Forest and Stream Management in the Oregon Coast Range

Forest and Stream Management in the Oregon Coast Range

edited by

Stephen D. Hobbs, John P. Hayes, Rebecca L. Johnson, Gordon H. Reeves, Thomas A. Spies, John C. Tappeiner II, and Gail E. Wells

Oregon State University Press

Corvallis

The paper in this book meets the guidelines for permanence and durability of the Committee on Production Guidelines for Book Longevity of the Council on Library Resources and the minimum requirements of the American National Standard for Permanence of Paper for Printed Library Materials Z39.48-1984.

Library of Congress Cataloging-in-Publication Data
Forest and stream management in the Oregon coast range /
edited by Stephen D. Hobbs ... [et al.].
p. cm.
Includes bibliographical references.
ISBN 0-87071-544-5 (alk. paper)
1. Ecosystem management--Oregon. 2. Forest ecology--
Oregon. 3. Stream ecology--Oregon. I. Hobbs, Stephen D.
QH76.5.O7 F67 2002
333.75'09795--dc21

2002002680

First edition 2002
Printed in the United States of America

Oregon State University Press
101 Waldo Hall
Corvallis OR 97331-6407
541-737-3166 • fax 541-737-3170
http://oregonstate.edu/dept/press

Contents

Acknowledgments

This book is a product of the Coastal Oregon Productivity Enhancement (COPE) Program (1987-1999). A cooperative research and education effort between public and private organizations, the program sought greater understanding of the forest and stream resources in the Oregon Coast Range and their management through fundamental and applied research. Although several hundred people were involved in the COPE Program, some deserve special recognition for the roles they played in the success of the program and the preparation of this book.

We would like to express our appreciation to George Brown, Dean Emeritus of the College of Forestry at Oregon State University, for his vision and leadership. Without them the COPE Program and this book would not have been possible. We would also like to thank Barte Starker, Chair of the COPE Advisory Council, for his support and encouragement throughout the COPE Program and the preparation of this book. Other members of the COPE Advisory Council during preparation of the book were: John Christie, Jim Clarke, Mike Collopy, Ray Craig, Dennis Creel, Don Davis, Gina Firman, Jim Furnish, Mike Hicks, Nancy MacHugh, Van Manning, Joyce Morgan, Steve Morris, Carrie Philips, Mike Propes, Al Tocchini, and Gary Varner. Bob Ethington and Charles Philpot, former directors, and Tom Mills, current director, USDA Forest Service Pacific Northwest Research Station, provided important leadership throughout the COPE Program. In addition, 60 organizations provided financial support to the COPE Program and the preparation of this book. These included federal and state agencies, county, city, and tribal governments, the forest products industry, and small woodland owners. In particular, the Forest and Rangeland Ecosystem Science Center of the USGS Biological Resources Division, the USDA Forest Service Pacific Northwest Research Station, and the College of Forestry at Oregon State University provided major financial support for preparation of the book.

It was our good fortune to have worked with a dedicated and tireless support staff without whose help this book would not have been possible. Karen Bahus, Laurel Grove, and Caryn Davis put in many long hours to provide much-needed copyediting. Tristan Bahr and Cindy Rossi provided skillful and rapid word processing and Leah Rosin contributed skilled proofreading. Don Poole prepared many of the illustrations and standardized the figures throughout the book. Thanks also to Gretchen Bracher, Laurie Houston, Christina Kakoyannis, Alissa Moses, and Kathryn Ronnenberg for their work on the illustrations.

In addition to Editorial Board review, outside experts evaluated individual chapters and provided many valuable suggestions that contributed significantly to the overall quality of the book. We recognize these reviewers for their contribution to this book. They were:

Dr. Robert Bilby, Weyerhaeuser Company
Mr. Jerry Beatty, USDA Forest Service
Mr. Stephen Dow Beckham, Lewis and Clark College
Dr. Gordon Bradley, University of Washington
Dr. Carol Chambers, Northern Arizona University
Dr. Dean DeBell, USDA Forest Service
Dr. Ted Dyrness, USDA Forest Service
Dr. Greg Filip, Oregon State University
Dr. Donald Goheen, USDA Forest Service
Dr. Everett Hansen, Oregon State University
Dr. Miles Hemstrom, USDA Forest Service
Dr. Norm Johnson, Oregon State University
Dr. Alan Kanaskie, Oregon Department of Forestry
Dr. David Marshall, USDA Forest Service
Mr. Jim Mayo, USDA Forest Service
Dr. William McComb, University of Massachusetts
Dr. Stephen McCool, University of Montana
Dr. Phil McDonald, USDA Forest Service
Dr. Maureen McDonough, Michigan State University
Dr. Thomas McMahon, Montana State University
Dr. Scott Reed, Oregon State University
Dr. Carlton Yee, Humboldt State University

S.D.H.
J.P.H.
R.L.J.
G.H.R.
T.A.S.
J.C.T.
G.E.W.

Authors

Kelly M. Burnett, USDA Forest Service, Pacific Northwest Research Station, Corvallis, Oregon
William H. Emmingham, Department of Forest Science, College of Forestry, Oregon State University, Corvallis, Oregon
Ellen Michaels Goheen, USDA Forest Service, Southern Oregon Forest and Disease Technical Center, Medford, Oregon
Stanley V. Gregory, Department of Fisheries and Wildlife, College of Agricultural Sciences, Oregon State University, Corvallis, Oregon
Joan Hagar, Department of Forest Science, College of Forestry, Oregon State University, Corvallis, Oregon
John P. Hayes, Department of Forest Science, College of Forestry, Oregon State University, Corvallis, Oregon
David E. Hibbs, Department of Forest Science, College of Forestry, Oregon State University, Corvallis, Oregon
Stephen D. Hobbs, Department of Forest Science, College of Forestry, Oregon State University, Corvallis, Oregon
Rebecca L. Johnson, Department of Forest Resources, College of Forestry, Oregon State University, Corvallis, Oregon
Julia Jones, Department of Geosciences, College of Science, Oregon State University, Corvallis, Oregon
Richard F. Keim, Department of Forest Engineering, College of Forestry, Oregon State University, Corvallis, Oregon
Loren D. Kellogg, Department of Forest Engineering, College of Forestry, Oregon State University, Corvallis, Oregon
Ginger V. Milota, Department of Forest Engineering, College of Forestry, Oregon State University, Corvallis, Oregon
Janet Ohmann, USDA Forest Service, Pacific Northwest Research Station, Corvallis, Oregon
Robert J. Pabst, Department of Forest Science, College of Forestry, Oregon State University, Corvallis, Oregon
Laurie Parendes, Department of Geosciences, Edinboro University, Edinboro, Pennsylvania
Gordon H. Reeves, USDA Forest Service, Pacific Northwest Research Station, Corvallis, Oregon
Barbara Schrader, Department of Forest Resources, College of Forestry, Oregon State University, Corvallis, Oregon
Arne E. Skaugset, Department of Forest Engineering, College of Forestry, Oregon State University, Corvallis, Oregon
Thomas A. Spies, USDA Forest Service, Pacific Northwest Research Station, Corvallis, Oregon
George Stankey, USDA Forest Service, Pacific Northwest Research Station, Corvallis, Oregon
Ben Stringham, Department of Forest Engineering, College of Forestry, Oregon State University, Corvallis, Oregon
Fred Swanson, USDA Forest Service, Pacific Northwest Research Station, Corvallis, Oregon
John C. Tappeiner II, Department of Forest Resources, College of Forestry, Oregon State University, Corvallis, Oregon
Walter G. Thies, USDA Forest Service, Pacific Northwest Research Station, Corvallis, Oregon
Beverly C. Wemple, Department of Geosciences, College of Science, Oregon State University, Corvallis, Oregon
Cathy Whitlock, Department of Geography, University of Oregon, Eugene, Oregon

Introduction

Stephen D. Hobbs and Thomas A. Spies

A Wealth of New Information

We have learned much about the ecology and management of forest and stream resources in the Oregon Coast Range since the Coastal Oregon Productivity Enhancement (COPE) Program was begun in 1987. The COPE Program was a cooperative research and education effort between federal and state agencies, the forest products industry, city, county, and tribal governments, and small woodland owners. Administered by the College of Forestry at Oregon State University and the USDA Forest Service Pacific Northwest Research Station, the COPE Program had the goal of developing objective, science-based information for more informed decision making about the management of forest and stream resources in the Coast Range. During its 12-year history, the COPE Program sponsored 60 studies involving 130 researchers from 14 organizations. This effort produced over 300 publications. When this information is considered in aggregate with research done by other organizations, the wealth of new knowledge developed since 1987 is impressive. Although there will always be uncertainty and new questions, research results produced by COPE scientists and others provide a solid information base for decisions and future research.

This book was written to provide resource managers, specialists, technicians, and policymakers with a synthesis of information needed to meet the challenges of forest and stream management in the Coast Range. Many of the issues facing managers and policymakers today require broad perspectives and integration of multiple disciplines. Consequently, each chapter is written for a diverse audience. For example, the wildlife chapter (Chapter 5) is not aimed narrowly at wildlife biologists; its intent is to provide silviculturists and others with a better understanding of wildlife habitat needs and opportunities for silvicultural operations to improve habitat quality. The chapter also contains new information that will be of particular interest to wildlife biologists. Thus each chapter is written broadly. The intent is to give people a greater understanding and appreciation for disciplines other than their own, and how these disciplines interact within the context of forest and stream management.

This introductory chapter sets the context for the book by providing an overview of Coast Range resources. Perhaps most importantly, it describes key principles or emerging "truths" fundamental to understanding and managing aquatic and terrestrial ecosystems in the Coast Range. These principles form the foundation for the book and have shaped the synthesis of information found in each chapter.

Overview of the Oregon Coast Range

The Coast Range Physiographic Province in Oregon is a range of relatively low but steep mountains that extends from the Columbia River south to about the Middle Fork of the Coquille River in Coos County (Franklin and Dyrness 1988) (Figure 1-1). The Province is bounded on the south by the Siskiyou Mountains, which consist of older (pre-Tertiary)

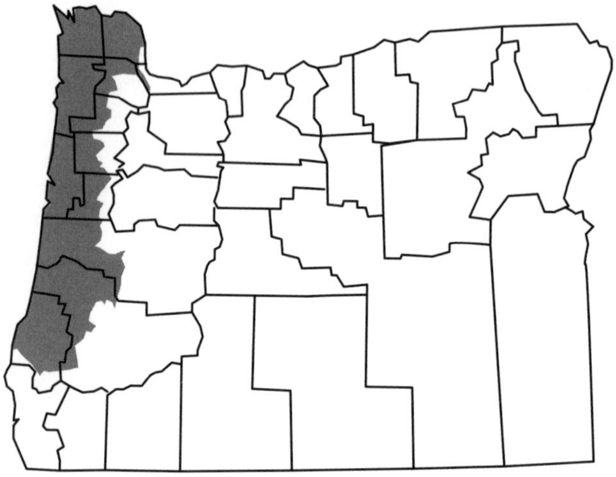

Figure 1-1.The Coast Range Physiographic Province in Oregon (superimposed over county boundaries) is bounded on the north by the Columbia River, on east by the Willamette Valley, and on the south by the Siskiyou Mountains.

rocks than those of the Coast Range (Tertiary), and on the east by the Willamette Valley.

The diverse environments of the Oregon Coast Range set the context for ecological processes and human experiences. For example, the terrain of the Coast Range can be thought of, quite simply, as having two states: low, flat areas and steep uplands. Where it is low and relatively flat, water typically accumulates in the form of streams, rivers, and wetlands, hardwood trees are common, and people typically graze cattle, build roads, and live in settlements. Where it is steep, forests of conifers typically develop and people build roads, harvest the trees for wood fiber, or hike trails and camp to experience the beauty of the forest. The Coast Range encompasses all or part of 20 major watersheds that are sources of drinking water for many cities and habitat for anadromous fish.

Spies and others (Chapter 3) describe the Oregon Coast Range as containing some of the world's most productive forest ecosystems. They note that the maritime climate, with its moderate temperatures and abundant precipitation, provides ideal growing conditions, particularly for coniferous forests. However, there are variations in climate. The northwestern part of the Coast Range is wetter and cooler than areas in the southeastern Coast Range. This differentiation is particularly important during the late spring and summer months when most plant growth occurs. There is also variation in the soils of the Coast Range that reflects the interaction of parent material, climate, vegetation, and physiography through time. For example, soils in the drier southeastern part of the Coast Range have more distinct horizons, while those in the wetter northwestern part generally have surface horizons high in organic matter with a high cation exchange capacity.

From a landscape perspective, Spies and others (Chapter 3) note that broad-scale patterns of vegetation generally follow a gradient of annual precipitation, summer moisture stress, and seasonal temperatures that runs from the north coast to the southwest interior of the Coast Range. Conifers are the dominant type of vegetation, but differences in conifer species composition and associated vegetation are obvious along the gradient. Whereas Sitka spruce (*Picea sitchensis*) and western hemlock (*Tsuga heterophylla*) may be found in abundance with Douglas-fir (*Pseudotsuga menziesii*) on the north coast, Oregon white oak (*Quercus garryana*) and Pacific madrone (*Arbutus menziesii*) are common associates with Douglas-fir in the interior southwest of the Coast Range. The broad-scale patterns we see today also reflect the influence of disturbances such as timber harvesting and wildfire. Disturbances have a profound effect on stand structure and composition and when considered in aggregate, they can influence broad-scale vegetation patterns. Today, the Tillamook State Forest is composed of relatively young forests resulting from wildfires in the mid-twentieth century. Likewise, stand structures on federal and industrial lands are on divergent courses, largely because of new federal environmental policies and the concomitant reduction in timber harvests from federal lands. As federal stands continue to grow older and industrial stands continue to be harvested on relatively short rotations, type of ownership and the types of management practices associated with them will be increasingly obvious in broad-scale vegetation patterns.

How forest and stream resources are valued by society has a profound influence on broad-scale vegetation patterns. Johnson and Stankey (Chapter 2) describe how values associated with the Oregon Coast Range have changed and how they continue to evolve today. Prior to Euro-American settlement in the early 1800s, Native Americans utilized Coast Range resources for daily subsistence and as a source of spiritual support. In contrast, Euro-American settlers viewed the forests as an impediment to agriculture and as an unending source of wood to help develop their communities. By the late 1800s, forest and stream resources were viewed almost exclusively as commodities for extraction. This was particularly true following World War II, when timber harvests increased and intensive forest management practices were put into place on both private and public lands. When considered in aggregate, these changes in the biophysical environment have substantially altered the landscape. How we value forest and stream resources continue to evolve today. As Oregon's population changes and becomes less connected to the wood products industry, an increasing number of Oregonians are concerned about the environment and are less willing to forgo noncommodity values for timber harvests and the production of wood products. Whether these changing values will continue and how they will affect broad-scale vegetation patterns are unknown. (also see Chapters 2 and 10).

Choices

The information in this book, considered in aggregate, presents clear messages about choices in the Oregon Coast Range. One of the most important messages is that there are options for management for multiple benefits, within limitations—ecological, socioeconomic, and legal/political. These limitations largely define society's decision space or choices about how resources will be managed and utilized (also see Chapter 10). The good news is that new research information has increased the probability that Coast Range forests and streams may be managed in ways that sustain both desired ecological and socioeconomic values and benefits. Research has improved our capability to manage sustainably in important ways, and this increased capability has presented us with more choices about how resources will be managed.

We now have better information about the inherent dynamics of forests and streams that allows us to design management systems that more closely match those of native ecosystems, with the goal of sustaining species and processes that are dependent on those disturbance regimes (see Chapters 3 and 4). It is unrealistic to seek a return to conditions that existed prior to Euro-American settlement, but understanding pre-settlement conditions and the disturbances that shaped them provides a yardstick against which we can measure change, either positive or negative. Equally important, this knowledge enables us to develop management strategies and practices to provide for society's needs in ways that more closely resemble the spatial and temporal patterns of natural processes such as disturbance. Such management should contribute to the conservation of biological diversity and ecosystem resilience, factors essential to the sustainability of natural resources.

Research has provided indicators with which we can assess current conditions and set clear measures of success, a process that has evolved to the point where regional and state multi-resource assessments are possible. Examples include the Coastal Landscape Analysis and Modeling Study (CLAMS) (Spies et al. in press), Oregon State of the Environment Report 2000 (Science Panel 2000), and Oregon's First Approximation Report for Forest Sustainability (Oregon Department of Forestry 2000). These efforts are important because they let us examine aquatic and terrestrial ecosystems across ownerships so that a composite picture of current conditions, as well as more spatially explicit conditions, can be developed. This ability enables policymakers and managers to identify areas of concern, relate conditions to management activities, and identify relationships important to policy decisions. We also have information that helps us to better understand watershed and landscape linkages to avoid undesirable or unintended consequences of actions.

Decision support tools to evaluate the effects of current and alternative policy and management scenarios at broad spatial and temporal scales are now becoming available. These tools offer an opportunity to significantly improve the quality and quantity of information used in policy decisions. Continued improvements in remotely sensed data, faster and more powerful computers, and models that better represent biophysical conditions and processes (e.g., forest growth, harvest schedules, input of large wood to streams) and also incorporate socioeconomic conditions will enable us to explore the potential implications of future policies. Although much work remains to be done in multi-scale, multi-resource modeling applied to broad landscapes, significant progress has already been achieved (e.g., CLAMS, Bettinger et al. 2000, Spies et al. in press). As we gain additional understanding of aquatic and terrestrial ecosystems and how human activities affect them, confidence in our ability to predict future outcomes will improve. Continued development of these tools presents an opportunity for improved policy and scientific analyses.

Research has provided us with information needed to expand our choices for management in ways that may provide greater overall compatibility among seemingly exclusive uses of forests. The use of natural disturbance regimes as a guide for timber management balanced with ecological values represents a significant philosophical shift in our approach to management. Likewise, we know how to use silvicultural practices to increase vegetation structural diversity in stands and watersheds where management history has simplified stand structure. We are also learning to design forest management strategies at landscape, watershed, and regional scales to increase overall benefits to both ecological and socioeconomic systems.

Key Principles

The following key principles about the Coast Range and the management of its resources emerge from the book:

Forest and stream ecosystems of the Oregon Coast Range are spatially and temporally dynamic and are shaped by human and natural disturbance. The Coast Range contains a broad array of environments that provide habitat for terrestrial and aquatic animals and plants and a context for ecosystems and their management (Chapter 3). The types and quality of habitats form a vegetation mosaic across the landscape that is constantly changing at different spatial and temporal scales in response to human and natural disturbances. The vegetation mosaic of today is quite different from the one observed when Euro-American settlers first set foot in the Coast Range. For example, prior to Euro-American settlement, acres of old-growth forest were spatially and temporally variable ranging from 25 to 75 percent of the Coast Range area. Currently old-growth forests occupy less than 5 percent of the area (Chapter 3).

Aquatic ecosystems are also very dynamic (Chapter 4). Historically, a wide range of conditions existed in Coast Range streams. In terms of fish, these varied from productive to non-productive streams. The time elapsed since a major disturbance, generally a catastrophic landslide following a wildfire, was the primary determinant of stream conditions. Streams that were 80-150 years away from such disturbances had the best habitat for salmonids. Currently, many streams have fish habitat that is in relatively poor condition because of changes in the type, frequency, and spatial extent of disturbances and the effects of disturbances on vegetation and ecosystem processes.

The resiliency of forest and stream ecosystems is influenced by the type of disturbance, and its intensity, distribution, and frequency. Forest disturbances come in a variety of types, sizes, and frequencies. Both human and natural disturbances affect biological diversity and ecosystem resiliency. Understanding natural processes of disturbance, including forest diseases (Chapter 8) and landslides (Chapter 9) in the context of forest development can provide a foundation for sustainable forest management. This understanding can contribute to management both of lands designated primarily for timber production and those designated primarily for conservation values (Chapter 3).

A dynamic mosaic of different habitat types and stages of forest development across the landscape is central to maintaining biological diversity and ecological processes in the Oregon Coast Range. The full array of forest stand diversity can take many centuries to develop. For example, forests 1, 10, 100, 200, and 500 years old have different structures and compositions. Each forest age-structure class contributes to the diversity and productivity of the region's landscape (Chapter 3). How the different patches of the vegetation mosaic are spatially and temporally distributed across the landscape will strongly influence ecosystem resilience and biological diversity. The composition and configuration of the vegetation mosaic can also influence the distribution and abundance of wildlife (Chapter 5).

Ecosystems are interconnected in space and time. Landscape scale phenomena, including movement of water and sediment along stream and road networks and interception of light by tall trees, create important links among terrestrial and aquatic ecosystems. These linkages may not be readily apparent, but they can have strong impacts on the function of Coast Range forests and streams (Chapter 3). For example, landslides and debris flows can transmit wood and sediment from one site to another over hundreds and sometimes thousands of feet. Road networks can serve as corridors for invasion of exotic species and can increase the rate at which water moves through a watershed.

Humans are an integral part of forest and stream ecosystems and their influence will continue to evolve. Population growth in the Oregon Coast Range is concentrated mostly along the coastal strip and valley margins. Changing demographics are altering the character and values of coastal communities. A decline in lumber and wood products employment and payroll and increased reliance on transfer payments, dividends, and rents make coastal residents less connected with the forest industry, which affects public acceptability of forest practices. People have a shifting conception of coastal forest resources, which are increasingly valued for non-commodity outputs and services such as habitat for threatened and endangered species (see Chapters 2 and 10). How forests and streams are managed in the future will be increasingly influenced by the complex interplay of such things as globalization, worldwide human population growth, increasing demand for wood, improved efficiencies in manufacturing, and other changing socioeconomic factors (Chapter 10).

Silviculture, including timber harvest, can be used to improve fish and wildlife habitat. Wildfire was once the major disturbance agent that created early-successional habitats and large accumulations of dead wood in Coast Range landscapes. Wildfires are less common since the advent of fire suppression and will be less common in the future, although some fires will still undoubtedly occur. Today, the dominant forest disturbance is cutting of trees for wood products and for fish and wildlife habitat enhancement. The disturbance regime of intensive forest management is not the same as that of wildfire regimes. Disturbances related to intensive timber management are more frequent and remove more forest structure than wildfire regimes. However, silvicultural practices during timber management can be modified to more closely approximate natural disturbances and in some cases silvicultural practices such as thinning can help restore forest structure to desired conditions faster than would occur without any management at all. Making timber harvest activities resemble natural disturbance would likely result in improved conditions in many Coast Range stream systems (Chapter 4). Harvesting operations can be conducted in ways that enhance fish and wildlife habitat, minimize impacts to soil and water, and remove wood from the forest for economic and social benefits (Chapter 6). Riparian-area enhancement can be successfully integrated with traditional upslope timber harvesting and may be more cost effective than stand-alone operations following harvesting.

There are many acres of young forests on both private and public lands. The focus of silviculture is shifting to include the management of these young conifer and hardwood stands for purposes that include both timber harvesting for wood products and fish and wildlife habitat enhancement. These young forests are generally quite dense because they were started for timber production and because reforestation was so successful. They can both produce timber and develop a wide range of forest structures including mixed species and multiple layered stand characteristics of old-growth forests. However, on many sites careful thinning will be needed to produce old-growth characteristics at a more desired rate (Chapter 7).

Collaboration among diverse groups of people, including landowners, will be necessary to make progress toward sustainability in the Oregon Coast Range. Natural resource management professionals must confront the challenges of developing new institutions and new ways to encourage informed dialogue among themselves and all interested stakeholders (Chapter 2). Achieving sustainability will require cooperation among diverse interest groups. A key ingredient of success will be continued improvement in our ability to integrate, evaluate, and communicate the interplay of biophysical, socioeconomic, and policy factors as they affect the quality of life for current and future generations (Chapter 10).

It is important to keep these principles in mind while reading the book. An understanding of the principles and appreciation of their significance, will help the information presented in individual chapters become more meaningful. The principles provide a big-picture context and foundation upon which successful and lasting management strategies can be built.

Literature Cited

Bettinger, P., K. N. Johnson, J. Brooks, A. A. Herstrom, and T. A. Spies. 2000. *Phase I Report on Developing Landscape Simulation Methodologies for Assessing the Sustainability of Forest Resources in Western Oregon.* CLAMS Simulation Modeling Report. Oregon Department of Forestry. Salem OR.

Franklin, J. F., and C. T. Dyrness. 1988. *Natural Vegetation of Oregon and Washington*. Oregon State University Press. Corvallis.

Oregon Department of Forestry. 2000. *Oregon's First Approximation Report for Forest Sustainability*. Salem.

Science Panel. 2000. *Oregon State of the Environment Report 2000: Statewide Summary.* Oregon Progress Board, Salem.

Spies, T. A., G. H. Reeves, K. M. Burnett, W. C. McComb, K. N. Johnson, G. Grant, J. L. Ohmann, S. L. Garman, and P. Bettinger. In press. Assessing the ecological consequences of forest policies in a multi-ownership province in Oregon, in *Integrating Landscape Ecology into Natural Resource Management*, J. Lui and W.W. Taylor, eds. Cambridge University Press.

Forest and Stream Management in the Oregon Coast Range: Socioeconomic and Policy Interactions

Rebecca L. Johnson and George Stankey

Introduction

The Coastal Oregon Productivity and Enhancement (COPE) project was developed to improve our understanding of the forest and stream ecosystems of the coastal region. Such understanding, it is presumed, is fundamental to the improved, rational, and scientific management of this region and for the long-term, sustainable flow of goods and services to society.

These purposes are similar to the primary objectives that drive much natural resource management policy and action today. Yet, if any single attribute characterizes natural resource management today, it is *conflict*. Of course, there is nothing new about conflict in the arena of natural resources; debates about alternative uses and values for resource management trace back many years. At the dawn of the 20th century, disagreements over the role and future of the nation's forests led not just to a split between Gifford Pinchot and John Muir, but also and more fundamentally institutionalized the ideological debate between wise use and preservation (e.g., Dana and Fairfax 1980). More than any other attribute, conflict may characterize our pluralistic society, whether the topic of discussion is health care, education, or resource management. However, there is a current disquiet that debates over the goods and services to be derived from our forests—ranging from endangered species, old growth, and wilderness to increased yields of fiber, jobs, and a source of community stability—have grown so intense, partisan, and vitriolic as to have brought traditional forest management, with its primary emphasis on commodity production, to a virtual halt. Moreover, this is a conflict characterized by increasing vilification and intolerance that leave little or no room for negotiation, compromise, and resolution. Efforts to resolve this conflict span several years, and include the 1993 Forest Summit in Portland, Oregon, which featured face-to-face discussions between President Clinton and a wide range of industry, environmental, and community leaders from the region. The Forest Summit led to development of the Forest Ecosystem Management Assessment Team (FEMAT 1993) report. But despite the high level of political attention and the level of scientific investment in producing the FEMAT report, its successful implementation remains problematic.

Herein lies an irony. As the other chapters in this book reveal, our understanding of the biophysical features of the coastal forests and streams is substantial and increasing rapidly. While major questions obviously remain, the pool of knowledge at our disposal today regarding these systems—how they function, how they respond to stress, how they can be managed to produce an array of products—is relatively large. Yet, our ability to tap that pool of knowledge to develop a scientific base for forest management or to translate that knowledge into implementable programs and policies is often stymied by the lack of consensus among the various stakeholders in the region. This is perhaps nowhere more evident than in the coastal region of Oregon, and it forms the basis for the central thesis of this chapter: *the changing sociopolitical landscape of the region will have profound impacts on the nature, extent, and level of forest management on the Oregon coast in the 21st century.*

How profound are these changes? In a July 1997 three-part series entitled "Oregon's crowded coast," the Portland *Oregonian* newspaper profiled a region in rapid, even turbulent, change. Through anecdotes, interviews, polls, analysis of census data, and other sources, the series revealed how this region has changed dramatically in virtually every respect in recent years. Rural communities, many with few if any land-use controls, have grown by 50 percent or more in the past decade; lot prices reach well into six figures; and local infrastructure (including roads, water, and sewage) is strained and often broken. Although the series primarily focused on the immediate coastal strip adjacent to Highway 101, the changes and their implications will resonate throughout the wider coastal region, from the shore of the Pacific Ocean to the crest of the Coast Range.

These rapid and radical changes to the coast's sociopolitical landscape have three key dimensions that bear directly and profoundly on the management of coastal forests in the future. These are (1) shifting conceptions of forest resources, (2) changing judgments of the social acceptability of forest management practices and conditions, and (3) changes in the relationship between political power and organizational authority. We briefly discuss the implications of each below.

Shifting conceptions of forest resources

Some 50 years ago, the economist Erich Zimmermann (1950) observed that "resources are not, they become" (p. 11). His cogent comment reminds us that resources are fundamentally social constructs. Those things we describe as resources are defined by the value or utility they provide society. While our conception of value or utility typically tends to be framed in the economic marketplace, value can be communicated through other means as well (Koch and Kennedy 1991). An excellent example of this would be wilderness values; despite the lack of any well-defined economic market through which wilderness values can be measured or captured, these values have been manifested through a political marketplace (Caulfield 1975).

Natural resources are *social constructs*, tied to conceptions of value and importance. The idea of a social construct is simple: it is people who give meaning to the idea of resources—what they are, how important they are. Quite literally, resources are "in the eye of the beholder." One person may see a towering tree as an emblem of spiritual renewal and inspiration, while the next person sees it as raw material for homes. Neither is right, neither is wrong; it is a matter of perception, and the meanings we bring to our perceptions of the world around us are shaped by a host of complex factors, including personal history, education, and world view. Furthermore, the notion of what constitutes a resource can change, and such changes are driven, in turn, by two factors. First, features of the biophysical environment that once held no value [what Zimmermann (1950) called "neutral stuff"] can come to hold value. This change can be driven by various forces, including technological change, scientific knowledge, economic markets, and social perception. For example, the Pacific yew was considered a weed species until taxol, derived from yew bark, was found to have potential beneficial effects in the treatment of breast cancer; this weed species then became a valued forest resource.

Second, resources can change in terms of the specific goods and services associated with them. For example, old-growth forests have long been valued for the high-quality, high-value timber they produce. However, such forests have also come to be valued for their significant aesthetic, social, cultural, and scientific values, and much of the conflict over old-growth forests hinges on the relative importance of these alternative conceptions of value.

The conception of resources associated with the forests and streams of the Oregon Coast Range has been evolving for a long time and has several distinct components. For the early Native American occupants of the coast, the region was a rich environment, replete with valued resources for daily subsistence as well as spiritual sustenance. Euro-Americans who settled the area beginning in the early 1800s, however, had perceptions of the region that conflicted sharply with those held by Native Americans. Primarily because of security concerns, the same biophysical system that sustained Native Americans was in some ways perceived as a barrier to Euro-American civilization (Bunting 1997). The big trees were an impediment to early farmers, and the fires with which the Indians renewed the prairie grasses threatened settlers' homes and farms. As industrialization grew in the late 1800s and early

1900s in response to growing expansionism and capitalism, the resource values of the region were defined almost solely in instrumental terms, and human efforts focused on using resources in the vast forests to support settlement, trade, and development.

The forest landscape that contemporary society inherited was largely shaped by its earlier use and occupancy by Native Americans and western settlers. Although the changes wrought by these early residents of the coastal region were bounded by certain biophysical constraints, the vegetation in the coastal landscape was substantially altered by them. The use and cultivation of favored species, the use of fire, and the exploitation of forests to fuel industrial expansion changed the region's natural forests and associated processes and systems.

Today, we find yet another transition underway, driven at least in part by the rapid change in the region's population and a concomitant growth in concern (in some cases, codified in law, such as the Endangered Species Act) about coastal forests for their value as habitat for endangered species, as a source of biodiversity and ecological heritage, and for their aesthetic, recreational, and amenity values. The implications associated with changes in the region's population are exacerbated by major structural changes in its economy and by external policy shifts in response to value changes occurring across the wider society. Although it is not clear whether these changes represent the emergence of new values held by a new constituency, or whether these values have always been present but unarticulated and unrecognized, the result is that development of appropriate forest management policies and practices now takes place in a different arena for social evaluation.

The social acceptability of forest practices and conditions

A second major issue with which natural resource policymakers and managers must contend is social acceptability—the extent to which the various practices and programs associated with management are understood and supported by the public. The issue of social acceptability is inherently complex; it raises perplexing questions about what is judged to be acceptable and by whom (Brunson et al. 1996). Nonetheless, such judgments can have a major influence on public policies and practices.

The concept of social acceptability can be traced to work by the rural-sociologist Walter Firey and economist Marion Clawson (1975). Both Firey and Clawson were concerned with understanding why certain resource policies and practices persisted in different cultures and failed in others. Firey focused on three key, interdependent aspects of the problem. For a policy or practice to persist, he believed, it had to be (1) biologically possible (i.e., consistent with biophysical realities and constraints); (2) economically feasible (i.e., consistent with the prevailing economic order); and (3) culturally adoptable (i.e., consistent with prevailing cultural norms) (Firey 1960). Clawson (1975) described a similar but more detailed framework: to persist, practices must be (1) biologically and physically feasible; (2) economically efficient; (3) economically equitable; (4) culturally and socially acceptable; and (5) administratively practical. In both Firey's and Clawson's frameworks, each criterion is mutually constraining, meaning that a successful policy or practice must meet all conditions. Thus a silvicultural practice that may be both grounded in solid biophysical science and economically efficient is still likely to fail if it lacks public understanding and support. Considerable effort has been devoted to increasing the rigor of the scientific rationale underlying management decisions and to the maximization of monetary benefits over costs, but little attention has been given to the issue of social acceptability. Nevertheless, practices that are judged unacceptable in light of prevailing norms will not be sustainable, *irrespective of how sound the biophysical rationale or how rigorous their economic foundation.*

Thus, as we think about forest and stream management in the Coast Range, the practices employed by managers (public or private) not only must be grounded in sound scientific principles and in an economic system that operates rationally, but also must be consistent with social judgments of acceptability. When such judgments are negative, the management practices—perhaps even silvicultural treatments to restore old-growth conditions or riparian management to protect stream quality—may prove difficult or impossible to implement. For example, previous work has shown that the acceptability of forest practices is judged on the presence of some specific attributes: a diversity of tree species, the presence of big trees, and the capacity for visual penetration (Ribe 1989).

Social acceptability is obviously a function of who makes such judgments. In the Oregon coastal region over the past decade, the size and structure of the area's population have changed dramatically. The region was once dominated by people in timber and fishing, but today these industries are in decline while the region's population is increasingly characterized by people in the service industries and especially by retirees. The immigration of new people also brings different perspectives, perceptions, and levels of political sophistication. These recent immigrants are likely to have only limited contact and experience with primary resource production economies, such as timber harvesting or fishing. Their interest in the region's natural resources might well derive more from aesthetic and/or recreational concerns than from a commodity or economic perspective.

Although some evidence shows large differences in the values identified as important by rural and urban residents (e.g., Steel et al. 1994), it is incorrect to assume that conflicts between "newcomers" and "old-timers" are inevitable or that all rural (or urban) residents hold the same views about resource management. Often there is as much variability within such a group as there is between groups.

The two key dimensions discussed thus far, the shifting conceptions of resources and the changing social acceptability of forest and stream management practices and policies, constitute a major component of the changing sociopolitical landscape of the Oregon coast. Taken together, they will have a significant influence on which natural resource values managers will need to enhance and which they will need to protect, as well as how they will carry out such enhancement or protection. However, a third aspect of this landscape warrants attention; it is a dimension that derives from the political nature of natural resource management.

Political power and organizational authority

The third key dimension of the changing sociopolitical landscape of the Oregon coast relates to the distribution and exercise of political power. Given changes in the public perception of values and appropriate management activities associated with forests and other natural resources, forest managers will increasingly find their plans and decisions subject to critical public scrutiny and evaluation. Although state and federal agencies clearly hold certain authorities relative to the management of forests and streams in the region, these authorities are neither unlimited nor immune to change.

Potapchuk (1991) distinguished between the concepts of *political power*, which is the capacity to affect choice and behavior held by the body politic, and *authority* (sometimes referred to as "legal power"), which is delegated to organizations such as the USDA Forest Service or Oregon Department of Forestry (ODF) through the political process. Agencies and their staff clearly have the authority to take certain actions and, indeed, the public expects that they will responsibly exercise this authority. However, such authority derives from the political power held by the wider society and accorded the organization; what has been accorded can also be withdrawn.

The concept of authority in a democratic government holds potent implications. If agencies such as the Forest Service or ODF are perceived as failing to exercise their authority appropriately, the rules of the game can be changed. Thus, if such organizations fail to respond to changing values and changing public support associated with the forests they manage, they may not survive (Clarke and McCool 1996). Arguments supporting major reform of the Forest Service (O'Toole 1988) or the rule changes proposed by the Committee of Scientists (1999) to reform the National Forest Management Act reflect such pressures.

Fashioning a responsible and thoughtful response to calls for the abolition, reform, or restructuring of natural resource organizations can be difficult. The perception of organizational failure might be inaccurate, but in this case, the perception is the reality. Thus, natural resource management professionals must confront the challenge of developing new ways to encourage informed dialogue among themselves and all interested stakeholders. However, current institutional mechanisms to achieve such an objective are poorly developed (Yankelovich 1991). Moreover, the biophysical and technical issues confronting managers today are extraordinarily complex and the debate over appropriate goals [and associated objectives, strategies, and management techniques; see Thompson and Tuden (1987)] is increasingly divisive and contentious. Thus, there seems little choice other than to respond to the challenge of developing innovative institutional structures and processes.

These basic characteristics of the sociopolitical landscape of the Oregon coastal region shape much of our discussion and especially our thoughts about the future. In our comments, we further specify the nature and, where possible, the causes of the unfolding changes. We also identify some potential strategies for responding to these changes as well as some of the future challenges whose implications and consequences for forestry are as yet unclear.

Sociopolitical Organization of the Oregon Coast Range

Although the term "Oregon Coast Range" seems to imply a particular geographic region, in reality the term embraces a complex mix of jurisdictions, tenures, and conceptions of place that confound both understanding and efforts to promote consensus and direction. As the previous discussion suggested, the human meanings associated with the Coast Range have changed substantially over the past 150 years and still vary today across different populations. Other chapters will define the Coast Range along biological or physical boundaries, but the relevance of these boundaries for human populations is problematic. Although physical features such as the Coast Range mountains and the Pacific Ocean have profound impacts on human settlement patterns and commerce, other factors influence the way people form communities, meanings, and identities.

Geopolitical boundaries

Geopolitical boundaries provide one logical way to define communities of people. Most socioeconomic data are gathered at the city or county level, even though those boundaries rarely coincide with biological or physical boundaries. Choosing a group of counties to represent the "Oregon coast" is problematic for a number of reasons. Curry, Coos, Lincoln, Tillamook, and Clatsop counties are clearly coastal, while Lane and Douglas counties have only a small portion of their land area and population along the coast (Figure 2-1 in color section following page 52). Benton, Polk, Yamhill, Washington, and Columbia counties contain land in the Oregon Coast Range, but are also heavily influenced by the economy of the Willamette Valley. If the purpose of analysis is to describe demographic and socioeconomic changes along the Oregon coast, then it would be reasonable to include only the wholly coastal counties. However, if the purpose is to analyze policies that affect Coast Range forests, then counties on the eastern side of the Coast Range should be included as well, because the effects will extend beyond the coastal area.

Communities of interest

Although human communities can be described in terms of geographic place, they can also be linked through interests. For example, Carroll and Lee (1990) described loggers as composing an "occupational community" in that they "share a common (or community) life set apart from others in society" (p. 142). People in the fishing industry could be similarly identified; those linked to fisheries have industry associations as well as social organizations, such as the Fishing Families Project and the Fishermen's Wives Association. Recent crises in resource management for both timber workers (related to declines in northern spotted owl [*Strix occidentalis*] and marbled murrelet [*Brachyramphus marmoratus*] populations) and fisheries workers (related to declines in coho salmon [*Oncorhynchus kisutch*]) have served as catalysts for group organization. Most importantly, members of these communities support one another in the same way that residents of a geographic community might, even though they might be located far from one another.

Communities of interest can also organize around issues other than occupation, such as recreation resources. On the Oregon coast, recreational salmon and steelhead anglers are linked to both the ocean fishery and the many coastal streams. Deer and elk hunters are linked to big game populations in the Coast Range, and campers and hikers are tied to the recreation setting provided by Coast Range forests. Still others are linked through their common concerns with conservation issues, such as the preservation of old growth or the protection of biodiversity.

Obviously, identifying and mapping such communities of interest is extremely difficult. However, any socioeconomic analysis of the Oregon coast needs to include the complex array of communities and interests. Actions that threaten these interests can trigger community mobilization efforts, even when those communities are scattered well beyond the geographic boundaries of the Oregon coastal region.

Cultural boundaries

Many communities of interest, such as the forestry and fishing communities identified above, would claim that their cultures are distinct from those of their neighbors. Other cultural boundaries also exist in the Oregon Coast Range, the most obvious of which is the Native American culture. Five confederated tribes are currently located in the Coast Range: Lower Umpqua, Coos, Siletz, Coquille, and Grand Ronde. Although reservations have formal boundaries, many tribal members live off-reservation yet maintain their cultural identity.

The rights and interests of the confederated tribes hold special, if often neglected, importance for the future management of the forests in the coastal region and, potentially, for all forms of development. A large body of judicial and legislative action acknowledges the tribes as sovereign governments. As such, the tribes must be consulted on a government-to-government basis regarding policy development. Consultation embraces more than notification and coordination; it includes meaningful discussions and collaborations with tribal governments on policy development, planning design, and project formulation. Tribes must be consulted as legally recognized sovereign governments, as landowners who are potentially affected by natural resource policy decisions and actions, and as holders of treaty rights that take precedence over other uses (FEMAT 1993). These treaty rights can include reserved rights for fishing, gathering, hunting, and grazing and can extend beyond reservation boundaries. Thus, management actions (e.g., silvicultural prescriptions) that affect the exercise of these rights must be undertaken in ways that ensure that these rights are protected. Moreover, only Congress can modify treaty rights; federal courts have ruled that if these rights are adversely affected, tribes must be compensated.

Another cultural boundary is that between rural and urban cultures. While most of the Coast Range is considered rural, many people who move to Coast Range communities these days come from urban areas and bring their urban culture, values, and perspectives, such as idealized expectations of rural life, with them. This is one of the potential causes of conflicts in values for natural resources. Although we have limited capacity to identify and map such cultural boundaries, policymakers must be both aware of and responsive to them.

Institutional boundaries

As new issues and problems arise, new institutions emerge to cope with them (such as COPE). Some of these can become permanent fixtures of the region, such as conservation districts and special improvement districts. Others might be ephemeral, existing only for a short time, in a limited space, and for a particular purpose.

Examples of new institutional features in the Oregon Coast Range are watershed councils, formed to oversee habitat restoration in Coast Range watersheds. These new institutions represent yet another way that people are organized and will be important when analyzing impacts of policies that affect Coast Range forests. They may both deliver programs and represent new entities with which more established institutions, such as the Forest Service or ODF, will have to deal.

More could be said about boundaries, but it is sufficient at this point to support our contention that a wide variety of ways of social organization are both possible and present. A key feature of this variable pattern of social organization is the fact that different boundaries, formed for different purposes and at different times, are seldom coterminous. Consequently, as actions are taken within one boundary for one particular purpose, they may or may not be consistent with actions taken within another, perhaps overlapping, boundary. Moreover, most sociopolitical boundaries are poorly linked to the biophysical system over which they are imposed. Efforts to deal with critical natural resource management issues and challenges are often thwarted or handicapped because of the mismatch between the institutional structures and the issue requiring attention or action. This problem has certainly become apparent as efforts to implement the concept of ecosystem management have proceeded. However problematic the definition of an ecosystem might be, the links between it and the ways in which social systems are organized are often remote or nonexistent.

The existence of a web of formal and informal, temporary and permanent, explicit and implicit boundaries poses an especially perplexing and confounding challenge for policymakers, resource managers, and citizens. Simply defining the various boundaries—where they are, to whom they are important, how they might facilitate or thwart management actions—is a major challenge. On the other hand, and in a more positive light, the diversity of boundaries can also be seen as offering a wide range of opportunities through which management can be undertaken. We are not restricted to working within the federal or state systems, for instance. However, the extent to which we are able to capitalize on this opportunity remains to be seen.

The Changing Oregon Coast

We earlier stated the central thesis of this chapter, that *the changing sociopolitical landscape of the region will have profound impacts on the nature, extent, and level of forest management on the Oregon coast in the 21st century*. This section will detail some of the changes in the economic structure and social characteristics of the Oregon coast.

In the following description of socioeconomic characteristics, the Oregon Coast Range is usually defined as the wholly coastal counties of Curry, Coos, Lincoln, Tillamook, and Clatsop. In addition, some descriptions coincide with resource or cultural boundaries, such as the fishing and timber industries or the Native American tribes.

Changing population

Population growth

In 1998, Oregon's five coastal counties had a population of 188,467—only 6 percent of Oregon's 3,281,974 people. From 1970 to 1998, the coastal population grew by 32 percent, a relatively modest growth rate (Figure 2-2). By comparison, Oregon's population grew by 56 percent over this period, a faster rate than U.S. population growth. Most of the coastal counties had similar growth trends, although Coos County suffered a relatively greater loss of population during the recession of the early 1980s (Figure 2-3), which hit timber-dependent counties the hardest. The high growth rates of the early 1990s have ended in coastal Oregon, and in recent years population growth has been slower than in the rest of the state and the nation (Figure 2-4).

The relatively slow population growth in coastal counties masks a much higher growth rate in coastal cities. Because of the severe topography of the coastal area and the attractiveness of the coast relative to the inland portions of these counties, coastal cities have absorbed the majority of new residents. For example, over the last decade, Warrenton grew by 56 percent, Brookings grew by

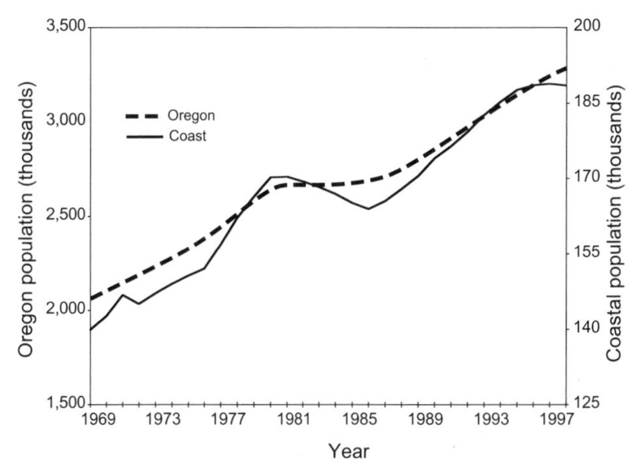

Figure 2-2. Population over time for Oregon and the Oregon coast, 1970–1998.

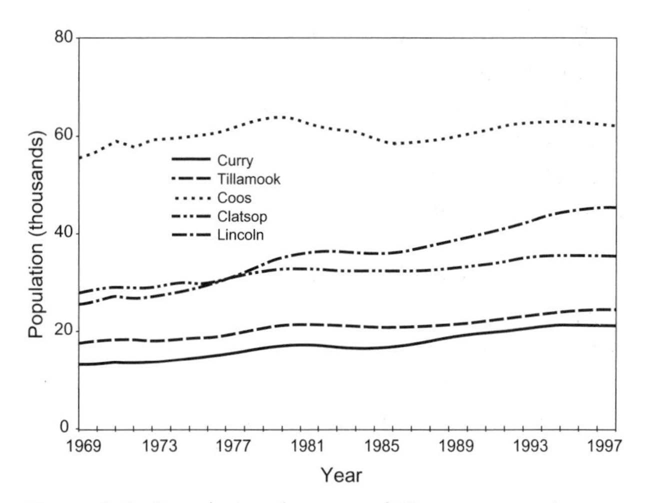

Figure 2-3. Population for coastal Oregon counties, 1970–1998.

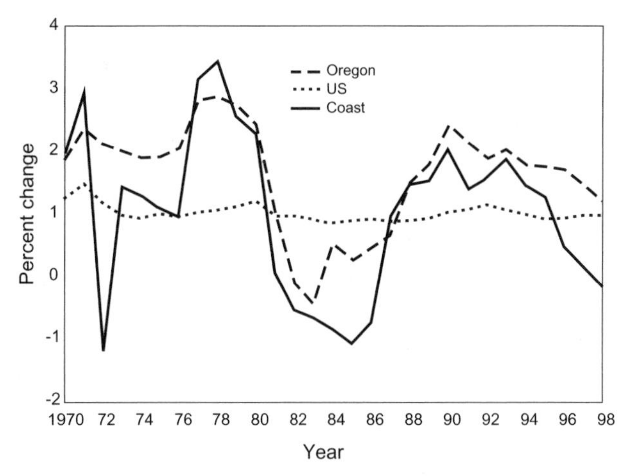

Figure 2-4. Percent change in population for the United States, Oregon, and the Oregon coast, 1970–1998.

49 percent, Manzanita grew by 48 percent, Florence grew by 29 percent, and Depoe Bay grew by 25 percent (*Oregonian* 1997).

The population growth figures also do not fully account for the increased pressure on coastal areas from second-home owners, as revealed by data on building permits. For example, the *Oregonian* (1997) reported that Seaside issued 1,013 building permits between 1990 and 1996, but the population grew by only 510 people. Florence issued 2,061 permits while its population grew by 1,229. Even building-permit numbers do not account for all the second homes built; because manufactured homes are not counted as "construction," these figures do not reveal the total number of new dwellings.

Population in coastal counties is expected to continue to grow into the foreseeable future (Office of Economic Analysis 1997), although differences are expected between counties. Curry, Lincoln, and Tillamook counties are expected to grow at over 1 percent per year through 2020. Clatsop County is projected to grow by around 0.8 percent and Coos County at around 0.45 percent per year over this time period.

It is clear that growth pressures in the far western portions of coastal counties are a significant issue for coastal residents. In addition to the problems caused by sheer numbers of new people, difficulties also arise when new residents have different characteristics and values from the current residents. One of the growing concerns along the coast is how to cope with the influx of retirees, who have special needs for health care and other services, yet desire to maintain a low cost of living (Corcoran et al. 1997).

Demographic structure

As in many parts of Oregon, population growth in the coastal region in the late 1980s and early 1990s was fueled by immigration. Figure 2-5 shows the net migration to coastal counties from 1988 to 1998. In the late 1980s and early 1990s, net migration was positive in all coastal counties, with Lincoln County showing the largest increase. In the late 1990s, however, both Clatsop and Coos counties had net out-migration, and in-migration in Lincoln and Tillamook counties slowed considerably. Such migration trends are likely to be influenced by economic factors in the different counties. For example, Lincoln County is more dependent on tourism and second-home owners, an economy that grew steadily in that period. Coos County is more dependent on the timber industry, and its population fluctuates more with changes in the timber economy.

In 1998, 43 percent of the immigrants to the Oregon coast were from other parts of Oregon, up from 37 percent in 1989 (Figure 2-6). The adjoining states of California and Washington accounted for about 12 percent, while around 45 percent came from other states. The only significant change between 1989 and 1998 was a reduction in immigrants from California, from 13 to 7 percent, probably as a result of the rebound in the California economy in the late 1990s.

The fact that population growth is fueled by immigration means that many coastal residents are new to their communities. A 1993 survey of coastal residents found that 47 percent of respondents had lived in their community less than 10 years, and 27 percent had lived there less than 5 years (Lindberg et al. 1994). The four most important reasons they gave for moving to the coast were natural environment (29 percent); job opportunity/closer to job (27 percent); lifestyle (small-town life, quality of life) (25 percent); and family reasons (be closer to family, moved with family) (25 percent). (Numbers add up to more than 100% because respondents were allowed to check more than one answer.) The natural and social amenities of the coast are clearly attractive to many people.

Many of the new migrants to the Oregon coast are retirees. The *Oregonian* (1997) reported that the fastest-growing sector of Florence's population is 75 years old or older, and a third of all residents are 60 years old or older. A 1993 survey found that 39 percent of residents in eight coastal communities were 60 years old or older (Lindberg et al. 1994). Population estimates from 1998 show that the percentage of population over 64 years old is significantly higher on the coast than in all of Oregon or the United States (Figure 2-7). Conversely, the percentage of the coastal population from age 18 to 34 is significantly lower, perhaps indicating an out-migration of young adults in search of employment opportunities. According to the 1990 census figures, Curry County had the highest percentage of population over 65 years old (24 percent), followed closely by Tillamook (21 percent) and Lincoln (20 percent), while Clatsop (16 percent) and Coos (17 percent) were closer to the overall Oregon percentage (14 percent).

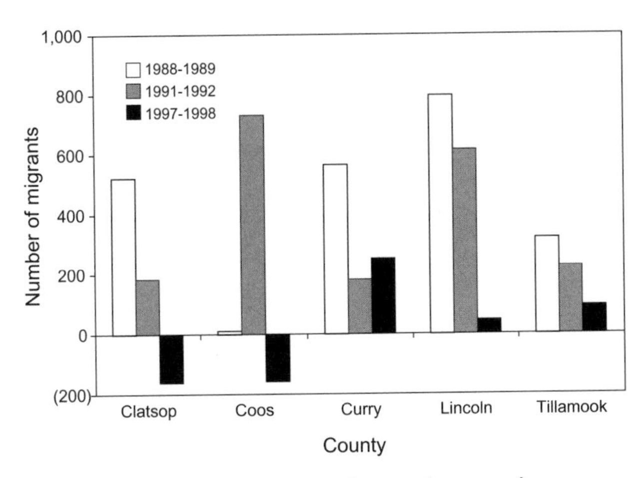

Figure 2-5. Net migration to Oregon's coastal counties.

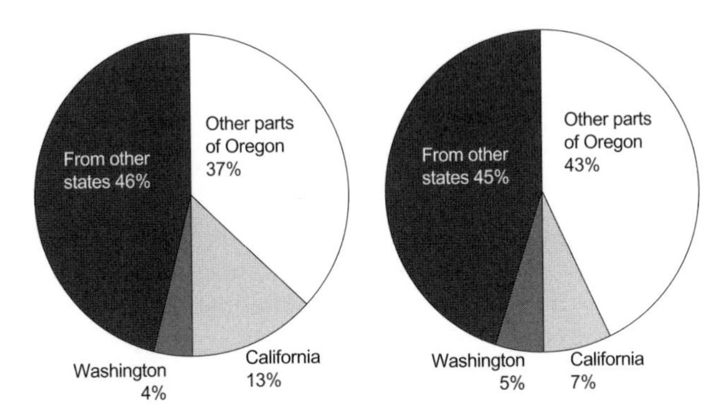

Figure 2-6. Origin of migrants to coastal Oregon during two tax years: (left) 1988–1989 and (right) 1997–1998.

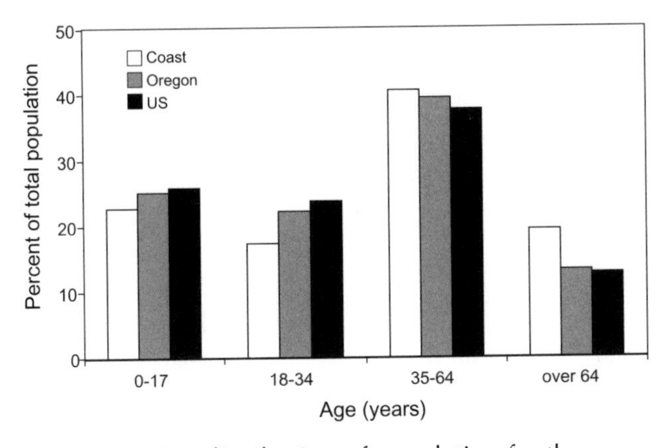

Figure 2-7. Age distribution of population for the Oregon coast, Oregon, and the United States in 1998.

The ethnic composition of the Oregon coast population is overwhelmingly dominated by whites, and this changed little between the census years of 1980 and 1990 (97 percent white in 1980 and 96 percent in 1990). The largest minority group is American Indians, who make up 2 percent of the coast population, ranging from 3 percent in Curry County to 1 percent in Clatsop County. While there is a perception that an increasing number of nonwhites are moving to the coast to work in the tourism and forest industries, the census numbers do not reflect this.

Changing patterns of land use

Between 1982 and 1997, the number of acres in this area dedicated to transportation and urban use increased significantly, while other uses decreased (Figure 2-8). However, most of the Coast Range acreage remains in forest use (Figure 2-9). Within the forest use category, there has been a net change in land ownership from nonindustrial private forest (NIPF to forest industry. Between 1985 and 1994, 19,012 acres (1.5 percent) changed from industry to NIPF, but 32,575 acres (4.6 percent) changed from NIPF to industry. While population and development pressures will continue to move land into transportation and urban uses, almost all of that change will take place along the narrow coastal strip bordering the Pacific Ocean. Undeveloped land along the coastal strip will become increasingly scarce and valuable, not only for future development, but as a source of open space and recreation opportunities for coastal residents. Moreover, the Coast Range foothills lying to the east will serve as an important scenic and aesthetic backdrop; this has important implications for the management of forests in this area.

The relative scarcity of land that could be developed in coastal counties has contributed to rising assessed property values (Figure 2-10). From 1978 to 1999, real assessed values (adjusted for inflation) in coastal Oregon grew 58 percent, more than in any other region in the state (Figure 2-11), and almost all of this increase has happened since 1990. From 1990 to 1999, three of the five coastal counties' real assessed values grew faster than the state average of 60 percent, with Tillamook County growing 82 percent, but Coos and Curry counties' real assessed values (which rose 41 and 50 percent, respectively) grew slower than the state's.

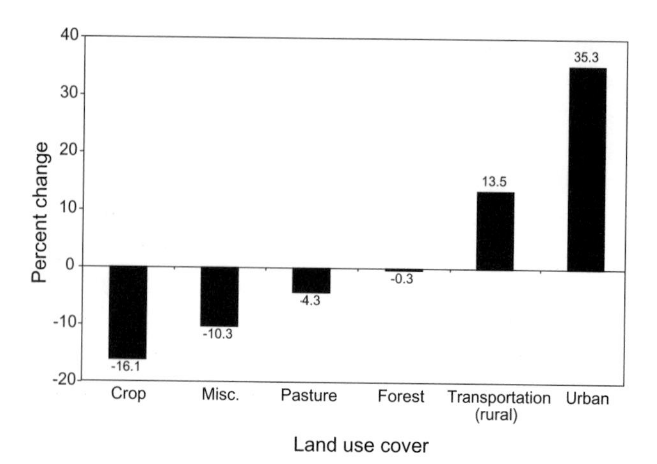

Figure 2-8. Percent change in land use acreage on the Oregon coast, 1982–1997.

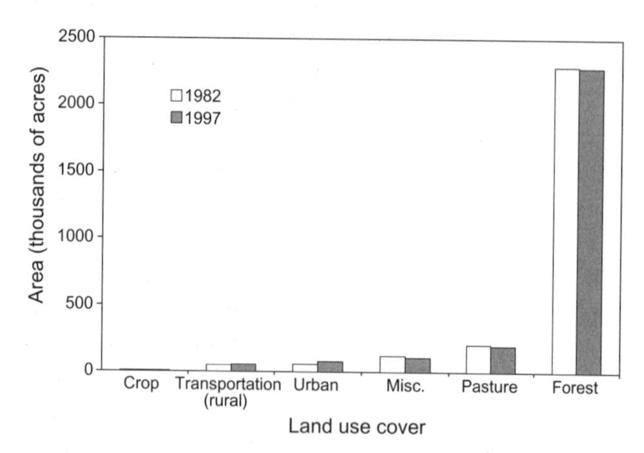

Figure 2-9. Land use cover on the Oregon coast in 1982 and 1997.

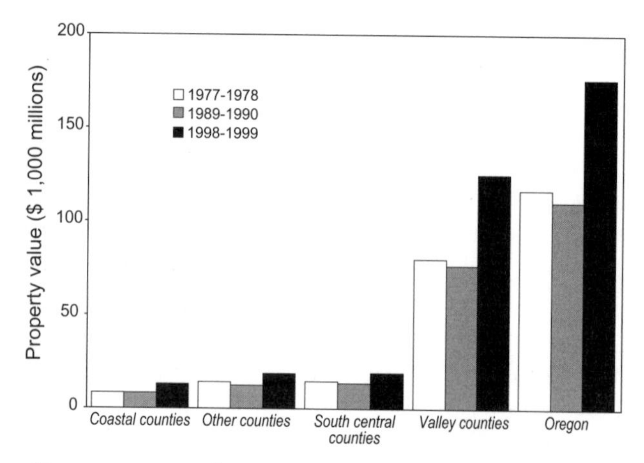

Figure 2-10. Real total assessed property values in Oregon, by region.

Changing economy

Income and employment

Perhaps the best measure of a region's overall economic well-being is the personal income of its residents. In nominal terms, per capita personal income has been rising steadily over the past three decades, although it is lagging behind income for Oregon and the United States. Coastal per capita income was over 90 percent of Oregon's per capita income from 1975 to 1988, getting as high as 94 percent in 1986, but it has been dropping steadily since 1988. When adjusted for inflation, 1997 per capita income in coastal Oregon was $3,950 less than in all of Oregon, and $5,318 less than in the United States (Figure 2-12). The real buying power of coastal residents declined significantly in the early 1980s and rose only slightly from 1986 to 1994, but it has grown more rapidly in recent years.

Real per capita income data for individual counties show similar trends over time, but differences in level between counties (Figure 2-13). Incomes in Clatsop County have been consistently higher than in the other counties, and Tillamook County has had the lowest incomes for the past 15 years. The gap between the two counties was almost $2,500 in 1994.

A significant source of income for coastal counties is transfer payments—that is, government payments to individuals, such as retirement, disability, medical, welfare, unemployment, veterans, and federal education and training payments. In recent years, this source represented about 26 percent of all personal income on the coast, up from 12 percent in 1969. Between 1989 and 1992, transfer payments increased by 10 percent or more per year for both the coast and all of Oregon. More recently they have risen by 5–8 percent per year. The composition of coastal Oregon transfer payments has changed significantly between 1969 and 1997 (Figure 2-14). While there has been a slight decrease in the percentage from retirement and disability, there has been a significant increase in medical payments (Medicare and Medicaid). This trend is true for Oregon as well (Figure 2-15), but to a lesser extent. Perhaps somewhat surprisingly, the percent of transfer payments from retirement and disability is virtually identical for the coast and all of Oregon. However, coastal residents received $2,857 in retirement and disability per capita in 1997, while the figure for all of Oregon was only $2,212.

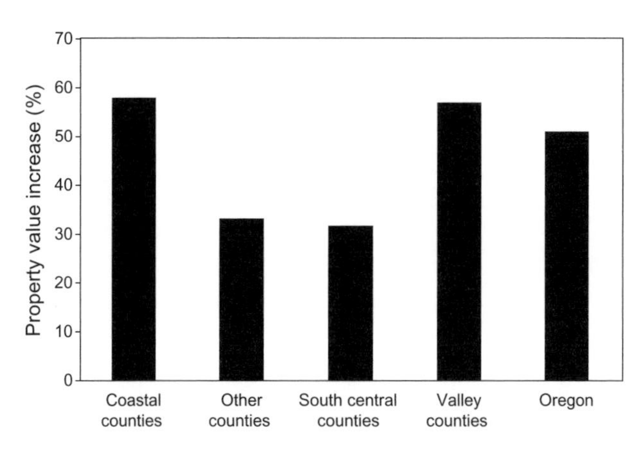

Figure 2-11. Percent increase in real total assessed property values in Oregon, by region, 1978–1999.

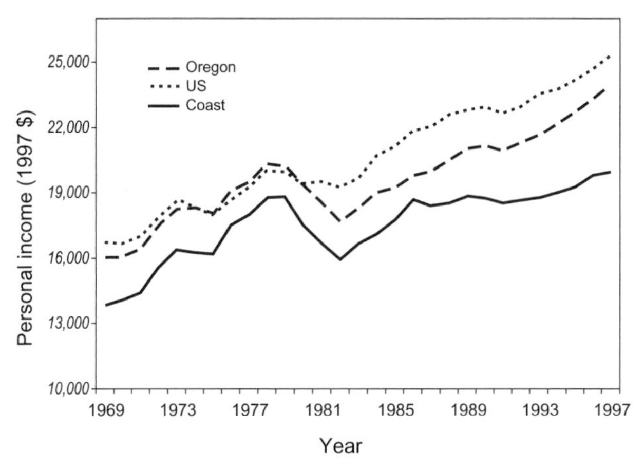

Figure 2-12. Real per capita personal income in coastal Oregon, Oregon, and the United States (1997 dollars).

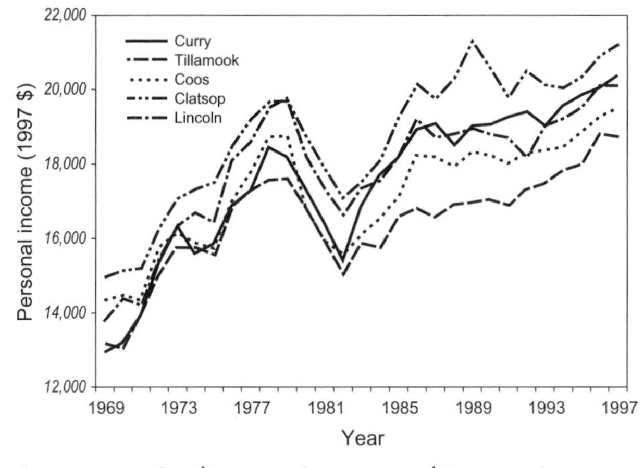

Figure 2-13. Real per capita personal income in Oregon's coastal counties (1997 dollars).

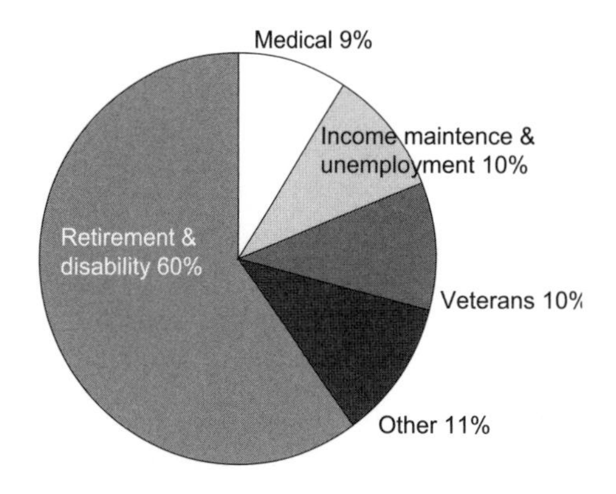

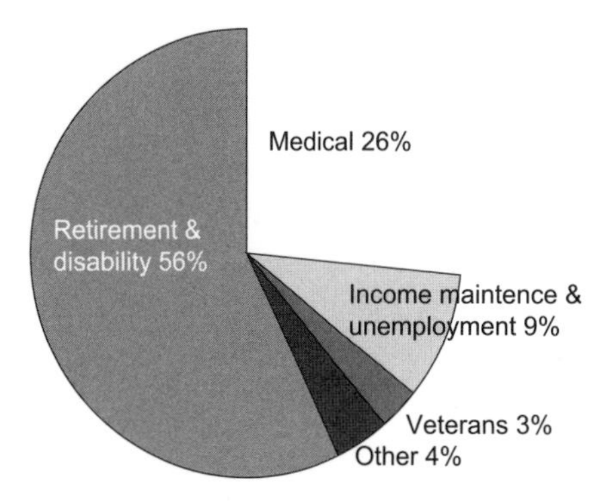

Figure 2-14. Transfer payments in coastal Oregon during 1969 (top) and 1997 (lower).

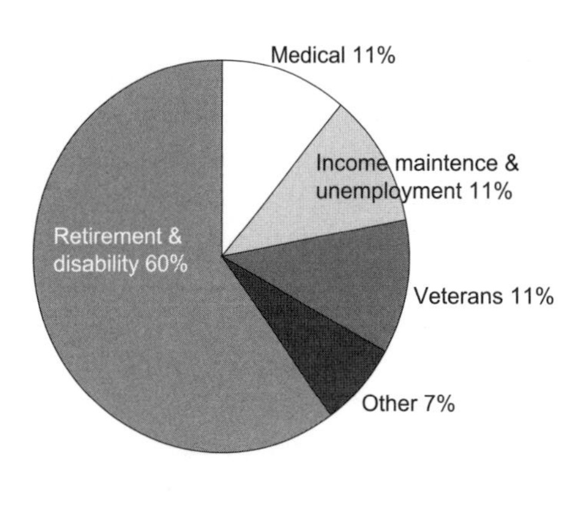

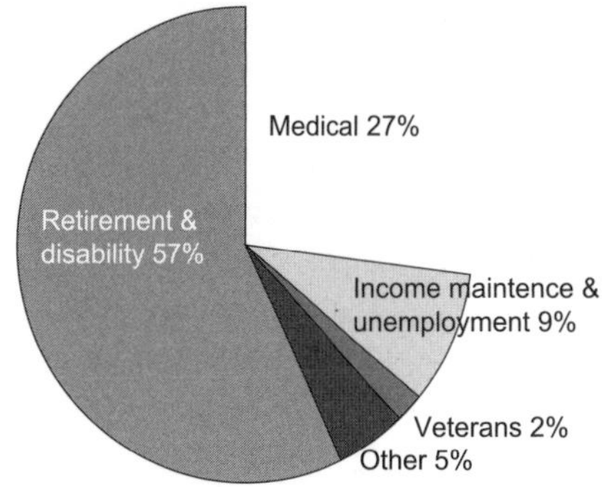

Figure 2-15. Transfer payments in all of Oregon during 1969 (top) and 1997 (below).

Another measure of the health of the economy is unemployment. In coastal areas, unemployment declined significantly from 1982 to 1990 and then increased slightly during the recession of the early 1990s and again in the mid-1990s (Figure 2-16). The coastal unemployment rate remains higher than both Oregon and U.S. rates, but is relatively low for a nonmetropolitan area. Among coastal counties, Tillamook County has had the lowest unemployment rate in recent years (5–7 percent) and Coos County has had the highest (8–10 percent).

Income levels have been slowly rising and unemployment rates have been slowly declining on the coast since 1982, although the long-term effects of the economic downturn of late 2001-early 2002 remain to be seen. Over the same period,the structure of the economy has been changing rather dramatically. Figure 2-17a shows employment by major sector in 1975, 1985, and 1995. Much of the manufacturing in coastal Oregon is in the lumber and wood products industry, which declined dramatically over this period. In 1975, lumber and wood products provided 18 percent of coastal Oregon employment, but by 1995 this had dropped to 6 percent, while retail trade grew from 19 percent to 26 percent (Figure 2-17b). Since the lumber and wood products sector historically paid higher wages than retail trade or services, the loss of jobs meant a significant loss in total payroll to coastal residents. Figure 2-18a shows that payroll for lumber and wood products, in real terms, has declined by 64 percent (from $292 million down to $106 million), while payroll in services has grown by 165 percent (from $86 million up to $229 million). Manufacturing payroll still makes up 20 percent of all coastal payroll, second only to government, which provides 26 percent (Figure 2-18b).

Payroll per employee has also been declining in coastal Oregon industries as the composition of industries and participation in the labor force change. In real dollars, payroll per employee declined in almost all industries from 1975 to 1995 (Figure 2-19), though it rose modestly in services and the "resource" industries (agriculture, fishing, mining, and forestry, not including wood processing). Since this measure counts full-time and part-time employees the same, the relatively low wages in retail trade and services can be misleading, since these industries have many part-time workers. The lumber and wood products industry still has some of the highest wages in the coastal area.

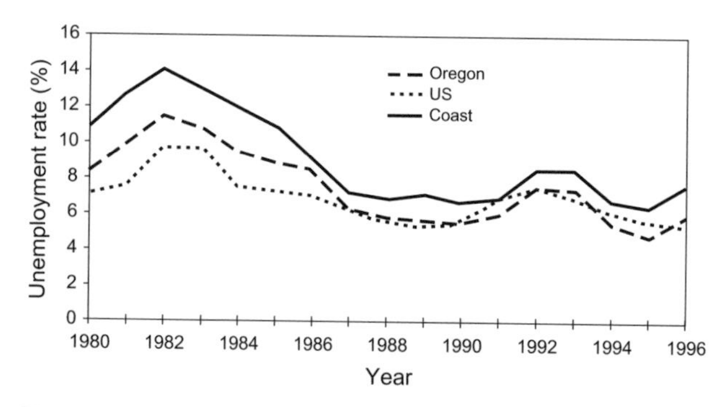

Figure 2-16. Annual average civilian unemployment rates in coastal Oregon, Oregon, and the United States.

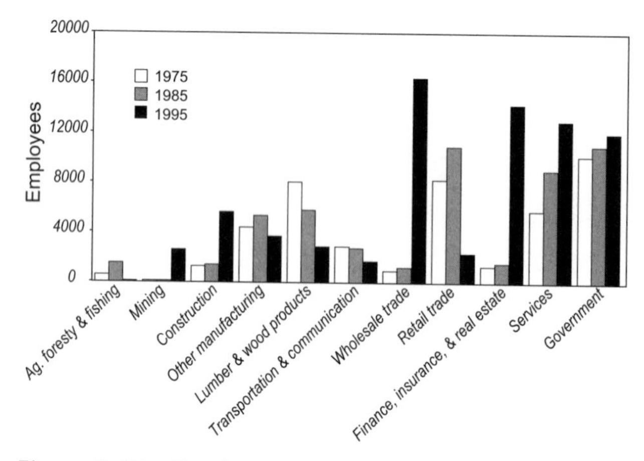

Figure 2-17a. Employment in coastal Oregon, by sector.

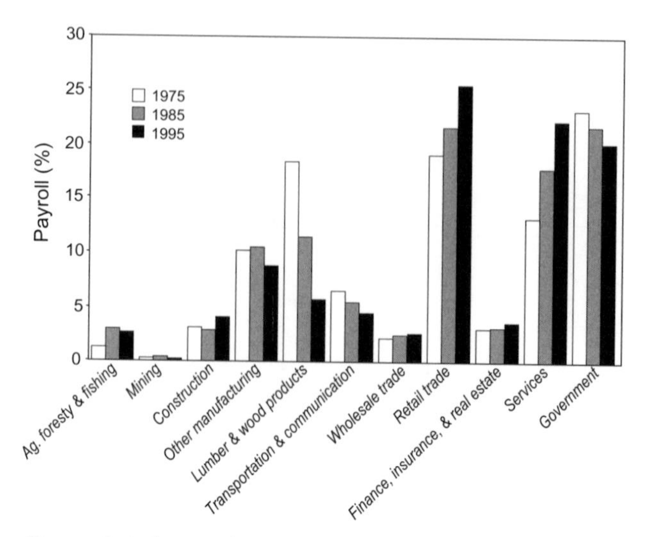

Figure 2-17b. Employment by sector, as a percentage of total coastal employment.

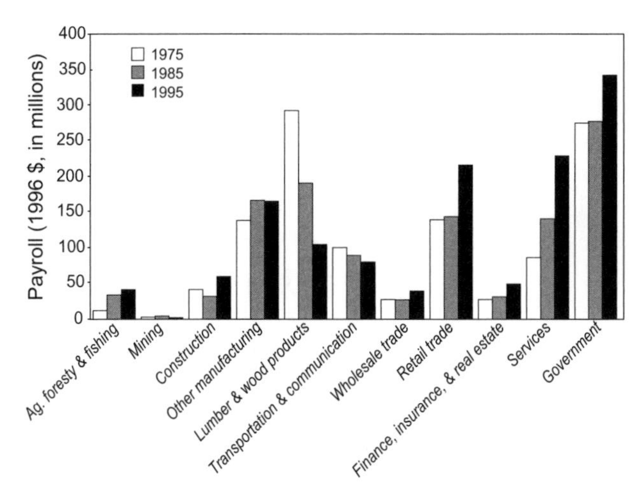

Figure 2-18a. Coastal payroll by sector, as a percentage of total coastal payroll.

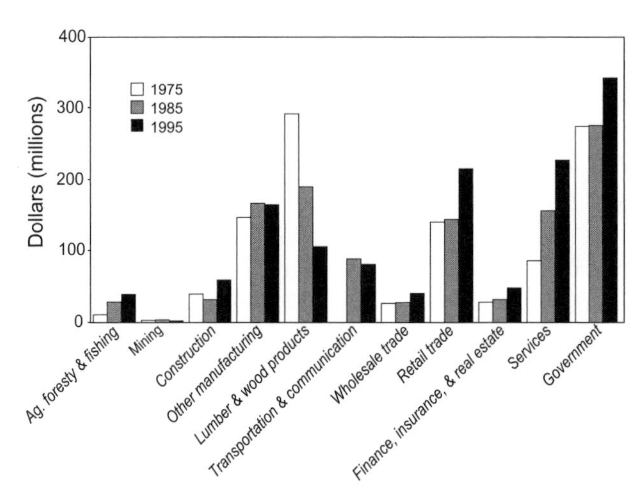

Figure 2-18b. Coastal payroll by sector, in 1995 dollars.

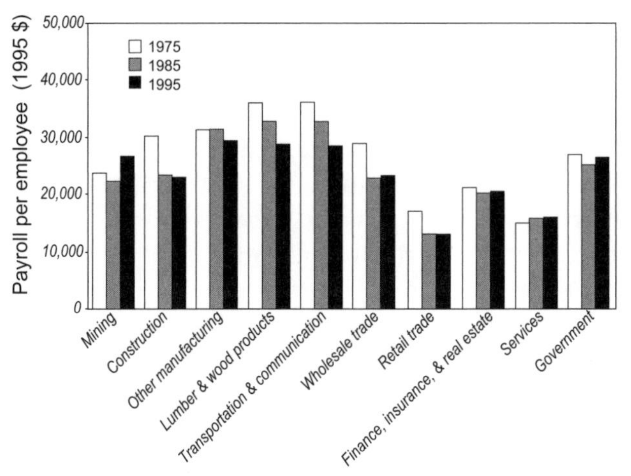

Figure 2-19. Payroll per employee, in 1995 dollars.

Over 75 percent of all employment in all coastal counties is in four sectors: manufacturing, retail trade, services, and government. All counties had significant increases in services and retail trade employment from 1975 to 1995, but other employment statistics for individual coastal counties are variable. The largest declines in lumber and wood products employment during these decades were in Lincoln County (–64 percent) and Coos County (–62 percent). Other manufacturing employment increased in all counties except Curry County (–3 percent), with Tillamook County increasing by 128 percent over this period. There were also differences in payroll per employee across counties, reflecting differences in the types of businesses that make up the broad sectors in each county. In Clatsop County, lumber and wood products payroll per employee (in 1995 dollars) declined from $42,005 (1975) to $29,137 (1995); the decline in Coos County was from $35,571 to $29,656 during this period. Services payroll per employee increased in real dollars in all counties, with the largest increase (from $13,051 in 1975 to $16,500 in 1995) in Tillamook County. Other manufacturing payroll per employee in 1995 differed significantly across counties, with Curry County being the lowest at $16,369, and Clatsop County the highest at $34,278.

Recreation and tourism industry

Oregon's coastal counties have 272 miles of ocean beaches and vast acreages of public lands available for outdoor recreation pursuits. The Bureau of Land Management is the major manager of public lands in Lincoln and Curry counties, and the Oregon Department of Forestry in Clatsop and Tillamook counties (Figure 2-20). Although the Oregon Parks and Recreation Department manages only a small amount of land in the Coast Range, the agency accounts for a relatively large share of coastal visitation, because so many state parks are located along the ocean and the main travel route of Highway 101. After growing for many years, attendance at coastal state parks declined through most of the 1990s (Figure 2-21). Although some of the decline may be due to transportation problems (e.g., landslides along major roads), poor weather, or increases in user fees, changes in the ways attendance is counted may be confounding the visitation figures.

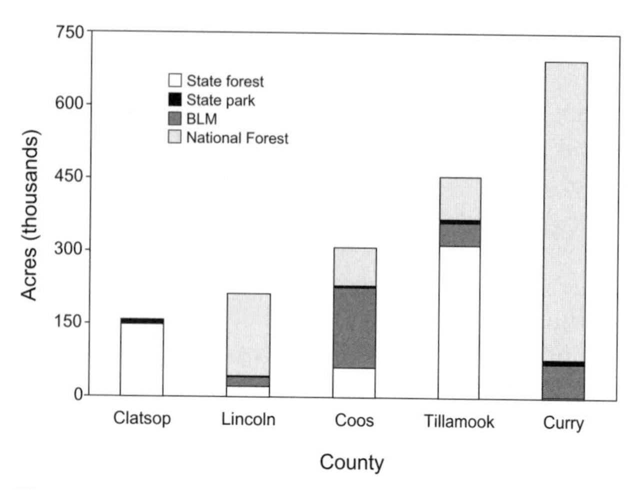

Figure 2-20. Acres of public recreation lands in Oregon's coastal counties.

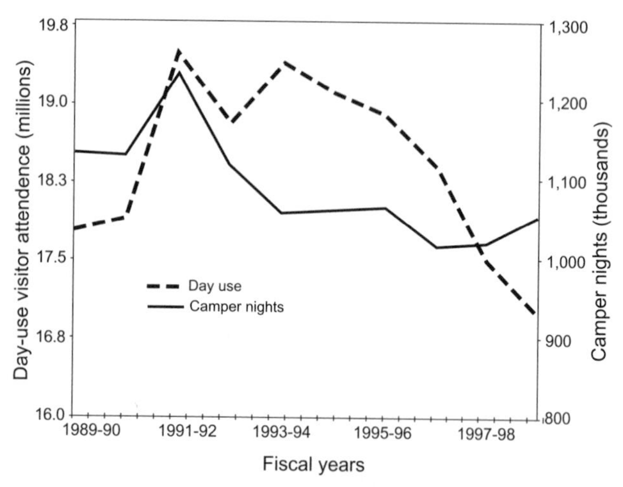

Figure 2-21. State park campground and day use visitation in Oregon's coastal counties, 1989–1999.

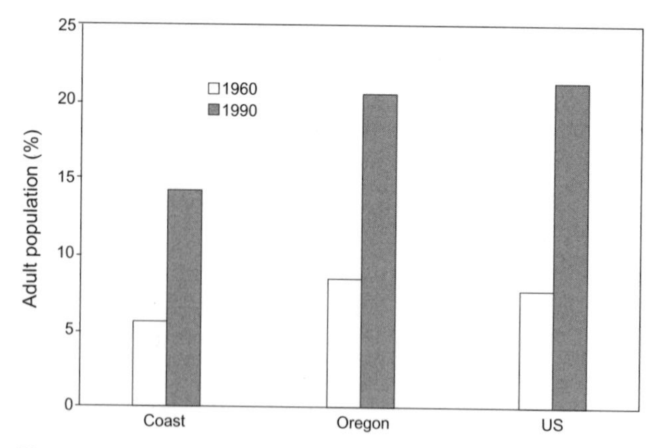

Figure 2-22. Percentage of the adult population (≥ 25 years old) with ≥ 4 years of college education.

Estimating the economic impact of the recreation and tourism industry on the coastal area can be difficult. The standard classifications used for economic data do not identify a single recreation or tourism industry. Instead, the tourism industry is made up of many different industries, none of which is solely associated with tourism. For example, the restaurant industry is clearly a part of the tourism industry, yet many local residents who are not tourists also eat at restaurants. Even motels, which predominantly serve tourists, are also patronized by business travelers and local businesses that host conventions or meetings. A study by Johnson et al. (1989) addressed this problem by surveying coastal tourism businesses to estimate the percentage of total sales that were to nonlocal households (tourists). In the north coast region, the estimate ranged from 22 percent in retail trade businesses to 90 percent in lodging establishments. In the central coast region, tourists accounted for only 13 percent of retail trade sales, 57 percent of restaurant sales, 73 percent of lodging sales, and 90 percent of amusement services sales. In the south coast region, only lodging had over 50 percent of sales to tourists, demonstrating the smaller magnitude of the tourism industry in that region. Total income generated from tourism sales was calculated through the use of input-output coefficients for each region. In the north, central, and south coast regions, annual local income generated from tourism sales was $42.06 million, $44.03 million, and $25.04 million, respectively. Since this study included only the six major economic sectors where tourists spend money (retail, lodging, restaurants, amusement, service stations, and marine services), total impacts of tourism were underestimated.

A more recent study of the coastal economy showed that tourism contributed 8 percent of total personal income to coastal residents (Davis and Radtke 1994). For comparison, contributions from other industries include 16 percent from the wood products industry (including paper), 5 percent from fishing, and 4 percent from agriculture. Transfer payments were by far the largest contributor to personal income at 24 percent. The figures for the whole coast can mask differences between coastal regions. A survey of residents in eight coastal communities showed that of all employed respondents, the percentage working in the tourism industry ranged from 4 percent in Coos Bay to 60 percent in Lincoln Beach (Lindberg 1995).

An alternative way to measure the economic impact of tourism is to estimate expenditures from tourist surveys (as opposed to business surveys), and then run those expenditures through an input-output model to estimate the multiplier effects. This procedure was used in a 1995 study of the economic impacts of outdoor recreation in Oregon (Johnson et al. 1995). The coast was divided into two regions. On the north coast, total personal income generated from outdoor recreation was $335.64 million, which supported 15,323 jobs. On the south coast, the figures were $157.47 million in income and 7,140 jobs. The drawback to this methodology is that agency estimates of visitor use must be relied upon to calculate total expenditures. While good estimates are available for developed recreation facilities, such as state parks, estimates for dispersed recreation use, such as that found on most Forest Service lands, are less reliable.

Regardless of methodology, it is clear that recreation and tourism play an important role in the economy of the coastal region. Annual estimates from the Oregon Tourism Commission show that the coast is the primary tourist destination in Oregon, far outdistancing any other region in the state. Travel expenditures including both business travelers and tourists in the coast region ($968.88 million) are second only to the Portland Metro region ($1.35 billion) (Dean Runyan Associates 1996). As population continues to increase in Oregon, especially in the Willamette Valley, coastal tourism will continue to grow.

A new industry that has grown tremendously in recent years is Indian casino gaming. Three casinos currently operate in the coast region: The Mill in Coos Bay, Chinook Winds in Lincoln City, and Spirit Mountain in Grand Ronde. Their 1997 gaming revenues (the difference between money bet and won by players, not the casino's net profits) were $20 million for The Mill, $29 million for Chinook Winds, and $82 million for Spirit Mountain (ECONorthwest 1997). These three casinos accounted for 62 percent of all Indian gaming revenues from the seven casinos operating in Oregon in 1997 and increased 26 percent between 1995 and 1997, showing the increasing popularity of this activity among both tourists and residents. It remains to be seen what impact casinos will have on other tourism businesses, which may be competing for the tourist dollar or may be complementary to casinos (e.g., lodging, service stations).

Changing education levels

Similar to trends in all areas of the United States, the education levels of Oregon coast residents have been rising over time. However, the percentage of population over 25 years old with 4 or more years of college education is considerably less for coastal residents than for all of Oregon or the United States (Figure 2-22). The percentages are highest in Clatsop and Lincoln counties (17 percent) and lowest in Coos County (12 percent). Both pull and push factors are likely to be at work here. Some of the coastal industries provide low-skilled jobs and attract workers with lower education levels. On the other hand, some people with higher education levels are unable to find employment opportunities in the coastal areas and leave to seek jobs in the Willamette Valley or out of state.

Public values about natural resources

Any discussion of natural resources leads quickly, and inevitably, to a discussion of values. This linkage is to be expected; as suggested earlier, natural resources are defined by society's assessment of worth and importance of some feature of the biophysical environment. Values, although subject to many definitions and conceptualizations, are at their root a reflection of importance.

The debate and conflict about the values of natural resources are driven largely by the fact that these values can be defined, measured, and captured in many different ways. For example, Stankey and Clark (1992) described a typology of values, including commodity (e.g., timber), amenity (e.g., scenery), and ecological (e.g., biodiversity). Each of these constitutes a value, but the manner in which those values are described, measured, and reflected in land-use decisions varies considerably. Typically, commodity values are measured in the economic marketplace and expressed in financial terms. Ecological or spiritual values, however, are much more difficult to identify and measure. Yet one should not conclude that these values are not real or that they are important only to those who espouse them.

Based on anecdotes as well as public opinion polls and scientifically based surveys, mounting evidence suggests that society's values regarding natural resources are in a major transition. One measure of the significance of these changes is that they are occurring at a host of spatial scales—globally, nationally, regionally, and locally. The changes are especially apparent in the challenges confronting natural resource management in the Pacific Northwest. "At the heart of this debate," Steel and Lovrich (1997) write, "are differing values and interests concerning the natural environment and the proper relationship of humans to their ecological surroundings" (p. 3).

Understanding the nature of these values and the factors that shape and alter them is essential to effective natural resource management. As implied in the opening discussion in this chapter, values shape our conceptions of resources and the goods and services they provide; they also shape our conceptions of their appropriate management. When values conflict—such as among different sectors of society (e.g., urban vs. rural) or between resource managers and citizens—the stage is set for conflict.

While the general consensus is that values are changing, the causes for and implications of these changes are not as well understood. In general terms, such factors as changes in population, technology, and economics are typically cited as being responsible for value changes. Also, while the specific nature of these changes is often difficult to document, the following trends in values about the environment are generally recognized (Dunlap 1991): public environmental concern increased dramatically in the late 1960s, coinciding with the rise of other social movements (e.g., civil rights); and concern for the environment declined in the 1970s, followed by a significant and steady increase in both public awareness of environmental problems and support for environmental protection efforts since the 1980s.

While the specific dimensions of environmental interest and concern might be problematic, polls and surveys in recent years have reported consistently high levels of public support. In a 1989 survey (Gallup Report 1989), Gallup reported that about 80 percent of the survey population described themselves as "environmentalists" (about half of these called themselves "strong" environmentalists). A 1993 national survey reported that only about 30 percent of respondents agreed that "plants and animals exist primarily for human use," while nearly 90 percent concurred with the statement that "humans have an ethical obligation to protect plant and animal species" (Steel and Brunson 1993). An especially revealing aspect of this survey relates to strong differences among age groups; for example, whereas about 60 percent of respondents over 50 years old agreed that "plants and animals exist primarily for human use," less than 10 percent of respondents in the 18–29 year age group agreed. More recently, a Gallup poll on the 30th anniversary of Earth Day in 2000 found that 83 percent of Americans agree with the goals of the environmental movement, and another Gallup poll in July 2000 reported that 71 percent of respondents said that candidates' positions on environmental issues were extremely or very important to them in the November 2000 presidential election.

In an assessment of national public values and attitudes about federal forest management practices, Steel et al. (1993) reported that only about 10 percent of respondents accorded "highest" priority to economic considerations, "even if there are negative environmental consequences." About 50 percent of respondents supported equal consideration of environmental and economic priorities in setting forest policy.

Compared with all respondents in a national survey, Oregon respondents held a more "anthropocentric" view about nature (i.e., they considered nature to be for human benefit) (Steel et al. 1997). In one sense, this is not surprising; Oregon respondents were about twice as likely as national respondents to be economically dependent upon the timber industry. However, Oregon respondents still espouse a biocentric view (i.e., a nature-centered perspective) more than an anthropocentric one; "the differences between the two samples (national and Oregon) lie in the *intensity* of value orientations and not in their direction" (Steel et al. 1997, p. 24; emphasis added).

Surveys have also examined links between location of residence (urban vs. rural) and values and attitudes toward forest management. Such analyses are driven by a hypothesis that anti-timber harvesting sentiments and pro-preservation support derive largely from urban settings, while "pro-industry" views are rural in origin. Brunson et al. (1997) tested this hypothesis, but also added the notion of dependency. They assessed whether respondents, irrespective of urban or rural residence, were dependent (e.g., connected by business, family) on the resource extraction industries. Their analysis reveals that "personal dependency is more influential on support for a resource management paradigm than rural or urban residency" (Brunson et al. 1997, p. 93).

Another conventional notion is that support for environmentalist perspectives is limited or nonexistent in rural, resource-dependent communities. However, research in California (Fortmann and Kusel 1990) and elsewhere suggests the contrary; such values are often widely held in rural communities (by long-term rural residents), including specific concerns with forest practices such as clearcutting (Blahna 1990).

The importation of values by newcomers has attracted much attention, but little research. On the Oregon coast, the increasing levels of immigration, especially by retirees, has raised concerns that local wishes, values, and concerns will simply be swept aside. Fortmann and Kusel (1990) suggested that this is not inevitable; both newcomers and "old-timers" might share many of the same concerns. In 1993, the Oregon Business Council conducted a survey explicitly focused on "values and beliefs." A key finding was that "the notion that newcomers are hostile to economic development or that they are more concerned about the environment is *not* born[e] out by the data in this study" (Oregon Business Council 1993, p. 26; emphasis added).

Survey results also cast doubt on the assertion that rural residents in general, and coast residents in particular, hold a "jobs at any cost" attitude. For example, Lindberg et al. (1994) reported that while coastal residents strongly supported programs and policies that create jobs, they strongly opposed options that either involved an increase in pollution or would permit the conversion of forest and farm land to residential use.

Assessing changes in the values held by residents of, and visitors to, the Oregon coast is difficult, because little research on these issues has been done. Much of what passes as fact is, in reality, apocryphal or anecdotal or must be inferred from data gathered elsewhere and at other scales. Still, the mounting evidence that major value changes are underway—globally, nationally, regionally, locally—cannot be dismissed or ignored. Although specific information on public values within the coastal region may be difficult to obtain, future natural resource management of the area must be undertaken within the larger social context. Forest management within the Oregon coastal region must be sensitive to the values and concerns held both within and outside that region. This is particularly important, given the powerful political influence of Portland and the Willamette Valley. As we shall discuss later, this will demand creative and innovative approaches and processes.

Interactions Between Society and Forest Management

Changing forest management and policy: implications for society

Forest management in the Coast Range has been changing in recent years, mostly due to changes in both state and federal policies. The sharp decline in timber harvests on federal forest lands, along with changes in markets and technology, are changing the forest industry, which then has associated impacts on society.

While overall income levels have been rising in coastal counties, this masks economic hardship in localized areas due to decreases in timber harvesting and processing. In small, timber-dependent communities in the Coast Range, unemployment and underemployment rates have risen, property values have declined, and workers have needed to be retrained or relocated. The decline of the timber industry can also have effects that do not show up in official statistics. Large companies and employees with high-paying jobs are often the only source of philanthropic activity for a community (Cheek 1996). Many aspects of community life are dependent on volunteers and donations in order to function. Another economic impact from reduced timber harvests is a decline in taxes and payments to counties. While school funding has been equalized across counties and there are no longer major impacts from timber harvest receipts, other functions of county and local government are affected.

In addition to the economic effects, forest management affects society in many other ways. Private forest companies in the Coast Range have increasingly been closing their lands to public use, which decreases the recreation opportunities in the Coast Range and puts pressure on public lands to provide more opportunities for recreation. Forest management practices and overall harvest levels can affect the quality of life for coast residents. Because income levels affect quality of life, individuals whose incomes decline when harvests are reduced will be negatively affected. Conversely, individuals and industries that value aesthetic characteristics of uncut forests will be positively affected. Changes in management practices such as increased use of thinning and longer rotations suggested by the draft plans on state forest lands (Oregon Department of Forestry 1998) will also contribute to aesthetic values but will decrease economic returns.

The changing mix of industries in the Coast Range might actually be contributing to population increases in the region. Capacity constraints on resources such as timber and fish stocks used to place bounds on growth in areas like the Coast Range. With the increase in other industries that are not tied to resource stocks such as tourism and retirees, there seem to be fewer bounds to growth.

Changing demographics: implications for forest management

The changing demographics of the Oregon Coast Range described in this chapter are likely to have major implications for future forest management in this area. Although most population growth is expected to occur directly adjacent to the coast, new residents (and visitors) will have many opportunities to interact with coastal forest environments and decision making.

Population growth fueled by immigration means that there will be many new residents in the coastal area with different interests and values. Many of these new residents will come from urban areas, and most will still live in the developed cities of the coast. This influx will create more demand for forest amenities (e.g., recreation), resulting in more people in the forest who will observe forest management practices. Increased population in the Willamette Valley will add to the demand for coastal forest amenities.

The influx of retirees to the coast means that more people will derive their income from transfer payments, and fewer will be dependent on resource extraction industries. With relatively fewer people working directly or indirectly in the timber industry, there might be less public support for traditional forest practices that conflict with recreation or scenic demands on coastal forests. Conversely, with relatively more people working in the tourism industry, there might be more demand for aesthetic forest landscapes, which are more attractive to visitors.

These demographic changes are not likely to be uniformly distributed along the Oregon coast. The north coast will be more influenced by the rapid population growth of the Portland metropolitan area, creating more demand for forest amenities and second homes. Although the south coast has a growing retirement industry, its tourism industry is less developed and the timber industry remains an important part of the economy (11 percent of employment in Curry County and 8 percent in Coos County). Therefore, we would expect more interactions between the public and forest management on the north coast than the south.

Value changes along the Oregon coast: implications for forest management

As our previous discussion indicates, population levels and structure along the Oregon coast have changed dramatically in recent years. These changes are driven by a host of factors, ranging from the "graying" of the population to changes in public policy to rapid changes in technology and the restructuring of primary industry. The coastal population today differs sharply from that of the past in terms of who the residents are, what they do, and why they live here.

The resident population is only a part of the changing picture, however. Although the Oregon coast has long been a destination for tourism, recreation, and amenity uses, the popularity of the region has reached unprecedented levels. Burgeoning traffic levels along Highway 101, overflowing campgrounds, and the development of a large service sector, including motels, restaurants, and whale-watching trips, all indicate the popularity of the region as a vacation venue.

The changes in resident population and the increase in tourism, as well as the associated uses, values, and preferences of both, will further affect the practice of forest management along the coast. First, there is strong evidence that major value changes are underway in American society. For example, Inglehart (1990) argued that the growth in economic well-being and security has contributed to the increased interest in a host of values associated with natural resources beyond commodity values. These include educational and scientific, spiritual and religious, as well as aesthetic and amenity values. Inglehart (1990) noted a growing shift from "an overwhelming emphasis on material well-being and physical security toward greater emphasis on the quality of life" (p. 5).

Our review of census and survey data for the coastal region provides some support for this shift. Growth, driven in large part by an influx of retirees, has created a resident population in the coastal region characterized by relative economic security, through various transfer payments, equity wealth, and so on. Education levels are high among these recent immigrants. Numerous studies report a strong correlation between educational attainment and support for environmental issues. There is also widespread concern that governments have failed to secure adequate environmental protection (Harris and Taylor 1990).

However, whether this rising environmental concern is widely held among Oregon coastal residents and visitors remains problematic. The data, from either longitudinal or routine cross-sectional surveys, are simply inadequate to make a definitive conclusion. Our assessment thus must be tentative and speculative; however, we would argue that the general patterns and relationships found elsewhere in more rigorous research are consistent with the limited, sometimes anecdotal, sources of data available in the coastal region.

Second, it seems likely that many of the so-called "new" values simply represent a change in the distribution of support for various types of values. Values are complex, and a full discussion lies well beyond the scope of this chapter. However, a relatively simple typology of values, such as that suggested by Stankey and Clark (1992), might include such things as:

- Commodity values—timber, range, minerals
- Amenity values—life style, scenery, wildlife, nature
- Environmental quality values—air and water quality
- Ecological values—habitat conservation, sustainability, threatened and endangered species, biodiversity
- Public use values—recreation, subsistence, tourism
- Spiritual values—re-creative, religious, inspirational

Such values have always been found in society, but the relative distribution of support for or judgments of importance of each of them might currently be shifting. For example, commodity values are still clearly important. However, Inglehart (1990) argued that support for other values, such as public use or amenity, may be growing.

Discerning the prevalence of and support for values other than commodity is difficult. The importance of commodity values has been relatively easy to determine, because these are set largely through well-developed market mechanisms, and prices that reflect public demands are available. When such prices are unavailable, the economic marketplace cannot readily provide information about values (Koch and Kennedy 1991). Most importantly, the lack of price signals constrains the ability of the marketplace to inform policymakers about shifts in priorities (Stankey et al. 1992) and, in fact, can lead to distortions in the priorities accorded to values and uses that are exchanged largely outside the market.

A related aspect of whether "new" values are emerging derives from a concern about fairness. Here, the argument is that the immigration of new people—retirees, second-home owners, telecommuters, or others—brings "new" values into existing communities and imposes these over traditional values. Because these newcomers often have high levels of political sophistication and capacity, their ability to influence policy and management may exceed that of the locals, giving the newcomers what might be seen as an unfair advantage.

Based on studies in northern California timber-dependent communities, Fortmann and Kusel (1990) put forward an alternative hypothesis. They suggested that new residents are not necessarily the source of "new" attitudes, but because they may be more able to act politically (e.g., participate in public involvement meetings, interact with politicians), they provide a voice for attitudes and values already present in these communities. On the Oregon coast, for example, we might find that residents have long been concerned about sustained production of salmon or the long-term productivity of Coast Range forests. However, these values may not have been explicitly articulated in the past. Today, new residents who share these values may also be better able and willing to raise such concerns in a public arena. Thus, long-held values might appear "new."

Third, an increasingly common feature of many environmental management issues today is that the support and opposition are spread geographically—local issues become state issues, state issues become national issues, and so on. Thus, management of forests along the Oregon coast will be assessed and evaluated by a larger body politic than just residents or tourists. Communication technologies contribute to the widespread concern about issues. The evening news, for example, frequently includes real-time and often graphic portrayals of environmental issues, be they demonstrations against the World Trade Organization, oil spills, landslides, or forest fires. Such events, and the media coverage they provoke, can quickly turn even the most local debate into a much wider concern. Thus, questions of appropriate forestry practices in the Coast Range will be debated as much in Portland and Eugene as in Coos Bay or Astoria.

Fourth, for a variety of reasons, the public is demanding greater levels of substantive involvement in natural resource decisionmaking. In part, this reflects a normative view that citizens in a democracy should be able to participate in decisions that have potential consequences for them. It also reflects the importance of the various goods and services associated with natural resource management. Simply put, these values are important to people and therefore decisions that might affect the production, maintenance, and distribution of these values are of concern to them. Finally, the push for greater public participation reflects many citizens' growing skepticism and distrust of governments, experts, and science. Increasingly, many people believe that public scrutiny and involvement are essential, even when the issues are highly technical.

For example, the Oregon State Board of Forestry recently proposed to codify the maximization of timber revenue from state lands. Some argue that such a policy already exists in statute and that it would not and does not disregard other forest values and uses; however, it is likely that this proposal will face considerable opposition (e.g., regarding effects on recreation or on salmon recovery). Another example is Tillamook State Forest, which is of increasing interest to the public as its trees approach harvesting age and as decisions are being made about the forest's direction and primary purpose. With the burgeoning urban population of Portland and its satellite suburbs only an hour away, it is likely that the question of future land use in the Tillamook State Forest will be sharply debated.

Summary and Conclusions

One interpretation of the preceding discussion might be that the changing sociopolitical landscape of the Oregon coastal region will result in a virtual end to production forestry, at least on public lands. We argue that this is *not* inevitable, but the changing sociopolitical landscape (which reflects the changes taking place across the nation) *does* require new and creative approaches to planning processes and especially to the involvement of citizens in those processes. Citizens who are deeply concerned with natural resource management issues, who are generally well educated, and who often bring well-developed political capacity to the arena are clearly important assets. Educated, caring, politically active citizens are essential to informed governance or civic science (Shannon 1991; Lee 1993).

The challenges and questions confronting forest management in the coastal region are complex. For example, in the Tillamook State Forest, any policy choice (including no decision) will carry important consequences and implications. Decisions to constrain or even eliminate timber production would affect timber availability, local employment, and prices for consumers; it might also have more complex effects on things such as fire danger or insect or disease infestations. Conversely, a sole focus on maximizing revenue through timber production could adversely affect a host of other values, despite assurances that these other values would be maintained and enhanced. It simply might prove impossible to do this.

How do policymakers, resource managers, and citizens contend effectively with the complex array of sectoral interests, such as commodities and recreation; spatial and temporal bounds; private-public interactions; high levels of scientific and technical complexity; national, regional, and local interests; competing values; and differing beneficiaries? Clearly, natural resource management professionals face a major challenge: developing effective institutional structures that can operate in the face of high levels of complexity, involving both technical-scientific uncertainty and the lack of social consensus about goals (Thompson and Tuden 1987). The recent emergence of ecosystem management has been both praised as long overdue and criticized for its lack of specificity. A central concern is that our current institutional structures (agencies, laws, policies) cannot achieve an ecosystem approach (e.g., Slocumbe 1993; Grumbine 1994).

At the outset of this chapter, we discussed new institutional structures for management. Watershed councils represent an example of new structures that are designed to link decision making to local conditions and contexts. Yet there is a two-fold challenge confronting such locally grounded structures. First, and perhaps ironically, these organizations have gained attention in part because of the growing recognition that an ecosystem approach is needed. However problematic the notion of ecosystem management might be, it likely involves broader spatial scales, longer time frames, and an integrated approach that embraces the biophysical, social, and economic dimensions. But this leads to what Lee and Stankey (1992) described as the "paradox of scale": while our view of the biophysical world is expanding to embrace larger scales, we are also concerned with enhancing decision-making roles at local levels. Many efforts have been made to establish collaborative decision-making structures that operate on the basis of consensus (e.g., Daniels and Walker 1996). Such approaches have also been criticized, however, as forsaking traditional democratic processes that are implemented through statutory means (McCloskey 1995).

A second challenge derives from the extent to which more locally grounded structures and processes can adequately represent the full range of values, interests, and uses present across society. For example, would a watershed council or similar structure located along the southern Oregon coast be able to effectively account for the interests, values, and concerns of people from Portland, Pendleton, or outside the state? If this proves problematic, what processes might be undertaken to provide a broader base from which local decisions could operate? Future institutional changes will need to address these questions in order to effectively minimize conflict and maximize societal benefits from forests.

Future Research Needs

Recent research efforts have recognized the need for an integrated approach in order to address the complex interactions of forest ecosystems. While some of these efforts include socioeconomic components, there is still a need to develop new ways to truly integrate the social system into an ecosystem framework. The Coastal Landscape Analysis and Modeling Study (CLAMS), which is currently underway in the Oregon Coast Range, represents one attempt to do this, but existing data and models do not adequately represent the linkages between the biological, physical, and social systems.

Economic relationships between forest commodities and human communities are the easiest to incorporate into an ecosystem model. Changes in the biological or physical systems that influence commodities can be readily converted into changes in economic value, as well as changes in jobs and income associated with those commodities. When the changes affect nonmarket goods and services (e.g., some types of recreation, species extinction, cultural values), it is much more difficult to estimate the changes in economic value. Many recreation values can be estimated with nonmarket valuation techniques, but much more research is needed to address the value of preserving species or cultural values. Concepts that are important to ecologists, such as biodiversity, need to be defined so that the public can reasonably understand them and think about their value in relation to other goods and services.

There is also a need to better understand the "production functions" for recreation opportunities that are supplied by forest ecosystems. For example, if forest managers increase the amount of thinning in the Coast Range and decrease the number of clearcuts, how will recreationists react? We need to collect better data on dispersed forest recreation use so that models can be developed to show how people respond to alternative forest management activities. In some cases, biological data are lacking. For example, we may know how anglers respond to increasing or decreasing fish populations in a stream, but we may not know how fish populations respond to changes in forest management activities.

Another area that requires future research is understanding how quality of life is related to forest ecosystems. Some people feel that forest management activities detract from the environmental aspects of quality of life and can hinder economic growth in other sectors of the economy, such as high-tech companies, tourism, and retirement (Power 1996). However, employment and income are certainly major components of quality of life, and timber harvesting and wood-processing jobs are often the preferred (and sometimes the only viable) alternative for some residents of the Coast Range.

While we have described a number of ways that forest management interacts with human communities, expressing those interactions in any type of systems model can be quite difficult. Many of the relationships are qualitative, and thus the challenge for future research is to integrate qualitative relationships into ecosystem models.

It will be important for future research to be truly interdisciplinary, and not just multidisciplinary, and yet many institutional barriers in our research system make integrated research difficult. Within any institution, researchers are usually aligned by discipline or function, making the lines of communication fairly narrow. There are also incentive systems, such as peer-reviewed publications, that do not fit well into an integrated research framework. Most researchers are rewarded for publishing in the top journals in their discipline, yet interdisciplinary research would seldom be published there. Also, an interdisciplinary effort and its concomitant institutional support and funding are difficult to sustain over the long periods of time it takes to accomplish the research. By necessity, interdisciplinary research requires multiple principal investigators; if some move on to other jobs during the course of a long project, the whole project can be set back months while replacements are sought.

For future research efforts to be successful, it will be necessary to develop appropriate incentive systems and institutions that make interdisciplinary research possible. Changes will have to take place throughout the research system. Funding sources will have to go beyond disciplinary lines, and agencies will have to pool resources to fund research that covers multiple forest resources. Research institutions will have to develop ways to streamline the process for people in different disciplines to work together. Journals and peer reviewers will have to recognize the scholarship of interdisciplinary research, or new journals will have to be started. Finally, individual researchers must broaden their outlook to try to understand how their discipline interacts with others in the context of forest ecosystems.

The central thesis of this chapter was that *the changing sociopolitical landscape of the region will have profound impacts on the nature, extent, and level of forest management on the Oregon coast in the 21st century.* If we are to make progress in understanding forest ecosystems and making wise forest policy choices, we must find new ways to integrate the social sciences into ecosystem research. We must begin to understand how changes in forest management affect human communities, and how changes in human communities impact forest management. This chapter has laid a foundation for future research to explore these issues in the Oregon Coast Range.

Literature Cited

Blahna, D. 1990. Social bases for resource conflicts in areas of reverse migration, pp. 159-178 in *Community & Forestry: Continuities in the Sociology of Natural Resources*, R. Lee, D. R. Field, and W. R. Burch Jr., ed. Westview Press, Boulder CO.

Brunson, M. W., L. E. Kruger, C. B. Tyler, and S. A. Schroeder, tech. eds. 1996. *Defining Social Acceptability in Ecosystem Management: A Workshop Proceedings.* General Technical Report PNW-GTR-369, USDA Forest Service, Pacific Northwest Research Station, Portland OR.

Brunson, M. W., B. Shindler, and B. S. Steel. 1997. Consensus and dissension among rural and urban publics concerning forest management in the Pacific Northwest, pp. 83-94 in *Public Lands Management in the West: Citizens, Interest Groups, and Values*, B. S. Steel, ed. Praeger, Westport CT.

Bunting, R. 1997. *The Pacific Raincoast: Environment and Culture in an American Eden, 1778-1900.* University of Kansas Press, Lawrence.

Carroll, M. S., and R. G. Lee. 1990. Occupational community and identity among Pacific Northwest loggers: implications for adapting to economic change, pp. 141-155 in *Community & Forestry: Continuities in the Sociology of Natural Resources*, R. G. Lee, D. R. Field and W. R. Burch, Jr., ed. Westview Press, Boulder CO.

Caulfield, H. P., Jr. 1975. *The Politics of Multiple Objective Planning.* Unpublished paper presented at the Multiple Objective Planning and Decision-Making Conference, Water Resources Research Institute, University of Idaho, Boise.

Cheek, K. A. 1996. *Community Well-being and Forest Service Policy: Reexamining the Sustained Yield Unit.* MS thesis, Oregon State University, Corvallis.

Clarke, J. N., and D. C. McCool. 1996. *Staking Out the Terrain: Power and Performance Among Natural Resource Agencies.* State University of New York Press, Albany.

Clawson, M. 1975. *Forests: For Whom and for What?* John Hopkins University Press, Baltimore MD.

Committee of Scientists. 1999. *Sustaining the People's Lands: Recommendations for Stewardship of the National Forests and Grasslands into the Next Century.* US Department of Agriculture, Washington DC.

Corcoran, P., F. Conway, L. Cramer, B. DeYoung, R. Johnson, and A. M. Morrow. 1997. *Coastal Recreation and Retirement Study: Perceptions of Residents.* Sea Grant Report, Oregon State University Extension Service, Corvallis.

Dana, S. T., and S. K. Fairfax. 1980. *Forest and Range Policy: Its Development in the United States.* McGraw-Hill Book Co., New York.

Daniels, S. E., and G. B. Walker. 1996. Collaborative learning: improving public deliberation in ecosystem-based management. *Environmental Impact Assessment Review* 16: 71-102.

Davis, S. W., and H. D. Radtke. 1994. *A Demographic and Economic Description of the Oregon Coast.* Oregon Coastal Zone Management Association, Inc., Newport.

Dean Runyan Associates. 1996. *Economic Impacts of the Oregon Travel Industry, 1991-1995.* Oregon Tourism Commission, Salem.

Dunlap, R. E. 1991. Trends in public opinion towards environmental issues: 1965-1990. *Society and Natural Resources* 4: 247-258.

ECONorthwest. 1997. *1997 Casino Revenue Survey.* Oregon State Lottery, Salem.

FEMAT (Forest Ecosystem Management Assessment Team). 1993. *Forest Ecosystem Management: An Ecological, Economic, and Social Assessment.* Report of the Forest Ecosystem Management Assessment Team, U.S. Government Printing Office 1973-793-071. U.S. Government Printing Office for USDA Forest Service; USDI Fish and Wildlife Service, Bureau of Land Management, and Park Service; U.S. Department of Commerce, National Oceanic and Atmospheric Administration and National Marine Fisheries Service; and the U.S. Environmental Protection Agency. Portland OR.

Firey, W. A. 1960. *Man, Mind, and Land: A Theory of Resource Use.* Free Press, Glencoe IL.

Fortmann, L., and J. Kusel. 1990. New voices, old beliefs: forest environmentalism among new and long-standing rural residents. *Rural Sociology* 55: 214-232.

Gallup Report. 1989. The environment, pp. 2-12 in *Gallup Report No. 285.* George Gallup International Institute, Princeton NJ.

Grumbine, R. E. 1994. What is ecosystem management? *Conservation Biology* 8(1): 27-38.

Harris, L., and H. Taylor. 1990. Attitudes to environment. *World Health Forum* 11: 32-37.

Inglehart, R. 1990. *Culture Shift in Advanced Industrial Society.* Princeton University Press, Princeton NJ.

Johnson, R., F. Obermiller, and H. Radtke. 1989. The economic impact of tourism sales. *Journal of Leisure Research* 21(2): 140-154.

Johnson, R., V. Litz, and K. A. Cheek. 1995. *Assessing the Economic Impacts of Outdoor Recreation in Oregon.* Oregon State Parks and Recreation Department, Salem.

Koch, N. E., and J. J. Kennedy. 1991. Multiple-use forestry for social values. *Ambio* 20: 330-333.

Lee, K. 1993. *Compass and Gyroscope: Integrating Science and Politics for the Environment.* Island Press, Washington DC.

Lee, R. G., and G. H. Stankey. 1992. Evaluating institutional arrangements for regulating large watersheds and river basins, pp. 30-37 in *Watershed Resources: Balancing Environmental, Social, Political, and Economic Factors in Large Basins,* P. W. Adams and W. A. Atkinson, compil. Forest Engineering Department, Oregon State University, Corvallis.

Lindberg, K. 1995. *Assessment of Tourism's Social Impacts in Oregon Coast Communities Using Contingent Valuation, Value-attitude, and Expectancy Value Models.* PhD dissertation, Department of Forest Resources, Oregon State University, Corvallis.

Lindberg, K., R. Johnson, and B. Rettig. 1994. *Attitudes, Concerns, and Priorities of Oregon Coast Residents Regarding Tourism and Economic Development: Results from Surveys of Residents in Eight Communities.* Oregon Sea Grant, Oregon State University, Corvallis.

McCloskey, M. 1995. *Report of the Chairman of the Sierra Club to the Board of Directors, November 18, 1995.* Sierra Club, San Francisco CA.

Office of Economic Analysis. 1997. *Long-term Population and Employment Forecasts for Oregon.* Department of Administrative Services, State of Oregon, Salem.

Oregon Business Council. 1993. *Oregon Values & Beliefs: A Summary.* Oregon Business Council, Salem.

Oregon Department of Forestry. 1998. *Draft Northwest Oregon State Forest Management Plan* (April). Oregon Department of Forestry, Salem.

Oregonian. 1997. Oregon's crowded coast: A three part series. July 6-8, 1997. Portland OR.

O'Toole, R. 1988. *Reforming the Forest Service.* Island Press, Covelo CA.

Potapchuk, W. R. 1991. New approaches to citizen participation: building consent. *National Civic Review* 80: 158-168.

Power, T. M. 1996. *Lost Landscapes and Failed Economies: The Search for a Value of Place.* Island Press, Washington DC.

Ribe, R. G. 1989. The aesthetics of forestry: What has empirical preference research taught us? *Environmental Management* 13(1): 55-74.

Shannon, M. A. 1991. Resource managers as policy entrepreneurs. *Journal of Forestry* 89(6): 27-30.

Slocumbe, D. S. 1993. Implementing ecosystem-based management. *BioScience* 43: 612-623.

Stankey, G. H., and R. N. Clark. 1992. *Social Aspects of New Perspectives in Forestry: A Problem Analysis.* Grey Towers Press, Milford PA.

Stankey, G. H., P. J. Brown, and R. N. Clark. 1992. Allocating and managing for diverse values of forests: the market place and beyond, pp. 257-272 in *Proceedings from Integrated Sustainable Multiple-use Forest Management Under the Market System, IUFRO International Conference, September 6-12, 1992,* N. E. Koch and N. A. Moiseev, compil. IUFRO and the Committee on Forest of the Russian Federation Ministry of Ecology and Natural Resources, Pushkino, Moscow Region, Russia.

Steel, B., and M. Brunson. 1993. *Western Rangelands Survey: Comparing the Responses of the 1993 National and Oregon Public Surveys*. Unpublished, Department of Political Science, Washington State University, Vancouver.

Steel, B. S., and N. P. Lovrich. 1997. An introduction to natural resource policy and the environment: changing paradigms and values, pp. 3-16 in *Public Lands Management in the West: Citizens, Interest Groups, and Values*, B. S. Steel, ed. Praeger, Westport CT.

Steel, B., P. List, and B. Shindler. 1993. Conflicting values about federal forests: a comparison of national and Oregon publics. *Society and Natural Resources* 7(2): 137-153.

Steel, B. S., P. List, and B. Shindler. 1997. Managing federal forests: national and regional public orientations, pp. 17-32 in *Public Lands Management in the West: Citizens, Interest Groups, and Values*, B. S. Steel, ed. Praeger, Westport CT.

Thompson, J. D., and A. Tuden. 1987. Strategies, structures, and processes of organizational decision, pp. 197-216 in *Comparative Studies in Administration*, J. D. Thompson, P. B. Hammond, R. W. Hawkes, B. H. Junker and A. Tuden, ed. Garland Publishing Co., New York.

Yankelovich, D. 1991. *Coming to Public Judgment: Making Democracy Work in a Complex World*. Syracuse University Press, Syracuse, NY.

Zimmermann, E. 1950. *World Resources and Industries*. Harper and Bros., New York.

The Ecological Basis of Forest Ecosystem Management in the Oregon Coast Range

Thomas A. Spies, David E. Hibbs, Janet L. Ohmann, Gordon H. Reeves, Robert J. Pabst, Frederick J. Swanson, Cathy Whitlock, Julia A. Jones, Beverly C. Wemple, Laurie A.Parendes, and Barbara A. Schrader

Introduction

Human relationships with the forests in the Oregon Coast Range and indeed everywhere are characterized by at least three paradoxes. First, many of the essential qualities of these forests (e.g. large trees and large accumulations of dead wood) arise from long periods of slow development, (by standards of human institutions and human life spans) but many of these same qualities cannot be sustained without periods of rapid, destructive change. Secondly, we humans and our institutions prefer order and predictability, yet forest ecosystems are marked by variability and unpredictability. Finally, the ecological consequences of our actions may not appear for many years, and some effects may appear far from the site of the initial action. Forest managers face a two-fold challenge: (1) to find a balance among the forces of growth and destruction, predictability and unpredictability, that produces desired values (goods and services); and (2) to know how actions at one point in time or space affect values at another point. These challenges, although great, pale in comparison to the challenge faced by society to determine just what mix of values are desirable in the first place. The determination of what is socially desirable, however, must be set within the context of what is biologically and economically possible. The information in this chapter is intended to help to set the ecological constraints on the output of social and economic values of the Coast Range forests.

We discuss twelve major ecological themes (regional environment, ecosystem types and patterns, vegetation in geologic history, deciduous forests, riparian zones, productivity, disturbance, tree death and decomposition, forest development, human influences, road effects, and aquatic-terrestrial linkages) that we believe form the foundation of ecologically based forest management in the Coast Range. These are divided into three general categories: (1) ecosystem patterns and history, (2) disturbance and vegetation development, and (3) landscape interactions. We conclude with a discussion of how an understanding of natural processes can contribute to reaching ecosystem goals. We draw primarily on information developed in the Coast Range but include information from other parts of Oregon and other regions where appropriate.

Ecosystem Patterns and History

Regional environment

The diverse environments of the Oregon Coast Range provide the context for ecological processes and constrain the range of management options for the region's forests and streams. The Coast Range is one of 10 physiographic provinces—large areas of relatively similar geology and climate—in Oregon (Franklin and Dyrness 1988). Except along the coast and the major rivers, the terrain of the province is quite rugged, with sharp ridges and steep slopes. The Coast Range mountains are the major topographic and climatic divide in the province (Figure 3-1a-c in color section following page 52). Elevations of main ridge summits range from about 450 to 750 meters, and reach a high of 1,249 meters on Marys Peak. The Coast Range province encompasses all or part of 20 major watersheds (Figure 3-1c).

Most of the geologic strata underlying today's coastal landscapes (Figure 3-1b) were laid down during the Tertiary period, from 75 to 2.5 million years ago. Geologic formations are primarily marine sandstones and shales, with basaltic volcanic rock, and related intrusive igneous rocks. Before the sea floor was uplifted to form the Coast Range, vast marine sedimentary beds were deposited 60 to 40 million years ago. Volcanic rocks also were formed over extensive areas northeast of present-day Tillamook during this period, and pillow basalts were deposited near present-day Alsea. From 40 to 26 million years ago, sedimentary beds were laid down near Vernonia and along the Nehalem River, extensive basalt flows occurred near what is now the Columbia River, and scattered igneous intrusions capped many of the prominent peaks. During the Pleistocene, which began 2.5 million years ago, rising sea levels deposited sands along the coast and drowned the mouths of coastal rivers.

Soils of the coastal province vary widely because of complex interactions among rock type, climate, vegetation, and physiography over time. Most soils are well drained and have poorly developed horizons, dark surface horizons high in organic matter, and high capacity to hold exchangeable cations. Such soils develop in areas of high winter precipitation and moderate winter temperatures, conditions typical of most of the province. In the southeastern portion of the province, where summers are warmer, soils generally are more strongly weathered, with more fully differentiated horizons. Characteristics of soils that have developed on the extensive sandstone deposits of the coastal province vary widely. Soils on steep, smooth mountain slopes tend to be shallow, with a stony-loam texture, whereas soils on uneven, benchy, and unstable slopes are deeper and derived from colluvium. Sandstone soils on broad ridgetops tend to be deep, with thick surface horizons high in organic matter. Soils developed from siltstone or shale are finer-textured than those developed from sandstone, with a silt loam surface horizon and a silty clay or clay-textured subsurface horizon. Soils developed from basalt tend to be shallow and stony. Soils on old, stabilized dunes along the coast range from excessively well drained to poorly drained and comprise loamy sand or fine sand. Soils derived from alluvium along the major streams are variable, ranging from well drained to poorly drained silt loams and silty clay loams.

The moderate, moist, maritime climate of the coastal province supports some of the most productive forest ecosystems in the world. In general, the maritime climate is characterized by mild temperatures; a long frost-free season; prolonged cloudy periods; narrow seasonal and diurnal fluctuations in temperature; mild, wet winters and cool, relatively dry summers; and heavy precipitation. Most precipitation falls as rain from October to March, resulting from cyclonic storms that approach from the Pacific Ocean on the dominant westerlies. Storm tracks shift northward during summer, and high-pressure systems bring fair, dry weather to much of the province, although coastal fog is common during this period. Large-scale patterns of variation of the coastal climate are associated with proximity to the Pacific Ocean, orographic effects, and latitudinal gradients. To varying degrees, the coastal mountains, oriented north-south, block maritime air masses and the moderating effect of the Pacific Ocean from the eastern slopes of the Coast Range and the interior valleys. From west to east rainfall decreases (Figure 3-2a in color section following page 52), winters become colder and summers warmer, and annual variability in temperature increases (Figure 3-2c in color section following page 52). In addition, precipitation decreases (Figure 3-2a) and temperature increases from north to south. During the summer, a critical period for plant establishment and growth in the coast, climatic conditions range from hot and dry in the southeast to cool and moist in the northwest (Figure 3-2b in color section following page 52).

In conclusion, the coastal province encompasses a broad array of environments that provide habitat for terrestrial and aquatic plants and animals, and a template for ecosystem processes. The sources of this ecological diversity are many, but the most important at a regional scale include landform, geology, soils, and climate. These broad-scale patterns of environment must be taken into account when managing these forests. Forest growth and dynamics will differ among climatic and geological zones and management plans and practices will, in many cases, differ as a result.

Forest ecosystem patterns

The ecosystems of the Coast Range vary over spatial scales ranging from ecoregions (e.g. the coastal Sitka Spruce [*Picea sitchensis*] Zone) (Franklin and Dyrness 1988) to microsites. This variability is important for several reasons. The diversity of environments and biological communities enables different species with different needs to coexist. Variation poses challenges for forest management plans and monitoring activities which must be tailored to different environments (coastal, interior, riparian, upslope) and different spatial scales (stands, landscapes, watersheds, ecoregions) if they are to achieve their objectives. Knowledge of how ecosystems vary with spatial scale or extent is important because scale affects management potential: a greater diversity of ecosystem values can be produced in an area the size of the entire Coast Range than in an area the size of an individual stand or a small watershed.

Broad-scale vegetation patterns follow climatic patterns in the Coast Range (Ohmann and Spies 1998). Forest vegetation, which is strongly dominated by conifer species, varies most strongly along a gradient of summer moisture stress, annual precipitation, and seasonal temperature, ranging from the southwest interior to the north coast. Woody plant species on the dry end of this gradient (southwest interior) include Oregon white oak (*Quercus garryana*), Pacific madrone (*Arbutus menziesii*), and poison oak (*Rhus diversiloba*) (Figure 3-3). Those on the moist end (north coast) include Sitka spruce, western hemlock (*Tsuga heterophylla*), false azalea (*Menziesia ferruginea*) and salmonberry (*Rubus spectabilis*). Additional regional variation in vegetation is influenced by soil and geological factors such as coastal dunes, which support communities of grasses, shrubs, and shore pine, and wet soils around the Willamette Valley floor which support communities dominated by Oregon ash (*Fraxinus latifolia*) (Ohmann and Spies 1998).

Ecologists recognize three major vegetation zones in the Coast Range, which are named after the dominant tree species of the late-successional stage of development in that environment (Franklin and

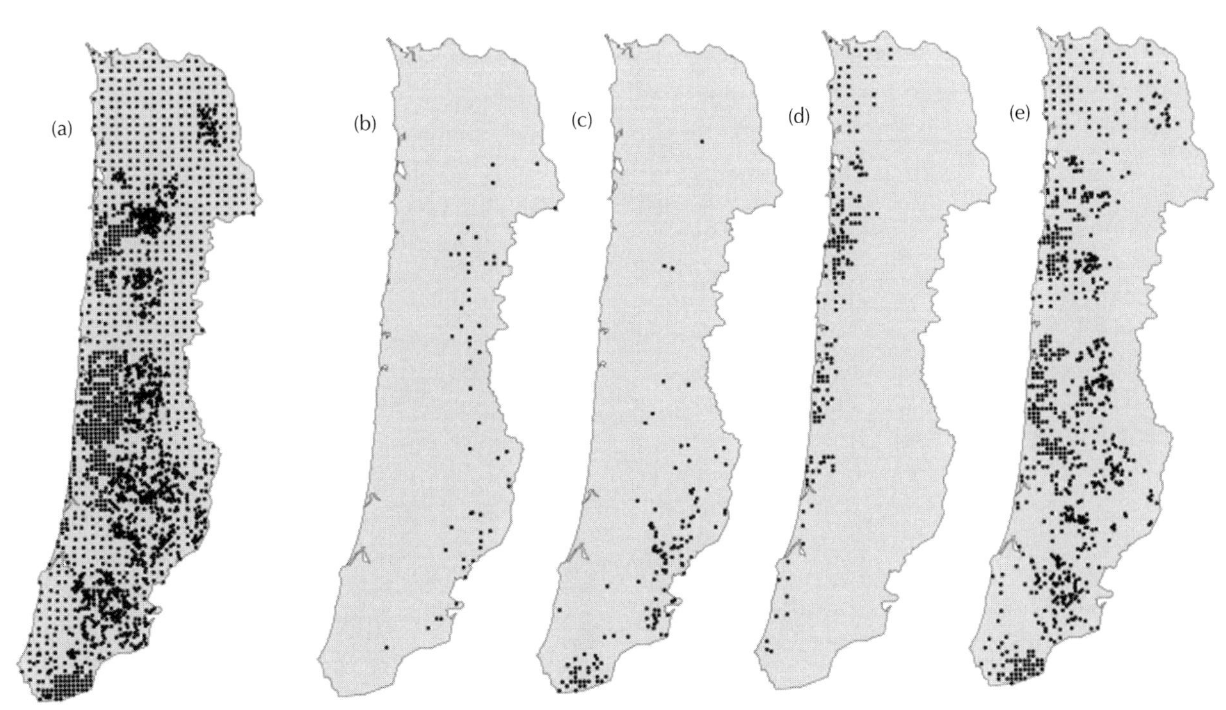

Figure 3-3. Occurrence and distribution of (a) all plots, (b) Oregon white oak, (c) madrone, (d) Sitka spruce and (e) western hemlock in forest plots in the Oregon Coast Range.

Dyrness 1988). These are the Sitka Spruce Zone on the coast, the Western Hemlock Zone in the central and eastern regions, and the Oak Woodlands of the Willamette Valley foothills on the eastern margin. A fourth zone, the Silver Fir (*Abies amabilis*) Zone, occurs sporadically on isolated mountain tops in the Coast Range. Each of these zones has a characteristic set of potential natural plant communities, that is, communities that would be found in later stages of forest succession and are indicative of the climatic and soil potential of a site.

Finer-scale variation in vegetation composition and structure has been imposed on these forest zones by disturbances such as wildfires and forest cutting, which have created a mix of younger and older forests over large areas. For example, much of the Tillamook State Forest is dominated by young forests that originated following the Tillamook burns of the 1930s, '40s, and '50s (Chen 1998). Today many of the broad-scale patterns of forest age and structure in the Coast Range correspond to large ownership blocks (Figure 3-4 in color section following page 52). For example, most of the mid- and older-aged forests are on federal lands of the Bureau of Land Management (BLM), and Siuslaw National Forest, and on the Elliott State Forest.

At scales such as small watersheds or landscapes, vegetation patterns are influenced by topographic position and proximity to streams (Figure 3-5 in color section following page 52). Plant community composition typically follows topographic trends (Figure 3-6 in color section following page 52). In lower topographic positions western hemlock/salmonberry plant associations are common types, whereas on drier, upper slopes, western hemlock/salal (*Galtheria shallon*) communities are common (Hemstrom and Logan 1986). Forest structure may also be influenced by topography. Within watersheds in the interior and valley-margin areas of the Coast Range, southwest-facing sites on upper slopes and ridges have fewer trees of shade-tolerant species, higher levels of shrub cover, and less coarse woody debris than north-facing sites on lower slope positions (Spies and Franklin 1991). Shade-intolerant hardwoods are frequently dominant over conifers in lower slope positions and along streams, where moisture stress is relatively low and landslide and flooding disturbance create sites favorable for their establishment (Pabst and Spies 1999). The spatial pattern of hardwoods and conifers along streams and topographic gradients can be quite complex, however (Figure 3-7). For example, well-developed conifer stands can be found near streams on higher benches and steep streamside slopes, and hardwoods can occur near ridgetops. Additional sources of variation in vegetation at this scale include frequency and severity of disturbance. Impara (1997) found more old Douglas-fir (*Pseudotsuga menziesii*) trees in lower slope positions than in higher ones, which suggests that fires in the Coast Range have been somewhat less frequent or less intense on lower slopes than on midslopes and ridgetops. Vegetation composition is also influenced by the size of a stream and distance from a stream (Pabst and Spies 1999; Nierenberg and Hibbs 2000).

The imprint of human activities on Coast Range is evident at the watershed scale. The level, streamside areas of many of the larger watersheds of the Coast Range are dominated by agricultural lands, many used for grazing and livestock production (Figure 3-8 in color section following page 52). The slopes along the watercourse are frequently a mosaic of young forests of different ages, reflecting widespread logging over the last 50 years on all ownerships.

Within stands, variation in forest composition and structure is influenced by several factors including fine-scale disturbance and environmental patterns. Forests of all ages that have developed with little or no human influence are frequently a mosaic of canopy gaps, patches of shrubs and tree regeneration, pockets of standing dead trees, and patches of dead and down trees. Where trees have been blown down recently, gaps may be scattered throughout the stand with a characteristic pattern of pit and mound topography.

Riparian areas are especially diverse and frequently consist of a mosaic of conifers, hardwoods, shrubs, down logs, and deposits of sediment left by floods and debris flows (Figure 3-9 in color section following page 52). Many patches along streams, especially those recently flooded, lack any trees at all (Pabst and Spies 1999; Nierenberg and Hibbs 2000). The size of canopy openings may influence the tree and shrub species that develop (Taylor 1990). Patterns of regeneration of hemlock and other shade-tolerant tree species are patchy, perhaps reflecting the distribution of rotten wood seedbeds as well as the presence of nearby (< 20 meters) hemlock parent trees (Schrader 1998). Stream junctions may be repositories of large dead wood where debris flows that began as upslope landslides stop at tributary junctions.

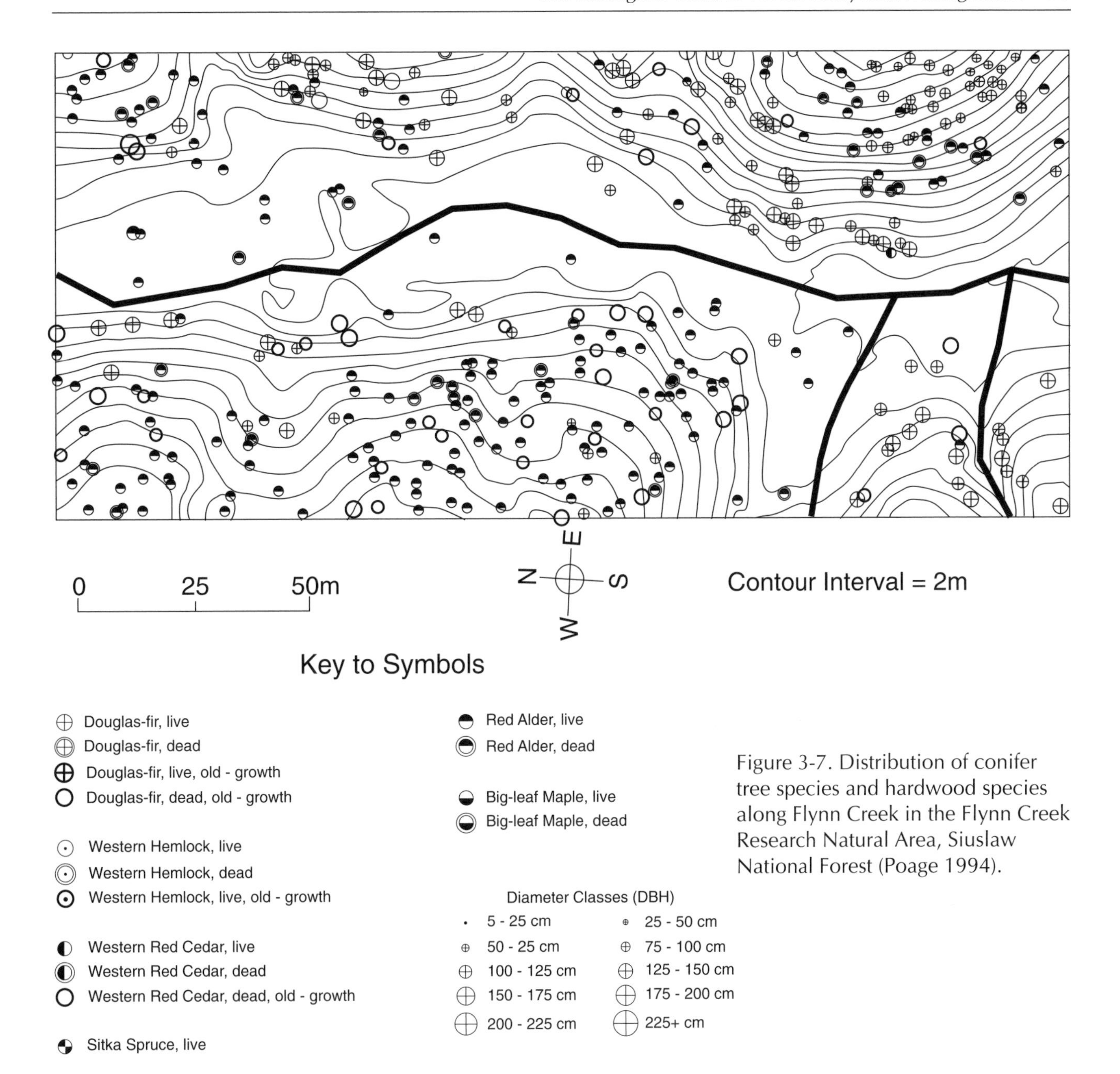

Figure 3-7. Distribution of conifer tree species and hardwood species along Flynn Creek in the Flynn Creek Research Natural Area, Siuslaw National Forest (Poage 1994).

Forests of the Coast Range may be thought of as a kaleidoscope of vegetation at different spatial scales. The patterns are constantly changing over time scales which span from decades to many centuries. Our ability to sustain desired levels of species diversity and ecosystem values is greatly dependent on the degree to which policymakers and managers take into account this dynamic mosaic in their plans and silvicultural practices. For example, at the regional level, it is clear that management practices appropriate for the interior and drier parts of the Coast Range will not be appropriate for the moist west slope and fog belt. Douglas-fir, which is the foundation of timber management throughout most of the Coast Range, will become a less important timber species in moist areas of the western slopes as Swiss needle cast (*Phaeocryptopus gaumannii*) drastically reduces growth rates in these areas (G. Filip, personal communication). Hemlock, on the other hand, is resistant to the disease and has relatively good growth rates on these sites. At landscape and watershed scales, dry sites may have a lower potential to develop large amounts of down wood than moist, lower slope positions but they may have a higher potential for developing shrubs which provide food and cover for wildlife species. The potential for conifer stands to develop near streams is somewhat lower than it is in mid- and upper-slope positions. Within stands, different habitat values are provided by canopy gaps, pockets of dead standing trees, and patches of multi-layered canopies. These structures increase the overall diversity of habitats and plant and animal species within forests.

In forest stands managed primarily for timber production, the goals are frequently to reduce spatial and species variability by filling as much of the available growing space with rapidly growing commercial timber crops. Although these efforts at homogenization are occasionally thwarted by regeneration failures, diseases, insects, and natural regeneration of noncommercial species, intensive forest management does result in less diverse stands and landscapes than might otherwise occur. The long-term consequences of these changes in diversity to ecosystem outputs are not well understood (Perry 1998); consequently, forest management in the Coast Range should be viewed as a large experiment, and as such should not proceed without controls, monitoring, and a willingness to change as new information becomes available.

Coast Range vegetation in geologic history

In the last 20,000 years, the Earth has undergone a shift from glacial conditions to the present interglacial period, the Holocene. In the course of this transition, the vast ice sheets disappeared, sea levels rose, and atmospheric carbon dioxide increased. This climatic change also triggered widespread biological reorganizations as species adjusted their ranges and abundance to form new biomes. In the Coast Range, subalpine forests and tundra, which were widespread during the last glacial period, were replaced by closed forests composed of plant species adapted to temperate conditions. This transition was most pronounced between 14,000 and 11,000 years ago, but major changes in vegetation also occurred before and since that time. Associated with, and perhaps triggering, these vegetational changes were alterations in the disturbance regime, that spatial and temporal pattern of events such as fire, windstorm, and floods that destroy vegetation and create new sites and resources for establishment and growth of new organisms. What can we learn from these past events that is relevant to understanding the modern Coast Range? What does the biotic response to previous environmental changes suggest about the sensitivity of present forests to future global change? Paleoecology, the study of past ecological interactions, provides an opportunity to answer these questions. With a long time perspective, one can consider the role of prehistoric events in shaping present-day vegetation composition and pattern. One can also evaluate whether pre-Euro-American conditions are an appropriate reference point for management, as some people have suggested (Aplet and Keeton 1999).

Our understanding of the long-term environmental history of Coast Range forests comes primarily from an analysis of the fossils preserved in the sediments of natural lakes and wetlands (Whitlock 1992). Pollen and plant macrofossils analyzed at closely spaced intervals in sediment cores are the primary tools used to infer the local vegetation history. When several sites show similar patterns of vegetational change, that is considered evidence of regional climate change.

Information on past fire occurrence comes from two sources. The first source is the forest of the present day; specifically, the study of fire-scarred tree rings and forest stand-age classes (Agee 1993; Impara 1997). These data provide fire histories with high spatial and temporal resolution, but they are only as old as the oldest living trees.

The second source of information is the analysis of particulate charcoal from cores taken in wetland sediments in which layers of abundant charcoal particles (so-called charcoal peaks) are considered evidence of a fire in the watershed (see Clark 1990; Long et al. 1998). The time between charcoal peaks in the core provides an estimate of the number of years between fires, or the mean fire interval. Charcoal records lack the spatial and temporal precision of dendrochronologic data, but they can disclose the changes in fire regime that accompany major reorganizations of vegetation over several thousand years. The interpretation of paleoecologic data is based on studies of the relationship of present-day pollen rain (deposition of pollen) and charcoal accumulation with modern vegetation, climate, and fire regimes. The better these relationships are known, the more confident we may be of the paleoecological reconstruction.

Another source of information is offered by general atmospheric circulation models or GCMs. These complex computer models, which are used routinely in present-day climate forecasting, depict the large-scale features of the Earth's climate that arise from a particular configuration of the large-scale controls of the climate system. In paleoclimatic research, the large-scale controls of interest are the variations in the seasonal cycle of insolation (produced by variations in the Earth's orbit around the Sun), the area covered by land and sea ice, atmospheric composition, and surface albedo (energy of reflected light) that occurred during the last 20,000 years (COHMAP Members 1988). The GCM simulations of full-glacial and Holocene climate conditions incorporate different configurations of these controls to show their influence on the position of the jet stream and the strength of atmospheric circulation features in northwestern North America for particular time periods (Thompson et al. 1993; Bartlein et al. in press).

The Oregon Coast Range has only a few records that describe past vegetation and climate based on pollen analysis (Heusser 1983; Worona and Whitlock 1995). An intensively studied site is Little Lake in Lane County, a landslide-dammed lake at 210 meters elevation and 45 kilometers inland from the Pacific Ocean. A 17-meter-long core from Little Lake spans the last 42,000 years, based on a series of radiocarbon dates and the known age of the Mazama ash, which is present in the core. Pollen studies from this core identify the nature of vegetation changes in the central Coast Range before, during, and after the last glacial maximum (Worona and Whitlock 1995; Grigg and Whitlock 1998). A detailed charcoal analysis of the sediments provides a record of fire history for the last 9,000 years (Long et al. 1998).

The Holocene Epoch, the last 11,000 years, has the greatest bearing on our understanding of modern ecosystem dynamics, because plant taxa (species or genera) that grow today in the region have been dominant throughout this period. Between 11,000 and 6,400 years ago, the Little Lake record revealed Douglas-fir, alder, oak, and bracken fern (*Pteridium* sp.) in higher proportions than occur in modern (2,400 years ago to present) pollen samples from the site. The charcoal record, which begins at 9,000 years ago, indicates that fires in the early Holocene were relatively frequent, with a

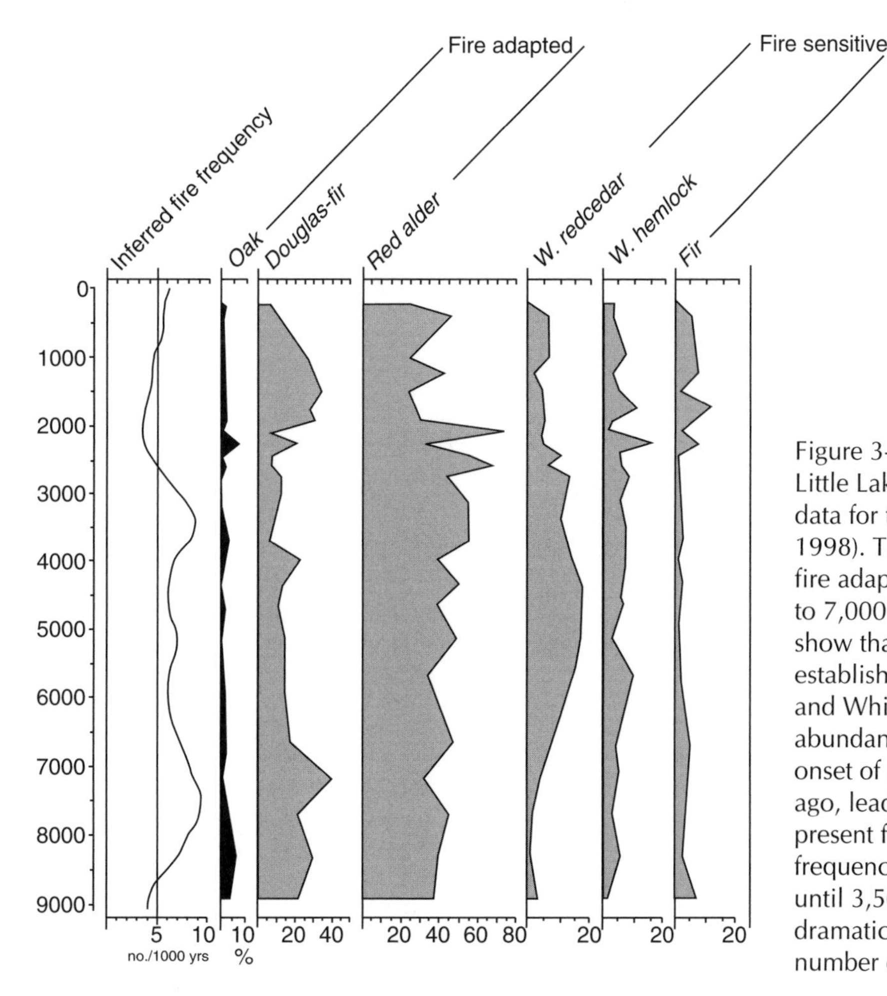

Figure 3-10. Fire and vegetation history at Little Lake based on charcoal and pollen data for the last 9,000 years (Long et al. 1998). The data show warm dry conditions, fire adapted species and frequent fires prior to 7,000 years ago. Longer pollen records show that these warm conditions were established about 11,000 years ago (Worona and Whitlock, 1995). Increases in the pollen abundance of fire-sensitive taxa imply the onset of cool wet conditions at 7,000 years ago, leading to the development of the present forest 2,400 years ago. Fires frequency, however, continued to be high until 3,500 years ago, when it dropped dramatically. In the last 2,000 years the number of fires has increased.

mean fire interval of < 175 years (Figure 3-10). The forest composition and fire regime probably resembled present-day conditions at low elevations in the western Cascade Range and at the eastern margin of the Coast Range. The Little Lake data are consistent with other records from the Pacific Northwest which suggest that the climate of the early Holocene was warmer and drier than today's climate. The dry conditions probably resulted from higher-than-present summer insolation, which in turn increased summer temperature, decreased effective precipitation, and strengthened the eastern Pacific subtropical high pressure system.

In the middle Holocene, changes in the pollen layer pattern suggest that the Coast Range became progressively cooler and more humid. This shift was part of a regionwide trend ascribed to decreasing summer insolation and the weakening of the subtropical high-pressure system in the last 7,000 years. Western redcedar (*Thuja plicata*) and western hemlock became more abundant at Little Lake, and the forest apparently had fewer openings than before, as indicated by the decline of bracken fern and oak. The charcoal record suggests that fires continued to be frequent (mean interval of < 175 years) until 3,500 years ago, despite the shift in regional climate toward cooler, wetter conditions. The fire regime may have been maintained by episodes of drought or possibly by the activities of Native Americans. Between 3,500 and 2,400 years ago, fires became less frequent, with the intervals increasing to 275-300 years. The great abundance of charcoal in the sediments at this time suggests that fires were stand-replacing events that consumed large amounts of woody biomass. Because of the wetter climate, landslides probably also increased in frequency, and they undoubtedly carried charcoal along with other sediment to streams and lakes (Long et al. 1998). About 2,400 years ago, the vegetation at Little Lake corresponded to that of the present day, being composed of Douglas-fir, red alder, western hemlock, western redcedar, and true fir (*Abies* spp.). In the last two millennia, the fire return interval has decreased to 160-190 years, suggesting that fires have become more frequent.

The paleoecologic record has three important implications for forest management. First, the composition and dynamics of Coast Range forests have not been static, but have changed along with climate over thousands of years. The Little Lake record shows changes in forest composition in the past that were most likely responses to shifts in summer drought and winter precipitation, which in turn were driven by changes in the seasonal amplitude of insolation and position of winter storm tracks (Worona and Whitlock 1995). How rapidly can forest communities in this region adjust to climate change? One episode of rapid vegetational change at Little Lake 14,850 years ago suggests that major changes in composition can occur in a matter of decades (Grigg and Whitlock 1998). During this period, spruce forest was replaced by one dominated by Douglas-fir in less than a century. Douglas-fir forest persisted for about 400 years, and then the area reverted back to spruce forest. The increase in Douglas-fir at Little Lake was accompanied by a prominent charcoal peak, suggesting that a fire, or perhaps two or three, closely spaced, helped trigger the vegetation change by killing spruce and creating soil conditions suitable for establishment of Douglas-fir. Warmer conditions allowed Douglas-fir to remain competitive for several decades before spruce returned. Other records of comparable resolution are required in order to determine whether this event is of regional significance. Nonetheless, the Little Lake data suggest that vegetation changes can occur rapidly, especially when the disturbance regime is altered. Levels of carbon dioxide and other so-called greenhouse gases are predicted to result in a global temperature rise of 2°-5°C in the next century (Intergovernmental Panel on Climate Change 1996). Paleoecological data suggest strongly that rising temperatures will lead to changes in the rates of growth of forest trees, seed production, and seedling mortality. Indirectly, they will influence the disturbance regimes of fire, insect infestation, and disease (Franklin et al. 1992). The fossil record suggests that climate change will trigger changes in disturbance regimes, creating disequilibrium in Coast Range forests as species adjust to new conditions.

The second implication of paleoecologic records for forest management is that they reveal the relatively ephemeral nature of modern plant communities. Present-day forests represent an association that has existed for less than three millennia, and in the Coast Range only a few generations of the forest dominants have been present. Plants apparently have responded to Holocene environmental changes as species rather than as whole communities, and in the process plant associations and competitive interactions have been dismantled and reformed. Our assessment of the present ecological condition of our forests should not focus on their geological age, which is a few thousand years and relatively short in terms of species generation times. Rather, we should consider their compositional diversity and structural complexity, which arose from a long history of adjustment of the species to changes in climate.

Finally, the records from Little Lake and other sites indicate that fire occurrence is closely tied to climate. The frequency of fires has fluctuated according to climatic shifts throughout the Holocene. In restoring natural fire regimes to forested ecosystems, we should recognize that studies based on tree ages and fire scars (Poage 1994; Impara 1997) sample only a small part of the prehistoric record and provide information on no more than the last stand-replacing event. The Little Lake record suggests that fires have become more frequent in the last 2,000 years, occurring about every 160-190 years. This fire regime would have produced a forest of mixed age classes at the time of Euro-American settlement. Prior to 2,000 years ago, a mean fire interval of 275-300 years in the Little Lake area suggests more extensive old-growth forests.

It would be impossible to recreate the coastal forests that existed at some point in the past, even if we could describe previous forest conditions perfectly (which we cannot) or knew the role of Native American activity precisely (which we do not). Restoration of forests based on some distant past mix of species and structures is untenable because the current set of climate conditions is unique on both centennial and millennial time scales. We cannot reconstruct the forests of the Little Ice Age, Medieval Warm Period, or the early Holocene, because the climate and disturbance regimes now are different than then. The paleoecologic record argues instead that natural ecosystems are best managed as dynamic systems, which in the Coast Range includes the possibility of changes in species' ranges and abundance and the occurrence of large stand-replacing fires because of global climate change.

The role of deciduous vegetation

One's first impressions of the Coast Range forests may be of conifers—Sitka spruce, western hemlock, and especially Douglas-fir. The landscape is dominated by conifers, and most research and management has focused on them. Yet the Coast Range also contains 14 species of broadleaf hardwood trees, each with a distinct ecology and ecological role (Niemiec et al. 1995). They are red alder (*Alnus rubra*), big-leaf maple (*Acer macrophyllum*), black cottonwood (*Populus trichocarpa*), bitter cherry (*Prunus emarginata*), Pacific madrone, Oregon white oak (*Quercus garryanna*), Pacific dogwood (*Cornus nuttalii*), white alder (*Alnus rhombifolia*), black oak (*Quercus kelloggii*), tanoak (*Lithocarpus densiflorus*), golden chinkapin (*Chrysolepis chrysophylla*), Oregon ash (*Fraxinus latifolia*), California-laurel (*Umbellularia californica*, and cascara (*Rhamnus purshiana*).

Two of these species, red alder and big-leaf maple, have quite large ranges, occurring over almost all of western Oregon. The others have much more limited ranges, although in specific locations each can be numerous. For example, Pacific madrone can be relatively common on dry sites in the southeastern portion of the Coast Range. Commercial markets exist for most of these species although only red alder occurs in enough volume to support a commercial industry.

Historically, the ranges of these species have been relatively stable for several thousand years although the abundances appear to have fluctuated, sometimes dramatically. Pollen records show that the present species composition has been in place for about 6,000 years (see below and Worona and Whitlock 1995) but abundance of each species appears to have varied by as much as 50 percent over periods of a few centuries. No one is sure of the cause of these variations but variation in fire regimes is a likely explanation.

The area covered by red alder probably increased greatly in the twentieth century as early forestry practices opened up favorable regeneration sites on slopes (Figure 3-11 in color section following page 52). Before this logging, some alder was seen on slopes, but most of it was probably restricted to riparian areas, where more frequent disturbance provided sites for regeneration (Figure 3-12 in color section following page 52). Before Euro-American settlement, Coast Range riparian areas appear to have contained a lot of alder with the amount increasing with stream size.

In southwest Oregon, forestry practices in the first two-thirds of the twentieth century increased the dominance of tanoak, madrone, and oaks (Hobbs et al. 1992). All these species either were left during logging or sprouted back from cut stumps to dominate many areas. Thus, their relative abundance, but not necessarily their absolute abundance, has increased in this area.

Coast Range hardwoods play a variety of ecological roles. Perhaps best known is alder's ability to fix nitrogen (Binkley et al. 1994). Alder uses bacteria housed in nodules on its roots to convert atmospheric nitrogen into ammonia at rates as high as 200 kilograms per hectare per year (approximately 200 pounds per acre per year; Figure 3-13 in color section following page 52). This nitrogen is used by the alder and later recycled through root decay and litter fall into the soil where it is accessible to other organisms. Most of the soils of the central and northern Coast Range have high levels of nitrogen (Bormann et al. 1994), suggesting that alder has grown on these sites at times in the past.

It has been suggested that a mix of deciduous tree leaves and conifer needles decompose faster than conifer needles alone. Such a mix would then have higher rates of nutrient cycling (Fried et al. 1988) and thus increase soil fertility (Figure 3-14 in color section following page 52). Hardwoods have also been identified with lichen diversity "hot spots" in predominantly coniferous forests (Neitlich and McCune 1997).

Hardwood trees make distinctive and important contributions to wildlife habitat. A study found that Oregon white oak provides 10 times the cavity habitat provided by Douglas-fir trees of the same diameter (Gumtow-Farrior 1991; Figure 3-15 in color section following page 52). The large fruits of oaks, madrone, chinkapin, and tanoak are a major food source for many animals and birds. The fact that the breeding territory of spotted owls in the mixed hardwood-conifer forest of southwest Oregon is smaller than in the coniferous forests to the north may be linked to high productivity in the south of the owls' prime food source, the seed-eating rodents such as the wood rat. Some wildlife species show strong and very specific connections to hardwood communities, like the white-breasted nuthatch (*Sitta carolinensis*) in oak woodlands and the white-footed vole (*Arborimus albipes*) in riparian alder (McComb 1994).

The role of hardwoods, especially red alder, in riparian areas is often underappreciated with today's emphasis on growing large conifers for in-stream structure. The abundance of alder normally increases from minor patches on most intermittent streams, to a narrow strip along first-order streams, to large patches along second-order and larger streams. Even where patches of conifers are found,

Figure 3-16. Landforms and geomorphic features of riparian areas.

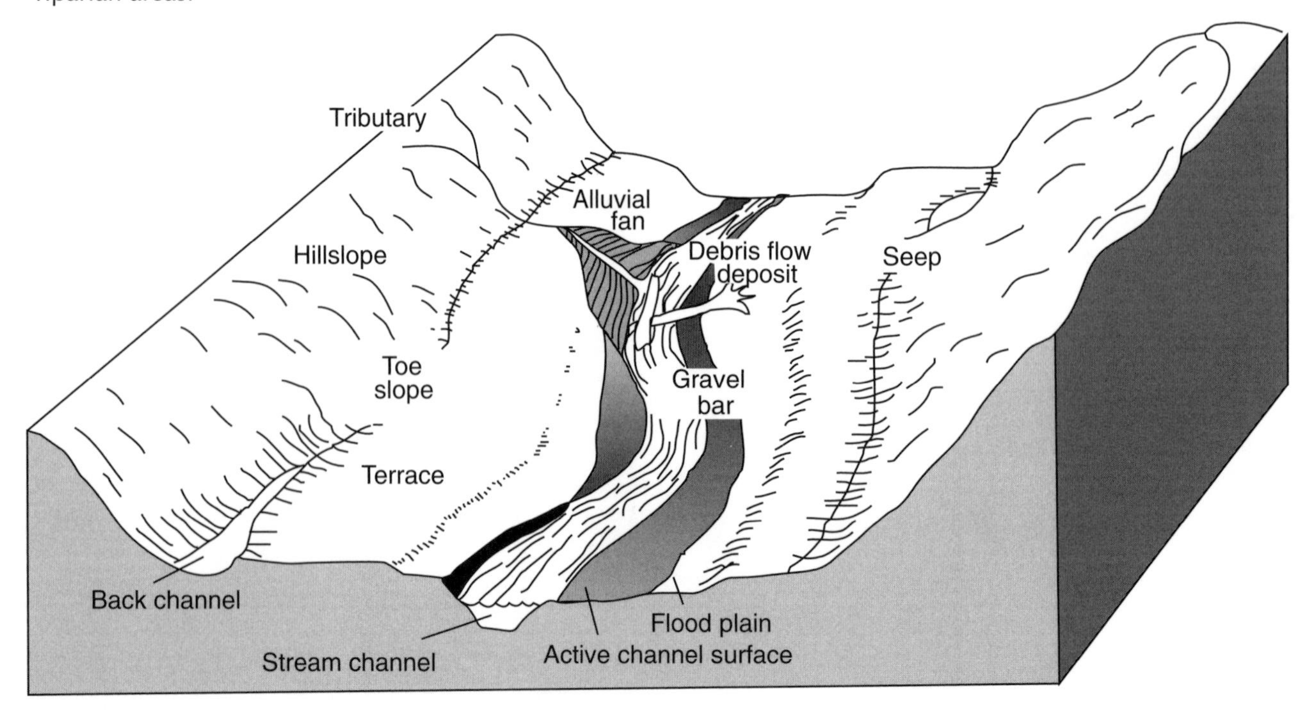

there is often a strip of alder trees immediately adjacent to the stream (D. Hibbs, personal observation). This alder is the primary source of bank-stabilizing roots; Douglas-fir is not sufficiently flood tolerant to play this role.

The nitrogen fixed by alder enters the stream in litter and groundwater, playing a role in regulating stream productivity (Gregory et al. 1991). Because deciduous hardwood leaf litter is readily decomposed, it is a major energy source for invertebrates. Alder produces more litter than Douglas-fir and therefore is a better source of food to the aquatic food chain (Zavitkovski and Newton 1971). Being deciduous, alder allows light to reach the stream in the spring and fall for in-stream photosynthesis and its leaves maintain shade in summer, when stream-temperature concerns are greatest. Hardwoods often provide the branches and small logs that transform a large conifer log into an effective sediment trap or complex hiding cover for wildlife.

Riparian forests

Riparian forests are distinctive ecosystems that thread through the coniferous landscapes of the Coast Range. The vegetation of these forests is shaped by forces originating within and beyond stream corridors. Floods, debris flows, and channel migration create landforms and expose or deposit substrates along streams (Figure 3-16). Landslides and other processes deliver sediment, wood, water, and nutrients to riparian areas from hillslopes. Other disturbances such as fire, wind, and pathogens influence both riparian and upland environments. The diversity of disturbances and unique environmental conditions that typify riparian areas are reflected in the patterns of vegetation. Riparian forests differ from upland forests in species composition, structural attributes, and consequently, in the way they function. In this section, we highlight what is known about the structure and composition of riparian forest vegetation in the Coast Range, and discuss the reasons why these patterns exist.

Riparian forests in the Coast Range are characterized by a mixture of hardwoods, conifers, and shrub-dominated openings. Tree densities are usually lower near the stream than further upslope, and conifers increase in dominance with distance and height from the stream (Figure 3-17) (McGarigal and McComb 1992; Minore and Weatherly 1994; Pabst and Spies, 1999). Conifer basal areas in mature, unmanaged riparian forests average 8-12 square meters per hectare within 5 meters of the stream, and 25-40 square meters per hectare within 30 to 50 meters of the stream. Hardwood basal areas range from 8 to 18 square meters per hectare over the same distances, but do not show clear trends with distance from the stream. These patterns of distribution are strongly related to landform over much of the Coast

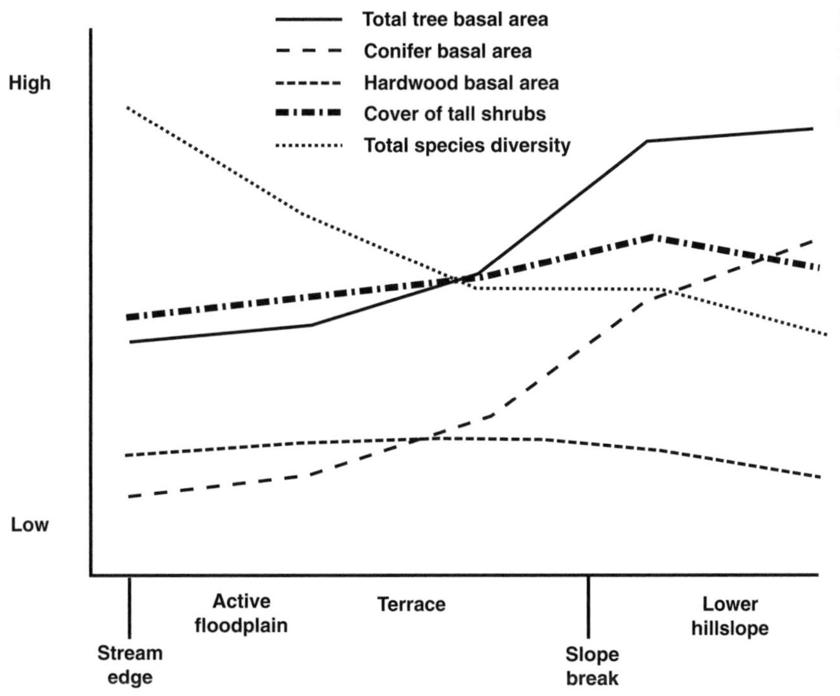

Figure 3-17. Generalized relationships between riparian forests attributes and distance from stream and landform.

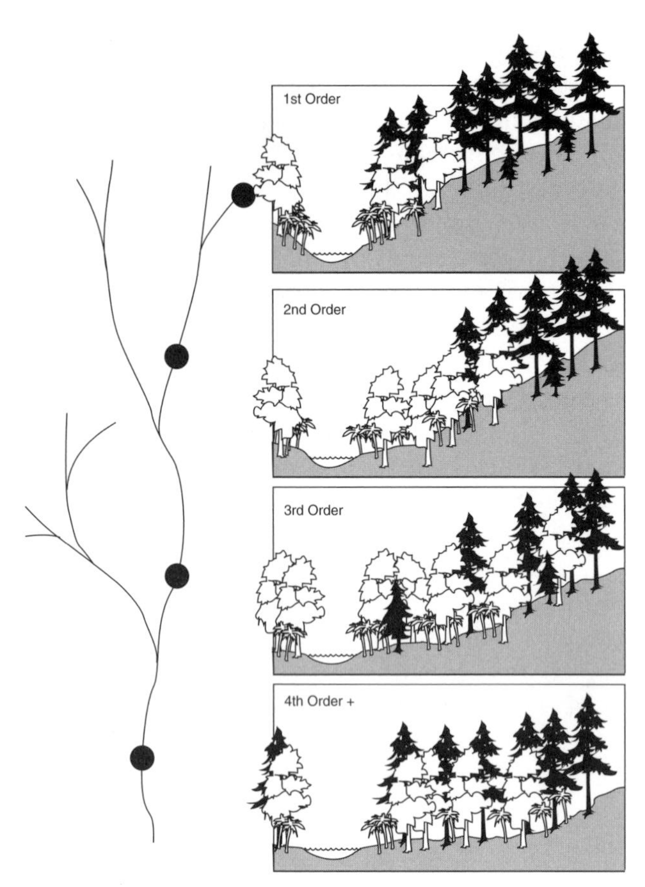

Figure 3-18. Stream valley cross-sections from first to fourth-order streams.

Range; for example, conifers are more likely to be found on hillslopes than on valley floors. The occurrence of snags reflects the distribution of live trees. Snag densities in unmanaged riparian forests range from 8 to 18 per hectare near the stream compared to about 48 per hectare upslope (Andrus and Froehlich 1988; McGarigal and McComb 1992; Pabst and Spies 1999).

A stand's location in the stream network also influences its structure and composition (Figure 3-18). Along steep-gradient headwater streams, riparian forests tend to be dominated by conifers, except where unstable slopes or competition with shrubs preclude trees from developing (Pabst and Spies 1998, 1999). The likely reason for this pattern is that small streams have less influence on the streamside environment than higher-order streams, allowing an upland environment to prevail.

Riparian forests along higher-order streams of the Coast Range generally have a mix of species, including an abundance of hardwoods. Valley floors of these larger streams are characterized by broader floodplains, greater hydrologic and topographic complexity, larger canopy gaps, and less constraint on the stream course compared to low-order streams (Naiman et al. 1992; Pabst and Spies 1999; Figure 3-18). Species composition in these locations may be affected by soil pH, soil moisture and depth to water table, soil texture and size of coarse fragments, successional status, landform, and the severity and frequency of flooding. Red alder, the dominant tree in Coast Range riparian areas, is well adapted to this environment. It germinates readily on newly deposited or exposed mineral substrates; can fix nitrogen in nitrogen-limited locations such as gravel bars; has a rapid juvenile growth rate that allows it to outcompete shrubs in some situations; can tolerate poorly drained soils and brief inundation, even during the growing season; and can sprout or produce adventitious roots (roots growing out of the stem) when the bole is buried with sediment (Harrington 1990). Bigleaf maple, western redcedar, and Sitka spruce tolerate moist soils and inundation to various degrees (Minore and Smith 1971; Walters et al. 1980), but their distribution along streams may be limited by competition, herbivory, lack of seed source, or the scarcity of favorable microsites for germination (Fried and Tappeiner 1988; Minore 1990; Emmingham and Maas 1994). Douglas-fir, which is intolerant of wet soils (Minore 1979), can grow along larger streams and rivers that are constrained by hillslopes or bedrock substrates, or where terraces are well drained and high above the water table.

Understory plant communities of Coast Range riparian forests are dominated by shrubs such as salmonberry, stinking black currant (*Ribes bracteosum*), red elderberry (*Sambucus racemosa*), and vine maple (*Acer circinatum*). These species have reproductive and physiological advantages allowing them to thrive in valley bottoms and on lower slopes (Tappeiner et al. 1991; O'Dea et al. 1995; Pabst and Spies 1998). In general, understory species composition is associated with landform and distance from the stream. For example, species diversity and the occurrence of non-native species are greater near streams and on valley floors than on adjacent lower hillslopes (Hibbs and Giordano 1996; Pabst and Spies 1998). There are two likely reasons for this: erosion and deposition provide germination sites in a relatively high, light environment for opportunistic species, and microsites such as seeps, back channels, and boulders provide unique environments that are uncommon on hillslopes.

Species composition is also influenced at broader scales by factors such as climate. For instance, riparian forests in or near the coastal fog belt (within a few kilometers of the ocean) have more Sitka spruce than elsewhere in the Coast Range, whereas those at the eastern and southern fringes have more grand fir (*Abies grandis*). In addition, there appear to be less red alder and salmonberry in riparian forests at the southern extreme of the Coast Range than in the north.

Coast Range riparian areas are difficult sites for tree establishment because of competition from shrubs (especially salmonberry), herbivory, and seed source limitations for some species. Where conifer regeneration does occur, it is usually at sites with large conifers (Minore and Weatherly 1994), particularly shade-tolerant conifers (Pabst and Spies 1999). Shade-tolerant species dominate conifer regeneration, whereas Douglas-fir (a shade-intolerant species) is uncommon (Minore and Weatherly 1994; Hibbs and Giordano 1996; Pabst and Spies 1999). Conifer regeneration is generally absent under dense cover of shrubs or hardwood trees. It is most prevalent on large, rotting, conifer logs, which provide an elevated, and possibly less-competitive, environment than the forest floor. Hardwood regeneration is dominated by red alder, which is found most frequently on newly exposed mineral substrates of floodplains and gravel bars.

In summary, compared to upslope areas, riparian forests of the Coast Range are characterized by lower densities and basal areas of live and dead trees, greater cover of shrubs, less tree regeneration, and higher diversity of plant species. Many of these stand attributes vary with landform, proximity to the stream, and location within both the stream network and the region. Hardwoods and shrubs seem better adapted than conifers to the disturbance regimes and soil moisture conditions of most riparian areas; thus, the competitive balance is tipped in their favor. For management purposes, it is important to recognize how the factors determining vegetation patterns are connected in the landscape, not only along stream networks, but between streams and adjacent hillslopes.

Productivity of Coast Range Forest and Stream Ecosystems

Ecosystem productivity is the source of the values and the goods and services, both market and nonmarket (Daily 1997), that we obtain from forests. In general, the more productive the ecosystem, the more goods and services it can supply. The Coast Range is among the most productive forest ecosystems in the world; it has the capacity to produce huge quantities of wood, the most common and easily quantified measure of forest productivity. However, the Coast Range ecosystem also produces large quantities of vegetation other than wood, as well as wildlife, fish, and water. This high output is a result of the climate and soil of the region. The high forest productivity of the Coast Range makes it a desirable location to produce timber for economic and social benefits (Figure 3-19 in color section following page 52). It also provides opportunities to produce nontimber values more rapidly and in greater quantity than most forest ecosystems in the Pacific Northwest. For example, large trees, which are important as the habitat of some wildlife and fish species, can develop more rapidly in the Coast Range than almost anywhere else in the region. In this section, we examine the basis of the high productivity of the Coast Range for both timber and nontimber resources.

Productivity of ecosystems can be defined in several ways. Net ecosystem productivity is the accumulation of matter in a given period of time by green plants (net primary productivity) plus the secondary accumulation of matter by other organisms including arthropods, insects, and vertebrates (net secondary productivity) that consume the primary production (Perry 1994; Barnes et al. 1998). In forests, net primary productivity, the most easily measured of the two, consists of the growth increment of plants, the fine and coarse litter (leaves, branches etc) they shed, the trees that die, and the plant parts consumed by other organisms. In stream ecosystems, algae, rather than trees, are the photosynthetic factories that capture sunlight and produce organic matter. Primary productivity in small streams is strongly controlled by conditions of the surrounding forest through its influence on light and nutrient inputs. As streams become larger, and are less shaded by vegetation, their primary productivity increases.

The above-ground net primary productivity of the moist coniferous forests of the Coast Range is higher than that of most forests in the world, including many tropical forests (Gohlz 1982; Kimmins 1987; Prince and Goward 1995). In the temperate zone, only the coastal redwood and evergreen broadleaf forests of northern California and southeastern Oregon may have higher above-ground productivity (Waring and Franklin 1979). Hemlock-spruce forests on the coast can produce 15 metric tons of aboveground biomass per hectare per year, whereas Douglas-fir-western hemlock stands in the Oregon Cascades produce less than 10 metric tons per hectare per year (Gohlz 1982). Trees are not the only source of productivity. In some stands in the Pacific Northwest, production by shrubs, herbs, and mosses is as much as 17 percent of the total net aboveground productivity (Long 1982). The cover of shrubs in the Coast Range, which is an indicator of biomass of shrubs and may be an indicator of shrub productivity, is higher than in any other province in Oregon (Ohmann 1996).

The most common indicator of potential productivity used in forest management is the forester's measure: site index or site quality (the total height of a tree at either 50 or 100 years). On this basis, the Coast Range is also quite productive. Much of the Coast Range is site quality II or higher for Douglas-fir (more than 170 feet in height at 100 years), although lower site quality is more common in the eastern portion of the province (Figure 3-20 in color section following page 52). The Pacific Northwest coastal region (west of the Cascade Mountains of Oregon and Washington), which includes the Oregon Coast Range, contains 7 percent of the forest land in the United States but 21 percent of the highest productivity lands, capable of producing 120+ cubic feet per acre per year of MAI (mean annual increment) at 50 to 70 years of age (Powell et al. 1993). Coast Range counties with the most productive timber lands (from highest to lowest) are Clatsop, Lincoln, Tillamook, Columbia, Coos, and Curry. The average forest land in all of these counties can potentially produce more than 100 cubic feet per acre per year of MAI.

The high rates of tree growth in the Coast Range also mean that large trees, which are important components of forest and stream habitat, can grow more rapidly than in almost any other region of the Pacific Northwest. In the Cascades, it may take over 200 years to produce trees over 170 feet tall, which is the height of the upper canopy at which many characteristics of old forest habitat begin to develop. Trees in the Coast Range can grow this tall in as little as 70 to 100 years (McArdle et al. 1961). Similarly, relatively large-diameter trees (24 inches) can grow in as little as 50 years on many sites in the Coast Range; such trees may require over 100 years to develop on many sites in the Cascade Range.

Intensive management to maximize production of wood and financial return on a site in the Coast Range (i.e., rotation lengths of 40 to 50 years) will not produce high levels of some other ecosystem goods and values including wildlife habitat associated with large trees (more than 30 inches in diameter) and large snags and down wood, carbon storage, and recreational opportunities (McComb et al. 1993; Curtis 1994). Timber productivity, as measured by MAI, may peak relatively early (50 to 70 years) but can increase or stay high for relatively long periods of time (Curtis 1994). Consequently, harvest ages of 40 to 50 years result in reduced total volume production and less than the trees' potential (Curtis 1994). Extending rotations could provide benefits not only in terms of total timber volume produced but to other forest values as well.

Why are Coast Range forests so productive? Climate is the main factor. Climate plays a major role because productivity is determined to a large degree by the constraints on growth set by climatic extremes, such as drought and freezing temperatures, that limit the ability of green plants to photosynthesize and fix carbon in plant tissues. Climatic constraints on the ability of forests to use light energy for photosynthesis are lower in the Coast Range than other regions in Oregon. For example, freezing temperature, soil drought, and vapor pressure deficit reduce the annual capture of photosynthetically active radiation (PAR) by only 8 percent compared to the potential maximum at Cascade Head on the Oregon Coast. Annual capture of PAR is reduced by 13 to 42 percent in the western Cascades and by more than 69 percent in the eastern Cascades and Juniper Zone of central Oregon, where colder winters and seasonal drought limit the time that plants can photosynthesize (Runyon et al. 1994).

In contrast to forests, where the social interest in productivity focuses on the primary producers (i.e. the trees and other vegetation), in streams the interest is on secondary productivity (i.e., consumers, especially fish, which are further up the food chain). Although the primary producers in the streams, the algae, play an important role in fish productivity, fish production is also influenced by many other factors as well, including habitat, temperature, oxygen, and mortality rates. Net primary productivity of streams in the Coast Range and other areas of western Washington and Oregon is relatively low in contrast to streams in other parts of the world (Gregory et al. 1987). The low net primary productivity of forested streams may be a consequence of the high productivity of the forests because stream productivity is strongly related to how much sunlight reaches the water. Net primary production is higher in open stream reaches; net primary productivity within recent clearcuts, deciduous forests, and closed coniferous forests averaged 210, 58, and 26 milligrams of carbon per square meter of stream surface per day, respectively (Gregory et al. 1991). In contrast, net primary productivity in a coastal coniferous forest has been measured at 1,800 milligrams of carbon per square meter per day (based on Fujimori et al. 1976). In steep, narrow coastal streams with dense conifer shade, the light reaching the stream may be less than 5 percent of full sunlight (Figure 3-21 in color section following page 52; Naiman and Sedell 1980). In larger, more open streams where red alder is dominant and in watersheds where forest cover has been removed by some disturbance, net primary productivity increases. This increased production of organic matter may move up through the food chain to produce more fish (Murphy and Hall 1981). However, productivity of fish is also related to in-stream habitat structure and temperature and thus may decline following disturbances even if net primary productivity of the stream increases (Gregory et al. 1987). Stream productivity is also influenced by inputs of fine and coarse litter from streamside and upslope vegetation. Leaves from herbs and deciduous shrubs and trees are important sources of energy and nutrients (especially nitrogen) to streams and are more readily decomposed than conifer needles (Gregory et al. 1991). Large wood contributes to stream productivity by forming dams that retain litter and sediment in the stream ecosystem and creating complex channels that are important in forming prime salmonid spawning and rearing habitat.

Although the primary productivity of streams in the Coast Range is relatively low, the Coast Range watersheds are capable of producing large amounts of fish biomass. Around the year 1900, the number of adult salmonids returning annually to coastal Oregon streams was estimated at 1,385,000 coho (*Oncorhynchus kisutch*) and 305,000 chinook (*Oncorhynchus tshawytscha*) (Lichatowich 1989). This represents over 15 million pounds of fish, assuming an average weight of 10 pounds. In the past, when many of these fish made their way into coastal streams and ended up as carcasses on the shore, they provided a source of food and nutrients for predators and scavengers, thus contributing to the productivity of the ecosystem as a whole (Willson and Halupka, 1995).

In summary, the Coast Range is one of the most productive forest ecosystems in the world, making it an ideal place to grow trees for profit and to produce wildlife species that use large live and dead trees. In contrast to the forests, the streams have relatively low productivity, largely as a result of shading by dense coniferous canopies. The productivity that these streams do have is greatly dependent on the terrestrial ecosystems that

Table 3-1. Examples of different kinds of natural disturbance agents in the Coast Range, by ecological categories.

Fire	Canopy gaps from wind, diseases, insects or animal damage	Soil disturbances	Inundation by water and sediment
Surface fires	Wind snap, breakage	Wind throw	Flooding
Understory fire	Root-rot mortality	Landslides	Deposits left by floods and debris flows
Crown fire	Death and damage from Douglas-fir bark beetle, Swiss needle cast, and Sitka spruce weevil	Debris flow	Inundation of riparian areas from beaver dams
		Stream bank failure	
		Erosion from floods	
	Herbivory by large vertebrates (beaver elk, bear)		

surround them. Consequently, what managers do in forests on the slopes will have long-lasting direct and indirect effects on various forms of productivity in the streams.

Ecological Forces of Disturbance and Development

Forest disturbances

Disturbances come in all types, shapes, and sizes in Coast Ranges forests and watersheds (Table 3-1). Ecologists define disturbances as discrete events that disrupt ecosystem function, species composition, or population structure, and change resources, sites for establishment, or the physical environment (Pickett and White 1985). Disruptions are integral to the productivity and biological diversity of the Coast Range ecosystems. For example, disturbances of one size or another control much of the timing and patterning of tree regeneration. Disturbances are often described in terms of regimes. A disturbance regime indicates the pattern of a given disturbance agent over long time periods and relatively large areas (Pickett and White 1985). A disturbance regime is described by measurements of the intensity, timing, and spatial distribution of a particular disturbance agent. Disturbances common in the Coast Range include: fire, wind, disease, insects, landslides, debris flows, flooding, and the activities of vertebrates such as beavers, bears, and humans. Although the natural disturbances of the Coast Range are diverse, they can be classified into four categories based on the general type and ecological effect (Table 3-1). These major categories are: (1) fire, (2) canopy gaps and patches resulting from forces such as wind, disease, insects, or beavers, (3) soil disturbances from landslides and floods, and (4) inundation from floods. Two or more of these can happen simultaneously; for example, wind may blow a tree over, creating a canopy gap and disrupting the soil. Humans create a fifth type of disturbance, either directly through cutting and removal of trees in small to large patches, or indirectly through activities such as fire suppression, road building, and introduction of exotic pests (e.g. gypsy moth) that affect natural disturbance regimes. In this section, we will highlight what is known about natural disturbance regimes in the Coast Range forests and briefly discuss how they influence ecosystem function and biological diversity.

Ecological responses of vegetation to these disturbances will depend on the individual species and the kind of disturbance. For example, surface and understory fires may kill the cambium of thin-barked species such as western hemlock and western redcedar, but not that of Douglas-fir, which has thick bark. On the other hand, Douglas-fir may be more susceptible than western redcedar or red alder to wet soils created by flooding

Until the advent of large-scale logging and effective fire suppression in the middle part of the twentieth century, wildfires were the dominant disturbance in Coast Range forests. The return interval for fires in the Coast Range over the last 2,000 years prior to Euro-American settlement probably ranged between 90 and over 400 years, depending on the ecoregion (Agee 1993; Long et al. 1998). The moist spruce zone in coastal and northerly areas experienced longer intervals between fires, while fire was a more frequent visitor to the valley margin areas in eastern and south-eastern parts of the Coast Range. Fire intensity ranged from severe, in which fire killed more than 70 percent of the canopy trees (Figure 3-22 in color section following page 52), to light, in which surface fires killed few canopy trees. Where fires were more frequent, they may have been less severe (Agee 1993). This general pattern was observed by Impara (1997), who found that fires were more frequent in the drier, eastern portions of the central Coast Range than in the central and coastal areas, but that more old-growth trees were present in the eastern areas. Fires were typically quite large in size. Impara (1997) estimated that mean fire sizes ranged from 66 square kilometers (16,309 acres) to over 190 square kilometers (46,930 acres) in one study area in the central Coast Range. Many fires were larger than these mean values. A large fire that occurred between 1845 and 1849 is estimated to have burned over 500,000 acres between the Siuslaw and Siletz rivers (Morris 1934). These large fires in the mid to late 1800s set the stage for extensive 100- to 150-year-old Douglas-fir and hemlock forests seen today in many areas of the Coast Range.

The estimates of fire sizes and return intervals suggest that before Euro-American settlement, the Coast Range was a slowly shifting mosaic of large and small patches of forest vegetation, ranging from shrubby areas to dense old forests dominated by conifers. Although fire has been the major force in shaping the forests in previous centuries, the Coast

Range would not ever have been totally covered by recently burned and young forests. In fact, it is estimated that the amount of old forest (more then 200 years old) probably ranged from 25 to 75 percent during the last 3,000 years (Wimberly et al. 2000).

Although wildfires have often killed trees over large areas, giving rise to landscapes of early successional forests, localized, more-frequent disturbances of canopy trees have played and continue to play important roles in advancing forest succession and creating habitat for animal species that use dead trees and multiple-layered forests. The death of individual trees and small groups of trees during stand development is a common and necessary process. During the first 100 years in the life of a moderately well-stocked stand of Douglas-fir, more than 90 percent of the trees will die from competition, disease, insects, and other causes. It may take another 400-900 years for the remaining Douglas-firs to die if there is no intervening stand-replacing disturbance. As the large trees in the canopy die, especially in middle-to-late stages of stand development, canopy openings are created that do not close from expansion of crowns of the surrounding trees. The resulting gaps are sites of increased light and moisture availability in the forest (Gray and Spies 1997) and they offer potential sites for invading shrubs and trees.

In western Oregon, canopy gaps in mature and old forests range between 0.0025 and 0.25 hectares in size (Spies et al. 1990; Taylor 1990; Figure 3-23 in color section following page 52). Near the coast, where high winds are more common, gap sizes are probably larger. In mature and old-growth forests, canopy gaps may cover 13 to 29 percent of the forest. Gaps can form at an annual average rate of 0.2 to over 1.0 percent. Thus, a point may experience a gap event every 100 to 500 years, or about the same average interval as wildfires of the past. The frequency of gap-forming events varies by topography, especially near the coast, where gaps caused by wind disturbance are more common on ridgetops and upper slopes than in lower slope positions (Ruth and Yoder 1953; Taylor 1990).

Tree species composition of coastal forests is influenced as much by gap dynamics as it is by larger-scale disturbances. For example, Taylor (1990) found that hemlock could regenerate in both small and large gaps, but that Sitka spruce could regenerate only in large gaps from 800 to 1,000 square meters in size. The occurrence of disturbances that create large gaps is apparently frequent enough to maintain spruce in the canopy of these forests. Douglas-fir probably requires very large gaps (more than 1,000 square meters) to regenerate and grow into the canopy (Spies et al. 1990). Gap formation, however, does not always lead to tree regeneration. Schrader (1998) found no relationship between abundance of gaps and hemlock regeneration in a survey of Douglas-fir stands 40 to 200 years old in the Coast Range. Other factors such as presence of seed sources, density of the shrub layer, and presence of rotted woody debris were more important than abundance of gaps in explaining the pattern of hemlock regeneration (Figure 3-24 in color section following page 52). In addition, seedling regeneration in gaps does not necessarily mean the trees will reach the canopy. In many forests, small canopy gaps will close after a few years and young trees will reach the canopy only if subsequent gap-forming events occur near the site of the initial tree regeneration.

Other disturbances that disrupt the forest floor and soil layers, such as windthrow, landslides, and floods, also are important to forest dynamics in the Coast Range, but rates, frequencies, and consequences of these types of disturbance are less well known. Landslides within intact, mature forests are typically less than 100 cubic meters in volume or 100 square meters in area (Swanson et al. 1981; Robison et al. 1999) and are confined to steep narrow low-order stream drainages. Landslides and other disturbances that expose mineral soil may be important to the maintenance of deciduous shrub and tree species, including salmonberry and alder, by providing establishment and colonization sites within the dense coniferous forests. Beavers (*Castor canadensis*) are also an important and often overlooked force in shaping the structure and dynamics of riparian and aquatic habitats. These rodents were quite common in the Coast Range in the past and are increasing in abundance again. Although beavers kill trees, they also can facilitate development of rearing habitat for coho salmon by creating large pools behind their dams (Leidholt-Bruner et al. 1992; Naiman et al 1992). In addition, the breaching of beaver ponds exposes bare substrates for establishment of herbs, shrubs, and trees in riparian areas.

In sum, the structure and composition of the forests of the Coast Range are shaped by disturbances of many kinds and at many scales. Where management goals seek to maintain the range of native forest conditions and species, the challenge for managers will be to incorporate natural disturbances of different frequencies and sizes into management plans. This could be done in several ways. For example, young stands could be thinned in small groups to simulate gap disturbances or older stands could be harvested with a green tree retention approach to simulate low- to moderate-severity wildfire.

Tree death and decomposition

Every year more than 1.5 million trees greater than 25 centimeters in diameter probably die in the Coast Range from natural causes, based on inventory estimates of 322 million live trees (J. Ohmann, personal communication) and assuming an average annual mortality rate of 0.5 percent. The accumulations of dead trees can be quite high in coastal forests of all ages. For example, the mass of dead-wood accumulations in old-growth forests of the Coast Range can exceed 25 percent of the live-tree biomass. The ratio of dead wood to live biomass in young forests that originate following wildfire or blowdown is generally much higher (100 to over 1,000 percent). When trees die, the boles and large branches continue to influence ecosystem processes and biological diversity in many ways. In this section, we will briefly review the overall dynamics of dead wood in stands, including the processes of wood decomposition, and their ecological functions.

The amount of coarse woody debris (CWD) in unmanaged stands varies according to disturbance history, stand development, and site conditions (Spies et al. 1988; Figure 3-25). Consequently, it is difficult to establish standards and guides that represent the "natural" state, because there are many natural states. Amounts of CWD are typically highest not in old-growth forests but in young forests that originated following severe disturbances that killed most of the live trees of the previous mature or old-growth stands. However, amounts of CWD can also be very low in young stands in the Coast Range if the predisturbance stands were young, with little carryover of wood into the new stand, or if the stand was logged. Carryover of dead wood from the previous stand is a legacy that can persist for many decades or centuries. It can play an important role in providing habitat for species that use dead wood if the current stand is young or

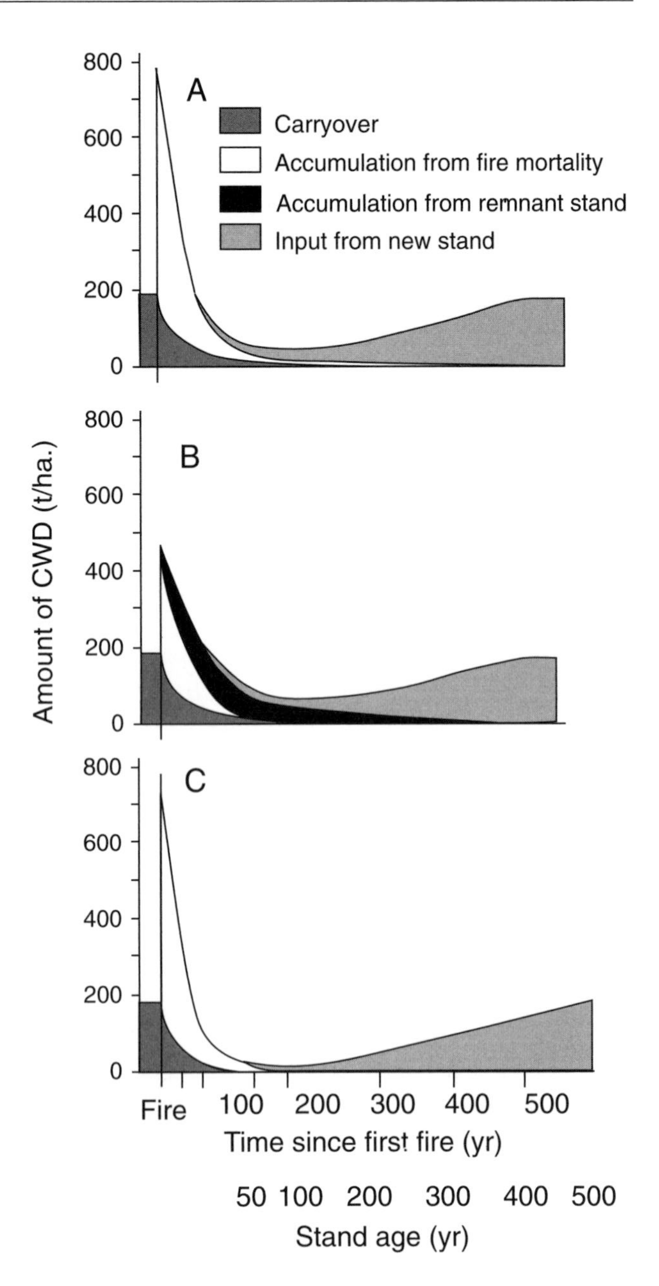

Figure 3-25. Dynamics of dead wood as a function of disturbance history (modified from Spies et al. 1988).

managed in a way that produces very little large dead wood. Many stands, watersheds, and streams in the Coast Range today contain a dead wood legacy of large conifer trees from the past that will not be replaced unless large trees are allowed to grow, die, and decay in the future.

When a tree dies, a series of decay actions transforms the physical structure and chemical composition of the tree, changing its potential as habitat for other organisms and its role in nutrient and water cycling and soil productivity. Trees decompose and are slowly incorporated into the soil through a number of processes including leaching (water percolating through the wood), fragmentation (breakage from falling, erosion by water in streams, boring and chewing of invertebrates), and consumption by microbes and fungi (Harmon et al. 1986). This sequence of rotting can take more than three or four centuries, depending on the climate and the size and species of tree.

Our understanding of the ecological role of large dead wood in forests and streams of the Coast Range is relatively poor, since most scientific interest in the subject is less than 20 years old (Harmon et al. 1986). However, we do know that CWD has several functions in forest ecosystems, including fixation of nitrogen, storage of carbon and water, development of soil structure, and the furnishing of habitat for cavity-nesting and forest-floor vertebrates, and for invertebrates, plants, and fungi (Harmon et al. 1986; Corn and Bury 1991).

Nitrogen fixation is the process by which nitrogen from the atmosphere or decomposing organic matter enters the soil and is made accessible to living plants. It is "fixed," or converted into a form that plants can use, with the help of soil microbes that sometimes interact symbiotically with the roots of certain plants. Nitrogen fixation occurring in CWD can add 1 kilogram per hectare per year of nitrogen to conifer ecosystems in the Pacific Northwest. Although this is a relatively small amount, it can be an important contribution in some nitrogen-limited conifer forests of western Oregon. In the Coast Range, where nitrogen levels in soils and inputs of nitrogen from precipitation are relatively high (but availability for plant growth is still limiting) the effect of input of nitrogen via fixation may be relatively small (Grier 1976; Harmon et al. 1986). Generally, CWD has a low nutrient content and a slow decomposition rate, suggesting that it plays a minor role in nutrient cycling during forest development, except perhaps following catastrophic disturbances, when large amounts of CWD are produced (Harmon and Hua 1991). Unlike its role in habitat, the role of CWD in site productivity in the Coast Range is unclear. In one modeling effort for a site containing old-growth spruce and hemlock near the coast, merchantable yield was 5 percent higher after six 30-year rotations when CWD was left after clearcutting of the initial stand than when the CWD was removed (Harmon et al. 1986). The effect of wood removal on productivity might actually be greater than indicated by the model because the model assumed that CWD actually made nitrogen unavailable during early stages of decomposition, whereas recent research indicates that the nitrogen leaches out of CWD throughout decomposition (M. Harmon, personal communication).

The role of standing and down dead wood as habitat for cavity-nesting birds such as pileated woodpeckers (*Dryocopus pileatus*) is well known (Maser et al. 1988), but its role in habitat for mammals, amphibians, plants, fungi, and invertebrates is less well known (Figure 3-26 in color section following page 52). Large decayed tree boles on the ground can function as seedling establishment sites in dense shrub layers in riparian forests (Pabst and Spies 1999). In the Coast Range, clouded salamanders (*Aneides ferreus*) are most often found under the bark of large Douglas-fir logs, while ensatina salamanders (*Ensatina eschscholtzii*) are associated with well-decayed logs (Corn and Bury 1991). The abundance of some fungal species differs with decay state and species of down logs in western Oregon (Harmon et al. 1994). It has been reported that species of mosses and lichens are associated with large down logs in boreal coniferous forests in Sweden (Esseen et al. 1997), but it is not known whether these associations also occur in the Coast Range. Several invertebrate species in European forests prefer dead tree boles of a particular size and state of decay (Heliovaara and Vaisanen 1984), but these relationships also are poorly known in the Pacific Northwest. In general, it is thought that large pieces of down wood (50 to more than 100 centimeters in diameter) are more biologically valuable than small pieces because of their longevity and the diversity of their substrates (thick bark, sapwood, and heartwood). However, accumulations of small down wood may also provide important habitat features for some wildlife species such as small mammals (A. Carey, personal communication).

In summary, the CWD created by the death of trees plays many roles in Coast Range ecosystems. On moist, highly productive sites the role of CWD in forest growth appears relatively small. However, large dead wood appears to play a very important role in creating habitat for several species of plants and terrestrial and aquatic vertebrates, and for many more species of fungi and invertebrates. Our understanding of the ecological role of dead wood in the Coast Range comes from a few studies on a very limited number of sites and for a small number of species. We have only a qualitative grasp of the implications to forest management of a decline in stocks of dead wood or reduction in sizes of pieces in managed stands. In other words, we can state whether effects of management are positive or negative on stocks of dead wood and habitat potential, but we do not have quantitative ways to predict how ecological benefits will change as amounts of dead wood change. Consequently, managers face considerable uncertainty in making site-specific decisions regarding amounts and kinds of dead wood to leave or produce in these forests.

Forest development

Disturbance and succession go hand in hand. While they differ in many characteristics, all disturbances have the effect of killing or removing some or all of the existing vegetation and freeing up resources for the establishment and growth of new plants. Plant growth in the forest is limited by one or more resources, most commonly light, water, and nitrogen. From the viewpoint of a seedling trying to get started on the forest floor, both light and water are usually in very short supply. Disturbance, by removing some or all of the living plants, makes light and moisture more available and creates mineral or organic substrates for establishment.

After a disturbance, new plant growth is usually abundant in the Coast Range (Franklin and Dyrness 1988; Stein 1995). New plants may be residuals from the previous forest, or they may grow from seed that was already in the soil, or seed that is carried in by wind or animals after the disturbance. Dispersal distances for plants in the latter group are quite variable, from several kilometers for fireweed (*Epilobium angustifolium*) to just a few tens of meters for hemlock in closed forests (Schrader 1998). In the residual-origin group are most of the mature-forest herbs and shrubs as well as broadleaf trees; they sprout from underground rhizomes (e.g., salmonberry; Tappeiner et al. 1991), roots (blackberry [*Rubus* spp.]), and stumps (maple). Many of the early-successional herbs and shrubs such as ceanothus (*Ceanothus* spp.) (Gratkowski 1962) come from seed that has been in the soil since the last big disturb-ance, often many decades. Many early-successional herbs (e.g. fireweed) and both early- and late-successional trees (e.g., alder and hemlock, respectively) rely on the wind to disperse seed into new areas. Plants that produce berries, like cherry and salmonberry, rely on animals for dispersal.

The complement of species present or arriving after a disturbance is variable, depending on the nature of the disturbance (what residual propagules are left) and the presence of nearby seed sources. This initial complement determines the early course of succession.

Competition among plants for resources after a disturbance can quickly become intense. The window in time when resources are available in excess—the best time for new plants to get started—is very short. Many early-successional plant species grow quickly, filling empty space and using all available resources.

Salmonberry is one nearly ubiquitous shrub species in the Coast Range that has been well studied (Tappeiner et al. 1991; Zasada et al. 1994). It is a strong competitor with tree regeneration, and its growth habitat is similar to that of several other common clonal plants (e.g. salal, bracken fern, and snowberry [*Symphoricarpos albus*]). Although salmonberry produces seed that are dispersed by birds, it regenerates after disturbance most often by sprouts from an underground rhizome. In a vigorous stand of salmonberry, a square meter of ground can have 10 meters of rhizome with buds every few centimeters. Salmonberry, well adapted to surviving disturbance, is usually among the first plants to colonize disturbed areas. The salmonberry rhizome is protected from most fire, and it contains lots of stored energy for quick growth after a disturbance. And because the plant contains such an abundance of buds on the rhizome, a disturbance such as a landslide, which breaks up a salmonberry patch and moves it around, rapidly spreads plants to new areas.

After a disturbance in a forest initiates a new plant community, the process of succession begins. This long sorting process is determined largely by the different abilities of species to colonize after a

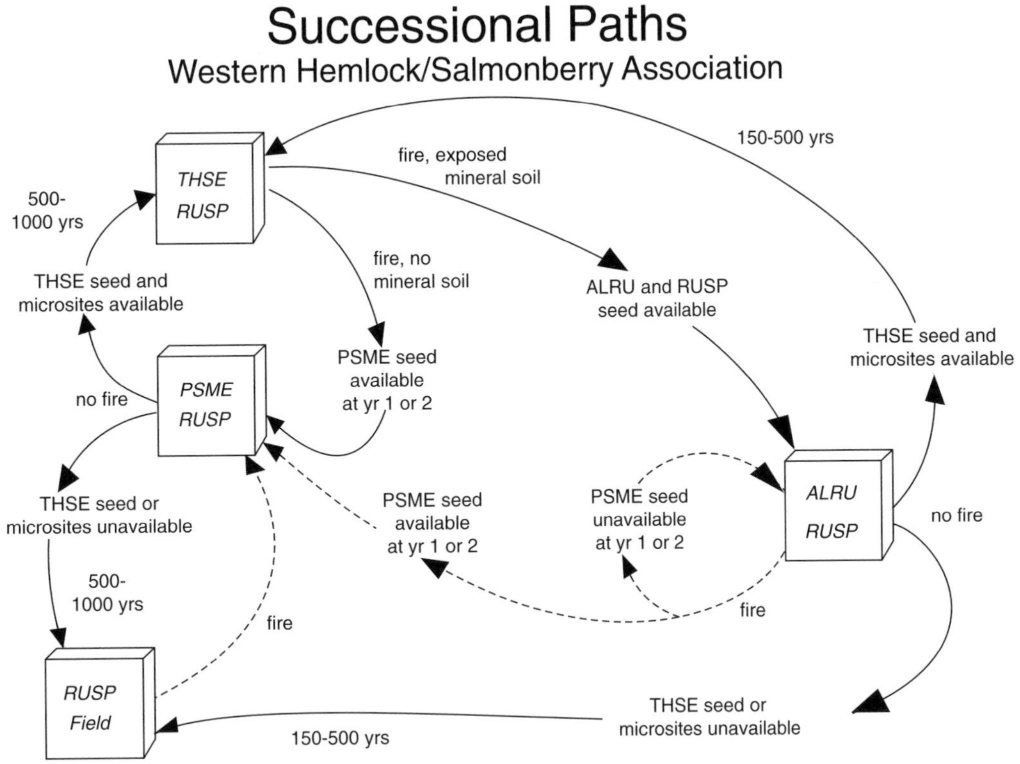

Successional Paths
Western Hemlock/Salmonberry Association

Figure 3-27. Depending on a variety of disturbance, site and seed source factors, the pathways of succession in the Oregon Coast Range can be quite varied. This figure represents the multiple pathways found in the western hemlock/salmonberry association on the Siuslaw National Forest (Hemstrom and Logan 1986).

disturbance and to compete with their neighbors. It is difficult to describe post-fire or post-logging succession because of the diversity of pathways that can occur (Figure 3-27). A common sequence following stand-replacement disturbance begins with dominance by invading and residual forest herbs and grasses such as fireweed, woodland groundsel (*Senecio sylvaticus*), candy flower (*Claytonia sibirica*), and bentgrass (*Agrostis capillaris*). After a few years, shrubs such as blackberry and salmonberry dominate the vegetation cover. Trees such as red alder, Douglas-fir, or western hemlock may become established in the first few years after disturbance, but do not form closed tree canopies for 10 to 25 years depending on the site (Henderson 1978; Hemstrom and Logan 1986).

Because many of today's Coast Range forests are young, the successional process now at work is a competitive sorting of species that colonize an area within a fairly short time after disturbance. After 50-75 years, the forest overstory begins to let through more light and shade-tolerant species begin to invade the forest understory. Later, canopy gaps are created as overstory trees age and die. This entire developmental process can take several hundred years.

The tree composition of late-successional forests varies throughout the Coast Range as a function of fire severity and return interval. In the moist near-coast zone, where fires were least frequent and seed sources of shade-tolerant conifers were abundant, mid to late stages of forest succession are dominated by hemlock and moisture-sensitive Sitka spruce. Douglas-fir is relatively less common in these areas than other areas of the Coast Range. On the other hand, in southwestern Oregon and the Willamette Valley margin, where fires were relatively frequent and less intense, fire-tolerant species like Douglas-fir, incense cedar (*Calocedrus decurrens*), ponderosa pine (*Pinus ponderoda*), madrone, and oaks are common in late successional forests. The central and northern parts of the Coast Range fall somewhere

Table 3-2. Expected percentages of landscape in different age classes in relation to fire interval (Van Wagner 1978).

Interval between wildfires (yrs)	*Age class of forest (yrs)*					
	> 80	*> 100*	*> 150*	*> 200*	*> 250*	*> 400*
	Expected percent of landscape					
50	20	14	5	2	<1	<1
100	45	37	22	14	8	2
150	59	51	37	26	19	7
200	67	61	47	37	29	14
250	73	67	55	45	37	20
300	77	72	61	51	44	26
350	79	75	65	56	49	32
400	82	78	69	61	53	37

in between these extremes. Fires tended to be large and stand replacing (Impara 1997). Shade-tolerant seed sources were probably patchy so succession to hemlock and redcedar forests may have been slow and sometimes truncated by succession to shrubs following senescence of short-lived shade-intolerant red alders.

If the frequency of stand-replacing disturbances is known, it is possible to estimate the amount of different successional stages or age classes of forest that have occurred in the past, or that might occur in the future landscape. Van Wagner (1978) developed a mathematical formula to calculate age-class distributions in landscapes based on the frequency of wildfires. In its simplest form, this model assumes that fire frequency is equally likely at any stand age. This assumption means that when a fire occurs in a landscape, some old stands may escape and some young stands may get burned, which is a scenario that has occurred frequently in the Coast Range. Under this scenario, a landscape will contain a percentage of forest that is older than the typical fire rotation. For example, in a landscape with a fire frequency of 150 years, about 30 percent of the landscape will be older than 200 years (Table 3-2).

A change in fire regime (e.g., frequency or intensity) can alter the development of forests. The control of fire in the drier parts of western Oregon has already resulted in some major changes in the composition and dynamics of forests and will certainly result in further changes. In southwest Oregon, fire control has increased the density of the understory and, in some places such as the eastern Oregon ponderosa pine forests, this understory growth is competing with overstory trees and competition is beginning to kill some trees (J. Tappeiner and T. Sensineg, personal observation). In the central and northern Coast Range, historic fire has made hemlock less common than it would be otherwise. Over the next century or more, fire control on lands managed for old-growth forest may result in an increase in hemlock and a concomitant reduction in regeneration of Douglas-fir.

Predicting compositional changes in succession in the Coast Range is not as easy as we might think. Figure 3-27 shows a complex web of pathways for vegetation change in the central and northern Coast Range. There is no single, linear pathway of change. Rather, depending on the disturbance type and intensity (i.e., the amount and distribution of available resources) and the available seed sources, several pathways are possible. For example, if fire and hemlock seed source are both lacking, succession in both alder and Douglas-fir stands can lead to a shrub-dominated community.

Influence of human activities

In the last 100 years, extensive human activities have become a dominant disturbance in the Coast Range. Presettlement activities by Native Americans and early patterns of Euro-American settlement and land use have left their mark on today's forested landscapes. Native Americans frequently burned the Willamette Valley grasslands and oak woodlands and the adjacent dry Douglas-fir forest of the foothills of the Coast Range (Boyd 1986; Agee 1993). However, according to Agee (1993), "...the case for widespread aboriginal fires throughout the wetter part of the Douglas-fir region is not convincing." Fire was thought to be used primarily to improve hunting and encourage growth of edible plants (Agee 1993; Robbins 1997). Burning by Native Americans stopped upon the arrival of Euro-Americans, who set fires of their own to clear land and to remove logging slash to prevent subsequent fires. Organized fire protection and regulation of

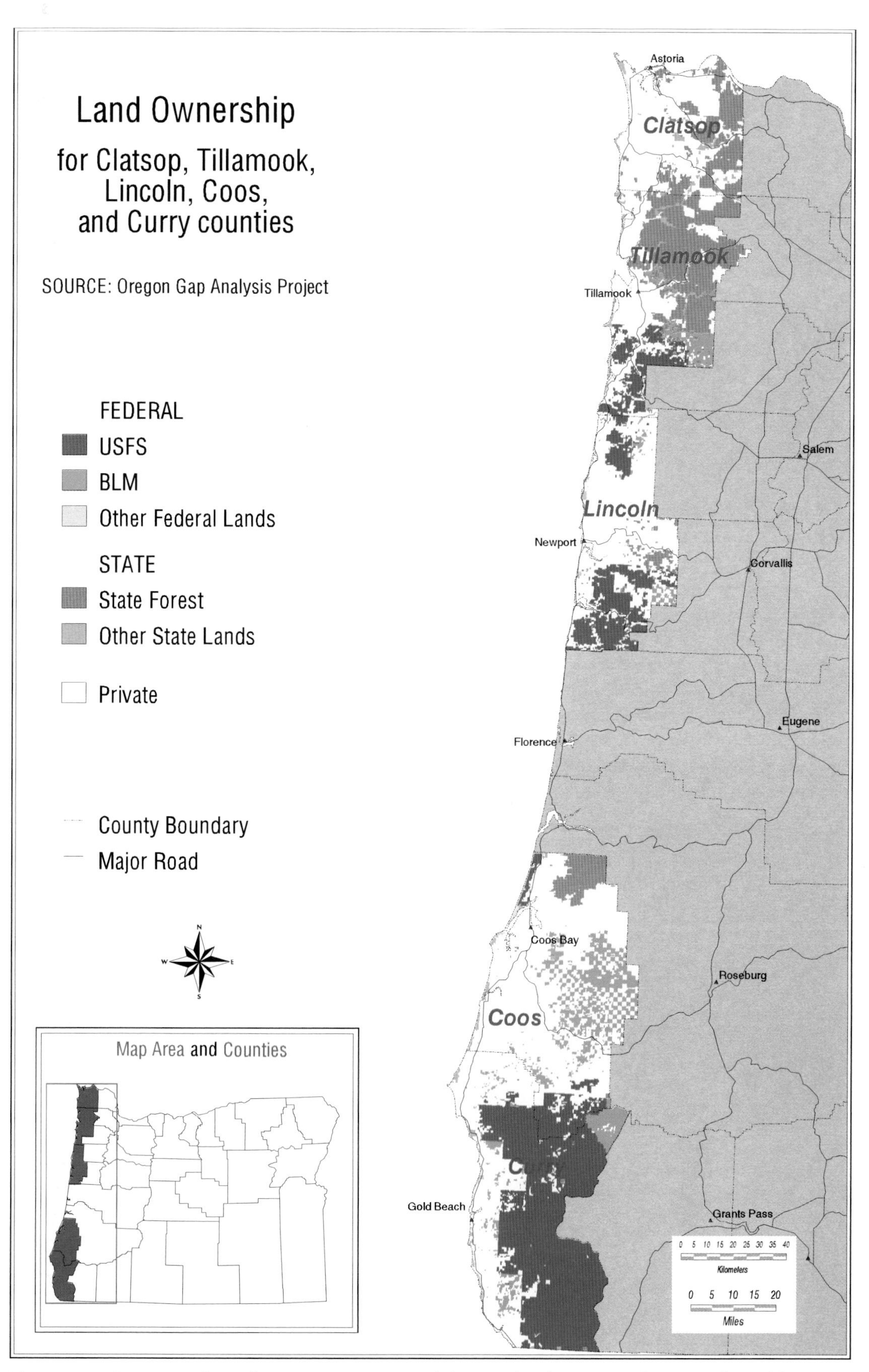

Figure 2-1. Counties along the Oregon coast, and those inland counties that are affected by policies toward Coast Range.

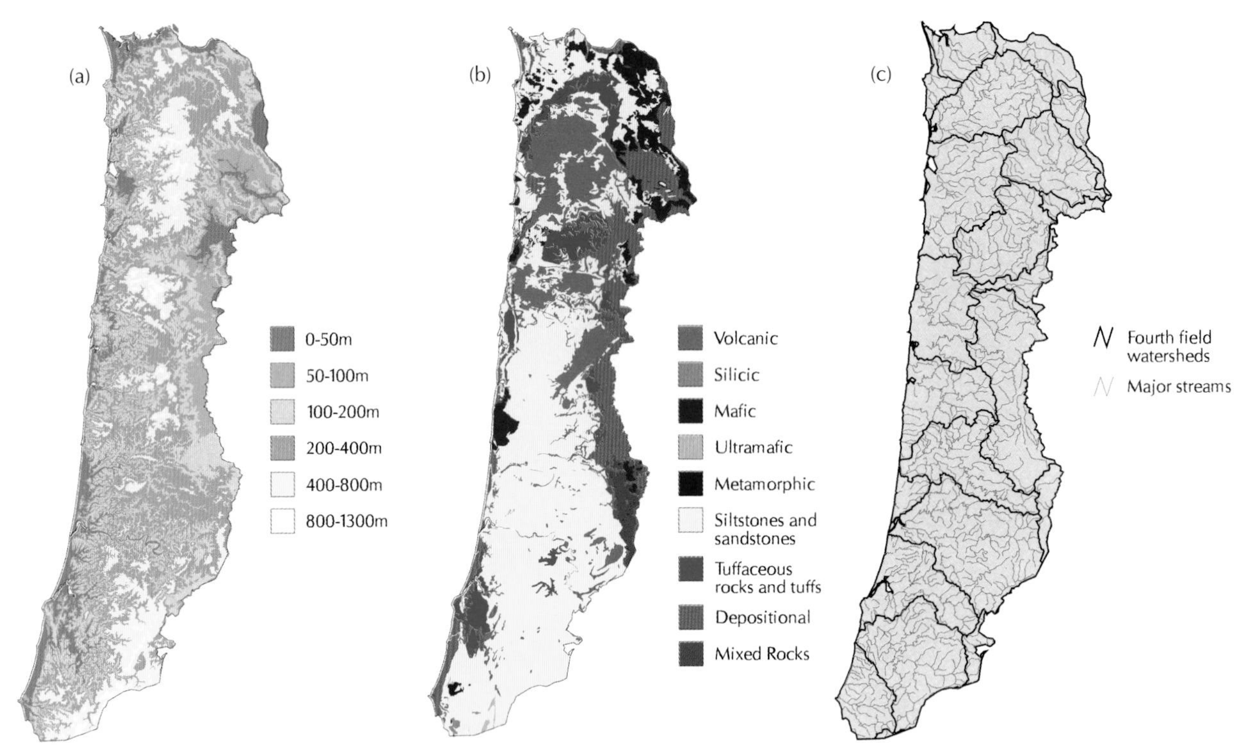

Figure 3-1 (a) Elevation (m) from digital elevation model of the Oregon Coast Range, 90-m resolution; (b) Geologic types, modified from Walker and MacLeod (1991). Red = igneous: volcanic and intrusive rocks; brown = igneous: silicic rocks (granite, diorite, rhyolite, dacite); violet = igneous: mafic rocks (basalt, basaltic andesite, andesite, gabbro); tan = igneous: ultramafic rocks (serpentine); black = metamorphic; yellow = sedimentary: siltstones, sandstones, mudstones, conglomerates; medium green = sedimentary: tuffaceous rocks and tuffs, pumicites, silicic flows—miocene and older; purple = depositional: dune sand, alluvial, glacial, glaciofluvial, loess, landslide and debris flow, playa, lacustrine, fluvial; dark green = mixed rocks (unspecified); and (c) Fourth-field hydrologic units (HUCs).

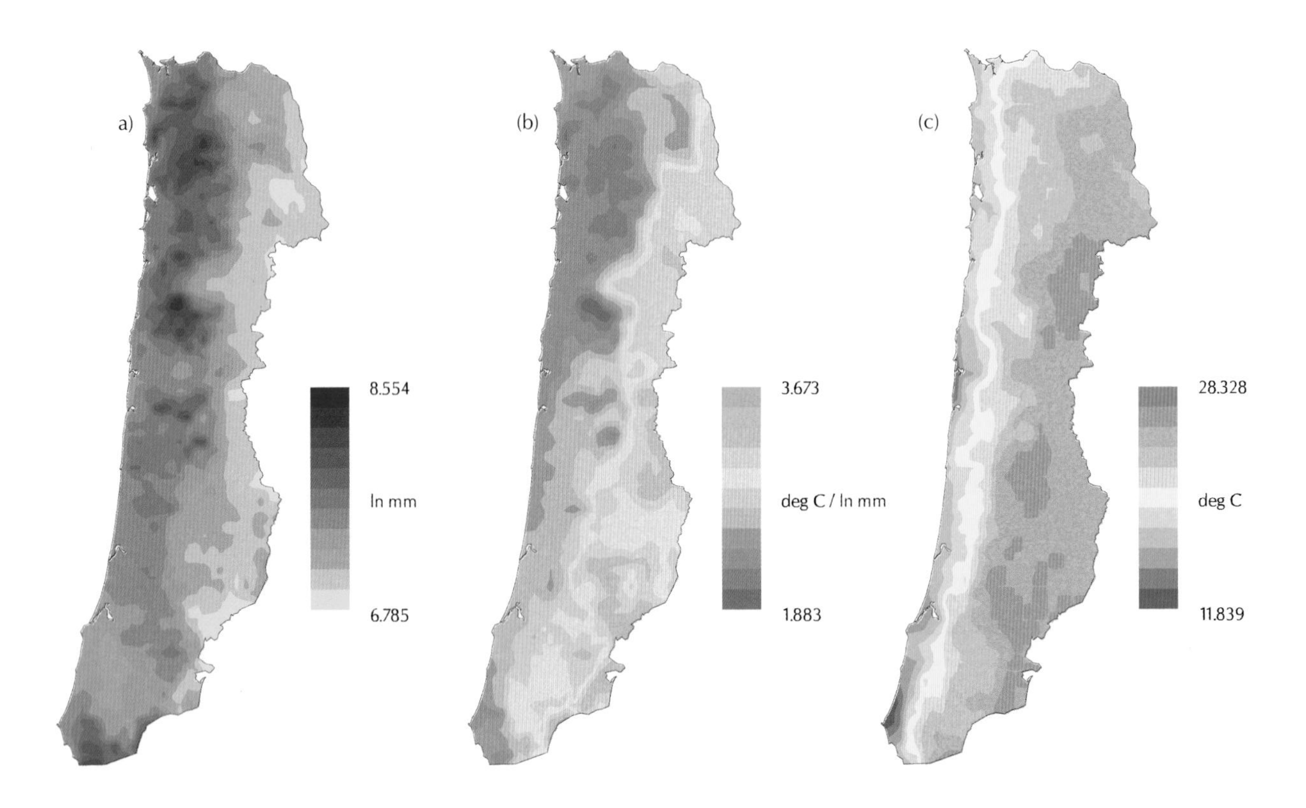

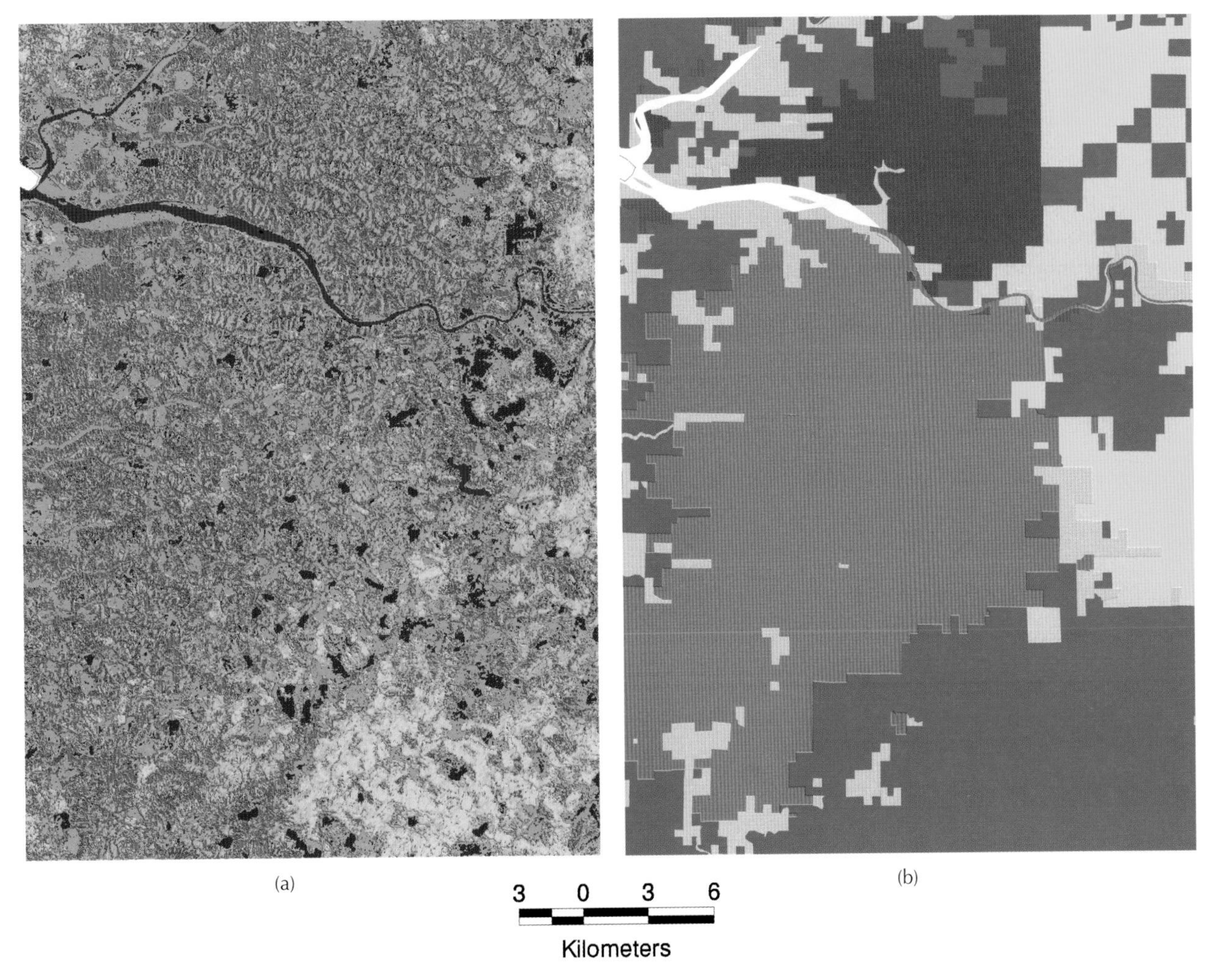

Figure 3-4. Land vegetation (a) and ownership (b) patterns (from 1988 Landsat TM satellite imagery) in the Siuslaw National Forest, Elliott State Forest, and industrial private lands of the central Oregon Coast Range. Imagery: dark blue = conifer forests > 50 cm tree diameter; light blue = conifer forests < 50 cm tree diameter; red = hardwood and shrub stands; tan = recently disturbed stands with vegetation cover < 70 percent; brown = recently disturbed stands with vegetation cover < 40 percent. Ownership: red = state; blue = private industrial; yellow = Bureau of Land Management; green = Forest Service.

Figure 3-2 (facing page) (a) Mean annual precipitation from PRISM model output (Daly et al. 1994); (b) index of moisture stress during the growing-season, computed as SMRTMP/SMRPRE, where SMRTMP = mean monthly temperature (degrees C) from May to September, and SMRPRE = mean monthly precipitation from May to September (in mm). Derived from PRISM model output (Daly et al. 1994); and (c) difference between mean August maximum temperature and mean December minimum temperature (degrees C), an index of seasonal variability and continentality. Derived from PRISM model output (Daly et al. 1994).

Figure 3-5. Patterns of vegetation (from Landsat TM satellite imagery) and stream drainages in the Cummins Creek Wilderness area, Siuslaw National Forest.

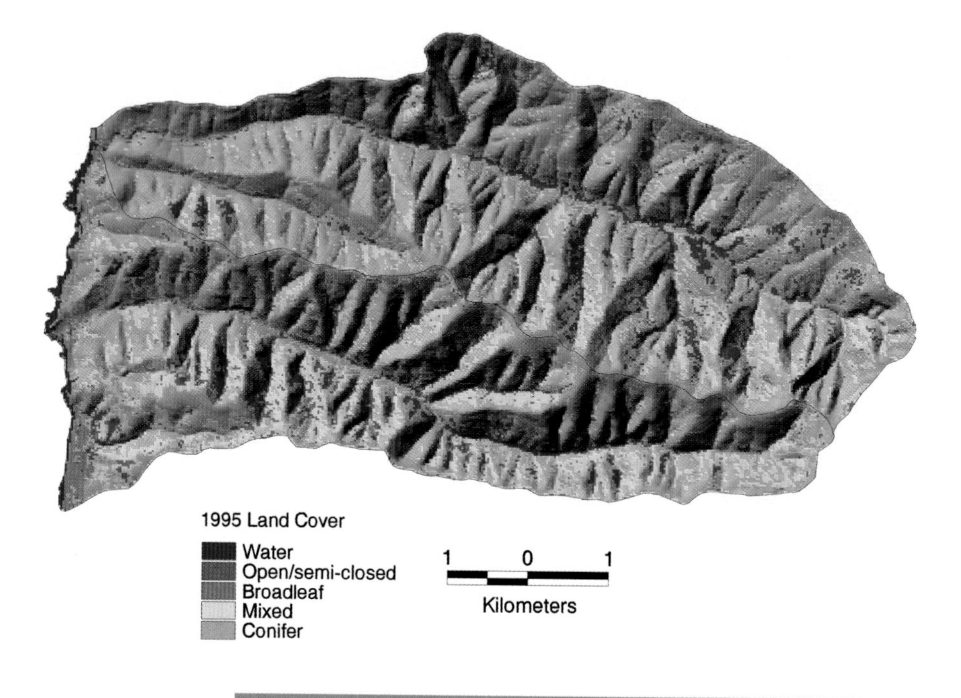

Figure 3-6. Distribution of plant communities in relation to topographic position and aspect in the western hemlock zone of the central Coast Range (adapted from Hemstrom and Logan 1986).

North aspect

South aspect

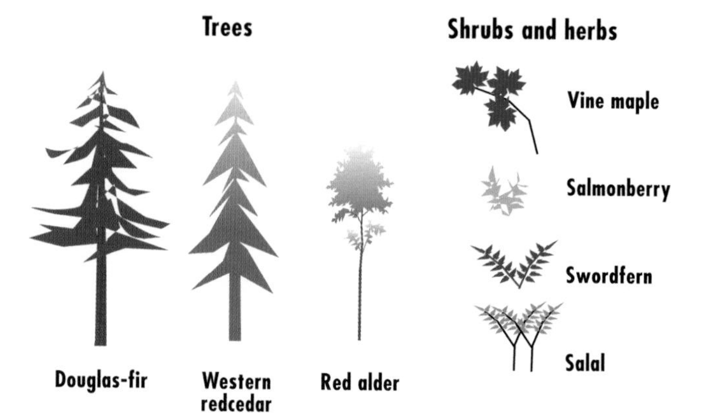

Figure 3-8. Landscape patterns of agricultural and forest lands along a river in the central Coast Range.

Figure 3-9. Debris flow deposit in the central Oregon Coast Range.

Figure 3-11. A hillside with intermixed patches of alder and Douglas-fir resulting from logging. Note linear patches of alder along skid roads.

Figure 3-13. The business end of nitrogen fixation, these root nodules of red alder contain the nitrogen-fixing bacteria *Frankia*.

Figure 3-12. A valley in early spring showing the historic pattern of alder in the bottom and Douglas-fir on the slopes. Note the bigleaf maple at the edge of the alder.

Figure 3-14. An older planted mix of red alder and Douglas-fir on this nitrogen-poor site resulting in Douglas-fir trees twice the normal size.

Figure 3-15. White oak in the foothills, a resource for wildlife and people.

Figure 3-19. 120-year-old Douglas-fir/western hemlock stand on productive site in the Drift Creek watershed.

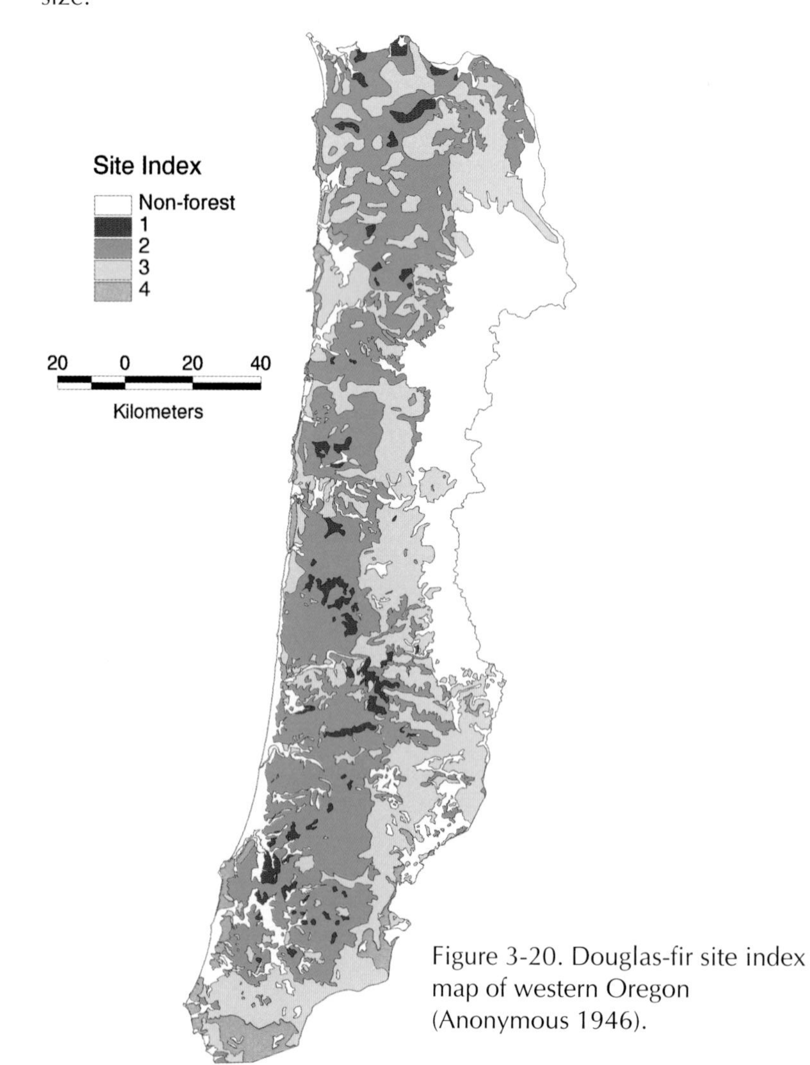

Figure 3-20. Douglas-fir site index map of western Oregon (Anonymous 1946).

Figure 3-21. Densely vegetated stream in the Oregon Coast Range.

Figure 3-22. Standing dead and fallen trees left after the Silver Fire in the southern Coast Range on the Siskiyou National Forest.

Figure 3-23. Canopy gap resulting from root rot in the Elk River watershed, southern Oregon Coast Range.

Figure 3-24. Shrub development in a canopy gap formed by breakage in hardwood canopy.

Figure 3-26. A well-decayed fallen tree bole in a Coast Range stand can serve as special habitat for insects, plants, and fungi. (Reprinted from *Ecology* v. 69, by permission of the publisher.)

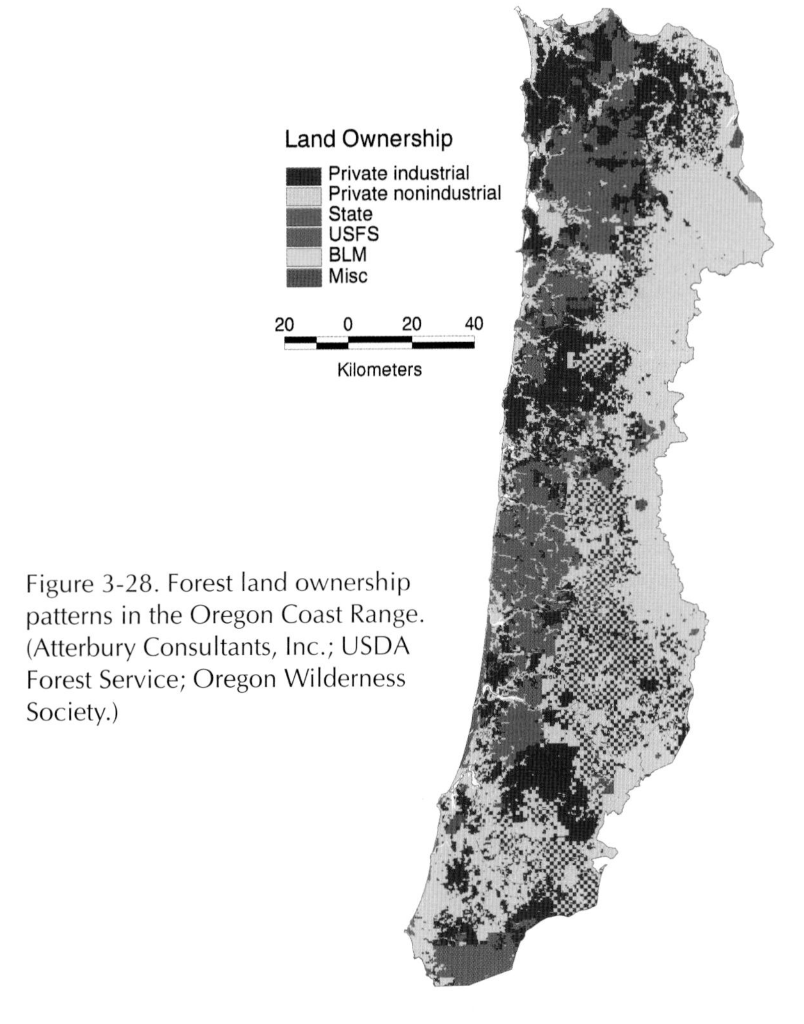

Figure 3-28. Forest land ownership patterns in the Oregon Coast Range. (Atterbury Consultants, Inc.; USDA Forest Service; Oregon Wilderness Society.)

Figure 3-29. Forest vegetation patterns in 1936 (U.S. Department of Agriculture Forest Service, Pacific Northwest Research Station).

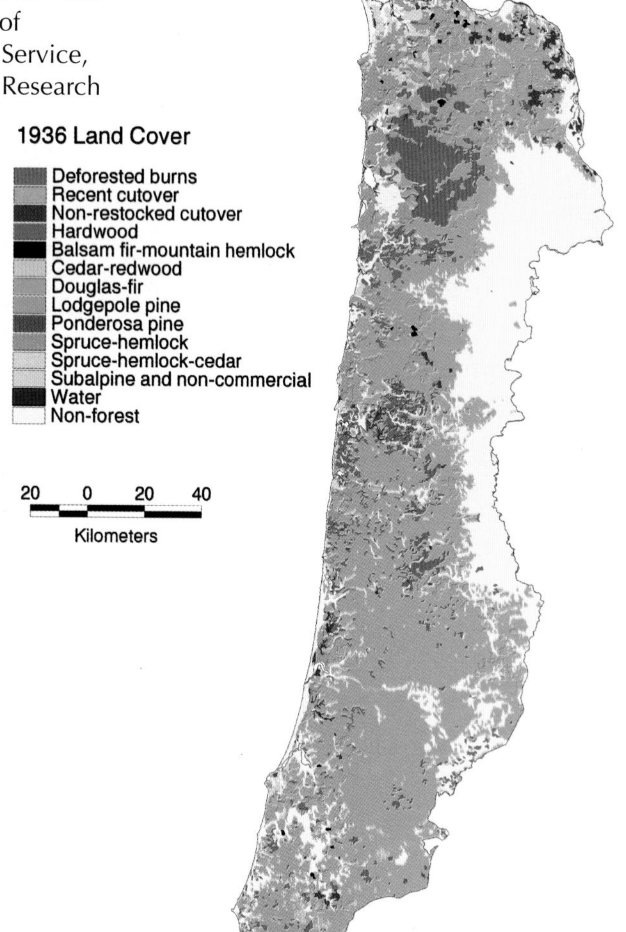

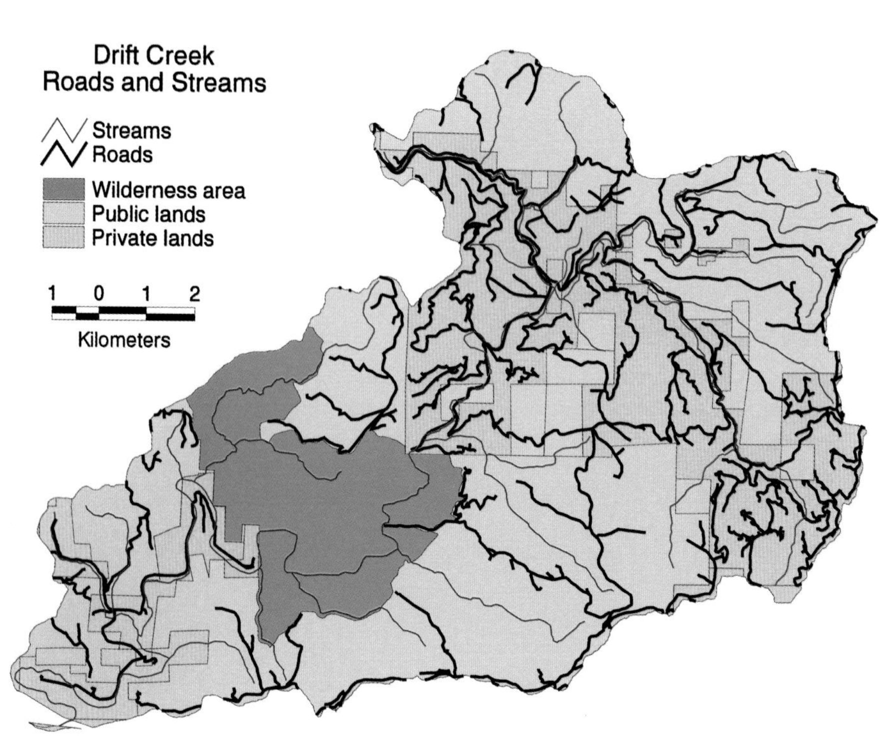

Figure 3-30. Three levels of road network development in the Drift Creek watershed, Oregon Coast Range: high level of development (road network to access complete forest harvest over entire area) (private lands), moderate (roads restricted to major ridges in area with less than 50 percent forest harvest) (public lands), and no road or forestry development (Wilderness Area).

slash burning in western Oregon began in the early 1900s.

The current structure and composition of coastal forests is most strongly influenced by logging and forest management activities which began in the 1800s and continue today. Grazing, fuelwood harvesting, introduction of nonnative plant species, and urbanization also have altered the composition of coastal forests and woodlands, especially near the population centers of the interior valleys. These fringe forests, as they have been called, have been highly fragmented by clearing for agriculture, roads, and buildings. Nonfederal timberland area in western Oregon has been declining at about 0.2 percent per year since 1960; about 60 percent of this has been converted to agricultural and urban land uses, and 40 percent to roads associated with timber harvest (MacLean 1990).

Current patterns of forest structure result primarily from historic and present-day logging and management activities, which are strongly associated with patterns of land ownership (Figure 3-28 in color section following page 52). The first stands in the Coast Range province to be logged, from the mid-1800s to the early 1900s, were the most accessible and productive lands in private ownership, primarily along the coast and major river drainages. These early logging operations left behind many more large live "cull" trees, snags, and logs than are commonly left today. Locations of recent cutovers and young restocked forests as of 1936 can be seen in Figure 3-29 (in color section following page 52). Some of these areas have now been harvested a second time, most notably in Clatsop and Columbia Counties in the northernmost portion of the Coast Range. Logging of older forest on federal lands began after World War II and accelerated in the 1970s; it was curtailed drastically after the listing of the northern spotted owl as a threatened species under the Endangered Species Act, and the implementation of the Northwest Forest Plan (FEMAT 1993). Currently, about 21 percent of the coastal province is federally owned, 9 percent is owned by the State of Oregon, and 68 percent is privately held (GIS data from Atterbury Consultants, Inc.). Seventy percent of the private timberland is owned by the forest industry (unpublished data, Forest Inventory and Analysis, Pacific Northwest Research Station, Portland, OR; Figure 3-28). These patterns contrast with forest land ownership in Oregon as a whole, where only 40 percent of timberlands are privately owned, and much more is federally owned.

Although landowners hold a wide variety of objectives for their land, timber production is the dominant land use. In general, timber management activities are more frequent and intense and less variable in size and intensity than natural disturbances, and they tend to simplify forest structure at both stand and landscape scales (Hansen et al. 1991). This is especially true of lands under intensive forest management, as practiced by the forest industry and other landowners seeking maximum financial returns from their land. At the stand level, intensive management consists of clearcutting most live trees and snags, site preparation by prescribed fire or herbicides to control competing vegetation, replanting with a single species (usually Douglas-fir), periodic thinning to maintain vigorous and evenly spaced crop trees, and harvesting at 40- to 100-year intervals (Hansen et al. 1991). Nonindustrial private lands in general are less intensively managed for timber, and partial harvesting is practiced more often. Where even-aged forest management is practiced on federal lands, management prescriptions generally call for (1) retaining more live trees, snags, and down logs after harvest than is typical in more intensively managed forests, (2) harvesting stands at older ages, and (3) managing competing vegetation without using herbicides. Management intensity on state lands falls somewhere in between that on federal and private lands.

Vegetation patterns at the landscape scale are governed primarily by the frequency, size, shape, and distribution of stand-level management activities. On industrial lands, which often consist of large ownership blocks, clearcut units tend to be larger (current maximum size is 120 acres) and more contiguous than on public or other private lands. Shorter rotation lengths (typically less than 50 years) result in landscapes dominated by young forests. In areas of nonindustrial private ownership, landscape patterns reflect the distribution of parcels owned by individual landowners, which typically are much smaller than those owned by the forest industry or managed by the federal agencies. Many National Forest landscapes retain the patterns created by three to four decades of staggering small (less than 16 hectares or 40 acres) harvest units in space and harvesting them at a constant rate. Many Bureau of Land Management (BLM) lands have a similar mosaic of clearcuts; this pattern is complicated by the section-by-section checkerboard of BLM land holdings, which results in a landscape of patchy older forests mixed with young forests on the adjacent private lands.

As of the mid-1980s, 12 percent of the nonfederal timberland in the coastal province was in an early-successional structural stage (approximately zero to 15 years old), 82 percent was mid-successional (approximately 15-80 years), and only 6 percent was late-successional (approximately more than 80 years; unpublished data, Forest Inventory and Analysis, Pacific Northwest Research Station, Portland, OR). Virtually all forest lands in private ownership have been harvested at least once in the past and current forests on these lands are less than 80 years old. Most of the scattered large, live trees, snags, and down logs that remain on nonfederal lands are the legacy of earlier logging of older forest. These stand structures are gradually decaying and are not being replaced (Ohmann et al. 1994). Most remaining late-successional forest in the coastal province is concentrated on federal lands, where it covers about 30 percent of the land area (T. Spies, unpublished data). Old-growth forests (more than 200 years old) cover only approximately 6 percent of the federal lands in the Coast Range (Siuslaw National Forest 1999). Under the Northwest Forest Plan, most of the early- and mid-successional forest on federal lands is being managed to accelerate the development of characteristics of late-successional forest.

Across the entire Coast Range, tree species are distributed along climatic gradients (Ohmann and Spies 1998). Within particular stands, however, timber management influences the relative abundance of species, although few species are totally eliminated from a site by logging (Ohmann and Bolsinger 1991; Ramey-Gassert and Runkle 1992; Halpern and Spies 1995). Intensive forest management aims to shift stand composition to the most valuable timber species: Douglas-fir across most of the province and western hemlock in the northwestern portions of the province. Harvested sites in the Coast Range that are not successfully regenerated to conifers are generally dominated by pioneer hardwood species, primarily red alder, or by shrubs such as blackberry, salmonberry, or salal. Planting of Douglas-fir over the last few decades on coastal sites historically dominated by western hemlock and Sitka spruce is believed to be contributing to spread of Swiss needle cast, a native forest pathogen. This foliage disease results in Douglas-fir needle loss, reducing tree growth and in some cases causing death. Incidence of laminated root rot (*Phellinus weirii*) is also believed to be increasing in coastal forests, in part due to forest management activities. This pathogen kills Douglas-fir in small patches, creating gaps in forest canopies. Spread of these tree diseases may force forest managers to convert many Douglas-fir plantations to hemlock or other conifers in the case of Swiss needle cast, or to broadleaf species such as red alder in areas where laminated root rot is a problem (A. Kanaskie, personal communication).

Watershed and Landscape-scale Processes

Influence of roads on ecosystem function at multiple scales

Forest roads have served many positive functions, including access for extraction of wood and other forest products, silvicultural activities, fire detection and suppression, and recreation. The unintended negative impacts of roads on forest ecosystems and watersheds include effects on water runoff (Harr et al. 1975; King and Tennyson 1984; Jones and Grant 1996; Wemple et al. 1996, Bowling and Lettenmeier 1997), surface erosion and attendant impacts on fish habitat (Reid and Dunne 1984; Duncan et al. 1987; Bilby et al. 1989; Foltz and Burroughs 1990), landslide initiation (Dyrness 1967; Swanson and Dyrness 1975; Megahan et al. 1978; Sessions et al. 1987), invasion of forest landscapes by exotic species (Forcella and Harvey 1983; Tyser and Worley 1992; Parendes 1997), spread of pathogens, wildlife dispersal, mortality due to collisions or hunting, and a range of other factors (Transportation Research Board 1997). Much is known about the layout, design, construction, and maintenance of roads in forest landscapes from engineering points of view. Various studies have focused on road location and improved engineering practices to reduce effects of roads on watersheds (Silen 1955; Burroughs et al. 1984, Swift 1984; Sessions et al. 1987, Piehl et al. 1988). In this discussion, we comment on effects of road networks on the ecosystems of steep forest landscapes with regard to movement of water, sediment, earth flows, exotic plants, and pathogens.

Extensive road networks are now widespread in Pacific Northwest landscapes. In developed parts of the Coast Range, for example, Freid (1994) observed road densities of 2.7 to 3.5 kilometers per square kilometer in sampled areas of federal and private forest management. An example is the Drift Creek watershed, which exhibits a variety of road densities and distributions, ranging from a roadless Wilderness Area, to Forest Service lands with road development limited to major ridges, to private timber lands where extensive roads are built during the first harvest cycle and then maintained indefinitely (Figure 3-30 in color section following page 52). Although roads commonly occupy less than 4 percent of the area of a forest landscape, they have distinctive properties that may disproportionately affect terrestrial and stream ecosystems. These properties include the wide distribution of roads in the full range of hillslope positions across a landscape (e.g. riparian, midslope, ridge); the persistent exposure of mineral soil, resulting in the invasion of a variety of plant species; compacted surface and subsurface materials; ditches and other drainage structures (e.g. culverts and water bars) that alter natural water flow paths; road cuts and fills on the hillside; vehicle and animal traffic; and linear openings in the forest canopy. Individually and together, these properties of roads affect many ecological and watershed processes.

A landscape or watershed perspective offers an important basis for evaluating effects of roads on forest and stream ecosystems (Jones et al. 2000). Assessing road effects at only the site level may miss important larger-scale effects. A landscape perspective includes the effects of interactions (e.g. routing of water and timing of peak flows) along road systems as well as interactions between roads and the adjacent forest and stream networks. Such interactions may be mediated by movement of water, other materials, and organisms. Assessing road effects from a landscape perspective is particularly critical now, when many managers and policymakers are focusing on protection of watersheds and species diversity.

Road networks interact with stream networks to influence the routing of water, sediment, and soil mass movements as they follow gravity flow paths through forest watersheds (Figure 3-31). Several recent studies have examined the possible roles of

Figure 3-31. Schematic view of interactions between road (pale gray) and stream (wave pattern) networks in terms of routing of water sediment, and mass movements down gravitational flow paths. The road network consists of a valley floor road segment parallel to a large stream, hillslope road segments perpendicular to streams, and near-ridge roads without streams (mostly). Water (wave pattern arrows) and debris slide/flow (gray arrows) flow paths are illustrated. Roads (1) intercept water in surface and subsurface flow paths, (2) alter water flow paths and extend the channel network, (3) initiate mass movements of sediment in unstable roadfills, (4) deposit sediment moved by mass movements on roads, and (5) on valley floors. Overall, roads function to divert water and sediment from paths followed in roadless landscapes, and they initiate multiple new, cascading flow paths.

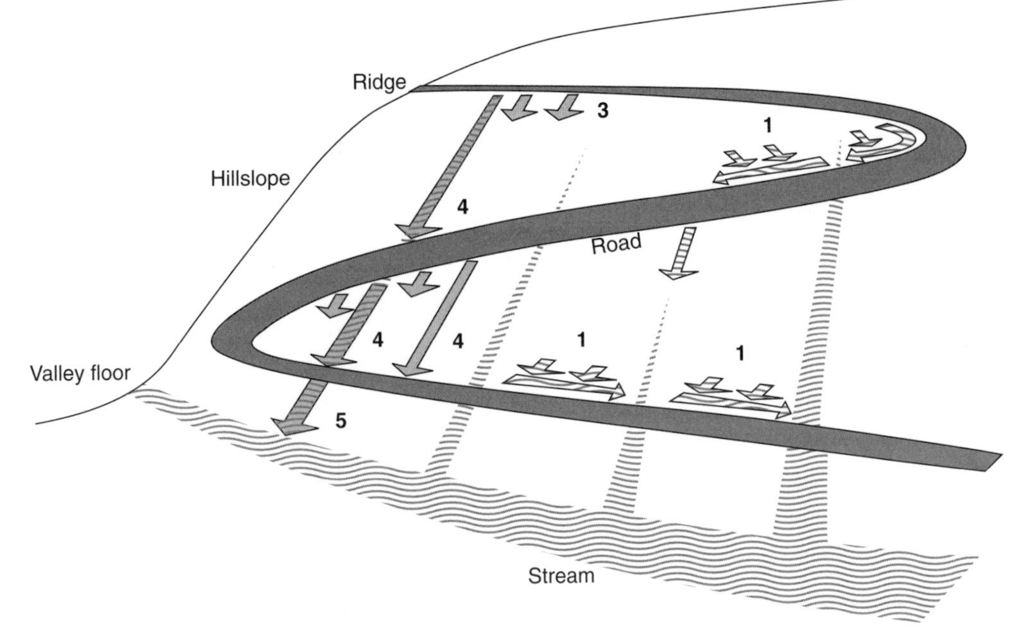

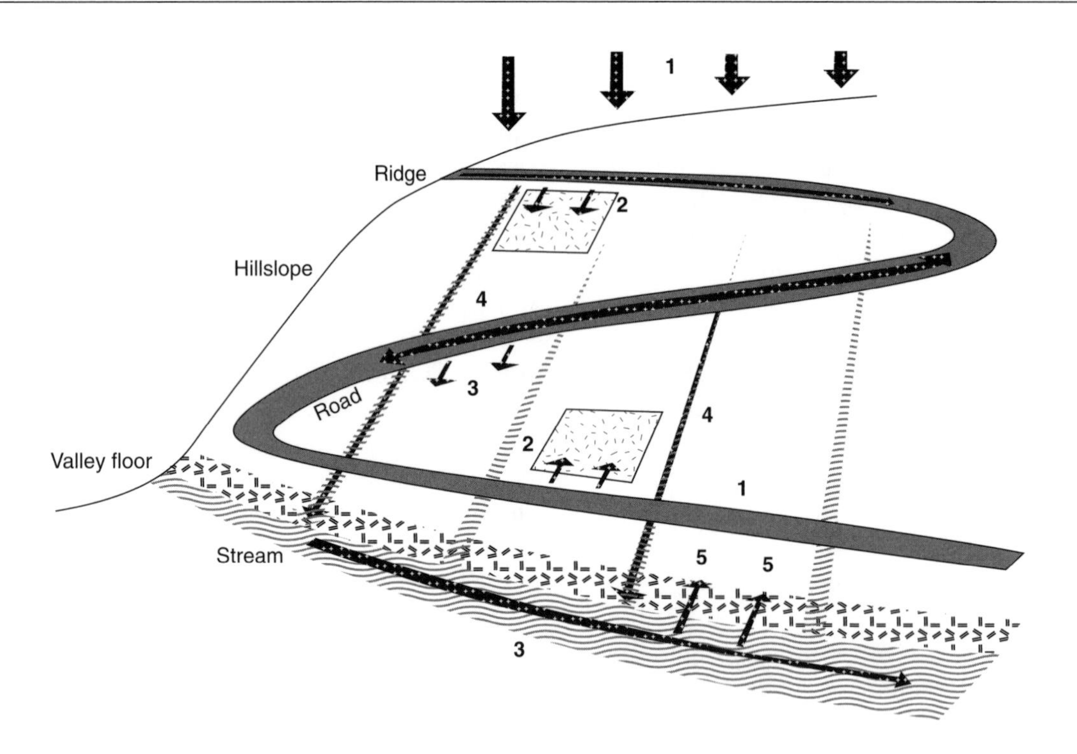

Figure 3-32. Schematic view illustrating movement of propagules via roads and associated air currents, animals and streams. Roads may function to (1) collect and transport propagules by air currents, animals or vehicles (cross pattern), (2) move propagules into clearcuts (confetti pattern) adjacent to roads, (3) move propagules into roadside forest (white), (4) move propagules into streams (wave pattern), and (5) deliver propagules to riparian forests (bird's foot pattern).

roads in increasing the number of flow paths through interception of subsurface water movement at roadcuts and conversion of it to surface flow along ditches that reach natural streams (Wemple et al. 1996; Bowling and Lettenmeier 1997). These changes in water routing through watersheds may alter the timing and magnitude of peak streamflows (King and Tennyson 1984; Jones and Grant 1996). Assessment of road-related erosion events during the February 1996 flood in the Cascade Range reveals that (1) the frequency and diversity of erosion events along road segments are greater the lower the road is situated on the hillslope, and (2) roads at near-ridge and especially midslope positions are net sources of sediment and channelized mass movements (debris flows); roads along valley floors tend to trap sediment and impede movement of debris flows before they can reach main channels and contribute to aquatic processes (B. Wemple, personal communication).

These and related observations of how roads function in landscapes have significant management implications. The position of a road segment relative to the flow paths of water, sediment, and mass movements can strongly influence site-specific engineering decisions. Roads near ridges in some surveyed landscapes have the lowest hydrologic and landslide impacts; hence, they may deserve lowest priority for erosion-control and watershed-restoration measures. On the other hand, locating roads near ridges in landscapes where sliding commonly begins in the oversteepened heads of stream channels, such as in the sandstone formations of the Coast Range, may cause increased sliding (Montgomery 1994). Managers should assess the potential for landslides to wash out bridges. If the potential is high, it may be prudent to design stream crossings to accommodate debris flows and minimize damage when they occur. Where the potential for debris flows is high, it may be prudent to design structures to accommodate debris flows and minimize damage when they occur. Intensive use of water bars may decrease slide damage to little-used roads by reducing the delivery of water to potential slide sites (Siuslaw National Forest 1997).

Roads appear to foster movement of some exotic plant and pathogen species into forest landscapes where seeds and spores can be dispersed by vehicles traveling the roads (Schmidt 1989; Lonsdale and Lane 1994) or by agents that do not follow roads (e.g., wind, birds, large mammals), but which capitalize on favorable habitats and canopy openings along roads (Wester and Juvik 1983; Parendes 1997; Figure 3-32). The role of roads in invasion of exotic species can be viewed in terms of opportunities for or barriers to dispersal and establishment. Dispersal occurs along the road network and from the road network into neighboring stands, streams, and riparian zones. Effects of exotic plants and pathogens can range from negligible to profound, as in the case of *Phytopthora laterella*, the root fungus that kills the high-value Port Orford cedar (*Chamaecyparis lawsoniana*) (Zobel et al. 1985). Road networks with extensive segments along ridgetops can greatly increase dispersal of pathogens such as *P. laterella*, which is transported into a forest landscape as spores in soil on vehicle tires, deposited along roads, and then spread downslope by groundwater and streams.

The role of roads in movement of exotic plant and pathogen species has several implications for managing forest landscapes. The initial introduction of pests and pathogens that could damage high-value resources should be minimized along road networks, even at the time of road construction. Measures to stop the spread of *P. lateralla*, for example, include thorough cleaning of equipment used in construction and maintenance (Zobel et al. 1985). Exotic plant species may be more likely to invade adjacent forest stands that have been opened by management activities (e.g., thinning) or other disturbances than to move into unmanaged stands (Hobbs and Huenneke 1992; DeFerrari and Naiman 1994). Useful strategies may include leaving a buffer of unthinned vegetation along roads to limit dispersal (Brothers and Spingarn 1992) or controlling or eradicating hosts of the pathogen along roadsides (Moody and Mack 1988).

Forest roads are important and possibly indelible parts of many Coast Range landscapes. Whether roads present serious, manageable, or negligible problems depends on the area and management objective being considered. By taking a comprehensive view of road functions and approaching roads from a landscape viewpoint, we can better plan forest and road management to minimize the undesired effects.

Aquatic-terrestrial linkages

Aquatic and terrestrial ecosystems of the Coast Range are linked through the many riparian zones that dissect the mountains and forests. The riparian zone extends upslope from the edge of the average high-water mark of the wetted channel (Gregory et al. 1991) and includes that portion of terrestrial areas where vegetation and microclimate are controlled by high water tables. The riparian "zone of influence" lies beyond the riparian zone. It is the

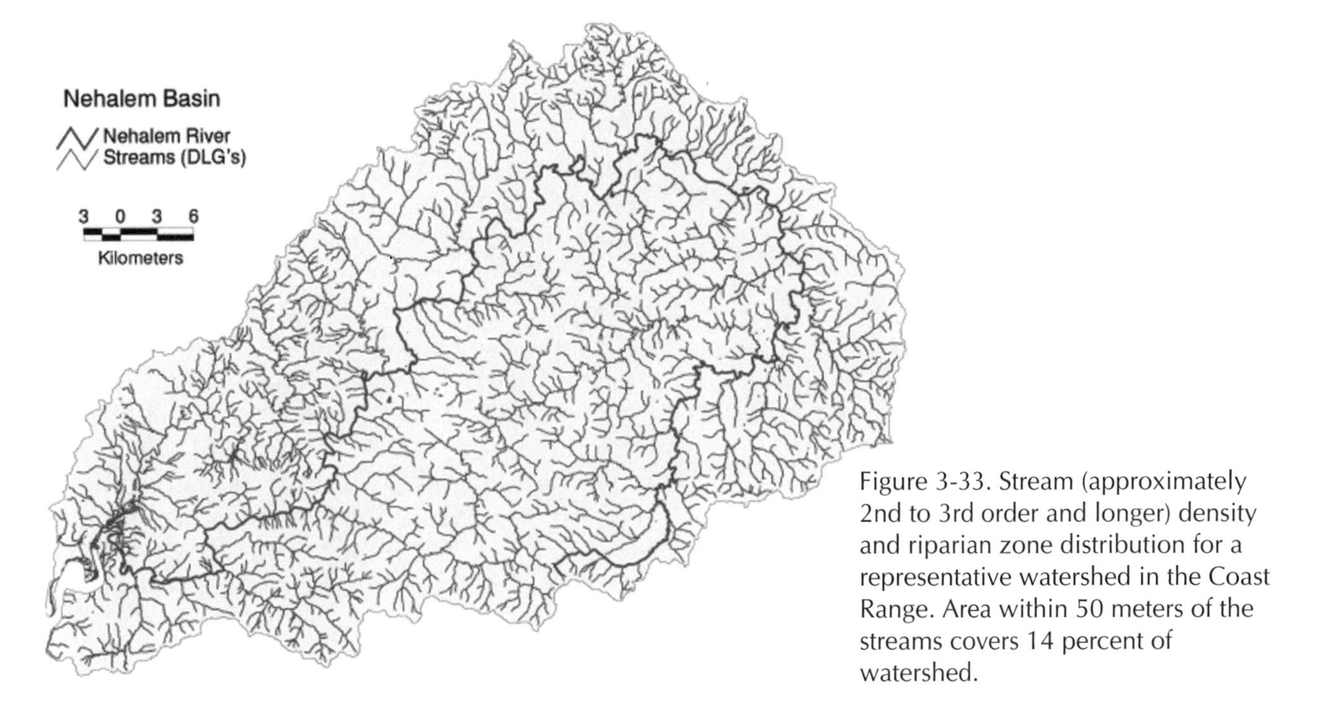

Figure 3-33. Stream (approximately 2nd to 3rd order and longer) density and riparian zone distribution for a representative watershed in the Coast Range. Area within 50 meters of the streams covers 14 percent of watershed.

transition area between the riparian zone and the upland area and is a zone where vegetation still influences the stream under some conditions (Gregory et al. 1991)

The width of the riparian zone and the extent of the zone of influence are related to stream size and valley morphology (Naiman et al. 1992). Small, headwater streams (i.e. first and second order) have relatively small riparian zones (Figure 3-18). However, while the amount of riparian zone for such a stream may be small, the total amount of riparian area in a watershed can be extensive because of the abundance of such streams (Figure 3-33). First- and second-order channels may compose more than 90 percent of the total stream network. These small channels are strongly influenced by terrestrial vegetation. Mid-order channels (third- to fifth-order) have larger riparian areas (Figure 3-18), which are determined by long-term channel dynamics, annual discharge, and valley morphology. Unconstrained reaches, those characterized by a low gradient and a wide valley, generally have larger riparian zones than more confined, steeper reaches. Larger rivers and streams (greater than sixth order) have well-developed floodplains and terraces that contain diverse riparian vegetation. The extent of the riparian zone is directly related to the size and complexity of the floodplain (Naiman et al. 1992).

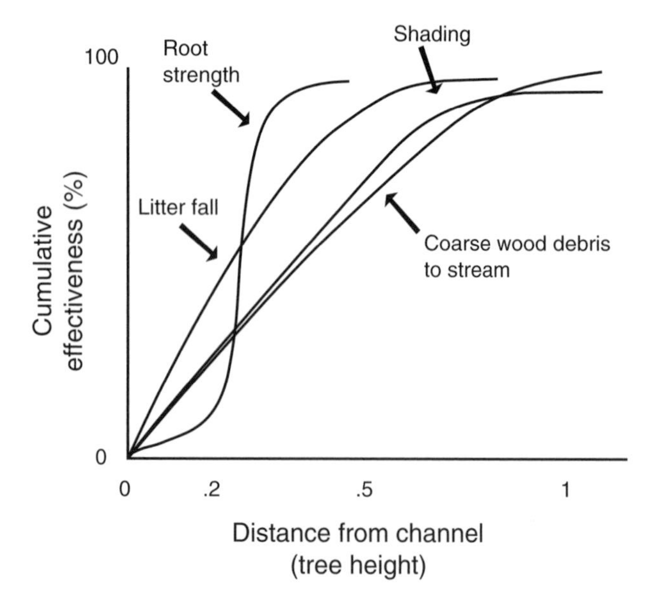

Figure 3-34. Generalized relation between distance from stream and ecological functions (FEMAT 1993).

Riparian vegetation has many functions in the aquatic ecosystem. It increases bank stability and resistance to erosion by two mechanisms: large and fine roots bind soil particles together, helping to maintain bank integrity during high flows (Swanson et al. 1982), and stems and branches which enter the stream increase roughness in the channel, reducing the erosion potential of flowing water (Spence et al. 1996). Litter from riparian vegetation provides major sources of energy for biota in streams. The quality, quantity, and timing of litter delivered to channels depend on vegetation type, stream orientation, valley topography, and stream morphology (Naiman et al. 1992). Deciduous and herbaceous materials decompose more quickly than coniferous inputs (Gregory et al. 1991). Riparian vegetation and downed wood in the riparian zone trap sediments and reduce the likelihood of landslides (Swanson et al. 1982). In addition, riparian vegetation controls the amount of solar radiation that reaches the channel, which in turn influences water temperatures and primary productivity (Beschta et al. 1987). These biotic and physical factors exert a strong influence on the structure and composition of biological communities in all sizes of streams.

The effect of riparian vegetation on the aquatic functions varies with distance from the channel (FEMAT 1993; Figure 3-34). Litterfall and bank stabilization are provided by vegetation closest to the stream. Shade and large-wood sources are influenced by a wider band of vegetation that extends further upslope.

Movement of materials from riparian areas to stream channels occurs both continuously and episodically. Litter input occurs throughout the course of the year, although input of deciduous materials occurs predominately in a 6-8 week period in the fall (Naiman et al. 1992). Large-wood input is both continuous and episodic. Single or small groups of trees fall periodically as a result of disease or wind and flooding. The relative importance of episodic and continuous inputs of wood and sediment into a given stream varies according to topography, geology, soil type and depth, and vegetation type. Large episodic inputs will be more pronounced in steeper, more unstable watersheds and less prominent in more stable areas.

Infrequent catastrophic disturbances may deliver large pulses of material to mid- and higher order channels. McGarry (1994) found that 48 percent of the large wood in Cummins Creek, a small wilderness-area stream on the Oregon Coast, was delivered to the channel via landslides. Historically, landslides and debris flows moved sediment and wood from riparian zones along lower-order channels and upslope areas and down stream to mid- and higher-order channels. In the past, these events were frequently associated with wildfires (Benda and Dunne 1997). The sediment and wood deposited in lower-gradient reaches had long-term impacts, from decades to centuries, on the physical and biological features of streams in the Coast Range (Reeves et al. 1995). The pattern of fire and landslides in the uplands gave rise to a changing mosaic of habitat conditions in streams. It has been proposed that streams followed a kind of "succession" or cycle of changes in sediment and wood that was closely tied to catastrophic disturbances and vegetation structure in the uplands (Reeves et al. 1995).

The pattern of human activities such as conversion of forest to agriculture and intensive forest management for timber production also affect the pattern and characteristics of inputs of wood and sediment. Riparian vegetation has been removed in all parts of watersheds, especially in the lower segments where forest was cleared for agriculture and other development. The riparian areas farther upstream in most watersheds have also been extensively altered. As a consequence, the amount of large wood in streams has been reduced, and landslide rates and stream temperatures have increased. These changes have led to a decline in the quality and quantity of freshwater habitats for anadromous salmonids (*Oncorhynchus* spp.) and are responsible, in part, for the current poor condition of many of these populations.

New management strategies for riparian areas are required if degraded aquatic ecosystems in the Coast Range are to be restored. Restoration programs will need to reestablish appropriate trees and shrubs in riparian zones in all parts of the watershed, from small headwater streams to valley bottoms along large rivers. Management strategies should be directed toward restoring and maintaining ecological processes that connect the riparian zone with the aquatic ecosystem (Reeves et al. 1995). These processes include the chronic and episodic delivery of sediments, wood, and organic litter. This conclusion does not mean that riparian zones become «no touch» areas, however. Many riparian areas need active management to accelerate the development of desired conditions and attributes. Strategies are needed that achieve the necessary goals and requirements of riparian zones and also provide for the production of wood and other products for human consumption.

Natural processes as a foundation for management

The production of goods and services from Coast Range forest ecosystems, including the maintenance of native species and natural ecological processes, is best achieved when we understand how natural disturbances and ecological processes have operated in the near and distant past and how they work today. We need this information not to recreate the past but to better understand the forces that created the forests we see today with all the diversity of their native species. Armed with this knowledge, we can better understand how the kinds of forest changes we are imposing either directly (through logging) or indirectly (through climate change or fire suppression) might affect the forests of the future.

As discussed above, fire has been a major force in past forest dynamics in the Coast Range. Today, few people want wildfires to burn as frequently or intensely as they did in the past. Managers are using prescribed burns to maintain certain habitat conditions, but wildfire will probably not be a major force in forest dynamics in the future. This does not mean that wildfires will not occur, but their extent and spread will probably be greatly restricted. Can we sustain desired native species, communities, and natural process in the absence of wildfire? Maybe. Prescribed fire could be used to thin understories and retain variable tree densities in old-growth forests where surface fires were relatively common in the past, such as in the southern and eastern parts of the Coast Range. Without low-severity fires, the structure of these forests will change as shade-tolerant, fire-sensitive trees such as western hemlock and grand fir invade and increase the density of the understory.

Other than prescribed fires, what substitutes are there for wildfire? Harvesting might be the first thing that comes to mind and, certainly, active management through the felling of trees has replaced fire as the dominant disturbance across the

Coast Range in the last 100 years. Silvicultural practices such as thinning, patch cutting and green-tree retention, and some clearcutting can substitute for some aspects of partial disturbances and large, severe fires, creating, for example, large patches of early-successional forest. However, we now know that traditional clearcutting disturbances have little similarity to wildfires. Although clearcutting removes the canopy and increases light at the forest floor, it differs from wildfires in many ways, most importantly in that clearcutting does do not leave the large live trees and dead wood that wildfires did. These legacies probably helped maintain habitat at stand and landscape levels for species such as pileated woodpeckers, ensatina salamanders, foliose lichens, fungi, and invertebrates. Consequently, new management practices such as green-tree retention (leaving one to 10 or more large trees per acre in cutting units), maintaining riparian buffer strips, and leaving live and dead conifers on a logging site make the effects of logging disturbances more similar to those of wildfires than traditional methods.

While it may be possible to imitate the dynamics and effects of wildfire at the stand level, it is far less clear how well we can do this at the landscape level. At this large a spatial scale we encounter the whole diversity of ecosystems and disturbance patterns. We know much less about the dynamics of forests at landscape scales and the effects of those dynamics on forest attributes such as upland and aquatic habitats. We do know that return intervals of natural disturbance have been much more variable (ranging from 5 to over 500 years) than are those determined solely by timber management objectives, which promote a single type of disturbance on a regular cycle of 40 to 60 years. The current mix of forest policies and practices in the Coast Range appears to provide for a certain range of disturbance intervals, with long rotations on federal and state lands and short rotations on private industrial lands. However, even with this greater temporal variation in disturbance regime, disturbances within large blocks of land will be relatively constant over time. One type of disturbance regime will be found on one area and a very different type on another area, according to ownership pattern. Can native species and processes be maintained if disturbance frequencies remain spatially relatively constant, between ownerships? Maybe. However, there is evidence that whole watersheds must occasionally cycle through a full range of successional stages if high-quality fish habitat is to be sustained (Reeves et al. 1995).

We now know that watershed disturbances such as landslides and debris flows help maintain aquatic habitats by delivering coarse sediments and large pieces of wood to streams. It is becoming increasingly clear that landslides originating in certain kinds of small, intermittent stream drainages have been a major source of large wood and coarse sediments in higher-order streams. Will streams provide the same quality of habitats they have in the past if landslide frequencies change (increase or decrease) and the wood delivered to streams declines in amount and size? Probably not. Will current forest practices on public and private land provide watershed processes and dynamics that meet aquatic conservation goals? Maybe. Are there alternatives to landslides for creating aquatic habitat? Yes, but they may not be practical over large areas or for long periods of time. Stream habitat can be engineered to some degree by placing large tree boles or boulders at selected points within stream networks. These structures probably can improve stream habitat quality in some localities for short periods of time. However, such structural enhancement of streams is difficult to do over large areas, and the added structures may function only until a flood transports them downstream or pushes them onto the floodplain. Developing management systems that incorporate landslides and debris flows containing large wood may be the best long-term solution.

Finally, it is clear that natural processes of vegetation development must be better understood if managers can hope to reach a goal of retaining native species and communities. Late-successional communities characterized by large trees, multiple canopy layers, and large amounts of dead wood require many centuries to develop. Some of these features, such as large trees and multiple layers, can be accelerated through silvicultural practices such as thinning of young plantations. This approach, which is practiced most commonly on federal and state lands, can probably be used in conjunction with long rotations (more than 120 years) to provide habitat for species that find optimum conditions in older forests. However, there may be a limit to the late- successional species or population numbers that can be maintained in stands created by accelerating habitat development. Some organisms such as canopy lichens or slow-moving salamanders may simply need more time to colonize forest than these strategies provide.

Another important natural process of vegetation development in the Coast Range is the competitive interaction between deciduous and coniferous woody plants. The outcome of the competitive struggle between these two life forms can determine the structure and function of forest ecosystems for many decades or even centuries. It is clear that the competitive ability of hardwoods relative to conifers is greater in riparian areas than upslope. Consequently, managers should not expect conifers to dominate the vegetative community near streams, except in low-order (headwater) streams. Efforts to regenerate conifers along streams must be tempered with the knowledge that conifer trees have probably always been patchily distributed in these environments because deciduous shrubs and hardwoods are naturally more competitive. If the goal is to increase the inputs of large conifer boles to streams, growing more large trees along headwalls and first-order streams may be more effective than trying to achieve conifer domination along higher-order (downstream) drainages.

Through focused research efforts, our knowledge of native species and natural processes of Coast Range forests has progressed tremendously over the last 10 to 20 years. We now recognize the ecological values of old-growth forests, large down wood in streams, and watershed-scale disturbances such as landslides and debris flows, and are beginning to understand how forest management practices affect resources other than wood-fiber production. We have a general knowledge of the habitat needs of many species of plants and animals that we lacked just a few years ago. We are beginning to understand how landscape-scale process might affect biological diversity. However, we lack knowledge of many of the details of ecological processes and habitat relationships that are essential to modern forest planning and the ability to predict the consequences of specific actions. We also lack the ability to see the big picture of forest management at the scale of the Coast Range and understand how individual management activities might add up to affect the ecological and social conditions as whole. Research will continue to provide answers to current questions but managers and the public should be prepared for some of the scientific "truths" of today to evolve into different forms as new information comes forth.

How should managers respond to uncertainty and new information and ideas? Certainly practicing some form of adaptive management (McLain and Lee 1996) is prudent, just as it is prudent to maintain a diversity of management strategies. Managers need to know details for managing particular sites and stands; however, it is these details that are slow in coming from research on ecosystems and it is these details that are most likely change over time and geography. General scientific and management principles are probably more durable. We hope this chapter provides some of the principles to empower managers to fill in the details as they struggle to sustain forest values.

Literature Cited

Agee, J.K. 1993. *Fire Ecology of Pacific Northwest Forests.* Island Press, Covelo CA.

Andrus, C., and H. A. Froehlich. 1988. Riparian forest development after logging or fire in the Oregon Coast Range: wildlife habitat and timber value, pp. 139-152. in *Streamside Management: Riparian Wildlife and Forestry Interactions*, K. J.Raedeke, ed. Contribution No. 59, University of Washington, Institute of Forest Resources, Seattle.

Aplet, G. H., and W. S. Keeton. 1999. Application of historical range of variability concepts to biodiversity conservation, pp. 71-86 in *Practical Approaches to the Conservation of Biological Diversity*, R. K. Baydack, H. Campa III, and J. B. Haufler, ed. Island Press, Washington D.C.

Barnes, B. V., D. R. Zak, S. R. Denton, and S. H. Spurr. 1998. *Forest Ecology*. 4th Edition. John Wiley and Sons, New York.

Bartlein, P. J., K. H. Anderson, P. M. Anderson, M. E. Edwards, C. J. Mock, R. S. Thompson, R. S. Webb, T. Webb III, and C. Whitlock. In press. Paleoclimate simulations for North America over the past 21,000 years: features of the simulated climate and comparisons with paleoenvironmental data. *Quaternary Science Reviews*.

Benda, L. E. 1994. *Stochastic Geomorphology in a Humid Mountain Landscape*. Ph.D. dissertation. University of Washington, Seattle.

Benda, L. E., and T. Dunne. 1987. Sediment routing by debris flows, pp. 213-223 in *Erosion and sedimentation in the Pacific Rim*, R. L. Beschta, T. Blinn, G. E. Grant, G. G. Ice, and F. J. Swanson, ed. International Association of Hydrological Sciences, Publication 165. Oxfordshire, United Kingdom.

Benda, L. E., and T. Dunne. 1997. Stochastic forcing of sediment routing and storage in channel networks. *Water Resources Research* 33:2865-2880.

Beschta, R. L., R. E. Bilby, G. W. Brown, L. B. Holtby, and T. D. Hofstra. 1987. Stream temperature and aquatic habitat: fisheries and forestry interactions, pp. 191-232 in *Streamside Management: Forestry and Fishery Interactions,* E. O. Salo and T. W. Cundy, ed. Contribution 57, Institute of Forest Resources, University of Washington, Seattle.

Bilby, R. E., K. Sullivan, and S. H. Duncan. 1989. The generation and fate of road-surface sediment in forested watersheds in southwestern Washington. *Forest Science* 35(2): 453-468.

Binkley, D., K. Cromack Jr., and D. D. Baker. 1994. Nitrogen fixation by red alder: biology, rates, and controls, pp. 57-72 in *The Biology and Management of Red Alder,* D. E. Hibbs, D. S. DeBell, and R. F. Tarrant, ed. Oregon State University Press, Corvallis.

Bolsinger, C. L., and K. L. Waddell. 1993. *Area of Old-growth Forests in California, Oregon, and Washington.* Resource Bulletin PNW-RB-197, USDA Forest Service.

Bormann, B. T., K. Cromack Jr., and W. O. Russell III. 1994. Influences of red alder on soils and long-term ecosystem productivity, pp. 47-56 in *The Biology and Management of Red Alder,* D. E. Hibbs, D. S. DeBell, and R. F. Tarrant, eds. Oregon State University Press, Corvallis.

Boyd, R. 1986. Strategies of Indian burning in the Willamette Valley. *Canadian Journal of Anthropology* 5:65-86. Reprinted 1999, pp. 94-138 in *Indians, Fire, and the Land in the Pacific Northwest,* R. Boyd, ed. Oregon State University Press, Corvallis.

Bowling, L. C., and D. P. Lettenmaier. 1997. *Evaluation of the Effects of Forest Roads on Streamflow in Hard and Ware Creeks, Washington.* Water Resources Series Tech. Rpt. No. 155, Dept. of Civil Engineering, University of Washington, Seattle.

Brothers, T. S., and A. Spingarn. 1992. Forest fragmentation and alien plant invasion of central Indiana old-growth forests. *Conservation Biology* 6: 91-100.

Burroughs, E. R., Jr., F. J. Watts, and D. F. Haber. 1984. Surfacing to reduce erosion of forest roads built in granitic soils, pp. 255-264 in *Symposium on Effects of Forest and Land Use on Erosion and Slope Stability.* Environment and Policy Institute, Honolulu, HI.

Chen, S. 1998. *Characterization of Fire Effects on Forest Ecosystems in the Tillamook Forest, Oregon.* MS thesis. Oregon State University, Corvallis.

Clark, J. S. 1990. Fire and climate change during the last 750 years in northwestern Minnesota. *Ecological Monographs* 60:135-159.

COHMAP Members 1988. Climate changes of the last 18,000 years: observations and model simulations. *Science* 241:1043-1052.

Corn, P. S., and R. B. Bury. 1991. Terrestrial amphibian communities in the Oregon Coast Range, pp. 305-318 in *Wildlife and Vegetation of Unmanaged Douglas-fir Forests,* L. F. Ruggiero, K. B. Aubry, A. B. Carey, and M. H. Huff, tech. coords. General Technical Report PNW-GTR-285, USDA Forest Service, Portland OR.

Curtis, R. O. 1994. *Some Simulation Estimates of Mean Annual Increment of Douglas-fir: Results, Limitations, and Implications for Management.* Research Paper PNW-RP-471, USDA Forest Service, Pacific Northwest Forestry and Range Experiment Station, Portland OR.

Daily, G. C. (ed.) 1997. *Nature's Services: Societal Dependence on Natural Ecosystems.* Island Press. Washington D.C.

Daly, C., R. P. Neilson, and D. L. Phillips. 1994. A statistical-topographic model for mapping climatological precipitation over mountainous terrain. *Journal of Applied Meteorology* 33:140-158.

DeFerrari, C. M. and R. J. Naiman. 1994. A multi-scale assessment of the occurrence of exotic plants on the Olympic Peninsula, Washington. *Journal of Vegetation Science* 5:247-258.

Duncan, S. H., R. E. Bilby, J. W. Ward, and J. T. Heffner. 1987. Transport of road-surface sediment through ephemeral stream channels. *Water Resources Bulletin* 23(1):113-119.

Dyrness, C.T. 1967. *Mass Soil Movements in the H. J. Andrews Experimental Forest.* Research Paper PNW-42, USDA Forest Service, Pacific Northwest Forest and Range Experiment Station, Portland OR.

Emmingham, W. H., and K. Maas. 1994. Survival and growth of conifers released in alder-dominated coastal riparian zones. *COPE Report* 7(2,3):13-15. Oregon State University, Corvallis.

Esseen, P., B. Ehnström, L. Ericson, and K. Sjöberg. 1997. Boreal forests. *Ecological Bulletins* 46:16-47.

FEMAT (Forest Ecosystem Management Assessment Team). 1993. *Forest Ecosystem Management: An Ecological, Economic, and Social Assessment.* Report of the Forest Ecosystem Management Assessment Team, U.S. Government Printing Office 1973-793-071. U.S. Government Printing Office for USDA Forest Service; USDI Fish and Wildlife Service, Bureau of Land Management, and Park Service; U.S. Department of Commerce, National Oceanic and Atmospheric Administration and National Marine Fisheries Service; and the U.S. Environmental Protection Agency. Portland OR.

Foltz, R. B., and Burroughs, E. R., Jr. 1990. Sediment production from forest roads with wheel ruts. *Watershed Planning and Analysis in Action, Symposium Proceedings of IR Conference,* ASCE, Durango CO.

Forcella, F., and S. J. Harvey. 1983. Eurasian weed infestation in western Montana in relation to vegetation and disturbance. *Madrono* 30:102-109.

Franklin, J. F. and C. T. Dyrness. 1988. *Natural Vegetation of Oregon and Washington.* Oregon State University Press, Corvallis.

Franklin, J. F., F. J. Swanson, M. E. Harmon, D. A. Perry, T. A. Spies, V. H. Dale, A. McKee, W. K. Ferrell, J. E. Means, S. V. Gregory, J. D. Lattin, T. D. Schowalter, and D. Larsen. 1992. Effects of global climatic change on forests in northwestern North America, pp. 244-257 in *Global Warming and Biological Diversity,* R. L. Peters and T. E. Lovejoy, ed. Yale University Press, New Haven CT.

Fried, J .S., J. C. Tappeiner II, and D. E. Hibbs. 1988. Bigleaf maple seedling establishment and early growth in Douglas-fir forests. *Canadian Journal of Forest Research* 18:1226-1233.

Freid, M. A. 1994. *The Effects of Ownership Pattern on Forest Road Networks in Western Oregon*. MS thesis. Oregon State University, Corvallis.

Fujimori, T., S. Kawanabe, H. Saito, C. C. Grier, and T. Shidei. 1976. Biomass and primary production in forest of three major vegetation zones of the northwestern United States. *Journal of the Japanese Forestry Society* 58 (10):360-373.

Gohlz, H. L. 1982. Environmental limits on aboveground net primary production, leaf area, and biomass in vegetation zones of the Pacific Northwest. *Ecology* 63(2): 469-481.

Gratkowski, H.J. 1962. *Heat as a Factor in Germination of Seeds of* Ceanothus velutinus *var.* laevigatus *T. and G.* Ph.D. dissertation. Oregon State University, Corvallis.

Gray, A. N., and T. A. Spies. 1997. Microsite controls on tree seedling establishment in conifer forest canopy gaps. *Ecology* 78(8):2458-2473.

Gregory, S. V., F. J. Swanson, W. A. McKee, and K. W. Cummins. 1991. An ecosystem perspective of riparian zones. *Bioscience* 41(8):540-551.

Gregory, S. V., G. A. Lamberti, D. C. Erman, K. V. Koski, M. L. Murphy, and J. R. Sedell. 1987. Influence of forest practices on aquatic production, pp. 233-256 in *Streamside Management: Forestry and Fishery Interactions*, E. O. Salo and T.W. Cundy, eds.Contribution 57, University of Washington, Institute of Forest Resources, Seattle.

Grier, C. C. 1976. Biomass, productivity, and nitrogen-phosphorus cycles in hemlock-spruce stands of the central Oregon coast, pp 71-81 in *Proceedings, Western Hemlock Management Conference*, W. A. Atkinson and R. J. Zasoski, eds. College of Forest Resources, University of Washington, Seattle.

Grigg, L. D., and C. Whitlock. 1998. Late-glacial vegetation and climate change in western Oregon. *Quaternary Research* 49:287-298.

Gumtow-Farrior, D. L. 1991. *Cavity Resources in Oregon White Oak and Douglas-fir Stands in the Mid-Willamette Valley, Oregon*. MS thesis. Oregon State University, Corvallis.

Halpern, C. B., and T. A. Spies. 1995. Plant species diversity in natural and managed forests of the Pacific Northwest. *Ecological Applications* 5(4):913-934.

Hansen, A. J., T. A. Spies, F. J. Swanson, and J. L. Ohmann. 1991. Conserving biodiversity in managed forests. *BioScience* 41:382-392.

Harmon, M. E. and C. Hua. 1991. Coarse woody debris dynamics in two old-growth ecosystems. *BioScience* 41(9):604-610.

Harmon, M. E., J. F. Franklin, F. J. Swanson, P. Sollins, S. V. Gregory, J. D. Lattin, N. H. Anderson, S. P. Cline, N. G. Aumen, J. R. Sedell, G. W. Lienkaemper, K. Cromack Jr., and K. W. Cummins. 1986. Ecology of coarse woody debris in temperate ecosystems, pp. 133-302, in *Advances in Ecological Research*, Vol. 15, A. MacFadyen and E. D. Ford, ed. Academic Press, Orlando FL.

Harmon, M. E., J. Sexton, B. A. Caldwell, and S. E Carpenter. 1994. Fungal sporocarp mediated losses of Ca, Fe, K, Mg, N, P, and Zn from conifer logs in the early stages of decomposition. *Canadian Journal of Forest Research*, 24:1883-1893.

Harr, R. D., W. C. Harper, J. T. Krygier, and F. S. Hsieh. 1975. Changes in storm hydrographs after road building and clearcutting in the Oregon Coast Range. *Water Resources Research* 11(3):436-444.

Harrington, C. A. 1990. *Alnus rubra*, pp. 116-123 in *Silvics of North America,* volume 2, Hardwoods, R. M. Burns and B.H. Honkala, ed. Agriculture Handbook 654, USDA Forest Service, Washington D.C.

Harris, A.S. 1990. *Picea sitchensis*, pp. 260-267 in *Silvics of North America,* volume 2, Hardwoods, R. M. Burns and B.H. Honkala, ed. Agriculture Handbook 654, USDA Forest Service, Washington D.C.

Heliovaara, K., and Vaisanen, R. 1984. Effects of modern forestry on northwestern European forest invertebrates: a synthesis. *Acta Forestalia Fennica* 189:1-32.

Hemstrom, M. A., and S. E. Logan. 1986. *Plant Association and Management Guide, Siuslaw National Forest*. R6-Ecol 220-1986a, USDA Forest Service Pacific Northwest Region.

Henderson, J. A. 1978. Plant succession on the *Alnus rubra/ Rubus spectabilis* habitat type in western Oregon. *Northwest Science* 52(3):156-167.

Heusser, C. J. 1983. Vegetation history of the northwestern United States, including Alaska, pp. 239-258, in *Late Quaternary Environments of the United States,* S. C. Porter, ed. University of Minnesota Press, Minneapolis.

Hibbs, D. E., and P. A. Giordano. 1996. Vegetation characteristics of alder-dominated riparian buffer strips in the Oregon Coast Range. *Northwest Science* 70(3):213-222.

Hobbs, R .J., and L. F. Huenneke. 1992. Disturbance, diversity, and invasion: implications for conservation. *Conservation Biology* 6:324-337.

Hobbs, S. D., S. D. Tesch, P. W. Owston, R. E. Stewart, J. C. Tappeiner II, and G. E. Wells. 1992. *Reforestation Practices in Southwestern Oregon and Northern California*. Forest Research Laboratory, Oregon State University, Corvallis.

Impara, P. C. 1997. *Spatial and Temporal Patterns of Fire in the Forests of the Central Oregon Coast Range*. Ph.D. dissertation. Oregon State University, Corvallis.

Intergovernmental Panel on Climate Change (J. T. Houghton, L. G. Meira Filho, B. A. Callander, N. Harris, A. Kattenberg, and K. Maskell). 1996. *Climate Change 1995; The Science of Climate Change,* Cambridge University Press.

Jones, J. A., and G. E. Grant. 1996. Peak flow responses to clearcutting and roads in small and large basins, western Cascades, Oregon. *Water Resources Research* 32(4):959-974.

Jones, J. A., F. J. Swanson, B. C. Wemple, and K. U. Snyder. 2000. Effects of roads on hydrology, geomorphology, and disturbance patches in stream networks. *Conservation Biology* 14 (1):1-10.

Kimmins, J. P. 1987. *Forest Ecology*. Macmillan Publishing Co., New York.

King, J. G., and L .C. Tennyson , 1984. Alteration of streamflow characteristics following road construction in north central Idaho. *Water Resources Research* 20(8):1159-1163.

Leidholt-Bruner, K., D. Hibbs, and W. McComb. 1992. Beaver dam locations and their effects on distribution and abundance of coho salmon fry in two coastal Oregon streams. *Northwest Science* 66:218-223.

Lichatowich, J. A. 1989. Habitat alternation and changes in abundance of coho (*Onchorhynchus kisutch*) and chinook salmon (*O. tshawytscha*) in Oregon's coastal streams, pp. 92-99 in *Proceedings of the National Workshop on Effects of Habitat Alteration on Salmonid Stocks*, C. D. Levings, L.B. Holtby, and M.A. Henderson, ed. Canadian Special Publication of Fisheries and Aquatic Sciences 105.

Long, C. J., C. Whitlock, P. J. Bartlein, and S. H. Millspaugh. 1998. A 9,000-year fire history from the Oregon Coast Range, based on a high-resolution charcoal study. *Canadian Journal of Forest Research* 28:774-787.

Long, J. N. 1982. Productivity of western coniferous forests, pp. 89-125 in *Analysis of Coniferous Forest Ecosystems in the Western United States*, R. Edmonds, ed. Hutchison Ross Publishing Co., Stroudsburg PA.

Lonsdale, W. M., and A. M. Lane. 1994. Tourist vehicles as vectors of weed seeds in Kakadu National Park, northern Australia. *Biological Conservation* 69:277-283.

MacLean, C. D. 1990. *Changes in Area and Ownership of Timberland in Western Oregon: 1961-86*. (Resour. Bull.) PNW-RB-170, USDA Forest Service, Pacific Northwest Research Station, Portland OR.

Maser, C., S. P. Cline, K. Cromack, Jr., J. M. Trappe, and E. Hansen. 1988. What we know about large trees that fall to the forest floor, pp. 25-46 in *From the Forest to the Sea: A Story of Fallen Trees*, C. Maser, R. F. Tarrant, J. M. Trappe, and J. F. Franklin, tech. eds. General Technical Report. PNW-GTR-229, USDA Forest Service, Pacific Northwest Research Station, Portland OR.

McArdle, R. E., W. H. Meyer, and D. Bruce. 1961. *The Yield of Douglas-fir in the Pacific Northwest*. Technical Bulletin No. 201. U.S. Department of Agriculture Washington D.C.

McComb, W. C. 1994. Red alder: interactions with wildlife, pp. 131-140 in D. E. Hibbs, D. S. DeBell, and R. F. Tarrant, eds. *The Biology and Management of Red Alder*. Oregon State University Press, Corvallis.

McComb, W. C., T. A. Spies, and W. H. Emmingham. 1993. Douglas-fir forests: managing for timber and mature-forest habitat. *Journal of Forestry* 91(12):31-42.

McGarigal, K., and W. C. McComb. 1992. Streamside versus upslope breeding bird communities in the central Oregon Coast Range. *Journal of Wildlife management* 56(1):10-23.

McGarry, E. V. 1994. *A Quantitative Analysis and Description of the Delivery and Distribution of Large Woody Debris in Cummins Creek, Oregon*. MS thesis. Oregon State University, Corvallis.

McLain, R. J., and R. G. Lee. 1996. Adaptive management: pitfalls and promises. *Environmental Management* 20:437-448.

Megahan, W. F., N. F. Day, and T. M. Bliss. 1978. Landslide occurrence in the western and central northern Rocky Mountain physiographic province in Idaho, pp. 116-139 in *Forest Soils and Land Use*: Proceedings of the 5th North American Forest Soils Conference, Colorado State University, Fort Collins.

Minore, D. 1979. *Comparative Autoecological Characteristics of Northwestern Tree Species: A Literature Review*. Gen. Tech. Rep. PNW-87, U.S. Department of Agriculture Pacific Northwest Forestry and Range Experiment Station, Portland OR.

Minore, D. 1990. *Thuja plicata*, pp. 590-600in *Silvics of North America*, volume 1, Conifers, R. M. Burns and B. H. Honkala, eds. Agriculture Handbook 654, USDA Forest Service, Washington D.C.

Minore, D., and C. E. Smith. 1971. *Occurrence and Growth of Four Northwestern Tree Species over Shallow Water Tables*. Res. Note PNW-160, USDA Forest Service, Pacific Northwest Forestry and Range Experiment Station, Portland OR.

Minore, D., and H. G. Weatherly. 1994. Riparian trees, shrubs and forest regeneration in the coastal mountains of Oregon. *New Forests* 8:249-263.

Montgomery, D. R. 1994. Road surface drainage, channel initiation and slope stability. *Water Resources Research* 30:1925-1932.

Moody, M. E., and R. N. Mack. 1988. Controlling the spread of plant invasions: the importance of nascent foci. *Journal of Applied Ecology* 25:1009-1021.

Morris, W. G. 1934. Forest fires in Oregon and Washington. *Oregon Historical Quarterly* 35: 313-339.

Murphy, M. L., and J. D. Hall. 1981. Varied effects of clearcut logging on predators and their habitat in small streams of the Cascade Mountains, Oregon. *Canadian Journal of Fisheries and Aquatic Science* 38:137-145.

Naiman, R. J., and J. R. Sedell. 1980. Relationships between metabolic parameters and stream order in Oregon. *Canadian Journal of Fisheries and Aquatic Science* 37: 834-847.

Naiman, R. J., T. J. Beechie, L. E. Benda, D. R. Berg, P. A. Bisson, L. H. MacDonald, M. D. O'Connor, P. L. Olson, and E. A. Steel. 1992. Fundamental elements of ecologically healthy watersheds in the Pacific Northwest coastal ecoregion, pp. 127-188 in *Watershed Management: Balancing Sustainability and Environmental Change*, R. J. Naiman, ed. Springer-Verlag, New York.

Neitlich, P. N., and B. McCune. 1997. Hotspots of epiphytic lichen diversity in two young managed forests. *Conservation Biology* 11:172-182.

Newton, M., B. A. El Hassan, and J. Zavitkowski. 1968. The role of red alder in western Oregon forest succession, pp. 73-84 in *Biology of Alder*, J. Trappe, J. Franklin, R. F. Tarrant, and G. M. Hansen, ed. USDA Forest Service, Pacific Northwest Forestry and Range Experiment Station, Portland OR.

Niemiec, S. S., G. R. Ahrens, S. Willits, and D. E. Hibbs. 1995. *Hardwoods of the Pacific Northwest.* Contribution 8, Forest Research Laboratory, Oregon State University, Corvallis.

Nierenberg, T. R., and D. E. Hibbs. 2000. A characterization of unmanaged riparian areas in the central Oregon Coast Range of western Oregon. *Forest Ecology and Management* 129 (2000): 195-206.

O'Dea, M. E., J. C. Zasada, and J. C. Tappeiner II. 1995. Vine maple clone growth and reproduction in managed and unmanaged coastal Oregon Douglas-fir forests. *Ecological Applications* 5(1):63-73.

Ohmann, J.L. 1996. *Regional Gradient Analysis and Spatial Pattern of Woody Plant Communities in Oregon*. Ph.D. dissertation. Oregon State University, Corvallis.

Ohmann, J. L., and C. L. Bolsinger. 1991. Monitoring biodiversity with permanent plots—landscape, stand structure, and understory species (abstract), pp. 525-526in *Wildlife and Vegetation of Unmanaged Douglas-fir Forests*, L. F. Ruggiero, K. B. Aubry, A. B. Carey, and M. H. Huff, tech. coords. General Technical Report PNW-GTR-285, USDA Forest Service.

Ohmann, J. L., and T. A Spies. 1998. Regional gradient analysis and spatial pattern of woody plant communities of Oregon forests. *Ecological Monographs* 68 (2):151-182.

Ohmann, J. L., W. C. McComb, and A. A. Zumrawi. 1994. Snag abundance for cavity-nesting birds on nonfederal forest lands in Oregon and Washington. *Wildlife Society Bulletin* 22:607-620.

Pabst, R. J., and T. A. Spies. 1998. Distribution of herbs and shrubs in relation to landform and canopy cover in riparian forests of coastal Oregon. *Canadian Journal of Botany* 76:298-315.

Pabst, R. J., and T. A. Spies. 1999. Composition and structure of unmanaged riparian forests in the coastal mountains of Oregon. *Canadian Journal of Forest Research* 29:1557-1573.

Parendes, L. A. 1997. *Spatial Patterns of Invasion by Exotic Plants in a Forested Landscape.* Ph.D. dissertation, Oregon State University, Corvallis.

Perry, D. A. 1994. *Forest Ecosystems.* John Hopkins University Press, Baltimore MD.

Perry, D. 1998. The scientific basis of forestry. *Annual Review of Ecological Systematics* 29:435-466.

Pickett, S. T. A., and P. S. White, eds. 1985. *The Ecology of Natural Disturbance and Patch Dynamics.* Academic Press, New York.

Piehl, B. T., R. L. Beschta, and M. R. Pyles. 1988. Ditch-relief culverts and low-volume forest roads in the Oregon Coast Range. *Northwest Science* 62(3):91-98.

Poage, N. J. 1994. *Comparison of Stand Development of a Deciduous-dominated Riparian Forest and a Coniferous-dominated Riparian Forest in the Oregon Coast Range.* MS thesis, Oregon State University, Corvallis.

Powell, D. S., J. L. Faulkner, D. R. Darr, Z. Zhu, and D. W. MacCleery. 1993. *Forest Resources of the United States, 1992.* GTR RM-234. Rocky Mountain Forest and Range Experiment Station. USDA Forest Service, Fort Collins CO.

Prince, S. D., and S. N. Goward. 1995. Global primary production: a remote sensing approach. *Journal of Biogeography* 22:815-835.

Ramey-Gassert, L. K., and J. R. Runkle. 1992. Effect of land use practices on composition of woodlot vegetation in Green County, Ohio. *Ohio Journal of Science* 92:25-32.

Reeves, G. H., L. E. Benda, K. M. Burnett, P. A. Bisson, and J. R. Sedell. 1995. A disturbance-based ecosystem approach to maintaining and restoring freshwater habitats of evolutionarily significant units of anadromous salmonids in the Pacific Northwest. *American Fisheries Society Symposium* 17:334-349.

Reid, L. M., and T. Dunne. 1984. Sediment production from road surfaces. *Water Resources Research* 20:1753-1761.

Robbins, W. G. 1997. *Landscapes of Promise—the Oregon Story 1800-1940.* University of Washington Press, Seattle.

Robison, E. G., K. Mills, J. Paul, L. Dent, and A. Skaugset. 1999. *Storm Impacts and Landslides of 1996.* Draft final report. Oregon Department of Forestry, Salem.

Root, T. L., and S. H. Schneider. 1995. Ecology and climate: research strategies and implications. *Science* 269:334-341.

Runyon, J., R. H. Waring, S. N. Goward, and J. M. Welles. 1994. Environmental limits on net primary production and light-use efficiency across the Oregon transect. *Ecological Applications* 4:226-237.

Ruth, R. H., and R. A. Yoder 1953. *Reducing Wind Damage in the Forests of the Oregon Coast Range.* Research Paper 7. USDA Forest Service, Pacific Northwest Forest and Range Experiment Station, Portland OR.

Schmidt, W. 1989. Plant dispersal by motor cars. *Vegetation* 80:147-152.

Schrader, B. A. 1998. *Structural Development of Late Successional Forests in the Central Oregon Coast Range: Abundance, Dispersal and Growth of Western Hemlock* (Tsuga heterophylla) *Regeneration.* Ph.D. dissertation. Oregon State University, Corvallis.

Sessions, J., J. C. Balcom, and K. Boston. 1987. Road location and construction practices: effects of landslide frequency and size in the Oregon Coast Range. *Western Journal of Applied Forestry* 2(4):119-124.

Silen, R. R. 1955. More efficient road patterns for a Douglas fir drainage. *Timberman* 56(6):82-88.

Siuslaw National Forest. 1997. *Road Survey*. Siuslaw National Forest, Corvallis OR.

Siuslaw National Forest. 1999. *Oregon Coast Range Province Effectiveness Monitoring Pilot Study*. Final draft. On file at Siuslaw National Forest, Corvallis OR.

Spence, B. C., G. A. Lomnicky, R .M. Hughes, and R .P. Novitzki. 1996. *An Ecosystem Approach to Salmonid Conservation.* TR-4501-96-6057, Management Technology, Corvallis OR.

Spies, T. A,. and J. F. Franklin. 1991. The structure of natural young, mature, and old-growth Douglas-fir forests, pp. 91-110 in *Wildlife and Vegetation of Unmanaged Douglas-fir Forests*, L. F. Ruggiero, K. B. Aubry, A. B. Carey, and M. H. Huff, tech. coords. General Technical Report PNW-GTR-285, USDA Forest Service, Portland OR.

Spies, T. A., J. F. Franklin, and T. B. Thomas. 1988. Coarse woody debris in Douglas-fir forests of western Oregon and Washington. *Ecology* 69:1689-1702.

Spies, T. A., J. F. Franklin, and M. Klopsch. 1990. Canopy gaps in Douglas-fir forests of the Cascade Mountains. *Canadian Journal of Forest Research* 20:649-658.

Stein, W. I. 1995. *Ten-year Development of Douglas-fir and Associated Vegetation after Different Site Preparation on Coast Range Clearcuts.* Research Paper PNW-rp-473, USDA Forest Service. Pacific Northwest Research Station, Portland OR.

Swanson, F. J., and C. T. Dyrness. 1975. Impact of clearcutting and road construction on soil erosion by landslides in the western Cascade Range, Oregon. *Geology* 3(7):393-396.

Swanson, F. J., M. M. Swanson, and C. Woods. 1981. Analysis of debris-avalanche erosion in steep forest lands: in example from Mapleton, Oregon, USA, pp. 67-75 in *Erosion and Sediment Transport in Pacific Rim Steeplands.* I.A.H.S. Publ. No. 132. Wallingford, U.K.

Swanson, F. J., S. V. Gregory, J. R. Sedell, and A. G. Campbell. 1982. Land-water interactions: the riparian zone, pPp. 267-291 in *Analysis of Coniferous Forest Ecosystems in the Western United States*, R. L. Edmonds, ed. US/IBP Synthesis Series 14. Hutchinson Ross Publishing Co., Stroudsburg PA.

Swanson, F. J., T. K. Kratz, N. Caine, and R. G. Woodmansee. 1988. Landform effects on ecosystem patterns and processes. *BioScience* 38(2):92-98.

Swift, L. W. Jr. 1984. Gravel and grass surfacing reduces soil loss from mountain roads. *Forest Science* 30(3): 657-670.

Tappeiner, J. C. II, J. C. Zasada, P. Ryan, and M. Newton. 1991. Salmonberry clonal and population structure in Oregon forests: the basis for a persistent cover. *Ecology* 72(2):609-618.

Taylor, A. H. 1990. Disturbance and persistence of Sitka spruce [*Picea sitchensis* (Bong) Carr.] in coastal forests of the Pacific Northwest, North America. *Journal of Biogeography* (1990) 17, 47-50.

Thompson, R. S., C. Whitlock, P. J. Bartlein, S. P. Harrison, and W. G. Spaulding. 1993. Climate changes in the western United States since 18,000 yr B.P, pp. 468-513 in *Global Climates Since the Last Glacial Maximum*, H. E. Wright, Jr., J. E. Kutzbach, T. Webb III, W. F. Ruddiman, F. A. Street-Perrott, and P. J. Bartlein, ed. University of Minnesota Press, Minneapolis.

Transportation Research Board. 1997. *Toward a Sustainable Future—Addressing the Long-term Effects of Motor Vehicle Transportation on Climate and Ecology.* Special Report 251. Transportation Research Board. National Research Council, National Academy of Sciences. Washington DC.

Tyser, R. W., and C. A. Worley. 1992. Alien flora in grasslands adjacent to road and trail corridors in Glacier National Park, Montana (USA). *Conservation Biology* 6:253-262.

USDA Forest Service. 1946. Map on file at the Forestry Sciences Laboratory, Corvallis, OR. USDA Forest Service, Pacific Northwest Research Station, Corvallis OR.

Van Wagner, C. E. 1978. Age-class distribution and the forest fire cycle. *Canadian Journal of Forest Research* 8:220-227.

Walker, G. W., and N. S. MacLeod. 1991. Geologic map of Oregon, scale 1:500,000, two sheets. United States Geological Survey. Reston VA.

Walters, M. A., R. O. Teskey, and T. M. Hinckley. 1980. *Impact of Water Level Changes on Woody Riparian and Wetland Communities.* Vol. VIII, Pacific Northwest and Rocky Mountain Regions. FWS/OBS-78/94. U.S. Fish and Wildlife Service.

Waring, R. H., and J. F. Franklin. 1979. Evergreen Coniferous Forests of the Pacific Northwest. *Science* 204:1380-1386.

Wemple, B. C., J. A. Jones, and G. E. Grant. 1996. Channel network extension by logging roads in two basins, western Cascades, Oregon. *Water Resources Bulletin* 32(6):1195-1207.

Wester, L., and J. O. Juvik. 1983. Roadside plant communities on Mauna Loa, Hawaii. *Journal of Biogeography* 10:307-316.

Whitlock, C. 1992. Vegetational and climatic history of the Pacific Northwest during the last 20,000 years: implications for understanding present biodiversity. *The Northwest Environmental Journal* 8:5-28.

Willson, M. F., and K. C. Halupka. 1995. Anadromous fish as keystone species in vertebrate communities. *Conservation Biology* 9:489-497.

Wimberly, M. C., T. A. Spies, C. J. Long, and C. Whitlock. 2000. Simulating historical variability in the amount of old forests in the Oregon Coast Range. *Conservation Biology* 14 (1):167-180.

Worona, M. A., and C. Whitlock. 1995. Late-Quaternary vegetation and climate history near Little Lake, central Coast Range, Oregon. *Geological Society of America Bulletin* 107:867-876.

Zasada, J. C., J. C. Tappeiner II, B. D. Maxwell, and M. A. Radwan. 1994. Seasonal changes in shoot and root production and in carbohydrate content of salmonberry (*Rubus spectabilis*) rhizome segments from the central Oregon Coast Range. *Canadian Journal of Forest Research* 24:272-277.

Zavitkovski, J., and M. Newton. 1971. Litterfall and litter accumulation in red alder stands in western Oregon. *Plant and Soil* 35:257-268.

Zobel, D. B., L. F. Roth, and G. M. Hawk. 1985. *Ecology, Pathology, and Management of Port-Orford Cedar* (Chamaecyparus lawsoniana). USDA Forest Service, Pacific Northwest Forest and Range Experiment Station, Portland OR.

Fish and Aquatic Ecosystems of the Oregon Coast Range

Gordon H. Reeves, Kelly M. Burnett, and Stanley V. Gregory

Introduction

As in other parts of the Pacific Northwest (PNW), (Stouder et al. 1997), fish are important elements of human cultural, social, and economic systems and of natural ecosystems of the Oregon Coast Range. Many species provide sources of income, recreation, and food as well as having cultural significance to a variety of people (Schoonmaker and von Hagen 1996). Several species are integral components of aquatic and terrestrial ecosystems. Some, particularly the anadromous salmonids, are considered indicators of environmental conditions in marine and freshwater ecosystems. Certain anadromous species also may be food sources for a large suite of aquatic and terrestrial vertebrates and invertebrates (Cederholm et al. 1989; Willson and Halupka 1995; Bilby et al. 1996), as well as nutrient sources for riparian vegetation (Bilby et al. 1996).

This chapter will review the general ecology and biology of fishes in Coast Range streams and rivers, examine patterns of distribution and abundance, and discuss impacts of land management activities on fish and their habitats. We then consider restoration and future management directions for these populations and their freshwater ecosystems.

The Fish Fauna of the Oregon Coast Range

Streams in the Coast Range have only a fraction of the fish species that are found in similar-sized streams elsewhere. This pattern is similar to that seen in other parts of the PNW and western United States, which have about half as many fish families and one quarter of the fish species found in the eastern United States (Smith 1981; Minckley et al. 1986). Primary reasons for the low number of fish species in the Coast Range are relatively recent tectonic activity, the inability of species to move across the Continental Divide, and lack of direct faunal connections with other continents.

The primary types of fishes found in Coast Range streams and lakes are lampreys (Petromyzontidae), salmon and trout (Salmonidae), minnows (Cyprinidae), and sculpins (Cottidae). Suckers (Catostomidae) and sticklebacks (Gasterosteidae) occur in a number of river systems, but they are not as widely distributed as those fish mentioned above. Fish occupy the full range of habitats available in Coast Range river systems; however, not all fish are found in all parts of the stream network. Community composition varies depending on location in the stream network (Reeves et al. 1998).

Although the number of species is low, there is a wide variation in phenotypic (physical attributes), genetic, and life-history features within and among populations of many fish in the Coast Range. Species or populations can exhibit multiple life-history patterns. Reimers (1973) identified five distinct life-history variations within a population of ocean-type chinook salmon (*Oncorhynchus tshawytscha*) in Sixes River; ocean-type juveniles rear in streams and estuaries for a relatively short period before entering the ocean. Individuals spend from one month, to five to six months, to one year in freshwater before moving to the estuary, where residence time also is variable. Cutthroat trout (*O. clarkii*) may have resident forms, which spend their entire life in freshwater, and anadromous forms, which make repeated trips to large rivers or to the marine environment. Resident forms often reside above migration barriers and are reproductively isolated from anadromous forms. However, in some situations resident populations may produce anadromous individuals (Griswold 1996). Non-salmonids, such as the Pacific lamprey (*Lampetra tridentata*), also have wide variation in life histories. Individuals may spend two to more than five years in freshwater before moving to the ocean (van de Wetering 1998).

Fish populations in coastal Oregon streams may also exhibit variation among populations. Nicholas and Hankin (1988) documented the variation in size and age of return to freshwater and in fecundity among coastal Oregon chinook populations. T. H. Williams (personal communication) found large genetic and morphological variation among populations of coastal cutthroat trout populations in Oregon. Locally variable populations have also been reported for speckled dace (*Rhinichthys osculus*; Zirges 1973), longnose dace (*R. cataractae*; Bisson and Reimers 1977), and sculpins (*Cottus* spp.; Bond 1963) in coastal Oregon streams. This diversity within and among populations is a response to varying environmental conditions (Healey and Prince 1995) and likely has genetic and environmental components. Such variation within and among populations is also common in other parts of the PNW (Snyder and Dingle 1989; Swain and Holtby 1989) but does not appear to be as extensive in other parts of the United States (Smith 1981; Mahon 1984).

Anadromous fish are the dominant component of the fish fauna in the streams and rivers of the Coast Range, particularly coastal systems. These are fish that begin life in freshwater, move to the marine environment to grow and mature, and then return to freshwater to reproduce. Native species with anadromous life histories include lampreys, Pacific salmon and trout (*Oncorhynchus* spp.), and candle fish (*Thaleichthys pacificus*). Introduced anadromous species include striped bass (*Morone saxatalis*) and American shad (*Alosa sapidissima*).

The predominance of anadromous life histories in this region is likely attributable to two factors. Gross and others (1988) argued that in areas where freshwater productivity is less than marine productivity, such as along the Oregon coast, anadromous life histories predominate. Oregon is in the transition zone between changes in relative productivity of the marine and freshwater environment. Moving to the more productive marine environment allows individuals to grow to a larger size sooner than if they had remained in freshwater. Larger fish have a greater chance of reproducing successfully and can produce more offspring than smaller individuals.

Pacific salmon and trout (*Oncorhynchus* spp.) are the best known of the anadromous fishes in Coast Range rivers and streams (Table 4-1). Their general life cycle is shown in Figure 4-1 (in color section following page 84). All Pacific salmon and trout begin life as eggs deposited in the freshwater gravel.

Table 4-1. Life history characteristics of Pacific salmon and trout (*Oncorhynchus* spp.) found in coastal Oregon.

Common name/species	*Other names*	*Anadromous*	*Resident/non-anadromous*	*Years in fresh water*	*Years in salt water*	*Repeat spawners*
Pink salmon/*O. gorbuscha*	Humpback salmon	X		0	2	No
Chum salmon/*O. keta*	Dog salmon	X		0 (but 1-3 mo. estuary)	2 - 4	No
Coho salmon/*O. kisutch*	Silver salmon	X	X	1	2	No
Chinook salmon/ *O. tshawytscha*	King salmon Tyee salmon Spring salmon Blackmouth	X	X	<1 ("ocean type") 1 - 2 ("freshwater type")	2 - 5	No
Steelhead/*O. mykiss*	Rainbow trout	X	X	1 - 3	2 - 3	Yes
Sea-run cutthroat trout *O. clarki*	Harvest trout Blueback	X	X	2 - 3	<1 - 2	Yes

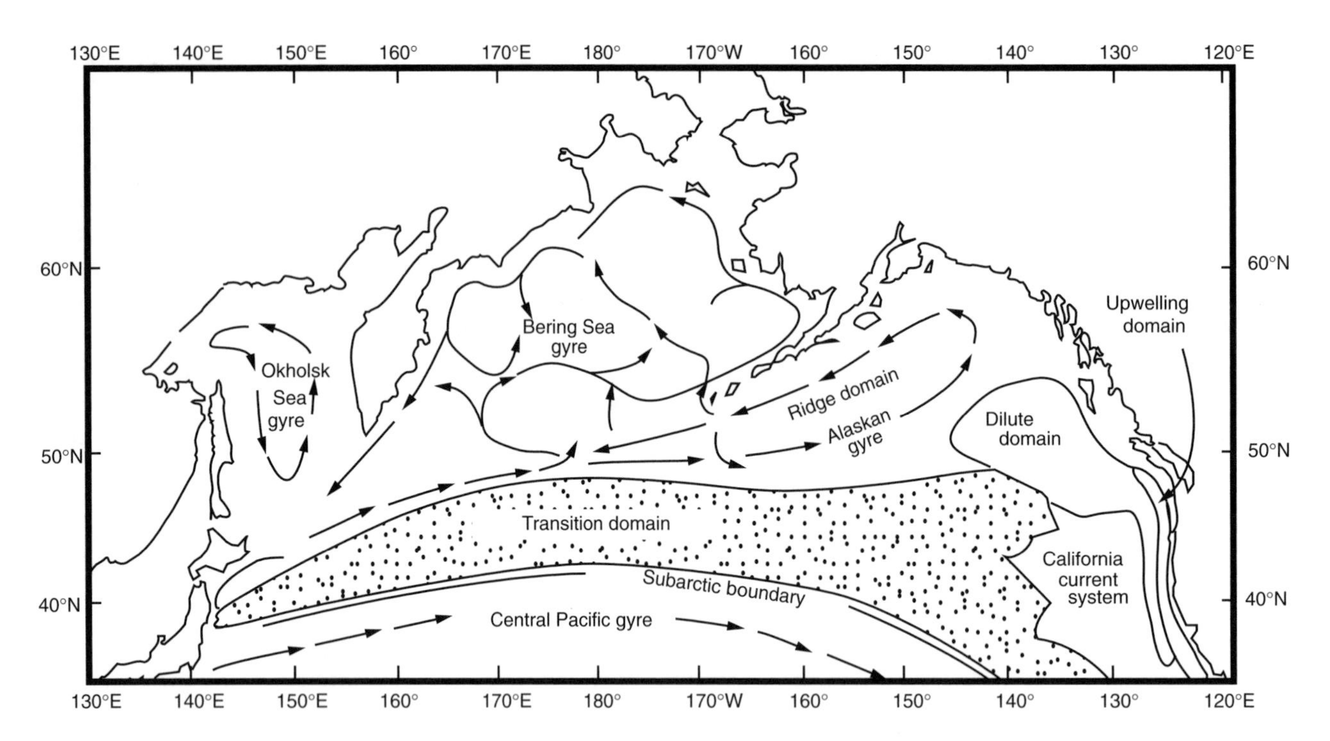

Figure 4-2. Location of Oregon relative to ocean transition zone between northern cool, nutrient-rich currents and southern warm, nutrient-poor currents (Fulton and LaBrasseur 1985). (Reprinted by permission from Washington Sea Grant Program, University of Washington.)

Alevins are the developing embryos. Upon emergence from the gravel, the small fish are known as fry or fingerlings. These fish are identified as "0+" individuals. Juveniles are older fry. Juveniles that have spent a winter in freshwater are identified as "1+" individuals. Smolts are individuals in the process of transitioning from freshwater to the ocean. This entails changes in such things as behavior (from defending territories to swimming in schools), body color (from various colors to silvery), and kidney operation (from retaining salt to excreting salt).

Each species has a unique life history (Table 4-1). Some, like pink (*O. gorbuscha*) and chum (*O. keta*) salmon, spend very little time in freshwater before moving to the ocean. In contrast, cutthroat trout and steelhead (*O. mykiss*) may spend two years or more in freshwater before migrating to the ocean. Time spent in the ocean also varies among species. Some, like pink salmon, spend a fixed amount of time (two years) in the ocean before returning to freshwater; for others, like chinook salmon, the time is quite variable (two to six years).

Ocean conditions for anadromous salmonids in Oregon are variable. Oregon is at the oceanic boundary between cool, nutrient-rich currents and warm, nutrient-poor currents (Figure 4-2) and productivity of the ocean off the Oregon coast depends on the location of these currents. During productive years, which are generally associated with a weak winter Aleutian low-pressure system (Hare et al. 1999), nutrient-rich currents move south towards the Oregon coast. Conversely, nutrient-rich currents move north, away from the Oregon coast, during a strong winter Aleutian low-pressure system and ocean productivity declines. This pattern generally occurs on a 20- to 30-year cycle (Mantua et al. 1997). The last productive period off the Oregon coast was from 1947 to 1976 (Miller et al. 1994); less productive conditions had occurred from 1925 to 1946 and have been prevalent since 1976. Survival rates of wild and hatchery populations of coho salmon (*O. kisutch*) are similar during more favorable conditions (Nickelson 1986; Coronado and Hilborn 1998). However, survival rates of wild populations are two to four times greater than hatchery stocks during times of poor ocean conditions (Nickelson 1986).

Estuaries are sites of early marine growth for anadromous salmonids, an important determinant of ocean survival (Pearcy 1992). They may be particularly important during times of low ocean productivity. However, there are few well-developed estuaries along the Oregon coast. The combination of sparse near-shore habitats and variable ocean conditions makes freshwater habitat more crucial for the survival and persistence of anadromous salmonid populations in Oregon than in more northerly areas.

Distribution of Fish in Coast Range Rivers and Streams

Organization of rivers and stream systems

River and stream systems consist of distinct spatial units that are organized hierarchically (Frissell et al. 1986). The units of organization dealt with in this chapter (from finest to coarsest spatial scale) are the habitat unit, the reach, and the watershed. The two primary types of habitat units are riffles and pools. Riffles, at base flow, are fast water units that are shallow and have a steep water-surface gradient. In contrast, pools are deeper and generally have a gentle surface slope with slow flow (O'Neil and Abrahams 1987). These units can be further classified into several types (Hawkins et al. 1993), which can be useful in some situations. They range in size from a few yards to 200 or more yards.

Habitat units may not always have clear physical boundaries, but they are distinct ecologically. Fish inhabiting them differ markedly in taxonomic composition and morphological, physiological, and behavioral traits. For example, pool dwellers, such as cutthroat trout and coho and chinook salmon, are often found in aggregations and are more active swimmers with more slender bodies and smaller paired fins (Figure 4-3a).

Fish that inhabit riffles, such as dace, are bottom-oriented fish, often possessing large pectoral fins to help maintain position (Figure 4-3b). Some, such as sculpins (Figure 4-3c), lack an air bladder or can adjust the air in the swim bladder to reduce buoyancy. Riffle dwellers are solitary or part of small, loose-knit groups.

A reach is an integrated series of habitat units that share a common landform pattern (Grant et al. 1990; Montgomery and Buffington 1997). Reaches are influenced by variation in channel slope, local side slopes, valley floor width, and riparian vegetation (Frissell et al. 1986). Gregory and others (1989) classified reaches as constrained (active channel to valley floor width ratio < 2; Figure 4-4a) and unconstrained (active channel to valley floor width ratio > 2; Figure 4-4b). Reaches vary in length from

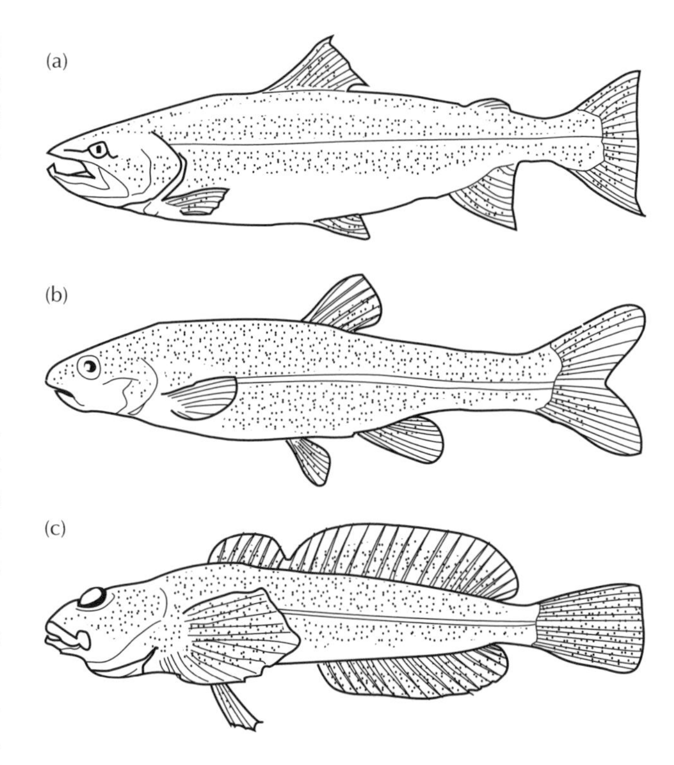

Figure 4-3. Fishes commonly found in coastal Oregon headwater streams: (a) cutthroat trout; (b) speckled dace; and (c) riffle sculpin (Reeves et al. 1998).

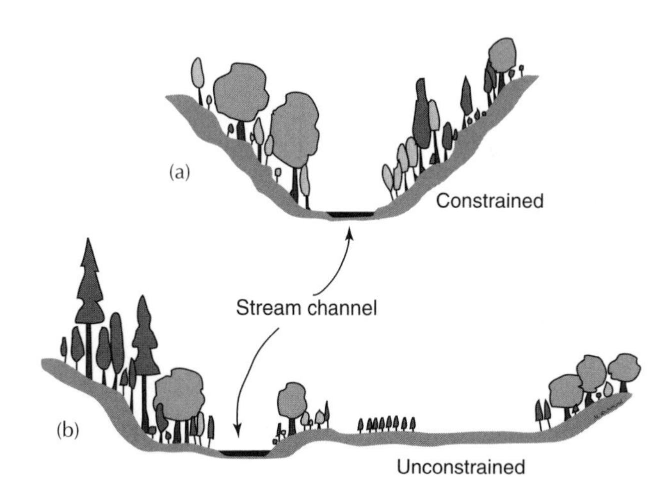

Figure 4-4. Reach types found in streams in the Oregon Coast Range: (a) constrained and (b) unconstrained (Reeves et al. 1998).

(Figures 4-3 and 4-4 reprinted from G. H. Reeves, P.A. Bisson, and J.M. Dambacher. 1998. Fish communities, pp. 200-234 in *River Ecology and Management: Lessons from the Pacific Coastal Ecoregion,* R.J. Naiman and R.E. Bilby, eds. Springer-Verlag, New York. By permission of the publisher.)

10^2 to 10^3 yards. A watershed contains a collection of reaches and the surrounding upslope areas to the drainage divide. The physical and biological characteristics of the watershed are determined to a large extent by the aggregate features of the reaches within it.

Sizes of watersheds can vary widely. For purposes of this chapter, we define watersheds as either being 4th or 5th order (Strahler 1957), or 6th or 7th field hydrologic units (HU). The size of watersheds, as used in this chapter, averages 4,500 to 5,000 acres. Understanding and explaining patterns of fish distribution and abundance and the factors influencing them are difficult, in part because of the hierarchical organization of streams and rivers. The pattern at a given level is influenced by factors at that level as well as factors at higher and lower organizational levels. For example, the assemblage of fishes in a reach varies with reach type. Abundance and distribution of fish within the reach are determined, in part, by these types, the condition of the habitat units, the position of the reach in the watershed, and the watershed condition.

Watersheds

Fish can be found throughout a watershed, and the structure and composition of the species assemblage change with location in the network (Figure 4-5 in color section following page 84). Generally, the number of fish species increases from the headwaters to lower portions of river systems in the Coast Range. As one moves downstream, some species are lost from the assemblage, while others are added; the latter process generally exceeds the former, resulting an overall increase in the number of species present. Reeves and others (1998) found a significant increase in the number of species moving from headwaters downstream in a northern and a southern coastal Oregon river system. This pattern has been observed in other parts of the PNW (Li et al. 1987) and the United States (Sheldon 1968; Horowitz 1978; Boschung 1987; Schlosser 1987).

Few fish species, primarily trout, sculpins, and dace, are present in headwater streams in the Coast Range. The trout are generally cutthroat and may include resident and anadromous forms. Resident fish do not move to the marine environment and often reside above migratory barriers. Such populations are generally small but nonetheless are ecologically important. Resident populations can in some circumstances produce offspring that can become anadromous, and some may contain very unique genetic information (Griswold 1996).

Reeves and others (1998) argued that physical conditions and not biological processes may be more important in determining the structure and composition of headwater assemblages in the PNW. Schlosser (1987) suggested that competition is generally the dominant process influencing the diversity of fishes in headwater streams. He reasoned that physical conditions are less influential because of the relatively low diversity of habitat types and environmental conditions in headwater streams. However, headwater streams, including those in the Coast Range, are very dynamic over multiple time scales. During the course of a year, stream flow fluctuates dramatically. Over longer periods (decades to centuries), stream channels may alter between bedrock following a debris torrent to large accumulations of sediment and wood. Also, the diversity of habitats (number of types) is relatively low. As a result of these variable conditions, the species pool is somewhat limited in these streams.

Many headwater streams may not contain fish, but they are important ecologically nonetheless. Fish may be precluded from these parts of the network because of geological barriers, such as waterfalls, or because they are too steep (> 10 percent slope). These streams, however, can be important habitats for several amphibians. Headwater streams also are sites where materials such as leaves and twigs, which form the food base for a myriad of aquatic insects, are stored behind pieces of large wood or large rocks and boulders. These materials and insects can be carried downstream to areas inhabited by fish, where they enter into the food chain.

Headwater streams are influenced by riparian and upslope vegetation (Naiman et al. 1992). These streams generally have relatively narrow riparian zones and they are strongly influenced by nearby upslope vegetation and surrounding topography. Headwater streams are generally in constrained reaches so the surrounding vegetation hillslopes shade the stream. Therefore, a relatively narrow riparian zone may effectively maintain water temperatures.

The reduction of light reaching headwater channels as a result of surrounding vegetation and steep canyon walls limits primary production, and production of fish is limited by this combination of cool waters and low primary production. Creating openings in the riparian vegetation may increase fish production (Murphy and Hall 1980). However, these increases may be offset by losses of production in downstream areas because of the cumulative effects of resulting temperature increases. Elevated temperatures in headwater streams can lead to higher temperatures in downstream areas, which may reduce habitat suitability and thus fish production.

Wood from surrounding riparian zones is important in headwater streams. Wood creates collection areas for sediment and organic materials, the latter providing the energy base for the food chain. Wood also creates pools, an important habitat for fish in these streams. Bilby and Ward (1989) found that relatively small pieces of wood could serve these functions.

Headwater streams and small tributaries may also be important sources of wood for mid-order channels. Headwater channels are frequently subjected to landslides and debris torrents that can deposit wood and sediment in fish-bearing headwater or middle-order streams. McGarry (1994) found that 49 percent of the wood volume in fish-bearing sections of Cummins Creek, a stream in a small wilderness area near Yachats, Oregon, was derived from landslides in headwater streams. Benda (unpublished data) examined some coastal streams following the 1996 floods and found up to 80 percent of the wood in a roadless watershed to be derived from landslides. May (1998) reported that landslides in managed streams may also be important sources of wood.

Not all headwater streams have the same potential to deliver wood to fish-bearing channels, however. Benda and Cundy (1990) identified the attributes of first- and second-order streams in the central Oregon coast that have the greatest potential for delivering desirable material to fish-bearing streams via debris torrents (Figure 4-6 in color section following page 84). These channels generally have gradients of 8 to 10 percent, depending on stream size, and junction angles of less than 45°. Leaving trees along potential debris-torrent routes obviously increases the potential for delivery of wood to higher-order, fish-bearing streams.

The number of fish species increases in the middle portion of the stream network (Figure 4-5). Species commonly found here include those found in headwaters as well as additional species that are morphologically and behaviorally similar to them. In coastal streams, this includes coho and chinook salmon and steelhead, along with other species, primarily minnows (Cyprinids) such as the red-side shiner (*Richardsonius balteatus*). Valley Coast Range streams draining to the Willamette River generally lack anadromous salmon and trout. Resident cutthroat trout and native minnows, such as the northern pike minnow (*Ptychocheilus oregonsis*), are found in these streams.

Reaches

One reason that the number of fish species increases in the middle portion of the stream network is that there are more types of reaches and habitats here than in headwaters. In Coast Range streams, salmonid assemblages differ in composition between the two reach types, constrained and unconstrained (Reeves et al. 1998). Coho and chinook salmon and age 1+ trout are generally more abundant in unconstrained than in constrained reaches (Figure 4-7a in color section following page 84). Age 1+ cutthroat trout and steelhead are numerically dominant in constrained reaches (Figure 4-7b in color section following page 84).

Unconstrained reaches generally have higher total numbers of fish than constrained reaches. In upper Elk River on the southern Oregon Coast, unconstrained reaches contain only about 15 percent of the total available habitat but account for 30 to 50 percent of the estimated number of juvenile anadromous salmonids (G. Reeves, unpublished data). Similar patterns have been observed in streams in southwestern Washington (Cupp 1989). Densities of cutthroat and rainbow trout (*O. mykiss*) in unconstrained reaches were more than twice those in constrained reaches of the McKenzie River in Oregon (Gregory et al. 1989).

The difference in the composition of the salmonid communities between the two reach types is attributable to two factors. Unconstrained reaches contain a greater diversity of habitat types than constrained reaches (Gregory et al. 1989; Schwartz 1990). Constrained reaches are typically dominated by fast-water habitats, such as rapids and cascades, whereas unconstrained reaches contain a variety of pool and riffle habitats and provide lateral refuges in backwaters and overflow areas of floodplains during floods. A greater variety of niches, which support more species, is created by more habitat types.

Fish community diversity and productivity of lower trophic levels are positively correlated at the reach scale. Trophic production is composed of primary production and secondary production. Primary production is the production of biomass by plants, which in Coast Range streams are algae and diatoms. Secondary production is production of biomass by organisms that feed on plants and other food sources. Secondary production in mid-order streams is supported by organic material produced in the channel, termed autochthonous, and organic matter derived either from the surrounding riparian zone and forest or from upstream areas, termed allochthonous (Vannote et al. 1980). Autochthonous production is primarily from algae, diatoms, and other plants that occur in the stream. These materials are eaten by aquatic insects, which in turn are eaten by fish. Allochthonous production is derived from materials such as leaves and needles that are colonized by fungi and bacteria. The bacteria and fungi in turn are eaten by aquatic insects that are eaten by fish.

Primary and secondary production are generally greater in unconstrained reaches than in constrained reaches. Unconstrained reaches are open and receive more light than constrained reaches, thus primary production is greater (Zucker 1993). The low gradient and wide floodplains of unconstrained reaches also result in the deposition and storage of allochthonous materials (Lamberti et al. 1989). The combination of high primary production and large amounts of allochthonous materials results in greater densities of aquatic insects and other organisms that are important food resources for fish (Zucker 1993).

Unconstrained reaches are also areas of greater hyporheic zone exchange (subsurface exchange of water between the floodplain and the stream channel) (Grimm and Fisher 1984; Triska et al. 1989; Edwards 1998). Hyporheic zones provide sites for storing and processing nutrients, insect production, and cool water, which contribute to the productivity of unconstrained reaches.

The presence of spawning anadromous fish can also contribute to the productivity of unconstrained reaches. Spawning, particularly by salmon, is generally greater in unconstrained reaches than constrained reaches (Frissell 1992). Juvenile salmonids and other fish may feed on eggs that drift in the water column during spawning activities. Eggs are nutrient rich and provide large amounts of energy. When salmon die after spawning, the carcasses are both colonized by fungi and bacteria and eaten directly by aquatic insects and fish (Bilby et al. 1996; Wipfli et al. 1999). Carcasses also provide food for a suite of mammals and birds (Cederholm et al. 1989; Willson and Halupka 1995) and nutrients for riparian vegetation (Bilby et al. 1996). Juvenile salmonids in streams with larger returning adult runs grow faster than do juveniles in streams with smaller runs (Bilby et al. 1998).

Unconstrained reaches are particularly susceptible to impacts from land management activities. These reaches are natural deposition zones because of their low gradient and wide valley. Accelerated erosion from activities such as timber harvest and road building in areas above unconstrained reaches can aggrade channels in these reaches, which can fill in, causing a reduction in complexity or loss of a pool. Roads and improperly constructed culverts located in valley bottoms reduce connectivity between channels and their floodplains. The consequence is the reduction or elimination of access to areas of habitat that may be particularly important during high flows.

Constrained reaches are sources of cool water for unconstrained reaches. Cooler water from upstream constrained reaches helps maintain water temperatures in unconstrained reaches in the range favorable to fish and other organisms. Increases in the temperature of water entering unconstrained reaches can affect the biota and have significant impacts on watershed productivity.

Constrained reaches provide several other important ecological functions. They are the transport portion of the stream network. Large wood in constrained reaches often forms temporary jams that often break during high flows; the wood, sediment, and organic material from them then move through these reaches to unconstrained reaches. Sediment and wood form the fundamental materials that create and maintain habitat for fish and other aquatic biota.

The influence of riparian zones differs with reach type. The zone of influence is generally narrower in constrained reaches than in unconstrained reaches. In constrained reaches, the ecological functions provided by riparian zones, such as sources of wood, shade, and allochthonous materials, comes from within the height of one site potential tree (i.e., the mean size of a 200-year-old tree that can be grown at the site) from the channel (Figure 4-8). The

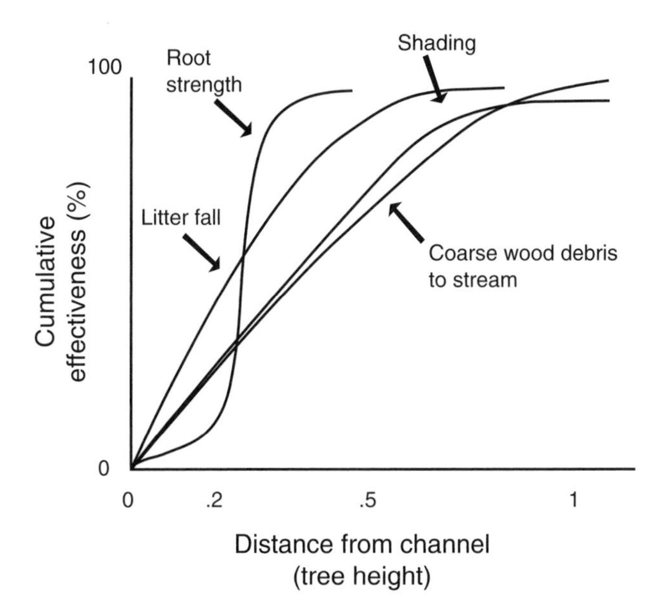

Figure 4-8. Zone of influence of selected ecological processes within riparian zones (FEMAT 1993).

role of riparian vegetation in providing shade may be reduced in constrained reaches with steep side walls that reduce the amount of solar radiation reaching the channel.

The riparian zone in unconstrained reaches is larger than in constrained reaches because it includes the valley floor as well as immediately adjacent side slopes. In intact unconstrained reaches, the stream channel often consists of multiple channels that migrate periodically across the valley floor. As it moves, the channel will interact with, and be influenced by, vegetation in all parts of the valley floor as well as on the adjacent hillsides. The extent of influence of the side-slope riparian zone is similar to that in constrained reaches (Figure 4-8).

Habitat units

The greater variety and production of fish in middle portions of the stream network compared with upper portions is due, in part, to the greater diversity of habitat types. Habitat types in headwater streams generally are restricted to pools, riffles, or cascades. Middle portions of the watershed have a greater variety of each habitat type. Units are also larger and deeper. There are, as a result, a greater number and variety of niches available. For example, deeper units allow species to partition habitat vertically, which cannot occur in shallower units. Within a habitat unit, structural features such as the amount of wood or boulders, substrate composition, flow velocity, and depth influence the number and biomass of species present (Sheldon 1968; Evans and Noble 1979; Angermeier 1987). Different combinations of these factors create an array of microhabitats.

Habitat complexity is directly related to the number and variety of microhabitats present in a unit, but it is difficult to explicitly quantify. It is usually determined by the amount of structural elements, such as wood or boulders, and the depth of a unit, and as a result, habitat complexity is usually expressed in relative terms.

Increased habitat complexity provides protection from predators, alters foraging efficiency (Wilzbach 1985), and influences social interactions (Fausch and White 1981; Glova 1986). In a Washington stream, Lonzarich and Quinn (1995) observed a general increase in the number of fish species and the abundance of each with increasing complexity of pools, measured by pool depth and amount of wood. However, each species responded to habitat features differently. Numbers of juvenile coho salmon, age 1+ steelhead, and cutthroat trout were directly correlated with pool depth. Numbers of Coast Range sculpin (*Cottus aleuticus*) did not respond to alteration of any habitat features. D. H. Olson (USDA Forest Service, Pacific Northwest Research Station, Corvallis, Oregon) found that numbers of salmonid species in a small coastal Oregon stream increased with maximum pool depth, pool surface area, and volume of wood. Cederholm and others (1997) found an increase in coho salmon smolt numbers with increased wood levels in a Washington stream, but steelhead showed no response.

The relative biomass of fish in Coast Range streams is also influenced by habitat complexity. More complex habitats result in more even distribution of biomass among species. Biomass of speckled dace, reticulate sculpin (*C. perplexus*), riffle sculpin (*C. gulosus*), and juvenile cutthroat trout increased with increasing levels of habitat complexity in a small Alsea River tributary (Fieth and Gregory 1993). In contrast, biomass of coho salmon showed no response to changes in habitat complexity in the summer, even though habitat complexity may influence coho salmon density at other seasons (Nickelson et al. 1992a; Quinn and Peterson 1996) and life history stages (McMahon and Holtby 1992). Densities of fish also decreased in southeastern Alaska streams (Dolloff 1986; Elliott 1986) and a midwestern stream (Berkman and Rabeni 1987) when habitat structure was reduced by the removal of wood and increased sediment deposition, respectively.

Stream fishes partition habitat by interactive and selective segregation (Nilsson 1967). In interactive segregation, species are capable of using the same niche, but one species is dominant and precludes the subordinate species from preferred habitats. The dominant species is generally more aggressive or more efficient at exploiting a particular resource. Therefore, the subordinate species will move into preferred habitat only if the dominant species is absent. Habitat use by juvenile coho salmon and age 1+ trout is influenced, in part, by interactions among species. Juvenile coho salmon are aggressive and preclude steelhead (Hartman 1965) and cutthroat trout (Glova 1978) from the heads of pools, where food resources are highest. Steelhead in turn dominate cutthroat trout and generally preclude them from habitats in larger stream systems (> 15 square kilometers) in British Columbia (Hartman and Gill 1968). Similar patterns of segregation have been observed between rainbow trout and cutthroat trout (Nilsson and Northcote 1981) and cutthroat trout and Dolly Varden (*Salvelinus malma*; Andrusak and Northcote 1971) in lakes. Reeves and others (1987) found that habitat use by redside shiner (*Richardsonius balteatus*) and juvenile steelhead was determined by interactive segregation. In cool water (< 20°C), steelhead excluded shiners from riffles, where food was most abundant, by aggressively driving the shiners away. Shiners then formed loose aggregations in pools in the presence of trout. When steelhead were absent, shiners moved to riffles.

Selective segregation involves differential use of available resources (Nilsson 1967). Each species uses different habitat for other resources such as food and occupies the same habitats whether alone or in the presence of the other species. Differences in habitat use arise from differences in behavior or body morphology. For example, selective segregation between juvenile summer steelhead and spring chinook salmon reduced interaction for space in Idaho streams (Everest and Chapman 1972). The fish use similar habitats when they are a given size, but because chinook salmon spawn in the fall and steelhead in the spring, chinook salmon emerge earlier and tend to be larger than co-occurring 0+ steelhead at all times. Consequently, interactions between juveniles for food and space are minimal. Similar patterns of segregation were observed between juvenile coho and chinook salmon (Lister and Genoe 1970). Differences in life history features or behavior, such as those described above, which lead to selective segregation are genetically encoded over time (Nilsson 1967).

Both types of segregation, particularly with regard to salmon and trout, may occur in streams in the Coast Range. Having knowledge of how fish partition habitat and other resources provides insight into how fish communities are organized and provides a basis for evaluating proposed management actions. For example, juvenile coho salmon and steelhead portion habitat by interactive segregation. The potential to increase numbers of one species, either through natural means or through supplementation, may be reduced if the other species currently occupies available habitat (Reeves et al. 1993). Conversely, if two species segregate selectively, as coho and chinook salmon do, the potential to increase numbers of one species may be greater. Understanding the type of segregation also provides insights in evaluating the response of the community to environmental alterations.

Seasonal distribution

Distribution of fish species and age classes throughout a watershed and within habitat units varies seasonally, particularly for anadromous salmonids. We will illustrate seasonal changes in distribution using the work of Sleeper (1993), who described this pattern for the portion of the watershed used by juvenile anadromous salmonids in Cummins Creek, a small watershed near Yachats, Oregon. Several other sources are used to describe seasonal changes in habitat use by the various species and age classes. Understanding these patterns is important for managing freshwater habitats, projecting potential impacts of land management activities, and developing restoration efforts.

Spring

In the spring, age 1+ coho salmon and age 2+ steelhead (a fish that has spent two or more winters in freshwater) are found primarily in the lower portions of the system (Figure 4-9 in color section following page 84). Many of these are pre-smolts, older fish that are preparing to move to the ocean. Larger cutthroat trout (> 20 centimeters) are most abundant in the lowest portion of the network, which is where these fish spend the winter. Larger, older cutthroat trout (1+) and steelhead also will be in the main channel during the spring (Figure 4-10 in color section following page 84). These fish are found in pools and are generally oriented to the bottom (Hartman 1965). Pools with the most complexity will have higher densities of 1+ steelhead (Lonzarich and Quinn 1995). Complexity, usually in the form of boulders and/or large wood, reduces current velocities and provides cover (Bisson et al. 1987). This is particularly important in the spring because the metabolic performance of the fish is reduced at lower water temperatures. Few fish are found in riffles and other fast water units at this time of year. Some age 1+ steelhead may use these units, but the high velocities associated with higher flow make these areas unfavorable for growth.

Recently emerged coho salmon in Coast Range streams are concentrated near spawning areas, which in Cummins Creek are in the upper portions of the watershed. The swimming abilities of recently emerged salmonid fry are limited because of their small size and, therefore, they use off-channel areas and quiet areas along stream margins (Figure 4-10). These areas provide low-velocity habitat and are created by boulders and pieces of large wood on and along stream margins (Moore and Gregory 1988). Small debris and detritus accumulate in these areas and, as a result, production of smaller invertebrates, such as midges (Chironomidae) and early life stages of other aquatic invertebrates, is high. These organisms are important food items for smaller salmonids. Loss of these off-channel and margin habitats will negatively impact the growth and survival of recently emerged fish. Species of recently emerged fish found in off-channel and stream-margin habitats will vary over the spring. In most coastal streams, coho and chinook salmon fry will be the first to emerge from redds. There may be little overlap of these two species, however, because they spawn in different parts of the watersheds (Figure 4-11 in color section following page 84) and at different times (Figure 4-12). Recently emerged steelhead will be found in these same habitats but later in the spring or in early summer because steelhead spawn after the salmon.

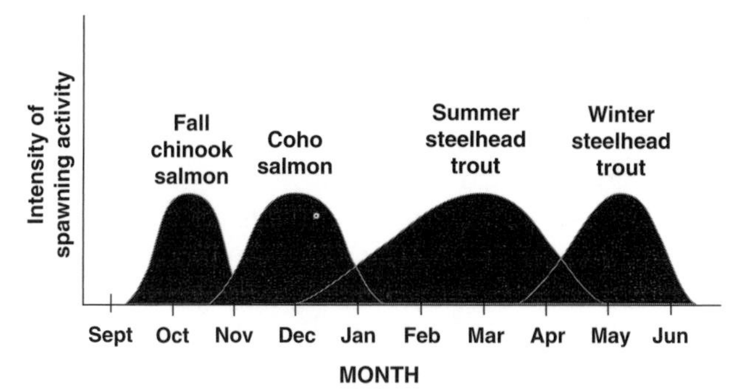

Figure 4-12. Spawning time of selected species of Pacific salmon (*Oncorhynchus* spp.) in coastal Oregon streams.

Summer

In early summer, the relative distribution pattern varies with species and age-classes. Coho that emerged this year are in the middle and parts of the upper portions of the system (Figure 4-13 in color section following page 84); recently emerged steelhead are more evenly distributed throughout the watershed. These differences are attributable, in part, to differences in spawning habitats. Coho salmon spawn in lower-gradient areas of the stream system, reaches 5 to 7 (these reaches are based in distance from mouth, not geomorphic settings). Steelhead spawn in a wider range of gradients and as a result spawn over a wider area. Numbers of larger cutthroat trout are also proportionately higher around coho spawning areas because they are likely preying on recently emerged coho salmon and steelhead. Both age-classes of steelhead are relatively evenly distributed throughout the watershed. There are no pre-smolts because these fish have migrated to salt water by this time. As summer flows decrease in Coast Range streams, so does the amount of off-channel habitat. Recently emerged steelhead use available off-channel habitat as well as stream margins (Figure 4-14 in color section following page 84).

Fish in backwater habitats can grow relatively rapidly at this time because of the large number of invertebrates present (Moore and Gregory 1988). As age 0+ steelhead grow, they shift to riffles in the main channels (Figure 4-15 in color section following page 84). Here food availability is high, consisting primarily of aquatic insects drifting in the current. However, the energetic cost to obtain food is also high because of the increased water velocities. Thus, fish hold in lower-velocity pockets, which are formed by larger substrates or wood in fast water units, and dart into the water column to capture drifting invertebrate prey. This strategy minimizes energy expenditure and maximizes growth potential. Age 0+ steelhead are territorial at this time.

Territory size is dependent on habitat complexity and food availability (Chapman 1966; Dill et al. 1981; Scrivener and Anderson 1982); the more complex the habitat and the greater the amount of food, the smaller the territories will be, and the greater the number of fish that can occupy a given area.

Coho salmon in coastal streams move from backwaters and stream margins into pools in summer (Figure 4-15). They tend to occupy the upper portions of the water column in the upstream portion of the pool (Hartman 1965), where they feed on a combination of aquatic and terrestrial insects (Chapman 1965). Coho salmon form dominance hierarchies within the pool (Mundie 1969). In this social system, the largest fish dominates all other individuals, the second-largest dominates all individuals except the largest, etc. Interactions among individuals are either through direct contact or through rather elaborate behavioral displays (Hartman 1965). The large, colorful fins of coho salmon are used extensively in these displays. Habitat complexity and food availability determine the strength of the hierarchy and the number of fish that a habitat unit can hold. The more complex the habitat and the greater the amount of food, the fewer interactions there will be among individuals. Older (≥ 1+) cutthroat trout and steelhead occupy the bottom of pools (Hartman 1965). Partitioning the habitat this way minimizes interactions with coho salmon. Lonzarich and Quinn (1995) found that trout numbers were directly related to the depth of pools.

Fall

The distribution patterns of fish shift downstream in early fall (Figure 4-16 in color section following page 84), primarily as a result of declining flows, which reduce available habitat, particularly in the upper portions of the watershed. Elevated water temperatures in Coast Range streams also may reduce habitat availability and suitability at this time. Fish may concentrate in areas of cool water, frequently in pools at or near cool-water tributaries. This situation is especially true for streams in the eastern part of the Coast Range. Habitat use will generally be similar to that seen in the summer. The primary difference is that larger 0+ steelhead will move from riffles into pools and occupy habitats similar to age 1+ steelhead (Hartman 1965; Figure 4-17 in color section following page 84). Pools, especially deeper ones, are more abundant in lower reaches. Additionally, channels in lower parts of the network may have less variable streamflows than those in upper areas, potentially providing better overwintering conditions.

Winter

Because of difficult working conditions, such as high water and turbidity, little research has been done in Coast Range streams at this time of year. As a result, little is known about the winter distribution of juvenile anadromous salmonids in the Coast Range. We believe that the winter distribution is probably similar to those observed in late fall and early spring. Fish, especially older age-classes, are in the middle to lower portions of the watershed. Nickelson and others (1992a) found the highest densities of coho salmon in off-channel habitats, which suggests that areas of the watershed with wide floodplains are probably the most heavily used in winter. The stream network, and with it the amount of habitat, expands with the onset of winter in Coast Range streams.

Small streams that are dry during other times of the year begin to flow. Intermittent valley-bottom streams that have water in them may provide overwintering habitat for juvenile salmonids. Studies in the Rogue River in southern Oregon (Everest 1977) and in northern California (Kralick and Sowerwine 1977) found that juvenile cutthroat trout and steelhead moved into small, low-gradient, intermittent streams when flows increased in late fall and early winter and remained there until flows dropped in the spring.

Juvenile salmonids also move from main channels to off-channel habitats in the winter (Tschaplinski and Hartman 1983; Nickelson et al. 1992a). Winter off-channel habitats include alcoves and side-channels, which may only be present during high flows, as well as permanent floodplain habitats, such as beaver ponds, which become connected to the main channel at high flows. Peterson (1982) found that juvenile coho salmon

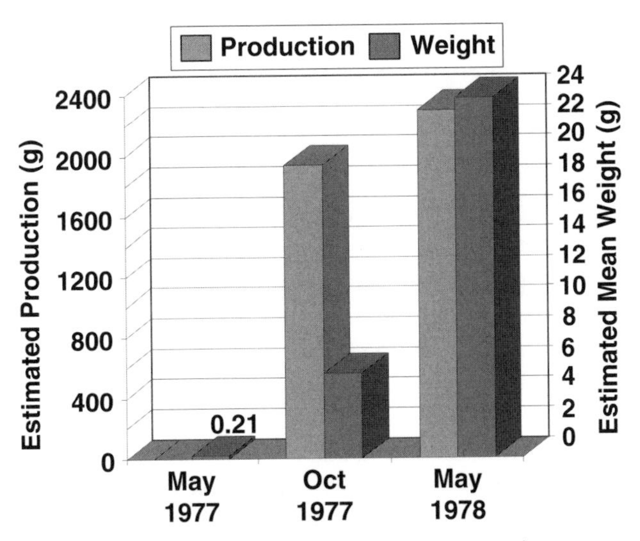

Figure 4-18. Estimated annual production and mean weight of young-of-the-year in the East Fork of the North Fork of the Mad River, California (Reeves 1978).

migrated as much as 20 miles before moving into off-channel ponds in a coastal Washington stream with the onset of high flows.

Direct loss or isolation of these habitats can reduce overwinter survival and production of salmonids (Tschaplinski and Hartman 1983). Fish that remain in the main channel in the winter concentrate in pools. Rodgers (1986) found coho salmon in Knowles Creek, a tributary of the Siuslaw River, in pools formed by debris-torrent deposits in the mainstem. Highest numbers were in deep pools with large amounts of wood. Such pools were formed at tributary junctions with small streams that had recently experienced debris torrents. Deep pools with large wood provide refugia for fish during high flows (McMahon and Hartman 1989; Harvey and Nakamoto 1998). Steelhead numbers are also greater in pools than riffles in winter (Grunbaum 1996).

In contrast to areas with harsh winters, winter may be a period of growth for fish in coastal Oregon streams. Water temperatures are mild, generally greater than 10°C for extended periods. Large amounts of organic materials that form the base of the food web, such as leaves and needles, are present in the channel. Carcasses of adults may also be present. These materials tend to accumulate in backwaters and behind concentrations of large wood (Moore and Gregory 1988) and provide food for invertebrates, which are in turn eaten by fish. Grunbaum (1996) found that age 1+ steelhead in coastal streams were active during the day in the winter and appeared to be feeding actively. Reeves (1978) found that more than 50 percent of the annual production of age 0+ steelhead in a small coastal northern California stream occurred over the winter (Figure 4-18). Mean size of individuals more than doubled at this time. Loss of winter production could have significant impacts on salmon and trout populations in Oregon streams.

Human Impacts on Fish and Fish Habitat

All species of Pacific salmon found in the Coast Range have been considered for listing under the Endangered Species Act. Species that have been listed include coho salmon coastwide; chum salmon in the lower Columbia River; steelhead on the southern Oregon coast; and spring chinook salmon in the Willamette Valley. Pink salmon were not listed primarily because they have already been extirpated from the coast. Fall chinook salmon coastwide and steelhead north of the Rogue River were not listed because of high population numbers. Lampreys are likely to be considered for listing in the near future.

The structure and composition of native fish communities in the Coast Range have been impacted by past and present human activities. A suite of factors is associated with the decline of native anadromous fishes in the Coast Range. These include loss or degradation of habitat, over-harvesting in sport and commercial fisheries, and influence of hatchery fish (Nehlsen et al. 1991). Variable ocean conditions also influence population numbers and may exacerbate the effects of these various human impacts. Habitat alteration is cited most frequently as being responsible for the decline of these fish (Nehlsen et al. 1991; Bisson et al. 1992).

Physical habitats in rivers and streams of all sizes throughout the Coast Range have been altered and simplified by human activities. Early settlers extensively channelized and diked the lower portions of larger rivers to facilitate transportation, control flooding, and develop pastures for agriculture by early settlers. The lower Coquille provides a typical example of this type of development. Prior to the arrival of Euro-Americans, the lower Coquille River had broad forested floodplains with well-developed side-channels and sloughs (Benner 1992). The main channel was sinuous and occupied the valley bottom. Beaver ponds were extensive and there were large accumulations of wood in the channel and estuary. As settlement

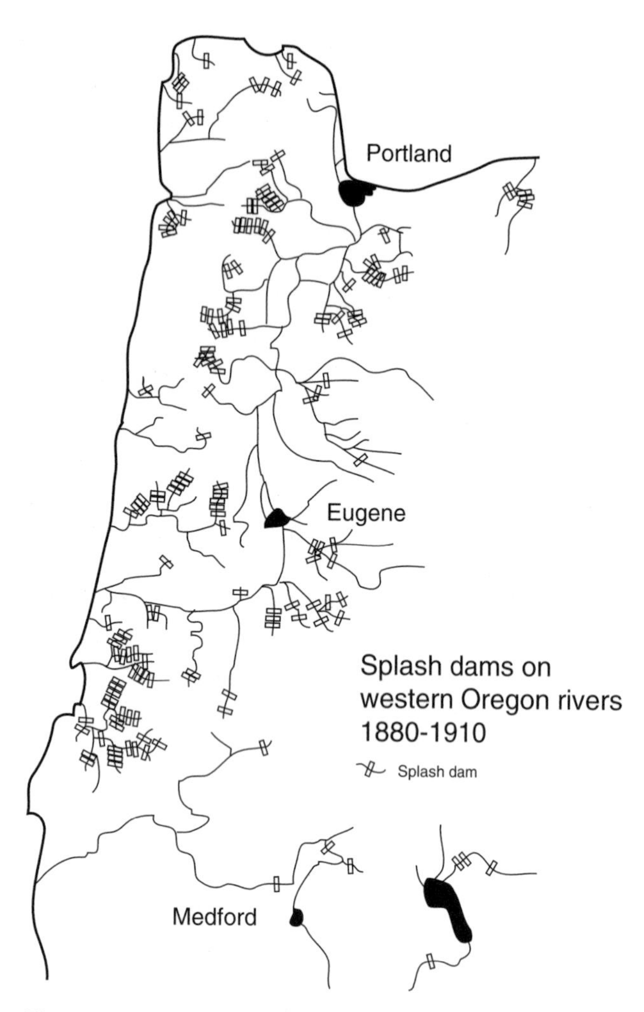

Figure 4-19. Location of permanent splash dams in the Oregon Coast Range, 1890-1910 (from J. R. Sedell).

proceeded, the channel was straightened and confined so that the floodplain is no longer connected to or interacts with the river. The floodplain was then often drained to create pastures for livestock. Large pieces of wood and rocks were removed from the channel to facilitate navigation and log transport. Secondary channels, backwaters, and oxbows, which are important habitats for many juvenile fishes, were lost as a result. In coastal streams, such habitats were historically important for coho salmon. Similar changes occurred on the Willamette River (Sedell and Froggatt 1984).

Smaller streams were also extensively affected by early development. Splash dams were built on many streams throughout western Oregon (Figure 4-19) to move logs from harvest sites to mills. These were permanent or temporary structures constructed across the channel to create a pond behind them, into which harvested trees were placed (Figure 4-20). During high flows, either the gate of a permanent dam was removed or the entire temporary structure was blown up, and the logs were carried downstream to processing facilities. Channels downstream of splash dams were straightened and obstructions removed to facilitate the movement of logs, resulting in very simplified channels, many of which persist today.

More recent activities have also altered habitat in many streams in the Coast Range. Timber harvest, urbanization, and agriculture have reduced the quantity and quality of habitat (Bisson et al. 1992). In the 1960s and 1970s, federal- and state-sponsored programs actively removed large wood from channels in the belief that accumulations of large wood impeded the upstream movement of fish. Additionally, management activities in riparian

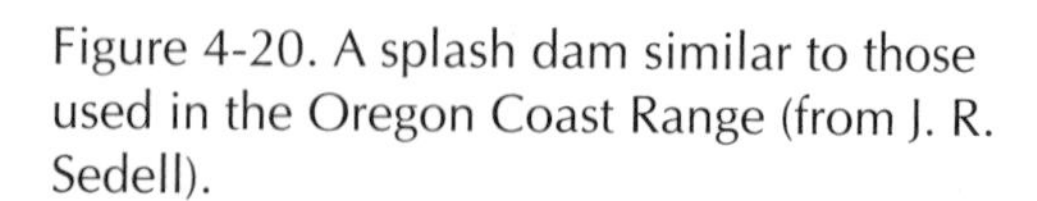

Figure 4-20. A splash dam similar to those used in the Oregon Coast Range (from J. R. Sedell).

areas, particularly timber harvest, increased along streams throughout the Coast Range. This reduced the amount of large wood that could be recruited to streams. The combination of loss of wood from active channel clearance and increased harvest, and increased levels of sedimentation in streams from timber harvest resulted in the loss of habitat, particularly pools. McIntosh and others (2000) found that the number of large pools (about 3 feet or more deep and 25 square yards or more in surface area), which are important habitats for many species and age-classes of fish, declined more than 75 percent in some coastal streams between the late 1930s to early 1940s and the mid 1980s.

Agricultural activities and growth of urban areas continue to impact aquatic systems and fish also. These activities occur primarily in the lower portions of watersheds, and reduce riparian habitats along main channels and smaller streams in the floodplain by diking and channelization. These impacts are particularly detrimental to coho salmon.

It is difficult to generalize about the response of fish to habitat alterations because responses vary with species, life-history stage, and location. Freshwater fish exist over a wide range of conditions, but the range that is generally most favorable for a species is relatively narrow (Larkin 1956). When environmental conditions change, relative abundances of the species in a community may shift. Those species favored by the new conditions increase, and those for which changes are less suitable decline. The result of this differential response is generally a decrease in diversity of the community, not usually from loss of species (the richness component of biodiversity) but rather because of changes in relative abundances of individuals of different species (the evenness component of biodiversity.

Reeves and others (1993) found that assemblages of juvenile anadromous salmonids in coastal Oregon watersheds where less than 25 percent of the basin was subjected to timber harvest and associated activities were more diverse than the assemblages in basins where more than 25 percent of the basin was harvested (Figure 4-21). The differences were primarily a result of increases in numbers of juvenile coho salmon and decreases in cutthroat trout numbers. Streams in systems with lower levels of timber harvest had more pools and more wood than those in systems with higher harvest levels. In steeper streams, coho salmon numbers declined as

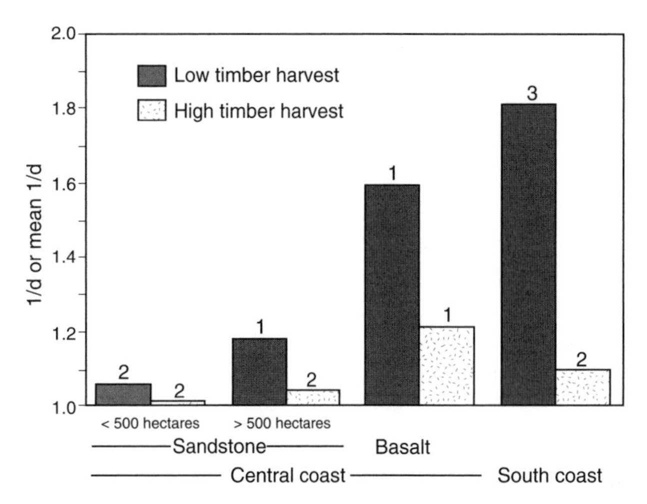

Figure 4-21. Mean diversity of juvenile anadromous salmonid assemblages in coastal Oregon watersheds with low levels (< 25 percent) and high levels (> 25 percent) of timber harvest (Reeves et al. 1993). (Reprinted by permission from Transactions of the American Fisheries Society.)

habitat complexity decreased. Cutthroat trout numbers decreased in lower-gradient systems. Coho salmon are more suited to slower water and trout to faster water (Bisson et al. 1988). Loss of structure in higher-gradient streams means the loss of slow-water habitats, and thus conditions were less favorable for coho salmon. On the other hand, loss of large wood in lower-gradient streams results in reduced cover and habitat suitability for cutthroat trout.

Similar responses have been observed in Carnation Creek, British Columbia (Hartman 1988; Holtby 1988), where cutthroat trout and steelhead numbers declined following timber harvest and coho salmon numbers increased. Chum salmon numbers also decreased following timber harvest (Scrivener and Brownlee 1989). Bisson and Sedell (1984) also noted a similar pattern of differential response in streams in Washington. In that case, 0+ steelhead were the dominant species in a stream following intensive timber harvest. Relative abundances of coho salmon and older than age 1+ cutthroat trout and steelhead were more even in a nearby pristine stream. The structure and composition of fish communities were similar in terms of species richness, but evenness was greater in a pristine stream than in an urban stream flowing into Puget Sound, Washington (Scott et al. 1986). A species' positive response to a change may not necessarily result in an increase in the species at a later life-history stage, however. Murphy and others (1986) found more 0+ coho salmon during the summer in stream sections in southeast Alaska with clearcuts than in sections where riparian zones were patch cut or not harvested. They attributed this increase to increased primary production resulting from increased sunlight reaching the stream in clearcut sections. However, overwinter survival in clearcut sections was less than in the other sections because habitat suitability decreased as a result of the reduction in the amount of large wood. Consequently, survival decreased and numbers in clearcut sections were less than in the other sections after the winter.

Coho salmon increased in numbers and grew faster following timber harvest in Carnation Creek, British Columbia (Holtby 1988). These changes were attributed to increased water temperatures and benthic invertebrate production. Following timber harvest, the majority of coho salmon smolts were age 1+ compared to a mixture of age 1+ and 2+ prior to harvest (in more northerly areas such as British Columbia and Alaska, coho salmon may spend two years in freshwater before moving to the marine environment compared to only one year in Oregon). However, increased adult returns were smaller than the increased smolt numbers because the timing of ocean migration was about 12 to 14 days earlier following timber harvest than it had been prior to harvest. It was believed that this earlier ocean entry resulted in increased predation by hake (*Merluccius productus*), which migrate along the coast at this time of year (Hartman and Scrivener 1990). Normally, coho salmon would enter the ocean after the hake had moved through the area. Therefore, the increase in smolt numbers was offset by decreased ocean survival, and adults returning to Carnation Creek declined over time.

Changes in environmental conditions that are not necessarily lethal can alter the structure and composition of fish communities by changing the outcome of interactions among potential competitors. Water temperature mediates interactions between redside shiners and juvenile steelhead (Reeves et al. 1987). At temperatures of 19° to 22°C (62° to 67°F), shiners displaced trout by exploitative competition; that is, they were able to obtain food more efficiently. Steelhead dominated at cooler temperatures, 12° to 15°C (51° to 56°F), because of interference competition; they prevented competitor access to food by establishing and defending territories. Dambacher (1991) attributed the distribution pattern of trout and shiners in Steamboat Creek, Oregon, to changes in competitive interactions associated with water temperature and reach gradient. Shiners dominate in warmer, low-gradient reaches and trout in cooler, higher-gradient reaches. Water temperature also mediated interactions for food and space between redside shiners and juvenile chinook salmon in the Wenatchee River, Washington (Hillman 1991).

Ecosystem Restoration

Current approaches

The most cited factor associated with the decline of anadromous salmonids in the PNW and northern California is alteration of freshwater habitats (Nehlsen et al. 1991). Freshwater habitats used by anadromous salmonids have been simplified by a suite of human activities (Hicks et al. 1991; Bisson et al. 1992; Spence et al. 1996). Simplification includes loss of habitat quantity, quality, diversity, and complexity. Lawson (1993) argued that the continuing decline of the quantity and quality of freshwater habitats must be reversed if anadromous salmonid populations are to be protected from further decline and extirpation.

Management agencies and private individuals and companies throughout coastal Oregon have in recent years implemented a number of restoration projects to improve the condition of freshwater habitats and to increase fish populations. These efforts have primarily focused on improving conditions in the stream channel. Reeves and others (1991) cite several examples of early restoration projects (late 1960s to early 1970s). These included blasting pools in bedrock stream reaches, placing gravel-filled wire gabions or baskets in streams to collect and retain spawning gravels, and placing boulders and large wood in channels. Many of these techniques, such as using wire gabions and blasting pools, have since been abandoned because they failed to produce the desired results.

Some recent in-channel restoration efforts in the Coast Range focused on increasing habitat complexity by placing wood in channels or creating off-channel habitats. The most successful projects created alcoves or small backwaters just off the main channel by excavation, or by creating a channel that connects naturally occurring alcoves with the main channel (Nickelson et al. 1992b; Crispin et al. 1993;

Solazzi et al. 2000). Pieces of large wood were placed in the alcove to enhance complexity. These projects increased the number of coho salmon smolts (House and Boehne 1985; House 1996) and migrant cutthroat trout and steelhead (Solazzi et al. 2000).

The likelihood that in-channel restoration efforts will be successful is increased by creating aggregations of wood at strategic locations in the stream network (Dewberry and Doppelt 1996). A watershed analysis (see Doppelt et al. 1996 and Cissel et al. 1998) provides the context for locating structures. Usually the best sites will be at or below tributary junctions in unconstrained reaches, which are natural deposition and collection areas. Aggregations should be anchored by "key pieces" of wood, which Dewberry and Doppelt (1996) recommend be larger than 2.5 feet in diameter and at least as long as the width of the active channel. The most effective pieces will be logs with the root wad attached. Structures should not be placed at every tributary junction; fewer larger structures are more effective than several small ones. In-channel restoration has produced mixed results at best. Frissell and Nawa (1992) found that the vast majority of restoration structures, many of which were in coastal Oregon and were primarily single pieces of wood, single boulders, and boulder clusters placed in streams on public lands, did not achieve their expected results. Structures failed because they were not properly placed in the stream or they were incorrect for a given situation (Nickelson et al. 1992b). For example, log structures placed in unconstrained reaches were often stranded when channels shifted after high flows. Kondolf (2000) provides guidelines for geomorphic aspects of habitat restoration.

Failure to include watershed-level considerations into restoration efforts greatly limits the contribution of in-channel work. Structure placement has been viewed as the end step in restoration efforts, but successful restoration of fish habitat is dependent on restoration of both in-channel and upslope conditions (Hartman et al. 1996). Reeves and others (1997) evaluated the effectiveness of habitat-forming structures, primarily boulders and large wood, to restoring pools for steelhead in Fish Creek, a tributary of the Clackamas River near Estacada, Oregon. Five years after completion of structures, the amount of pool habitat had increased to the desired level and steelhead smolt numbers had increased by 27 percent. Then the basin experienced major floods in November 1995 and February 1996, and 236 landslides occurred in the watershed, most originating from roads or timber harvest units. More than half of the structures were lost and the lower channel aggraded as a result.

Failure to consider large-scale watershed conditions and to halt activities that cause habitat degradation will negate any positive effects of in-channel manipulations (Jones et al. 1996). It is unlikely that focusing only on in-channel restoration will improve habitats and increase fish numbers to legally and socially demanded levels. The amount of degraded habitat that can be treated is a relatively small proportion of the total. It is estimated that currently the vast majority of streams in the Coast Range are deficient in large wood. Generally, treated areas are those that are readily accessible to needed equipment (Frissell 1997). As a result, sites that have the greatest need for restoration or have the greatest potential to respond to restoration are often ignored. In addition, many improvements are relatively simple and do not create the range or diversity of conditions required by multiple species (Cederholm et al. 1997). Streams in the Coast Range are dynamic. Consequently, in-channel structures have a relatively short life expectancy (10 to 20 years) and need to be replaced periodically. It is not clear whether responsible agencies, organizations, and businesses are willing to make the necessary long-term commitments to restoration programs.

Ecosystem approach

There is an emerging recognition that a more comprehensive ecosystem approach is necessary to assist in the recovery of imperiled fish populations and species. Williams and others (1989) reported that, between 1979 and 1989, no fish species listed under the Endangered Species Act were delisted because of improved status. They believed that the focus on restoring in-channel conditions and the failure to address ecosystem-level concerns were primary factors responsible for this lack of improvement. Recovery of anadromous salmonids in the Coast Range is doubtful unless restoration efforts are focused on ecosystems and watersheds.

Current efforts, particularly on private lands, continue to focus on in-channel restoration and are unlikely to make significant contributions to habitat or population recoveries. In-channel work should be considered a catalyst for, rather than the sole means of, achieving habitat and watershed recovery (Reeves et al. 1991). Structures may provide the initial recovery, particularly in more degraded situations, but there must also be efforts to restore watershed and ecological processes that create and maintain necessary habitat and environmental conditions. Having only an in-channel focus will result in short-lived results at best.

A primary goal of watershed/ecosystem restoration is to restore biological integrity, which is the ability of an ecosystem to recover from periodic disturbances and to express the historical range of ecological conditions (Angermeier 1997). Restoration policies and goals generally assume that societal demands, both material and nonmaterial, can be met from landscapes operating somewhere within their historical range of conditions. Thus, efforts to restore ecological conditions and processes, such as joining off-channel habitat connections and ensuring wood and sediment delivery and water flow, do not have to restore the full suite of conditions and processes found in natural systems. Rather, they need to restore the key ecological processes that will create the desired conditions across the landscape seasonally and over longer periods, such as decades to centuries.

Preventing initial degradation, which includes direct damage to habitat as well as the ecological processes that create and maintain habitat, is the most prudent and economical approach to ecological restoration. Repairing damage can be costly (Toth et al. 1997) and is not always successful; some impacts simply cannot be repaired. Mitigation measures, such as the credit provision in the Oregon State Forest Practices Rules that allow for harvest of trees from riparian areas if logs are placed in streams, can be expensive and less than effective. Clearly, past damages require restoration, but our efforts at habitat restoration and watershed management should put a strong priority on implementing land management policies and practices that protect and maintain the integrity of watersheds and their ecological processes. A primary principle of any successful watershed restoration program is that existing areas of good habitat must first be protected from degradation, particularly in the short term (McGurrin and Forsgren 1997).

Areas of good habitat are relatively rare in coastal Oregon today and future management could threaten many of these areas. Protecting these areas is essential in the short term to protect existing populations that supply colonists to other areas as they recover. This principle should be applied at all spatial scales from the site to the landscape. For example, at the site scale, emphasis should be placed on riparian areas that currently have desirable stand conditions. Management in these areas should be minimized to protect their integrity. Botkin and others (1995) concluded that the riparian management requirements on private lands in the Oregon State Forest Practices Act would do this. However, the Independent Multidisciplinary Scientific Team (1999) was less certain that riparian integrity is sufficiently protected by the act. The Forest Practices Act allows removal of larger trees over repeated entries, protects a relatively small area, and places primary emphasis on riparian zones along fish-bearing streams. These actions will likely compromise or eliminate necessary ecological functions and processes, such as large-wood recruitment.

Entire watersheds that currently are in good condition should also be protected from degradation. These watersheds should be 7th field HUs or larger. A watershed of this size is sufficiently large to allow anadromous salmonids to complete most of their freshwater life cycle, and could serve as a key source of migrants that can colonize new areas if they recover. A portion of the key watersheds established in the Northwest Forest Plan for federal lands fulfills this role. Unfortunately, these watersheds are generally limited to watersheds higher in river basins and are somewhat restricted in their distribution across the Coast Range. Therefore, their role in aiding the recovery of declining populations is restricted.

Because funds and time are always limited and the number of watersheds that are in good condition is small, priorities must be established for areas to be restored. Frissell (1996) presents a classification scheme for assigning such priorities:

(1) *Focal habitats* are critical refuge areas that have high-quality habitats. These can vary in size from a reach to a watershed. They have the highest priority because: (a) their protection will benefit multiple species; (b) potential biological/ecological benefits are high relative to costs of protection/restoration; and (c) the likelihood of near-term success is high.

(2) *Adjunct habitats* are areas immediately adjacent to focal habitats that have been degraded by human activities and do not currently support a high

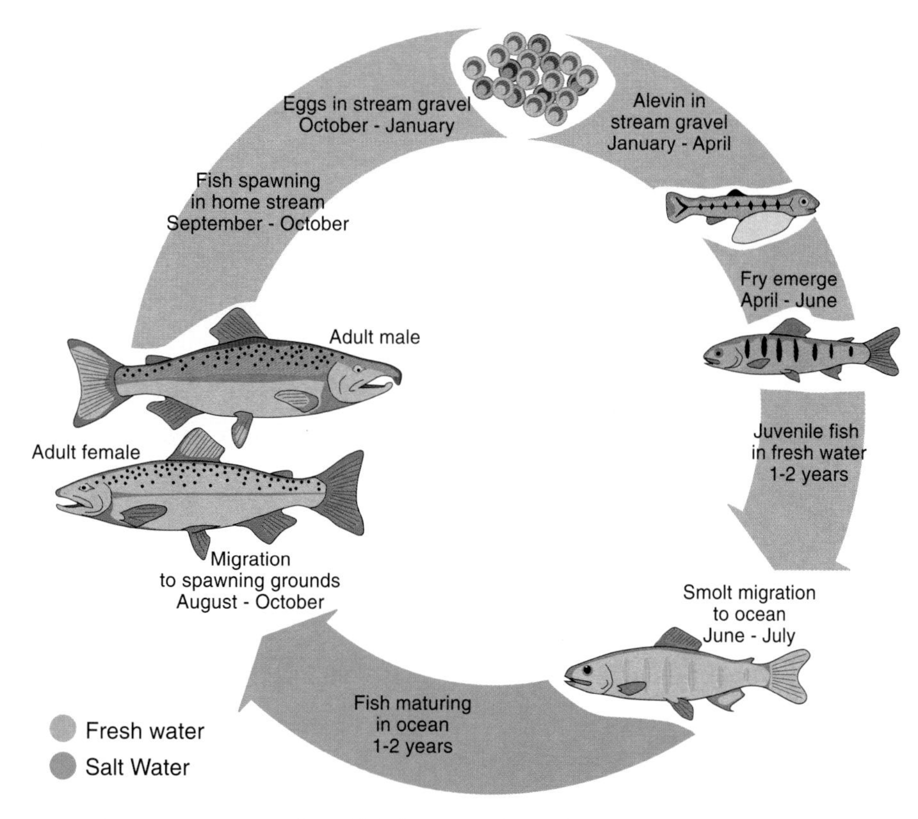

Figure 4-1. Generalized life history of anadromous Pacific salmon (*Oncorhynchus* spp.).

Figure 4-5. General summer rearing distribution of native fish in a coastal Oregon watershed.

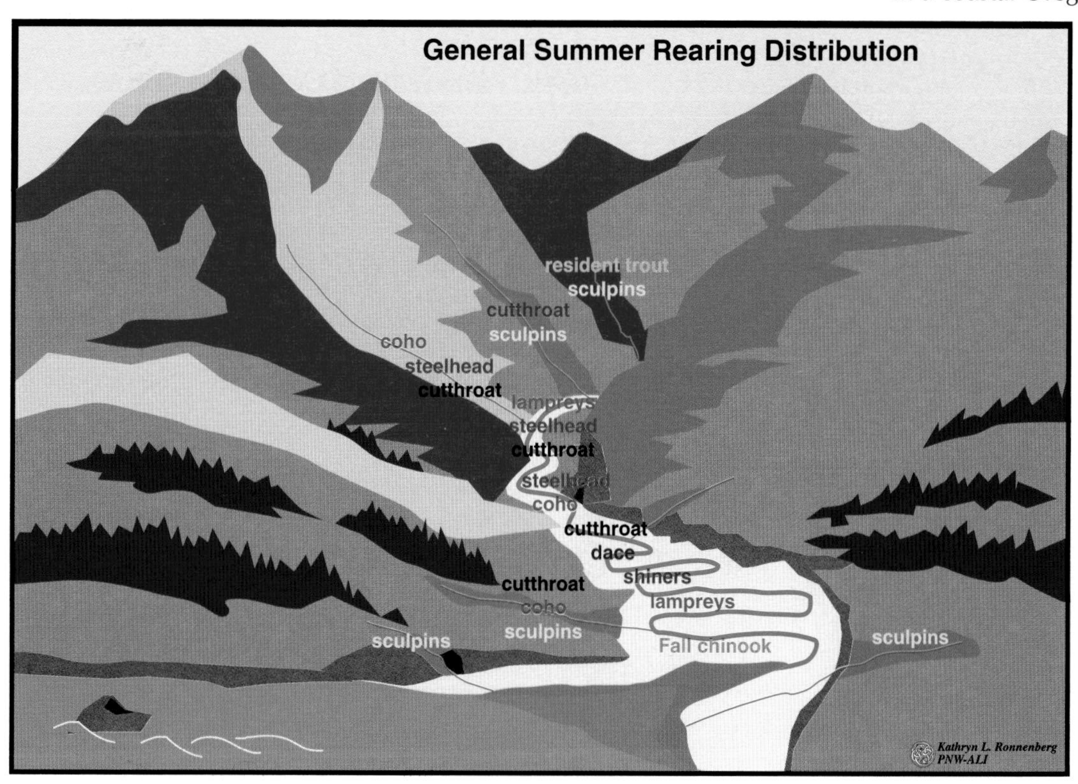

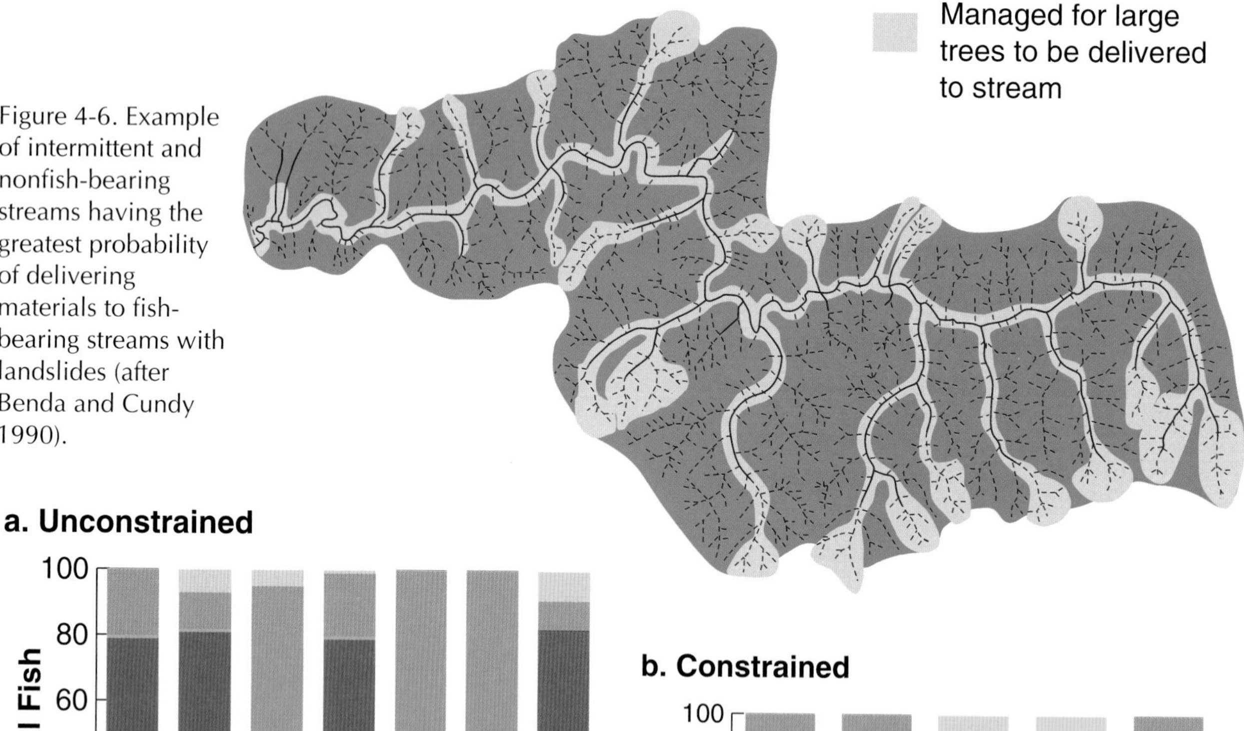

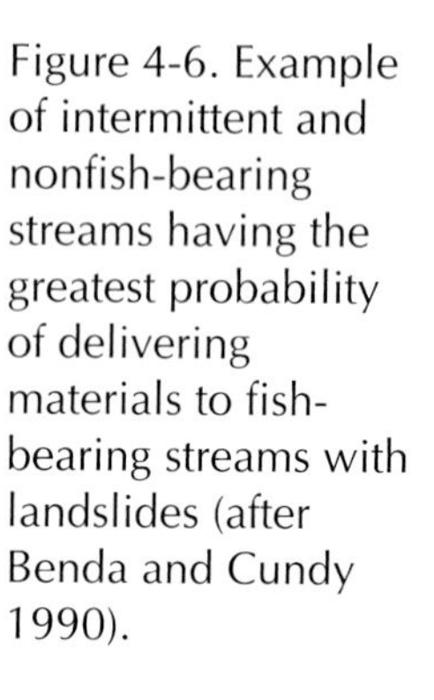

Figure 4-6. Example of intermittent and nonfish-bearing streams having the greatest probability of delivering materials to fish-bearing streams with landslides (after Benda and Cundy 1990).

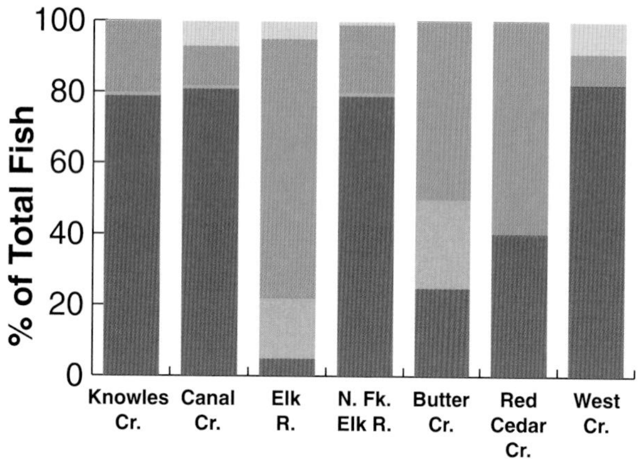

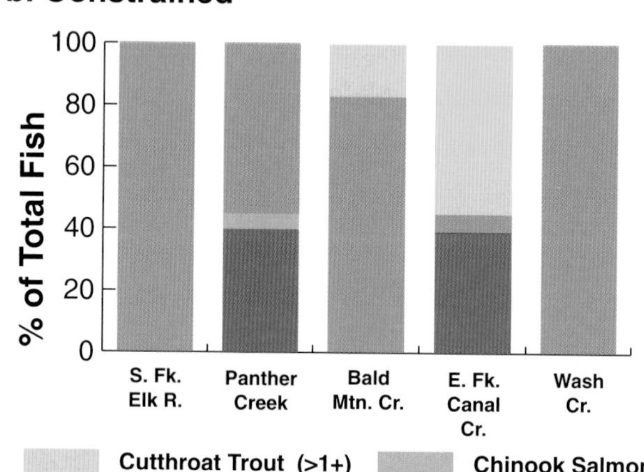

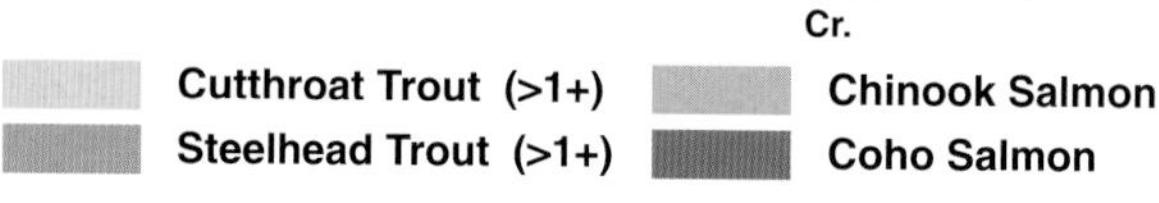

Figure 4-7. Composition of juvenile anadromous salmonid assemblage in different reach types, (a) constrained and (b) unconstrained, in streams in coastal Oregon. (Reprinted from G. H. Reeves, P. A. Bisson, and J. M. Dambacher. 1998. Fish communities, pp. 200-234 in *River Ecology and Management: Lessons from the Pacific Coastal Ecoregion,* R. J. Naiman and R. E. Bilby, eds. Springer-Verlag, New York. By permission of the publisher.

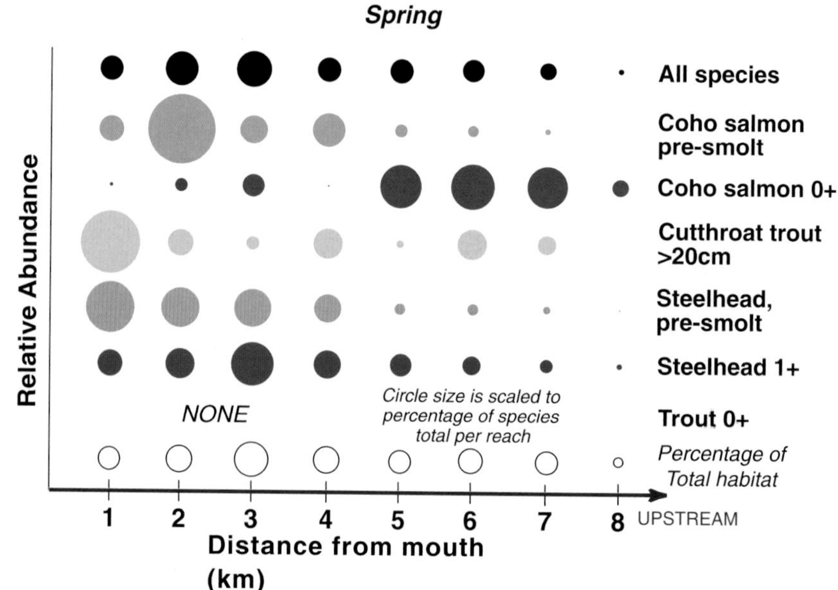

Figure 4-9. Seasonal distribution of juvenile anadromous salmonids in Cummins Creek, Oregon in spring (Sleeper 1993).

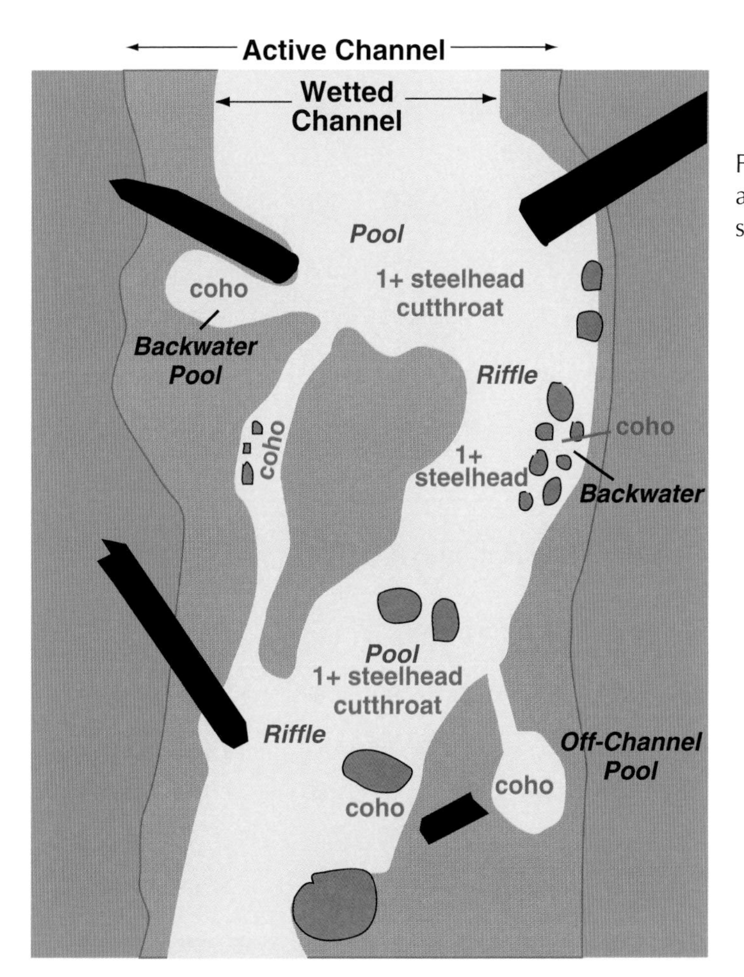

Figure 4-10. Generalized distribution of juvenile anadromous salmonids in coastal Oregon streams in spring.

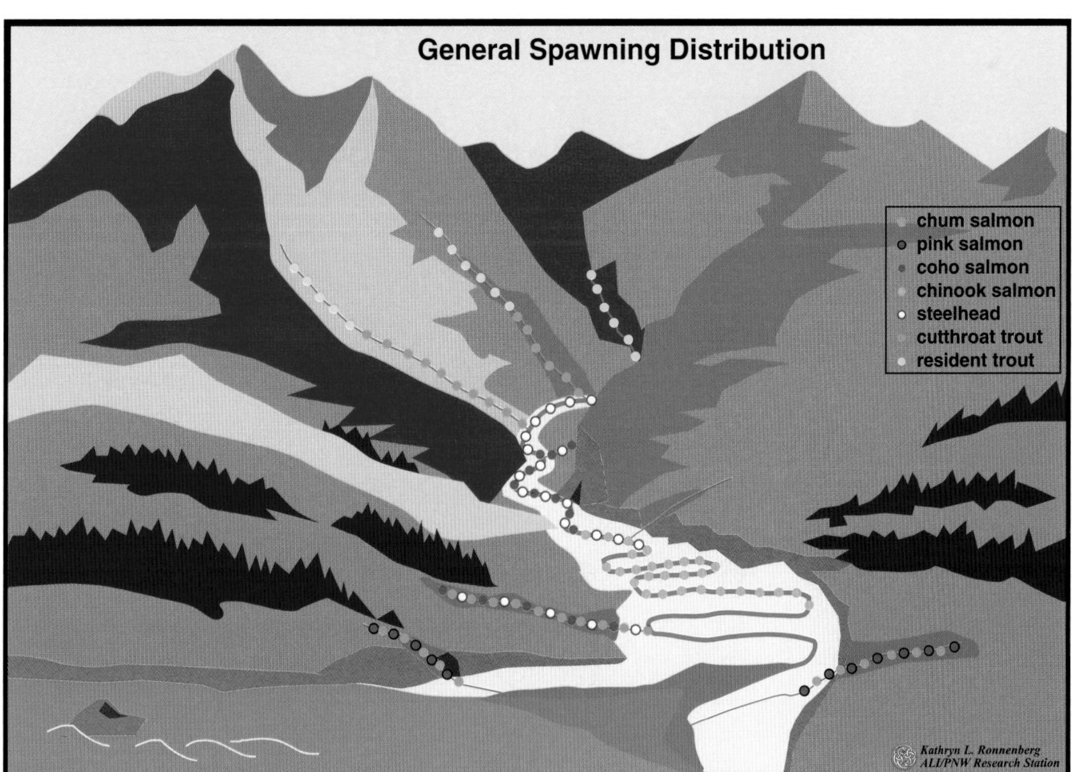

Figure 4-11. General spawning distribution of native fish in a coastal Oregon watershed.

Figure 4-13. Seasonal distribution of juvenile anadromous salmonids in Cummins Creek, Oregon in early summer (Sleeper 1993).

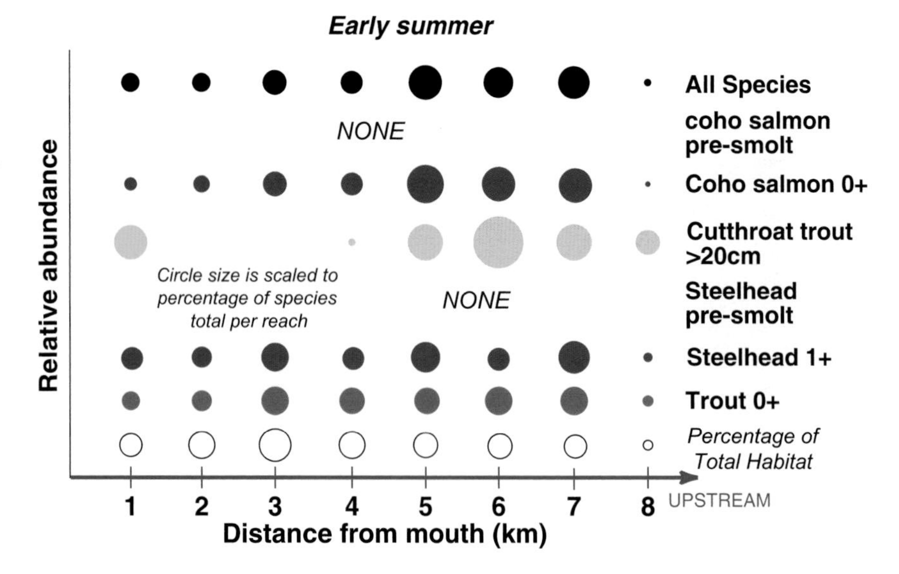

Figure 4-15. Generalized distribution of juvenile anadromous salmonids in coastal Oregon streams in mid-summer (Sleeper 1993).

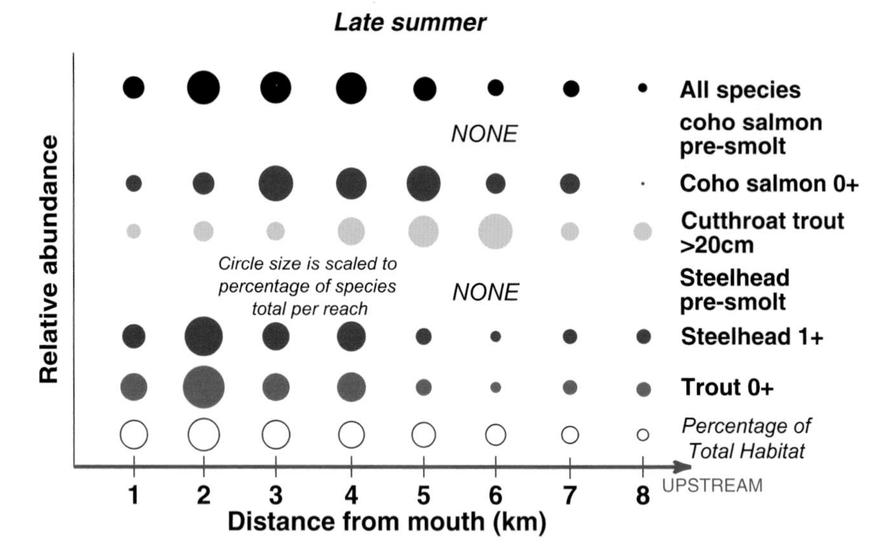

Figure 4-16. Seasonal distribution of juvenile anadromous salmonids in Cummins Creek, Oregon in fall (Sleeper 1993).

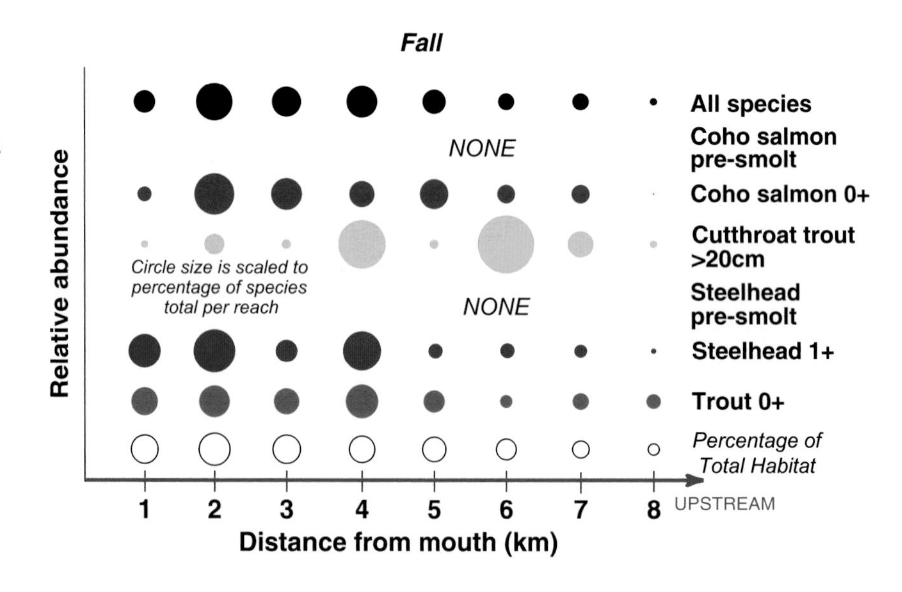

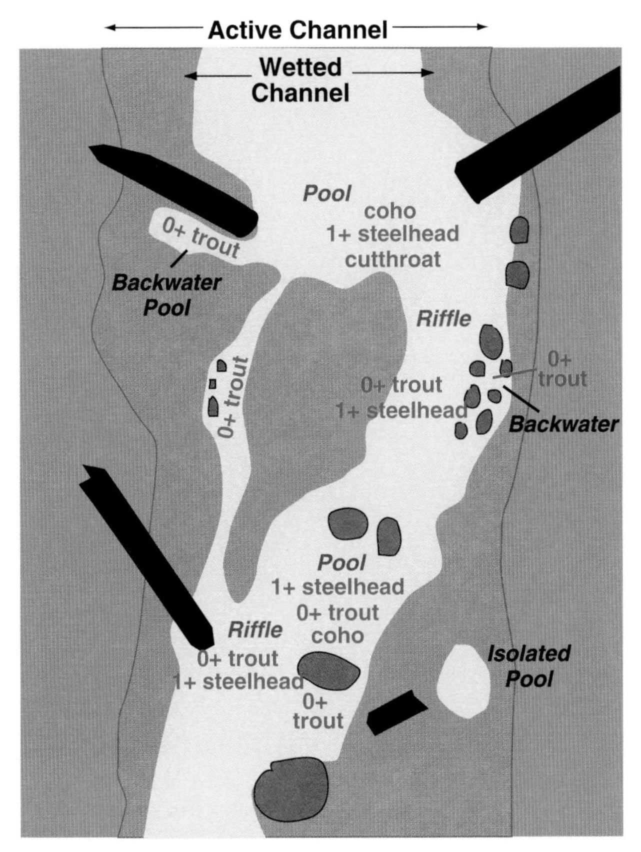

Figure 4-14. Seasonal distribution of juvenile anadromous salmonids in coastal Oregon streams in late summer.

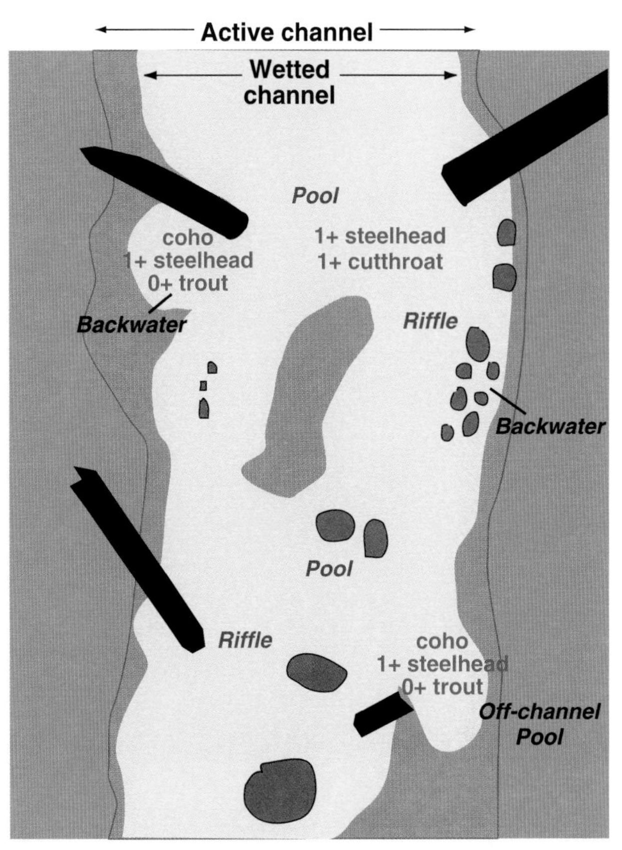

Figure 4-17. Generalized distribution of juvenile anadromous salmonids in coastal Oregon streams in fall.

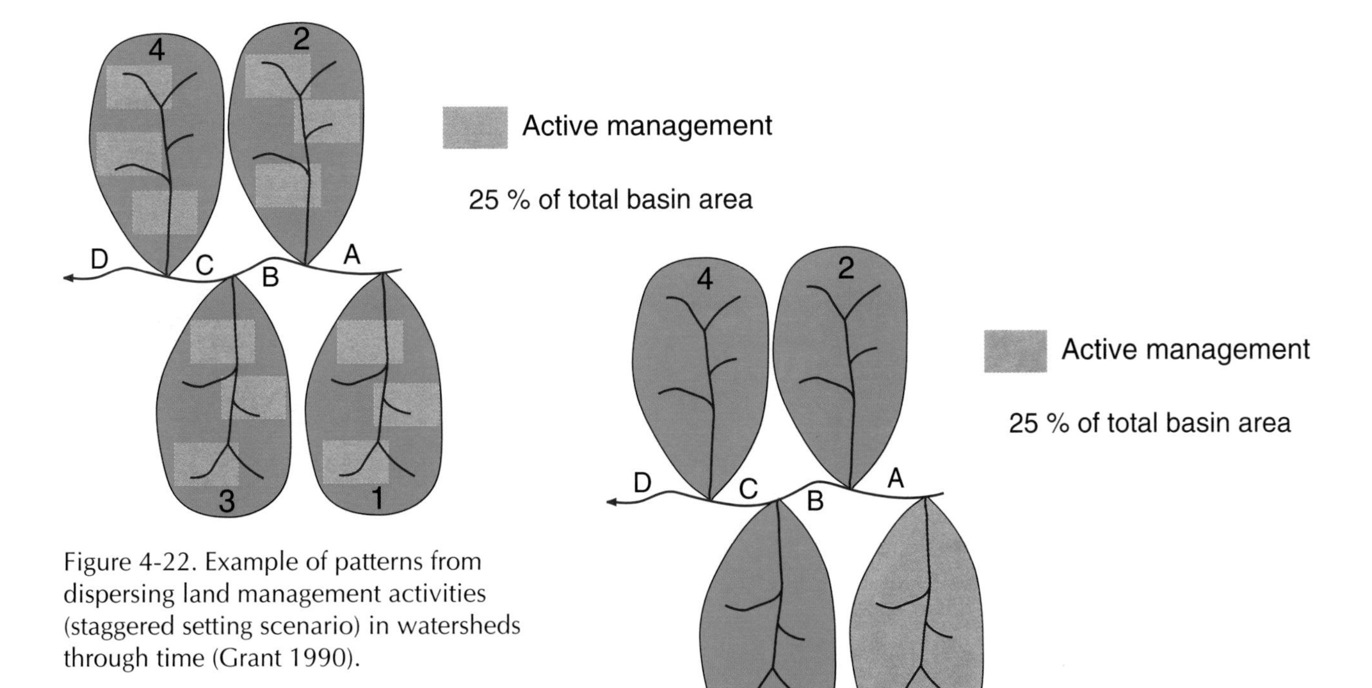

Figure 4-22. Example of patterns from dispersing land management activities (staggered setting scenario) in watersheds through time (Grant 1990).

Figure 4-23. Example of patterns from concentrating land management activities (minimum fragmentation scenario) in watersheds through time (Grant 1990).

Figure 5-4. Dead limbs and portions of oaks and other hardwoods provide valuable habitat for cavity-nesting birds and other species of wildlife (J. P. Hayes photo).

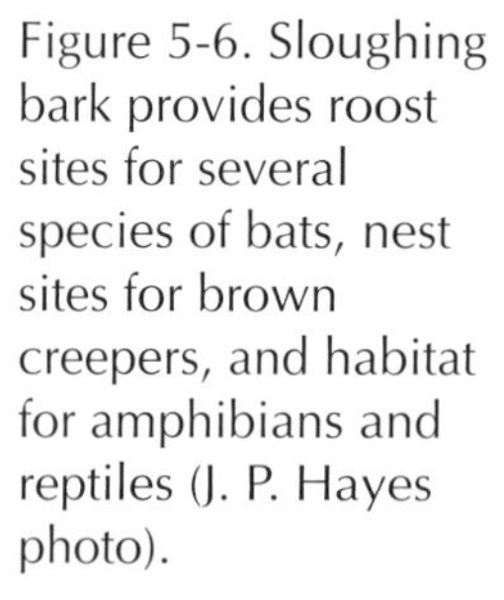

Figure 5-6. Sloughing bark provides roost sites for several species of bats, nest sites for brown creepers, and habitat for amphibians and reptiles (J. P. Hayes photo).

Figure 5-7 (a)

Figure 5-7 (b)

Figure 5-7 (c)

Figure 5-8. Large logs with hollow centers provide sites for nesting and provide den sites for larger species, such as black bear. (J. P. Hayes photo).

Figure 5-9. Artificially created snags, such as these snags made by topping live trees using chainsaws, can provide valuable habitat for species that use cavities in areas where few natural snags exist. (J. P. Hayes photo).

Figure 5-10. Deeply fissured bark, such as that on this tree, provides roosting habitat for some species of bats and habitat for arthropods that are fed upon by some species of wildlife. (D. Larson photo).

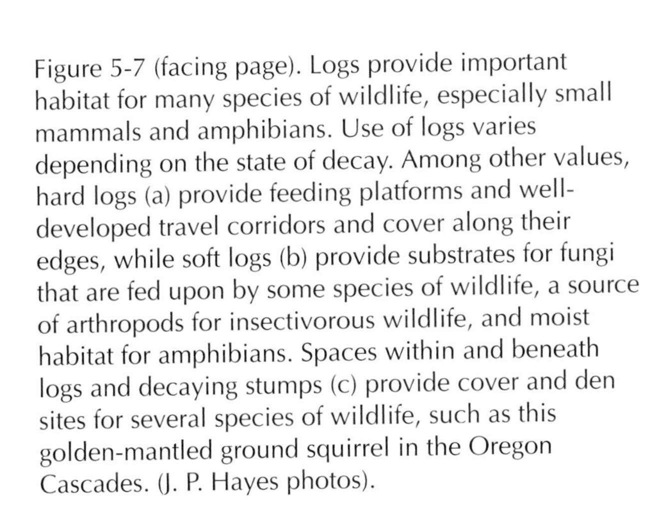

Figure 5-7 (facing page). Logs provide important habitat for many species of wildlife, especially small mammals and amphibians. Use of logs varies depending on the state of decay. Among other values, hard logs (a) provide feeding platforms and well-developed travel corridors and cover along their edges, while soft logs (b) provide substrates for fungi that are fed upon by some species of wildlife, a source of arthropods for insectivorous wildlife, and moist habitat for amphibians. Spaces within and beneath logs and decaying stumps (c) provide cover and den sites for several species of wildlife, such as this golden-mantled ground squirrel in the Oregon Cascades. (J. P. Hayes photos).

Figure 5-12. Thinning can dramatically influence the structural characteristics of a stand and influence wildlife use of an area. Differences in amount of light reaching the forest floor and some response of understory vegetation can be seen in these photos taken immediately before (a) and 4 years after (b) thinning. The large stump on the right-hand side of each photo provides a visual reference point. (M. Adam photos).

Figure 5-12 (a)

Figure 5-12 (b)

Figure 5-17. An alder-dominated riparian area in the Oregon Coast Range. (J. P. Hayes photo).

diversity or abundance of native species. They have high priority for restoration because the potential buffering effect and influence of adjacent focal habitats gives them a relatively good chance for recovery.

(3) *Nodal habitats* are areas with high species diversity or strong populations that are not connected to focal or adjunct habitats. They contain critical habitats required by certain life-history stages (e.g., spawning and early-rearing areas, overwinter habitats, etc.). These can be located throughout a watershed.

(4) *Critical contributing areas* are portions of the watershed that do not provide habitat directly but are important sources of materials that may create and maintain habitat or influence the quality of the existing environment. These may include tributaries that are sources of wood and sediment (see Figure 4-6) or cold water for fish-bearing streams.

(5) *Grubstake habitats* are areas where restoration costs will be high because much effort is generally required to restore them. However, because these habitats are often rare but essential for fish, the potential benefits may be large. Areas with grubstake habitats tend to be located in lower portions of the stream network, where alteration as a result of agriculture and urbanization has been extensive. Biotic and physical responses are likely to take a long time, perhaps decades. Examples include floodplains, estuarine marshes, and mainstems of lowland rivers.

(6) *Lost-cause habitats* are areas so severely degraded that recovery of any significance is unlikely. Examples include portions that are in heavily urbanized areas, or where activities such as timber harvest and agriculture are very intensive. Investing scarce resources in these habitats is simply not worthwhile. Biota are likely to benefit more from investment in other habitats.

Because social, political, and economic factors influence watershed and aquatic ecosystems, they must be considered in restoration efforts (McGurrin and Forsgren 1997). Issues associated with these factors include: (1) securing dependable long-term funding; (2) obtaining landowner interest and cooperation; (3) receiving understanding and support from local governments; (4) motivating agencies to work with citizens; and (5) influencing government planning decisions that potentially influence activities in the watershed. Overcoming these problems can be as formidable as the actual restoration work. McGurrin and Forsgren (1997) present several concepts for watershed groups to use in dealing with these concerns.

The Future: Ecosystem and Landscape Management

Our understanding of what constitutes the freshwater ecosystems of anadromous salmon and trout is continually evolving. To date, much of the focus has been on relatively small spatial scales, such as habitat units (Bisson et al. 1982; Nickelson et al. 1992a), reaches (Murphy and Koski 1989), and, to a lesser extent, watersheds (Reeves et al. 1993). There is an emerging need to move from these small spatial scales to the larger scale of ecosystems and landscapes because, in part, of the necessity to recover the freshwater habitats of anadromous salmonids with low or declining population numbers across the Coast Range.

A variety of sources, including interested publics, interest groups, scientific review and evaluation teams (National Research Council 1996; Independent Multidisciplinary Scientific Team 1999), regulatory agencies, and policy- and decision-makers, are calling for the development of policies and practices to manage the freshwater habitats of at-risk salmon and trout at ecosystem and landscape levels. Scientists are only beginning to work at these scales, so understanding of the broad-scale behavior of aquatic ecosystems over extended time periods is limited. Therefore, responsible agencies are struggling with the challenge to develop appropriate and effective policies and programs.

It is important to recognize that an ecosystem and a landscape are different entities and hence their management requirements differ. Ecosystems are vague entities with boundaries that may shift in space and time (Caraher et al. 1999). However, we consider the spatial scale of an ecosystem to be a watershed that is a 6th or 7th field HU, which is consistent with the definition of Hunter (1996). A landscape is a mosaic or collection of ecosystems (Hunter 1996) that occupy a relatively large area (2.47×10^5 to 2.47×10^7 acres; Concannon et al. 1999). Multiple watersheds that are contiguous are considered a landscape.

The foundation and principles for managing terrestrial systems and biota at the ecosystem and landscape levels are much more developed than they are for aquatic systems. Major paradigms of ecosystem management include (Lugo et al. 1999):

(1) Ecosystems are not steady state but are constantly changing through time.

(2) Ecosystems should be managed from the perspective of resilience, as opposed to stability.

(3) Disturbance is an integral part of any ecosystem and is required to maintain ecosystems.

Ecologists (Holling 1973; White and Pickett 1985) and managers recognize the dynamic nature of terrestrial ecosystems and how the associated biota and physical characteristics change through time. They are also aware that the range of conditions that an ecosystem experiences is determined to a large extent by the disturbances it encounters (a wildfire, hurricane, timber harvest and associated activities, etc.). Resilience is the ability of an ecosystem to recover to pre-disturbance conditions following a disturbance (Lugo et al. 1999). An ecosystem demonstrates resilience after a disturbance when the environmental changes caused by the disturbance are within the range of conditions that the system experienced before the disturbance. Reduced resilience results in a decrease in the diversity of conditions of a particular ecological state, the loss of a particular ecological state, or both (Lugo et al. 1999). Biological consequences of reduced resilience may include extirpation of some species, increases in species favored by available habitats, and an invasion of exotic species (Levin 1974; Harrison and Quinn 1989; Hansen and Urban 1992).

Yount and Niemi (1990) modified the definition of Bender and others (1984) and referred to a disturbance regime that maintains the resiliency of an ecosystem as a "pulse" disturbance. A pulse disturbance occurs infrequently, and there is sufficient time between disturbances to enable the ecosystem to recover to pre-disturbance conditions. A pulse disturbance allows an ecosystem to remain within its normal bounds to exhibit the same range of states and conditions that it does naturally. A "press" disturbance, on the other hand, reduces the resiliency of an ecosystem. It is a frequent or continuous impact that does not allow time for recovery to pre-disturbance conditions.

The less management actions resemble the disturbance regime under which an ecosystem evolved, the less resilient an ecosystem will be. The obvious challenge for ecosystem management is to make management actions resemble the natural disturbance processes and regime as closely as possible. The management disturbance regime should be more pulse-like and less like a press. Factors that must be considered in developing ecosystem management plans and policies are the frequency, magnitude (White and Pickett 1985; Hobbs and Huenneke 1992), and legacy (the conditions that exist immediately following the disturbance; Reeves et al. 1995) of disturbance regimes in the managed ecosystem. The impact on the ecosystem will depend on how closely the management disturbance regime resembles the natural disturbance regime with regard to these factors.

Landscape management strives to maintain a variety of ecological states in some desired spatial and temporal distribution. To do this, landscape management must consider (1) the development of a variety of conditions or states in individual ecosystems within the landscape at any point in time; and (2) the pattern resulting from the range of ecological conditions that are present (Gosz et al. 1999). Management must address the dynamics of individual ecosystems, the external factors that influence the ecosystems that comprise the landscape, and the dynamics of the aggregate of ecosystems. Obviously, understanding the dynamics of an individual ecosystem is demanding. Understanding the dynamics of the aggregate of ecosystems is much more challenging (Concannon et al. 1999).

Although a dynamic perspective of aquatic ecosystems is not widely held in the scientific community, the number of proponents is growing steadily (Minshall et al. 1989; Reeves et al. 1995; Benda et al. 1998). To establish a dynamic landscape perspective, the range of natural variability must be characterized at different spatial scales. Lower spatial scales (site, habitat unit) generally have a wider range of variation than do large scales (watershed, landscape). Wimberly and others (2000) demonstrated this for old-growth forests in the Coast Range. The amount of old growth at the finest spatial scale, the late-successional reserve (100,000 acres), ranged from zero to 100 percent over the time period of centuries. The variation was 15 to 80 percent at the scale of a national forest (7.27×10^5 acres). It was 25 to 73 percent at the province scale (5.6×10^6 acres).

The remaining challenge is to then determine how the location of each ecological condition moves across the landscape through time. The movement to ecosystem and landscape management for aquatic systems requires the articulation of principles and a conceptual basis to guide the development of policies and practices. However, there is little in the scientific literature to help with this. A major reason for this deficiency is that there is little or no consideration of time as an essential component of aquatic ecosystems. The major paradigms that shape the thinking about aquatic ecosystems, such as the River Continuum Concept (Vannote et al. 1980), do not consider time or its influence. Similarly, classification schemes such as that of Rosgen (1994) identify a single set of conditions for a given stream or reach type; no consideration is given as to how these conditions may vary over time. As a result, the dynamics of ecosystems and landscapes over long time periods (several decades to centuries) are not recognized, and the condition of aquatic ecosystems is expected to be relatively consistent through time. Therefore, a stream is expected to be in good condition at all times; any variation from this is considered to be unacceptable.

To develop effective ecosystem and landscape management policies for aquatic systems in the Coast Range, it is essential to understand the natural disturbance regime, how it has affected in-channel habitat within and among watersheds (6th and 7th HUs), and how it has been modified by human activities. To develop effective guidelines for management of aquatic ecosystems, it is critical to acknowledge that periodic disturbance is an integral part of these systems. Natural disturbances episodically delivered materials, sediment and wood, that formed the habitat over time. Also, suitability of the habitat for anadromous salmonids varied from good to poor through time in the past.

This perspective is not yet held very widely in the scientific community and is not even considered in setting management policies. We believe that it is necessary to assume that human activities, such as timber harvest and associated activities, have replaced wildfire and floods as the major disturbances in the Coast Range today. We must compare features of the natural disturbance regime and the human disturbance regime with regard to frequency, magnitude, and legacy. It should be made clear that we do not believe it is possible that the human disturbance regime will exactly mimic the natural regime. The challenge is to make the human disturbance regime more of a pulse disturbance than a press disturbance.

A case study

Understanding the natural disturbance regime is one of the first steps required for ecosystem and landscape management. The following is an example of the foundation for ecosystem and landscape management for timber harvest and associated activities in the central Coast Range. Reeves and others (1995) described the long-term dynamics of sandstone watersheds in the central Oregon coast. A brief synopsis of this follows. It should be noted that this case study is based on the examination of three watersheds that had little or no impact from human activities; however, they differed in the time since the last large wildfires and catastrophic landsliding event. Harvey Creek had most recently experienced landsliding and debris flows, perhaps 80-100 years before the study. Franklin Creek was estimated to have experienced such a disturbance 140 to 160 years before, and Skate Creek more than 200 years before.

We acknowledge that the small sample size—a consequence of the fact that watersheds in the Coast Range without a strong signal of human impact are very rare—may not fairly represent the central Oregon coast. Nonetheless, we believe that this case study provides an initial basis for understanding the natural disturbance regime of aquatic systems

Table 4-2. Physical conditions and composition of the juvenile assemblage in three streams in the Oregon Coast Range at different times from the last disturbance.

Stream	*Years since last major disturbance*	*Mean depth of pools (m)*	*Percent gravel*	*Mean pieces of wood (100 m)*	*Percent of assemblage*		
					Coho	*Steelhead*	*Cutthroat*
Franklin Creek	90-100	0.87	70	7.9	98.0	1.0	1.0
Harvey Creek	160-180	0.67	60	12.3	85.0	12.5	2.3
Skate Creek	> 300	0.08	10	23.5	100.0	0.0	0.0

Data from Reeves et al. 1995.

in the Coast Range. The historical natural disturbance regime in the central Oregon coast was dominated by infrequent wildfires and frequent, intense winter rainstorms (Benda 1994). Wildfires reduced the soil-binding capacity of roots. When intense rainstorms saturated the soils, there were catastrophic landslides and debris flows into the valley bottoms and streams. Such disturbances typically occurred on average every 300 years (Benda 1994).

Harvey Creek was the studied stream that most recently had experienced landsliding and debris flows. Its channels were aggraded with gravel-sized sediment 2+ meters deep. The result was that the stream had long expanses of riffle-like habitat with few pools (Reeves et al. 1995; Table 4-2). The few pools present were relatively deep, while the amount of large wood was relatively low (Table 4-2). Although large amounts of wood from upslope areas were delivered to the channels, much of it was buried in the sediments and did not immediately function to create habitat.

The likely immediate impacts of such a large disturbance event were direct mortality of fish, habitat destruction, elimination of or reduction in access to spawning and rearing areas, and temporary reduction or elimination of food resources. The diversity of fishes in Harvey Creek was relatively low (Reeves et al. 1995; Table 4-2). Coho salmon dominated the salmonid assemblage. Only a few juvenile cutthroat trout and steelhead were present, most likely because pools lacked large wood and the complexity it creates. Over time we would expect conditions in Harvey Creek to become more favorable to fish. Franklin Creek best approximated the complex habitat conditions that develop 80 to 140 years after a catastrophic disturbance. The amount of sediment in the channel would have declined over that period because of downstream transport and erosion (Benda 1994), exposing wood and large substrates that had been buried initially. There was also the recruitment of wood from the surrounding riparian zone, which now had trees that had become large enough to create habitat in the channel. Pools in Franklin Creek are shallower than those in Harvey Creek (Table 4-2), but they are more complex as a result of the increased amount of wood. Based on modeling the amount of sediment in channels as a result of natural wildfire and landsliding, Benda (1994) estimated that approximately 60 percent of the sandstone streams in the central Coast Range were in this condition at any point in time historically.

The diversity of the salmonid assemblage changed at this time as a result of the changes in physical conditions. Coho salmon still numerically dominate the assemblage in Franklin Creek (Table 4-2). However, cutthroat trout and steelhead make up a much larger proportion of the assemblage than they do in Harvey Creek. After an extended time since the last major disturbance (> 200 years), habitat conditions for juvenile anadromous salmonids were thought to decline), as in Skate Creek (Table 4-2). Old-growth forests in riparian zones would have delivered large amounts of wood to the channel, so the amount of wood in the channel of Skate Creek is now very high. However, the amount of gravel has declined because of erosion and downstream transport (Benda 1994). The result is that the stream has long expanses of bedrock. Pools are very shallow and not very suitable for fish. Juvenile coho salmon were the only fish found in this stream.

Life-history features of anadromous salmon and trout allowed them to persist in such a dynamic environment. Adaptations include straying by adults, relatively high fecundity rates, and movement by juveniles. Straying by adults is directly or indirectly genetically controlled (Quinn 1984) and aids the reestablishment of depressed or extirpated populations (Ricker 1989; Tallman and Healey 1994). The high fecundity of anadromous salmonids permits relatively quick establishment and growth in new areas with favorable conditions (Reeves et al. 1995). Juveniles may move from natal streams into unoccupied habitats and grow rapidly (Tschaplinski and Hartman 1983), thus helping to establish populations in the new areas.

Conclusions

We believe that human demands on aquatic ecosystems in the Oregon Coast Range will only continue to increase in the future. Given that, it will not be possible for the historical natural disturbance regime to operate, even in a relatively small number of watersheds. If aquatic ecosystems are to be conserved and restored, human activities will have to be viewed in the context of a disturbance regime that can sustain their long-term productivity. The challenge is to make the human disturbance regime resemble the natural regime as closely as possible; in other words, make human activities more of a pulse disturbance than a press disturbance (Yount and Niemi 1990). It is therefore necessary to identify those activities that can be modified to maintain required ecological processes and leave the legacy required for the resilience and persistence of the ecosystems.

A new disturbance regime

The recovery of degraded aquatic ecosystems in the Coast Range will be dependent on developing a new disturbance regime. One of the necessary steps is to compare the elements of the natural disturbance regime to those in the human disturbance regime, with respect to legacy, frequency, successional states, and size and spatial patterns of disturbance. The following example is for timber harvest. We have focused on timber harvest for this example because we feel that it offers the best opportunities for ecosystem and landscape management. Other human activities, such as agriculture and urbanization, need to be considered; however, they are very strong press disturbances and do not lend themselves to modification to the degree needed to develop disturbance regimes that are more pulse-like than press-like.

We recognize that timber harvest strategies vary with ownership and the owner's objectives. We will generalize about ecosystem management for aquatic systems in which timber management is the primary disturbance. The following compares timber harvest and the historic disturbance regime of aquatic ecosystems in the Coast Range. A more detailed discussion is presented in Reeves and others (1995). The timber harvest disturbance regime differs from the stand-replacing wildfires that affected aquatic ecosystems in several respects. One difference is the legacy of each. Wildfires left large amounts of standing and downed wood that was delivered, along with sediments, in landslide and debris torrents to fish-bearing streams in the valley bottoms (Benda 1994). The wood delivered via these hillslope processes can be a substantial amount of the total volume of wood found in streams (McGarry 1994). As described previously, once the amount of sediment declined to intermediate levels, high-quality habitats developed (Benda 1994). Landslides associated with timber harvest, on the other hand, contain primarily sediments, because trees along the landslide tract have been removed (Hicks et al. 1991). As a consequence, channels are simpler after timber harvest (Hicks et al. 1991; Reeves et al. 1993; Ralph et al. 1994) than they are after wildfire.

Timber harvest and wildfire also differ with regard to the frequency of disturbances. The interval between disturbances affects the range of conditions and ecological states that can develop within an ecosystem (Hobbs and Huenneke 1992). The extended time interval between natural disturbance events (300 years; Benda 1994) allowed a wide range of conditions to develop in aquatic ecosystems in the Coast Range. Timber harvest generally occurs at intervals shorter than this, generally 40 to 50 years on private lands and 80 to 100+ years on federal and state lands. The physical habitat conditions necessary to support the variety of fish naturally found in coastal streams may not develop in such relatively short periods, especially on private lands.

Another difference between the disturbance regime of timber harvest activities and of wildfire is the spatial distribution of successional stages under each regime. In the watersheds of the central coast area, Benda (1994) estimated that historically on average 15 to 25 percent of the forest would have been in early-successional stages. On federal lands in these watersheds approximately 35 percent are currently in early-successional stages (J. Martin, Siuslaw National Forest, personal communication). (The total percentage in early-successional stages would be larger if private lands were included.) This increase in early-successional stages has resulted in a concomitant decrease in the percentage of area of mid-successional forests, and those central coast watersheds with larger amounts of mid-successional forest appear to contain the most favorable habitats for anadromous salmonids (Botkin et al. 1995; Reeves et al. 1995).

A fourth difference between the natural disturbance regime and the current timber harvest disturbance regime is the size of the disturbance events and the landscape pattern created by the disturbance. Timber harvest activities generally occur in small individual actions that "disturb" areas of 40 to 120 acres and are distributed across the landscape. In contrast, wildfires resulted in larger but more concentrated areas of disturbance. In the central Coast Range, the mean size was 7,500 acres (Benda 1994).

Variation among watersheds in their suitability for fish is reduced under the timber harvest regime compared to the wildfire regime. Dispersal of timber harvest activities over relatively large areas subjects a greater number of watersheds to disturbance at any point in time and has degraded streams across the landscape, while the concentration of wildfire in a relatively small proportion of the landscape resulted in a variation in watershed conditions, ranging from poor to good, at any point in time (Reeves et al. 1995).

The legacy of timber harvest needs to include more large wood. Leaving large trees in riparian zones along selected landslide-prone channels will result in the delivery of more wood to fish-bearing streams and will increase the potential of aquatic ecosystems to develop conditions favorable for anadromous salmonids. The model of Benda and Cundy (1990) can be used to identify which landslide-prone channels have the greatest potential for delivering wood to fish bearing channels. This model was shown to be more than 90 percent accurate in identifying these channels in Coast Range watersheds impacted by the February 1996 storms (Robinson et al. 1999). Figure 4-6 shows an example of the distribution of these channels in a watershed.

Responsible agencies and landowners seem reluctant to extend riparian zones into intermittent and small nonfish-bearing streams on state and private lands. Opponents cite the scientific literature to argue that the vast majority of the wood found in fish-bearing streams is recruited from within a relatively short distance from the stream. However, we believe that cited papers do not support this contention. For example, McDade and others (1990) found that 90 percent of the wood came from within 66 feet of the stream. This result is based on using only a fraction (about 50 percent) of the total amount of the wood in the channel. The authors excluded wood pieces for which the origin in adjacent riparian zones could not be identified (McDade et al. 1990), thus excluding any landslide-delivered pieces. Additionally, only stream reaches that were not impacted to any extent by landslides were examined in the study (F. Swanson, Pacific Northwest Research Station, personal communication). Therefore, the results of this study showed the amount of wood coming from only one source—the immediately adjacent riparian zone. Riparian zone standards based on these kinds of results need to be re-evaluated in the development of ecosystem and landscape management plans and policies to ensure that all sources of wood are protected.

Extended time periods between disturbances from timber harvest could also be part of any ecosystem and landscape management plan. The interval between disturbances determines to a large extent the range and types of conditions that can develop in an ecosystem. Based on the limited observations of Reeves and others (1995), it appears that favorable conditions in central Oregon coast streams begin to appear 80 years or so following a disturbance; this is a rough approximation that requires more supporting research. The longer rotation between disturbances may not have to apply to an entire watershed if riparian zones are sufficiently large and include appropriate landslide-prone nonfish-bearing and intermittent streams.

Policies and practices for landscape management should consider allowing management activities to be concentrated at the ecosystem level rather than distributing activities over wider areas (Reeves et al. 1995). An example of this is shown in Figures 4-22 and 4-23 (in color section following page 84). The amount of activity allowable in a watershed is often limited by rules and regulations governing cumulative effects (Figure 4-22), but ecosystem and landscape effects may be lower if activities are concentrated in a given area rather than being dispersed (Figure 4-23). Grant (1990) modeled both scenarios to determine their effects on pattern of peak flow and found little difference between the two approaches. Concentrating rather than dispersing activities also confers benefits to terrestrial organisms that require late-successional forests (Franklin and Forman 1987).

Watershed reserves could also be considered in the development of ecosystem- and landscape-management policies, but only as short-term components. Reserves, such as the Key Watersheds identified in FEMAT (1993) or the Class I waters of Moyle and Yoshiyama (1994), are essential to protect watersheds that are currently in good condition. However, in dynamic environments reserves such as these simply act as holding islands that persist for relatively short ecological periods (White and Bratton 1980; Hales 1989; Gotelli 1991). Therefore, given the dynamic nature of aquatic ecosystems in the Coast Range, and elsewhere in the PNW, any single watershed reserve will not and should not be expected to provide high quality for extended periods. The challenge of ecosystem and landscape management is to manage for the future generations of "reserves" so that as good habitats become degraded, either through human or natural disturbances or through development of new ecological states (i.e., succession), others become available.

Many hurdles must be overcome to make viable ecosystem and landscape policies for aquatic systems work. Biologists, managers, and planners must consider much longer time frames than they are generally accustomed to using (Reeves et al. 1995). They need to acknowledge and understand the dynamics of ecosystems and landscapes in space and time. Legal and regulatory constraints may have to be reconsidered. For example, current cumulative effects regulations would prevent concentration of activities. Or, water quality standards for temperature or suspended sediment may be violated in some watersheds following disturbances, which may be a vital component of ecosystem and landscape management. There will also be social and economic barriers. In much of the Coast Range, there are multiple landowners with differing objectives. Coordinating objectives and timing of activities in such a mosaic so as to achieve ecosystem and landscape management will not be easy.

It is important that disturbance be recognized as an integral component of ecosystem and landscape management (Reeves et al. 1995; Gosz et al. 1999; Lugo et al. 1999). This will require helping managers, scientists, administrators, politicians, and the public to realize that periodic disturbances to aquatic ecosystems are essential to maintain long-term productivity and that they are not necessarily negative. Adjusting expectations that all watersheds should be in "good" condition at all times may be the biggest change required for successful ecosystem and landscape management of aquatic systems in the Oregon Coast Range.

Literature Cited

Andrusak, H., and T. G. Northcote. 1971. Segregation between adult cutthroat trout (*Salmo clarki*) and Dolly Varden (*Salvelinus malma*) in small coastal British Columbia lakes. *Journal of the Fisheries Research Board of Canada* 28: 1259-1268.

Angermeier, P. L. 1987. Spatiotemporal variation in habitat selection by fishes in small Illinois streams, in *Community and Evolutionary Ecology in North American Stream Fishes*, W. J. Matthews and D. C. Heins, ed. University of Oklahoma Press, Norman.

Angermeier, P. L. 1997. Conceptual rules of biological integrity and diversity, pp. 49-65 in *Watershed Restoration: Principles and Practices*, J. E. Williams, C. A. Wood, and M. P. Dombeck, ed. American Fisheries Society, Bethesda MD.

Benda, L. E. 1994. *Stochastic Geomorphology in a Humid Mountain Landscape*. PhD dissertation, Department of Geological Sciences, University of Washington, Seattle.

Benda, L. E., and T. Cundy. 1990. Predicting deposition of debris flows in mountain channels. *Canadian Geotechnical Journal* 27: 409-417.

Benda, L. E., D J. Miller, T. Dunne, G. H. Reeves, and J. K. Agee. 1998. Dynamic landscape systems, pp. 261-288 in *River Ecology and Management: Lessons from the Pacific Coastal Ecoregion*, R. J. Naiman and R. E. Bilby, ed. Springer-Verlag, New York.

Bender, E. A., T. J. Case, and M. E. Gilpin. 1984. Perturbation experiments in community ecology: theory and practice. *Ecology* 65: 1-13.

Benner, P. A. 1992. Historical reconstruction of the Coquille River and surrounding landscape, sections 3.2-3.3 in *The Action Plan for the Oregon Coastal Watershed, Estuaries, and Ocean Waters*. Near Coastal Waters Pilot Project, Oregon Department of Environmental Quality, Portland.

Berkman, H., and C. F. Rabeni. 1987. Effect of siltation on stream fish communities. *Environmental Biology of Fishes* 18: 285-294.

Bilby, R. E., and J. W. Ward. 1989. Changes in characteristics and function of woody debris with increasing size of streams in Washington. *Canadian Journal of Fisheries and Aquatic Sciences* 48: 2499-2508.

Bilby, R. E., B. R. Fransen, and P. A. Bisson. 1996. Incorporation of nitrogen and carbon from spawning coho salmon into the trophic system of small streams: evidence from stable isotopes. *Canadian Journal of Fisheries and Aquatic Sciences* 53: 164-173.

Bilby, R. E., B. R. Fransen, P. A. Bisson, and J. K. Walter. 1998. Response of juvenile coho salmon and steelhead to the addition of salmon carcasses in two streams in southwestern Washington, USA. *Canadian Journal of Fisheries and Aquatic Sciences* 55: 1909-1918.

Bisson, P. A., and P. E. Reimers. 1977. Geographic variation among Pacific Northwest populations of the longnose dace (*Rhinichthys cataractae*). *Copeia* 1977: 518-522.

Bisson, P. A., and J. R. Sedell. 1984. Salmonids populations in streams in clearcut vs. old-growth forests of western Washington, pp. 121-129 in *Fish and Wildlife Relationships in Old-growth Forests: Proceedings of a Symposium*, W. R. Meehan, T. R. Merrell Jr., and T. A. Hanley, ed. American Institute of Fishery Biologists, Morehead City NC.

Bisson, P. A., J. L. Nielsen, R. A. Palmason, and L. E. Grove. 1982. A system of naming habitat types in small streams, with examples of habitat utilization by salmonids during low stream flow, pp. 62-73 in *Acquisition and Utilization of Aquatic Habitat Inventory Information*, N. B. Armantrout, ed. Western Division, American Fisheries Society, Portland OR.

Bisson, P. A., R. E. Bilby, M. D. Bryant, C. A. Dolloff, G. B. Grette, R. A. House, M. L. Murphy, K. V. Koski, and J. R. Sedell. 1987. Large woody debris in forested streams in the Pacific Northwest: past, present, and future, pp. 143-190 in *Streamside Management: Forest and Fishery Interactions*, E. O. Salo and T. W. Cundy, ed. Contribution 57, Institute of Forest Resources, University of Washington, Seattle.

Bisson, P. A., K. Sullivan, and J. L. Nielsen. 1988. Channel hydraulics, habitat use, and body form of juvenile coho salmon, steelhead trout, and cutthroat trout. *Transactions of the American Fisheries Society* 117: 262-273.

Bisson, P. A., T. P. Quinn, G. H. Reeves, and S. V. Gregory. 1992. Best management practices, cumulative effects, and long-term trends in fish abundance in Pacific Northwest river systems, pp. 189-232 in *Watershed Management: Balancing Sustainability and Environmental Change*, R. J. Naiman, ed. Springer-Verlag, New York.

Bond, C. E. 1963. *Distribution and Ecology of Freshwater Sculpins, Genus Cottus, in Oregon*. PhD dissertation, University of Michigan, Ann Arbor.

Boschung, H. 1987. Physical factors and the distribution and abundance of fishes in the Upper Tombigbee River system of Alabama and Mississippi, with emphasis on the Tennessee-Tombigbee Waterway, pp. 178-183 in *Community and Evolutionary Ecology in North American Stream Fishes*, W. J. Matthews and D. C. Heins, ed. University of Oklahoma Press, Norman.

Botkin, D. B., K. W. Cummins, T. Dunne, H. Regier, M. Sobel, L. Talbot, and L. Simpson. 1995. *Status and Future of Salmon of Western Oregon and Northern California: Findings and Options*. Report #8, Center for the Study of the Environment, Santa Barbara CA.

Caraher, D. L., A. C. Zack, and A. R. Stage. 1999. Scales and ecosystem analysis, pp. 343-352 in *Ecological Stewardship: A Common Reference for Ecosystem Management, Volume II*, R. C. Szaro, N. C. Johnson, W. T. Sexton, and A. J. Malk, ed. Elsevier Science, Ltd., Oxford UK.

Cederholm, C. J., D. B. Houston, D. L. Cole, and W. J. Scarlett. 1989. Fate of coho salmon (*Oncorhynchus kisutch*) carcasses in spawning streams. *Canadian Journal of Fisheries and Aquatic Sciences* 46: 1347-1355.

Cederholm, C. J., R. Bilby, P. A. Bisson, T. W. Bunstead, B. R. Fransen, W. J. Scarlett, and J. W. Ward. 1997. Response of juvenile coho salmon and steelhead to placement of large woody debris in a coastal Washington stream. *North American Journal of Fisheries Management* 17: 947-963.

Chapman, D. W. 1965. Net production of juvenile coho salmon in three Oregon streams. *Transactions of the American Fisheries Society* 94: 40-52.

Chapman, D. W. 1966. Food and space as regulators of salmonid populations in streams. *American Naturalist* 100: 345-357.

Cissell, J. H., F. J. Swanson, G. E. Grant, D. H. Olson, S. V. Gregory, S. L. Garman, L. R. Ashkenas, M. G. Hunter, J. A. Kertis, J. H. Mayo, M. D. McSwain, S. G. Swetland, K. A. Swindle, and D. O. Wallin. 1998. *A Landscape Plan Based on Historical Fire Regimes for a Managed Ecosystem: The Augusta Creek Study*. General Technical Report PNW-GTR 422, USDA Forest Service, Pacific Northwest Research Station, Portland OR.

Concannon, J. A., C. L. Shafer, R. L. DeVelice, R. M. Sauvajot, S. L. Boudreau, T. E. Demeo, and J. Dryden. 1999. Describing landscape diversity: a fundamental tool for landscape management, pp. 195-218 in *Ecological Stewardship: A Common Reference for Ecosystem Management, Volume II*, R. C. Szaro, N. C. Johnson, W. T. Sexton, and A. J. Malk, ed. Elsevier Science, Ltd. Oxford UK.

Coronado, C., and R. Hilborn. 1998. Spatial and temporal factors affecting survival in coho and fall chinook salmon in the Pacific Northwest. *Bulletin of Marine Science* 62(2): 409-425.

Crispin, V., R. House, and D. Roberts. 1993. Changes in habitat, large woody debris, and salmon habitat after restructuring of a coastal Oregon stream. *North American Journal of Fisheries Management* 13: 96-102.

Cupp, C. E. 1989. *Identifying Spatial Variability of Stream Characteristics Through Classification*. MS thesis, University of Washington, Seattle.

Dambacher, J. 1991. *Distribution, Abundance, and Emigration of Juvenile Steelhead Trout (*Oncorhynchus mykiss*) and Analysis of Stream Habitat in the Steamboat Creek Basin, Oregon*. MS thesis, Oregon State University, Corvallis.

Dewberry, C., and B. Doppelt. 1996. Ecosystem-based in-stream channel modification from the whole watershed perspective, pp. 163-178 in *Healing the Watershed*, Second edition. The Pacific Rivers Council, Eugene OR.

Dill, L. M., R. C. Ydenberg, and A. H. G. Fraser. 1981. Food abundance and territory size in juvenile coho salmon (*Oncorhynchus kistuch*). *Canadian Journal of Zoology* 59: 1801-1809.

Dolloff, C. A. 1986. Effects of stream cleaning on juvenile coho salmon and Dolly Varden in southeast Alaska. *Canadian Journal of Fisheries and Aquatic Sciences* 47: 2297-2306.

Doppelt, B., J. E. Smith, R. C. Wissmar, and C. D. Williams. 1996. What is a watershed analysis? pp. 79-104 in *Healing the Watershed*, Second edition. The Pacific Rivers Council, Eugene OR.

Edwards, R. T. 1998. The hyporheic zone, pp. 399-429 in *River Ecology and Management: Lessons from the Pacific Coastal Ecoregion*, R. J. Naiman and R. E. Bilby, ed. Springer-Verlag, New York.

Elliot, S. T. 1986. Reduction of Dolly Varden and macrobenthos after removal of logging debris. *Transactions of the American Fisheries Society* 115: 392-400.

Evans, J. W., and R. L. Noble. 1979. Longitudinal distribution of fishes in an east Texas stream. *American Midland Naturalist* 101: 333-343.

Everest, F. H. 1977. *Ecology and Management of Summer Steelhead in the Rogue River, Oregon*. Fishery Research Report 7, Oregon State Game Commission, Corvallis.

Everest, F. H., and D. W. Chapman. 1972. Habitat selection and spatial interaction by juvenile chinook salmon and steelhead trout in two Idaho streams. *Journal of the Fisheries Research Board of Canada* 29: 91-100.

Fausch, K. D., and R. J. White. 1981. Competition between brook trout (*Salvelinus fontinalis*) and brown trout (*Salmo trutta*) for position in a Michigan stream. *Canadian Journal of Fisheries and Aquatic Sciences* 38: 1220-1227.

FEMAT (Forest Ecosystem Management Assessment Team). 1993. *Forest Ecosystem Management: An Ecological, Economic, and Social Assessment*. Report of the Forest Ecosystem Management Assessment Team, U.S. Government Printing Office 1973-793-071. U.S. Government Printing Office for USDA Forest Service; USDI Fish and Wildlife Service, Bureau of Land Management, and Park Service; U.S. Department of Commerce, National Oceanic and Atmospheric Administration and National Marine Fisheries Service; and the U.S. Environmental Protection Agency. Portland OR.

Fieth, S., and S. V. Gregory. 1993. *Influence of Habitat Complexity and Food Availability on the Fish Community in an Oregon Coastal Stream*. COPE Report 6(2): 6-8. Oregon State University, Corvallis.

Franklin, J. F. and R. T. T. Forman. 1987. Creating landscape patterns by forest cutting: ecological consequences and principles. *Landscape Ecology* 1: 5-18.

Frissell, C. A. 1992. *Cumulative Effects of Land Use on Salmon Habitat in Southwest Oregon Coastal Streams*. PhD dissertation, Department of Fisheries and Wildlife, Oregon State University, Corvallis.

Frissell, C. A. 1996. A new strategy for watershed protection, restoration, and recovery of wild native fish in the Pacific Northwest, pp. 1-25 in *Healing the Watershed*, Second edition. The Pacific Rivers Council, Eugene OR.

Frissell, C. A. 1997. Ecological principles, pp. 96-115 in *Watershed Restoration: Principles and Practices*, J. E. Williams, C. A. Wood, and M. P. Dombeck, ed. American Fisheries Society, Bethesda MD.

Frissell, C. A., and R. K. Nawa. 1992. Incidence and causes of physical failure of artificial fish habitat structures in streams of western Oregon and Washington. *North American Journal of Fisheries Management* 12: 182-197.

Frissell, C. A., W. J. Liss, C. E. Warren, and M. D. Hurley. 1986. A hierarchial framework for stream habitat classification: viewing streams in a watershed context. *Environmental Management*: 10: 199-214.

Fulton, J. D., and R. J. LaBrasseur. 1985. Interannual shifting of the Subartic Boundary and some of the biotic effects on juvenile salmonids, pp. 237-247 in *El Niño North: Niño Effects in the Eastern Subartic Pacific*, W. S. Wooster and D. L. Fluharty, ed. Washington Sea Grant Program, Seattle.

Glova, G. J. 1978. *Pattern and Mechanism of Resource Partitioning between Stream Populations of Juvenile Coho Salmon* (Oncorhynchus kisutch) *and Coastal Cutthroat Trout* (Salmo clarki clarki). PhD dissertation, University of British Columbia, Vancouver.

Glova, G. J. 1986. Interaction for food and space between juvenile coho salmon (*Oncorhynchus kisutch*) and coastal cutthroat trout (*Salmo clarki*) in a laboratory stream. *Hydrobiologia* 131: 155-168.

Gosz, J. R., J. Asher, B. Holder, R. Knight, R. Naiman, G. Raines, P. Stine, and T. B. Wigley. 1999. An ecosystem approach for understanding landscape diversity, pp. 157-194 in *Ecological Stewardship: A Common Reference for Ecosystem Management, Volume II*, R. C. Szaro, N. C. Johnson, W. T. Sexton, and A. J. Malk, ed. Elsevier Science, Ltd., Oxford UK.

Gotelli, N. J. 1991. Metapopulation models: the rescue effect, the propagule rain, and the core-satellite hypothesis. *American Naturalist* 138: 768-776.

Grant, G. E. 1990. Hydrologic, geomorphic, and aquatic habitat implications of old and new forestry, pp. 35-53 in *Proceeding of Symposium Forests Managed and Wild: Differences and Consequences*, A. F. Pearson and D. A. Challenger, ed. University of British Columbia, Vancouver BC.

Grant, G. E., F. J. Swanson, and M. G. Wolman. 1990. Pattern and origin of stepped-bed morphology in high-gradient streams, western Cascades, Oregon. *Geological Society of America Bulletin* 102: 340-352.

Gregory, S. V., G. A. Lamberti, and K. M. S. Moore. 1989. Influence of valley floor land forms on stream ecosystems. pp. 3-8 in *Proceedings of the California Riparian Systems Conference: Protection, Management, and Restoration for the 1990s, Conference held 22-24 September 1988, Davis, California*, D. L. Abell, ed. General Technical Report PSW-110, USDA Forest Service, Pacific Southwest Forest and Range Experiment Station, Berkeley CA.

Grimm, N. B., and S. G. Fisher. 1984. Exchange between interstitial and surface water: implications for stream metabolism and nutrient cycling. *Hydrobiologia* 111: 219-228.

Griswold, K. E. 1996. *Genetic and Meristic Relationships of Coastal Cutthroat Trout* (Oncorhynchus clarkii clarkii) *Above and Below Barriers in Two Coastal Basins*. MS thesis, Oregon State University, Corvallis.

Gross, M. R., R. M. Coleman, and R. M. McDowell. 1988. Aquatic productivity and the evolution of diadromous fish migration. *Science* 239: 1291-1293.

Grunbaum, J. B. 1996. *Geographical Variation in Diel Habitat Use by Juvenile (Age 1+) Steelhead Trout* (Oncorhynchus mykiss) *in Oregon Coastal and Inland Streams*. MS thesis, Oregon State University, Corvallis.

Hales, D. 1989. Changing concepts of national parks, pp. 139-149 in *Conservation for the Twenty-first Century*, D. Western and M. Pearl, ed. Oxford University Press, New York.

Hansen, A. J., and D. L. Urban. 1992. Avian response to landscape patterns: the role of species life histories. *Landscape Ecology* 7: 163-180.

Hare, S. R., N. J. Mantua, and R. C. Francis. 1999. Inverse production regimes: Alaska and west coast Pacific salmon. *Fisheries* 24(1): 6-14.

Harrison, S., and J. F. Quinn. 1989. Correlated environments and the persistence of metapopulations. *Oikos* 56: 293-298.

Hartman, G. F. 1965. The role of behavior in the ecology and interaction of underyearling coho salmon (*Oncorhynchus kisutch*) and steelhead trout (*Salmo gardneri*). *Journal of the Fisheries Research Board of Canada* 22: 1035-1081.

Hartman, G. F. 1988. Preliminary comments on results of studies of trout biology and logging impacts in Carnation Creek, pp. 175-180 in *Proceedings of the Workshop Applying 15 years of Carnation Creek Results*, T. W. Chamberlain, ed. Pacific Biological Station, Carnation Creek Steering Committee, Nanaimo BC.

Hartman, G. F., and C. A. Gill. 1968. Distributions of juvenile steelhead and cutthroat trout (*Salmo gairdneri* and *S. clarki*) within streams in south-western British Columbia. *Journal of the Fisheries Research Board of Canada* 22: 33-48.

Hartman, G. F., and J. C. Scrivener. 1990. Impacts of forestry practices on a coastal ecosystem, Carnation Creek, British Columbia. *Canadian Bulletin of Fisheries and Aquatic Sciences* 223.

Hartman, G. F., J. C. Scrivener, and M. J. Mills. 1996. Impacts of logging on Carnation Creek, a high-energy stream in British Columbia, and their implications for restoring fish habitat. *Canadian Journal of Fisheries and Aquatic Sciences* 53(Supplement 1): 237-251.

Harvey, B. C., and R. J. Nakamoto. 1998. The influence of large wood debris on retention, immigration, and growth of coastal cutthroat trout (*Oncorhynchus clarkii clarkii*) in stream pools. *Canadian Journal of Fisheries and Aquatic Sciences* 55: 1902-1908.

Hawkins, C. P., J. L. Kershner, P. A. Bisson, M. D. Bryant, L. M. Decker, S. V. Gregory, D. A. McCollough, C. K. Overton, G. H. Reeves, R. J. Steedman, and M. K. Young. 1993. A hierarchial approach to classifying stream habitat features. *Fisheries* 18(6): 3-12.

Healey, M. C., and A. Prince. 1995. Scales of variation in life history tactics of Pacific salmon and the conservation of phenotype and genotype, pp. 176-184 in *Evolution and the Aquatic Ecosystem: Defining Unique Units in Population Conservation*, J. Nielsen, ed. American Fisheries Society Symposium 17, Bethesda MD.

Hicks, B. J., J. D. Hall, P. A. Bisson, and J. R. Sedell. 1991. Response of salmonids to habitat change, pp. 483-518 in *Influences of Forest and Rangeland Management on Salmonid Fishes and Their Habitats*, W. R. Meehan, ed. Special Publication Number 19, American Fisheries Society, Bethesda MD.

Hillman, T. 1991. *The Effect of Temperature on the Spatial Interaction of Juvenile Chinook Salmon and the Redside Shiner and their Morphological Differences*. PhD Dissertation, Idaho State University, Pocatello.

Hobbs, R. J., and L. .F Huenneke. 1992. Disturbance, diversity, and invasion: implications for conservation. *Conservation Biology* 6: 324-337.

Holling, C. S. 1973. Resilience and stability of ecological systems. *Annual Review of Ecological Systems* 4: 1-23.

Holtby, L. B. 1988. Effects of logging on stream temperatures in Carnation Creek, British Columbia and associated impacts on coho salmon (*Oncorhynchus kisutch*). *Canadian Journal of Fisheries and Aquatic Sciences* 45: 502-515.

Horowitz, R. J. 1978. Temporal variability patterns and the distributional patterns of stream fishes. *Ecological Monographs* 48: 307-321.

House, R. 1996. An evaluation of stream restoration structures in a coastal Oregon stream, 1981-1993. *North American Journal of Fisheries Management* 16: 272-281.

House, R. A., and P. L. Boehne. 1985. Evaluation of instream enhancement structures for salmonid spawning and rearing in a coastal Oregon stream. *North American Journal of Fisheries Management* 5: 283-295.

Hunter, M. L. 1996. *Fundamentals of Conservation Biology*. Blackwell Science, Cambridge MA.

Jones, M. L., R. G. Randall, D. Hayes, W. Dunlap, J. Imhof, G. Lacroix, and N. J. R. Ward. 1996. Assessing the ecological effects of habitat change: moving beyond productive capacity. *Canadian Journal of Fisheries and Aquatic Sciences* 53 (Supplement 1): 446-457.

Kondolf, G. M. 2000. Some suggested guidelines for geomorphic aspects of anadromous salmonid habitat restoration proposals. *Restoration Ecology* 8(1): 48-56.

Kralick, N. J., and J. E. Sowerwine. 1977. *The Role of Two Northern California Intermittent Streams in the Life History of Anadromous Salmonids*. MS thesis, Humboldt State University, Arcata CA

Lamberti, G. A., S. V. Gregory, L. R. Ashkenas, R. C. Wildman, and J. D. McIntire. 1989. Influence of channel geomorphology on retention of dissolved and particulate matter in a Cascade Mountain stream, pp. 33-39 in *Proceedings of the California Riparian Systems Conference: Protection, Management, and Restoration for the 1990s, Conference held 22-24 September 1988, Davis, California*, D. L. Abell, ed. General Technical Report PSW-110, USDA Forest Service, Pacific Southwest Forest and Range Experiment Station, Berkeley CA.

Larkin, P. A. 1956. Interspecific competition and population control in freshwater fish. *Journal of the Fisheries Research Board of Canada* 13: 327-342.

Lawson, P. W. 1993. Cycles of ocean productivity, trends in habitat quality, and the restoration of salmon runs in Oregon. *Fisheries* 18(8): 6-10.

Levin, S. 1974. Dispersion and population interactions. *American Naturalist* 108: 207-228.

Li, H. W., C. B. Schreck, C. E. Bond, and E. Rexstad. 1987. Factors influencing changes in fish assemblages of Pacific streams, pp. 193-202 in *Community and Evolutionary Ecology in North American Stream Fishes*, W. J. Matthews and D. C. Heins, ed. University of Oklahoma Press, Norman.

Lister, D. B., and H. S. Genoe. 1970. Stream habitat utilization by cohabitating underyearlying chinook salmon (*Oncorhynchus tshawytscha*) and coho salmon (*O. kisutch*) salmon in the Big Quilicum River, British Columbia. *Journal of the Fisheries Research Board of Canada* 27: 1215-1224.

Lonzarich, D. G., and T. P. Quinn. 1995. Experimental evidence for the effect of depth and structure on the distribution, growth, and survival of stream fishes. *Canadian Journal of Zoology* 73: 2223-2230.

Lugo, A. E., J. S. Baron, T. P. Frost, T. W. Cundy, and P. Dittberner. 1999. Ecosystem processes and functioning, pp. 219-254 in *Ecological Stewardship: A Common Reference for Ecosystem Management, Volume II*, R. C. Szaro, N. C. Johnson, W. T. Sexton, and A. J. Malk ed. Elsevier Science, Ltd., Oxford UK.

Mahon, R. 1984. Divergent structure in fish taxocenes of north temperate streams. *Canadian Journal of Fisheries and Aquatic Sciences* 41: 330-350.

Mantua, N. J., S. R. Hare, Y. Zhang, J. M. Wallace, and R. C. Francis. 1997. A Pacific interdecadal climate oscillation with impacts on salmon production. *Bulletin of the American Meteorological Society* 78: 1069-1079.

May, C. 1998. *Debris Flow Characteristics Associated with Forest Practices in the Central Oregon Coast Range*. MS thesis, Oregon State University, Corvallis.

McDade, M. H., F. J. Swanson, W. A. McKee, J. F. Franklin, and J. Van Sickle. 1990. Source distances for coarse woody debris entering small streams in western Oregon and Washington. *Canadian Journal of Forest Sciences* 20: 326-330.

McGarry, N. 1994. A Quantitative Analysis and Description of the Delivery and Distribution of Large Woody Debris in Cummins Creek, Oregon. MS thesis, Oregon State University, Corvallis.

McGurrin, J., and H. Forsgren. 1997. What works, what doesn't, and why, pp. 459-471 in *Watershed Restoration: Principles and Practices*, J. E. Williams, C. A. Wood, and M. P. Dombeck, ed. American Fisheries Society, Bethesda MD.

McIntosh, B. A., J. R. Sedell, R. F. Thurow, S. E. Clarke, and G. L. Chandler. 2000. Historical changes in pool habitats in the Columbia River Basin. *Ecological Applications* 10: 1478-1496.

McMahon, T. E., and G. F. Hartman. 1989. Influence of cover complexity and current velocity on winter habitat use by juvenile coho salmon (*Oncorhynchus kisutch*). *Canadian Journal of Fisheries and Aquatic Sciences* 46: 1551-1557.

McMahon, T. E., and L. B. Holtby. 1992. Behavior, habitat use, and movements of coho salmon (*Oncorhynchus kisutch*) during seaward migration. *Canadian Journal of Fisheries and Aquatic Sciences* 49: 1478-1485.

Miller, A. J., D. R. Cayan, T. P. Barnett, N. E. Graham, and J. M. Oberhuber. 1994. The 1976-77 climate shift in the Pacific Ocean. *Oceanography* 7: 21-26.

Minckley, W. L., D. A. Hendrickson, and C. E. Bond. 1986. Geography of western North American freshwater fishes: description and relationship to intracontinental tectonism, pp. 519-614 in *The Zoogeography of North American Freshwater Fishes*, C. H. Hocutt and E. O. Wiley, ed. John Wiley & Sons, New York NY.

Minshall, G. W., J. T. Brock, and J. D. Varley. 1989. Wildfires and Yellowstone's stream ecosystems. *BioScience* 39: 707-715.

Montgomery, D. R., and J. M. Buffington. 1997. Channel-reach morphology in mountain drainage basins. *Geological Society of American Bulletin* 109(5): 596-611.

Moore, K. M. S. and S. V. Gregory. 1988. Summer habitat utilization and ecology of cutthroat fry (*Salmo clarkii*) in Cascade Mountain streams. *Canadian Journal of Fisheries and Aquatic Sciences* 45: 1921-1930.

Moyle, P. B., and R. M. Yoshiyama. 1994. Protection of aquatic biodiversity in California: a five-tiered approach. *Fisheries* 19(2): 6-19.

Mundie, J. H. 1969. Ecological implications of the diet of juvenile coho salmon in streams, pp. 135-152 in *Symposium on Salmon and Trout in Streams*, TG Northcote, ed. HR MacMillan Lectures in Fisheries, University of British Columbia, Vancouver.

Murphy, M. L., and J. D. Hall. 1980. Varied effects of clearcut logging on predators and their habitat in small streams of the Cascade Mountains, Oregon. *Canadian Journal of Fisheries and Aquatic Sciences* 38: 137-145.

Murphy, M. L., and K. V. Koski. 1989. Input and depletion of woody debris in Alaska streams and implications for streamside management. *North American Journal of Fisheries Management* 9: 427-436.

Murphy, M. L., J. Heifetz, S. W. Johnson, K. V. Koski, and J. F. Thedinga. 1986. Effects of clearcut logging with and without buffer strips on juvenile salmonids in Alaskan streams. *Canadian Journal of Fisheries and Aquatic Sciences* 43: 1521-1533.

Naiman, R. J., T. J. Beechie, L. E. Benda, D. R. Berg, P. A. Bisson, L. H. McDonald, M. D. O'Conner, P. L. Olson, and E. A. Steel. 1992. Fundamental elements of ecologically healthy watersheds in the Pacific Northwest coastal ecoregion, pp. 127-188 in *Watershed Management: Balancing Sustainability and Environmental Change*, R. J. Naiman, ed. Springer-Verlag, New York NY.

National Research Council. 1996. *Upstream: Salmon and Society in the Pacific Northwest*. National Academy Press, Washington DC.

Nehlsen, W., J. E. Williams, and J. A. Lichatowich. 1991. Pacific salmon at the crossroads: stocks at risk from California, Oregon, Idaho, and Washington. *Fisheries* 16(2): 4-21.

Nicholas, J. W. and D. G. Hankin. 1988. *Chinook Salmon Populations in Oregon Coastal River Basins: Life Histories and Assessment of Recent Trends in Run Strengths*. Information Report 88-1, Oregon Department of Fish and Wildlife, Portland.

Nickelson, T. E. 1986. Influences of upwelling, ocean temperature, and smolt abundance on marine survival of coho salmon (*Oncorhynchus kisutch*) in the Oregon production area. *Canadian Journal of Fisheries and Aquatic Sciences* 43: 527-535.

Nickelson, T. E., J. D. Rodgers, S. L. Johnson, and M. F. Solazzi. 1992a. Seasonal changes in habitat use by juvenile coho salmon (*Oncorhynchus kisutch*) in Oregon coastal streams. *Canadian Journal of Fisheries and Aquatic Sciences* 49: 783-789.

Nickelson, T. E., M. F. Solazzi, S. L. Johnson, and J. D. Rogers. 1992b. Effectiveness of selected stream improvement techniques to create suitable summer and winter rearing habitat for juvenile coho salmon (*Oncorhynchus kisutch*) in Oregon coastal streams. *Canadian Journal of Fisheries and Aquatic Sciences* 49: 783-789.

Nilsson, N. A. 1967. Interactive segregation between fish species, pp. 295-313 in *The Biological Basis of Freshwater Fish Production*, S. D. Gerking, ed. Blackwell Scientific, Oxford UK.

Nilsson, N. A., and T. G. Northcote. 1981. Rainbow trout (*Salmo gairdneri*) and cutthroat trout (*S. clarki*) interactions in coastal British Columbia lakes. *Canadian Journal of Fisheries and Aquatic Sciences* 38: 1228-1246.

O'Neil, M. P., and A. D. Abrahams. 1987. Objective identification of pools and riffles. *Water Resources* 20: 921-926.

Pearcy, W. G. 1992. *Ocean Ecology of North Pacific Salmonids*. University of Washington Press, Seattle.

Peterson, N. P. 1982. Immigration of juvenile coho salmon (*Oncorhynchus kisutch*) into riverine ponds. *Canadian Journal of Fisheries and Aquatic Sciences* 39: 1308-1310.

Quinn, T. P. 1984. Homing and straying in Pacific salmon, pp. 357-362 in *Mechanisms of Migration in Fish*, J. D. McCleave, G. P. Arnold, J. J. Dodson, and W. H. Neil, ed. Plenum Press, New York NY.

Quinn, T. P., and N. P. Petersen. 1996. The influence of habitat complexity and fish size on over-winter survival and growth of individually marked juvenile coho salmon (*Oncorhynchus kisutch*) in Big Beef Creek, Washington. *Canadian Journal of Fisheries and Aquatic Sciences* 53: 1555-1564.

Ralph, S. C., G. C. Poole, L. L. Conquest, and R. J. Naiman. 1994. Stream channel morphology and woody debris in logged and unlogged basins of western Washington. *Canadian Journal of Fisheries and Aquatic Sciences* 51: 37-51.

Reeves, G. H. 1978. *Population Dynamics of Juvenile Steelhead Trout in Relation to Density and Habitat Characteristics*. MS thesis, Humboldt State University, Arcata CA.

Reeves, G. H., F. H. Everest, and J. D. Hall. 1987. Interactions between the redside shiner (*Richardsonius balteatus*) and the steelhead trout (*Salmo gairdneri*): the influence of water temperature. *Canadian Journal of Fisheries and Aquatic Sciences* 43: 1521-1533.

Reeves, G. H., J. D. Hall, T. D. Roelofs, T. L. Hickman, and C. O. Baker. 1991. Habitat restoration, in *Influences of Forest and Rangeland Management on Salmonid Fishes and Their Habitats*, WR Meehan, ed. Special Publication Number 19, American Fisheries Society, Bethesda, MD.

Reeves, G. H., F.H. Everest, and J. R. Sedell. 1993. Diversity of juvenile anadromous salmonids assemblages in coastal Oregon basins with different levels of timber harvest. *Transactions of the American Fisheries Society* 122: 309-317.

Reeves, G. H., L. E. Benda, K. M. Burnett, P. A. Bisson, and J. R. Sedell. 1995. A disturbance-based ecosystem approach to maintaining and restoring freshwater habitats of evolutionarily significant units of anadromous salmonids in the Pacific Northwest. *American Fisheries Society Symposium* 17: 334-349.

Reeves, G. H., J. D. Hall, and S. V. Gregory. 1997a. The impact of land-management activities on coastal cutthroat trout and their freshwater habitats, pp. 138-144 in *Sea-run Cutthroat Trout: Biology, Management, and Future Considerations*, J. D. Hall, P. A. Bisson, and R. E. Gresswell, ed. Oregon Chapter, American Fisheries Society, Corvallis.

Reeves, G. H., D. B. Hohler, B. E. Hansen, F. H. Everest, J. R. Sedell, T. L. Hickman, and D. Shively. 1997b. Fish habitat restoration in the Pacific Northwest: Fish Creek, pp. 335-359 in *Watershed Restoration: Principles and Practices*, J. E. Williams, C. A. Wood, and M. P. Dombeck, ed. American Fisheries Society, Bethesda MD.

Reeves, G. H., P. A. Bisson, and J. M. Dambacher. 1998. Fish communities, pp. 200-234 in *River Ecology and Management: Lessons from the Pacific Coastal Ecoregion*, R. J. Naiman and R. E. Bilby, ed. Springer-Verlag, New York.

Reimers, P. E. 1973. *The Length of Residence of Juvenile Fall Chinook Salmon in the Sixes River, Oregon*. Research Reports of the Fish Commission of Oregon, Volume 4, Portland.

Ricker, W. E. 1989. History and present state of odd-year pink salmon runs of the Frasier River region. *Canadian Technical Report Fisheries and Aquatic Sciences* 1702: 1-37.

Robinson, E. G., K. Mills, J. Paul, L. Dent, and A. Skaugset. 1999. *Storm Impacts and Landslides of 1996: Final Report*. Forest Practices Technical Report Number 4. Forest Practices Monitoring Program, Oregon Department of Forestry, Salem.

Rodgers, J. D. 1986. *The Winter Distribution, Movement, and Smolt Transformation of Juvenile Coho Salmon in an Oregon Coastal Stream*. MS thesis, Oregon State University, Corvallis.

Rosgen, D. L. 1994. A classification of natural rivers. *Catena* 22: 169-199.

Schlosser, I. J. 1987. A conceptual framework for fish communities in small warmwater streams, pp. 17-24 in *Community and Evolutionary Ecology in North American Stream Fishes*, W. J. Matthews and D. C. Heins, ed. University of Oklahoma Press, Norman.

Schoonmaker, P., and B. von Hagen (editors). 1996. *The Rain Forests of Home: An Exploration of People and Place*. Island Press, Washington DC.

Schwartz, J. S. 1990. *Influence of Geomorphology and Land Use on Distribution and Abundance of Salmonids in a Coastal Oregon Basin*. MS thesis, Oregon State University, Corvallis.

Scott, J. B., C. R. Steward, and Q. J. Stober. 1986. Effects of urban development on fish population dynamics in Kelsey Creek, Washington. *Transactions of the American Fisheries Society* 115: 555-567.

Scrivener, J. C., and B. C. Anderson. 1982. Logging impacts and some mechanisms which determine the size of spring and summer populations of coho fry in Carnation Creek, pp. 257-272 in *Proceedings of the Workshop Applying 15 Years of Carnation Creek Results*, T. W. Chamberlain, ed. Pacific Biological Station, Carnation Creek Steering Committee, Nanaimo BC.

Scrivener, J. C., and M. J. Brownlee. 1989. Effects of forest harvesting on gravel quality and incubation survival of chum salmon (*Oncorhynchus keta*) and coho salmon (*Oncorhynchus kisutch*) in Carnation Creek, British Columbia. *Canadian Journal of Fisheries and Aquatic Sciences* 46: 681-696.

Sedell, J. R., and J. L. Froggatt. 1984. Importance of streamside forests to large rivers: the isolation of the Willamette River, USA, from its floodplain by snagging and streamside forest removal. *Vereinigung fur Theoretishe Limnologie Verhandlungen* 22: 1828-1834.

Sheldon, A. L. 1968. Species diversity and longitudinal succession in stream fishes. *Ecology* 49: 193-198.

Sleeper, J. D. 1993. *Seasonal Changes in Distribution and Abundance of Salmonids and Habitat Availability in a Coastal Oregon Basin*. MS thesis, Oregon State University, Corvallis.

Smith, G. R. 1981. Late Cenozoic freshwater fishes of North America. *Annual Review of Ecology and Systematics* 12: 163-193.

Snyder, R. S., and H. Dingle. 1989. Adaptive, genetically based differences in life history between estuary and freshwater three spine sticklebacks (*Gasterosteus aculeatus*). *Canadian Journal of Zoology* 67: 2448-2454.

Solazzi, M. F., T. E. Nickelson, S. L. Johnson, and J. D. Rodgers. 2000. Effects of increasing winter rearing habitat on abundance of salmonids in two coastal Oregon streams. *Canadian Journal of Fisheries and Aquatic Sciences* 57: 906-914.

Spence, B. C., G. A. Lomnicky, R. M. Hughes, and R. P. Novitzski. 1996. *An Ecosystem Approach to Salmonid Conservation*. TR4501-96-6057. ManTech Environmental Research Corp., Corvallis OR.

Stouder, D. J., P. A. Bisson, and R. J. Naiman. 1997. Where are we? Resources at the brink, pp. 1-12 in *Pacific Salmon and Their Ecosystems: Status and Future Options*, D. J. Stouder, P. A. Bisson, and R. J. Naiman, ed. Chapman & Hall, New York.

Strahler, A. N. 1957. Quantitative analysis of watershed geomorphology. *American Geophysical Union Transactions* (EOS) 38: 913-920.

Swain, D. P., and L. B. Holtby. 1989. Differences in morphology and behavior between juvenile coho salmon (*Oncorhynchus kisutch*) rearing in a lake and in its tributary stream. *Canadian Journal of Fisheries and Aquatic Sciences* 46: 1406-1414.

Tallman, R. F., and M. C. Healey. 1994. Homing, straying, and gene flow among seasonally separated populations of chum salmon (*Oncorhynchus keta*). *Canadian Journal of Fisheries and Aquatic Sciences* 51: 577-588.

Toth, L. A., D. A. Arrington, and G. Begue. 1997. Headwater restoration and reestablishment of natural flow regimes: Kissimmee River of Florida, pp. 425-444 in *Watershed Restoration: Principles and Practices*, J. E. Williams, C. A. Wood, and M. P. Dombeck, ed. American Fisheries Society, Bethesda MD.

Triska, F. J., V. C. Kennedy, R. J. Avanzino, G. W. Zellwegar, and K. E. Bancala. 1989. Retention and transport of nutrients in a third-order stream in northwestern California: hyporheic processes. *Ecology* 70: 1893-1905.

Tschaplinski, P. J., and G. F. Hartman. 1983. Winter distribution of juvenile coho salmon (*Oncorhynchus kisutch*) before and after logging in Carnation Creek, British Columbia, and some implications for overwinter survival. *Canadian Journal of Fisheries and Aquatic Sciences* 40: 452-461.

van de Wetering, S. J. 1998. *Aspects of Life History Characteristics and Physiological Aspects of Smolting in Pacific Lamprey*, Lampetra tridenta, *in central Oregon Coast Stream*. MS thesis, Oregon State University, Corvallis.

Vannote, R. L., G. W. Minshall, K. W. Cummins, J. R. Sedell, and C. E. Cushing. 1980. The river continuum concept. *Canadian Journal of Fisheries and Aquatic Sciences* 37: 130-137.

White, P. S., and S. P. Bratton. 1980. After preservation: the philosophical and practical problems of change. *Biological Conservation* 18: 241-255.

White, P. S., and S. T. A. Pickett. 1985. Natural disturbance and patch dynamics: an introduction, pp. 3-13 in *The Ecology of Natural Disturbance and Patch Dynamics*, S. T. A. Pickett and P. S. White, ed. Academic Press, Orlando FL.

Williams, J. E., J. E. Johnson, D. A. Henderson, S. Contreras-Balderas, J. D. Williams, M. Navarro-Mendoza, D. E. McAllister, and J. E. Deacon. 1989. Fishes of North America, endangered, threatened, and of special concern. *Fisheries* 14(6): 2-21.

Willson, M. F., and K. C. Halupka. 1995. Anadromous fish as keystone species in vertebrate communities. *Conservation Biology* 9: 489-497.

Wilzbach, M. A. 1985. Relative roles of food abundance and cover determining habitat distribution of stream dwelling cutthroat trout (*Salmo clarki*). *Canadian Journal of Fisheries and Aquatic Sciences* 42: 1668-1672.
Wimberly, M. C., T. A. Spies, C. J. Long, and C. Whitlock. 2000. Simulating historical variability in the amount of old forests in the Oregon Coast Range. *Conservation Biology* 14: 167-180.
Wipfli, M. S., J. Hudson, D. T. Chanoler, and J. P. Caouette. 1999. Influence of salmon spawner densities on stream productivity in Southeast Alaska. *Canadian Journal of Fisheries and Aquatic Sciences* 56: 1600-1611.
Yount, J. D., and G. J. Niemi. 1990. Recovery of lotic communities and ecosystems from disturbance-a narrative review of case studies. *Environmental Management* 14: 547-570.
Zirges, M. H. 1973. *Morphological and Meristic Characteristics of Ten Populations of Blackside Dace,* Rhinichthys osculus nubilus *(Girad) from Western Oregon*. MS thesis, Oregon State University, Corvallis.
Zucker, S. J. 1993. *Influence of Channel Constraint on Primary Production, Periphyton Biomass, and Macroinvertebrate Biomass in Streams of the Oregon Coast Range.* MS thesis, Oregon State University, Corvallis.

Ecology and Management of Wildlife and Their Habitats in the Oregon Coast Range

John P. Hayes and Joan C. Hagar

The Coast Range Supports a Diversity of Animal Species

The forests and streams of the Oregon Coast Range (including the Siskiyou Mountains) are home to more than 200 species of wildlife: 19 species of amphibians, 16 species of reptiles, 65 species of mammals, and more than 100 species of birds (Figure 5-1). Thirty-five of the species that currently or recently inhabited the Oregon Coast Range are considered threatened, endangered, or of special concern (Table 5-1). This tally excludes species that are not native, that do not breed in Oregon, and that inhabit primarily nonforested areas. Although many of the wildlife species that inhabit the Oregon Coast Range are geographically widely distributed, a few species, such as Dunn's salamander (*Plethodon dunni*), Del Norte salamander (*Plethodon elongatus*), Baird's shrew (*Sorex bairdi*), Pacific shrew (*Sorex pacificus*), fog shrew (*Sorex sonomae*), white-footed vole (*Arborimus albipes*), and red tree vole (*Arborimus longicaudus*), are found primarily in the forests of the Oregon Coast Range. The maritime climate profoundly influences habitats of the Coast Range, and the Pacific Ocean also contributes some unusual species to the forest fauna. For example, marbled murrelets (*Brachyramphus marmoratus*) and harlequin ducks (*Histrionicus histrionicus*) are seabirds that feed and winter in coastal waters, but reproduce in forests. A broad range of natural variability in habitat types is largely responsible for the diversity of Coast Range wildlife, because each species has unique habitat requirements.

The assemblages and distributions of wildlife species found in the Oregon Coast Range today are the result of both historical and current influences. Natural disturbances, such as wind, pathogens, and fire, have created a variety of habitat types and conditions across a range of spatial and temporal scales: from tree fall gaps that are rapidly reoccupied by forest vegetation, to shrub fields that persist for decades following a fire, to old-growth forests that may have covered thousands of acres for centuries. Large-scale disturbance events, such as the fires that swept through the Coast Range in the mid-1800s, have greatly influenced today's forest structure (see Chapter 3). People have influenced the distribution and abundance of wildlife in the Oregon Coast Range through patterns of settlement, alteration of vegetative communities, hunting, trapping, and introduction of species. We have inherited the legacy of past practices, activities, and events, which provide the constraints and opportunities that drive

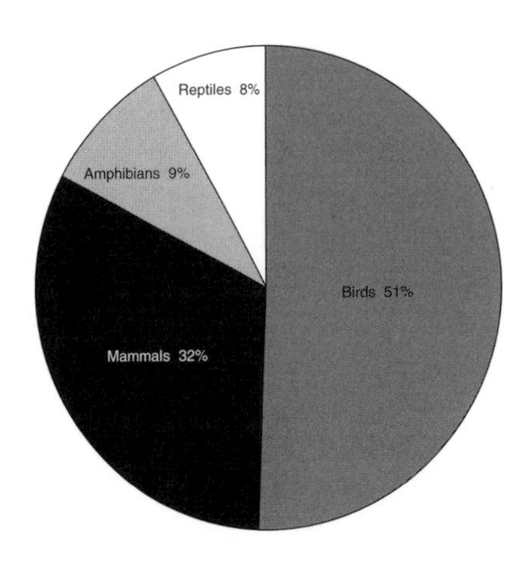

Figure 5-1. Composition of Coast Range wildlife by taxonomic group.

Table 5-1. Rare, threatened, endangered, and sensitive terrestrial, vertebrate wildlife species that use or have used forest habitats in western Oregon (adapted from Oregon Natural Heritage Program 1998).

Species	Federal Status[1]	State Status[2]	TNC Rank[3] Global	TNC Rank[3] State
Amphibians				
Clouded salamander	–	S	4	4
Tailed frog	SoC	S	4	3
California slender salamander	–	S	5	2
Western toad	–	S	4	4
Cope's giant salamander	–	S	3	2
Del Norte salamander	–	S	3	2
Northern red-legged frog	SoC	S	4	3
Foothill yellow-legged frog	SoC	S	3	3
Columbia torrent salamander	–	S	3	3
Southern torrent salamander	SoC	S	3	3
Reptiles				
Northwestern pond turtle	SoC	S	3	2
Sharptail snake	–	S	5	3
Birds				
Northern goshawk	SoC	S	5	3
Marbled murrelet	T	T	3	2
Olive-sided flycatcher	SoC	S	4	4
Pileated woodpecker	–	S	5	4
Willow flycatcher	SoC	S	5	U
American peregrine falcon	–	E	3	1
California condor	E	–	1	X
Bald eagle	T	T	4	3
Acorn woodpecker	–	–	5	3
Purple martin	–	S	5	3
Bank swallow	–	S	5	4
Black phoebe	–	–	5	3
Allen's hummingbird	–	–	5	3
Western bluebird	–	S	5	4
Northern spotted owl	T	T	3	3
Mammals				
White-footed vole	SoC	S	3	3
Ringtail	–	S	5	3
Gray wolf	E	E	4	X
Pacific western big-eared bat	SoC	S	4	3
California wolverine	SoC	T	3	2
Silver-haired bat	–	S	5	4
American marten	–	S	5	3
Pacific fisher	SoC	S	3	2
Long-eared bat	SoC	S	5	3
Fringed bat	SoC	S	5	3
Long-legged bat	SoC	S	5	3
Yuma bat	SoC	–	5	3
Columbian white-tailed deer	E	–	2	2
Western gray squirrel	–	S	5	4

today's management options. Through current management activities, we shape wildlife assemblages and their environments. Our actions create the habitats for future populations of wildlife that we pass down as an inheritance for future generations.

In recent years there has been increased attention to the influence of land-management activities on biodiversity. There are many reasons for this. For some, aesthetic considerations drive the concern over biodiversity; many people enjoy seeing animals and plants in their natural habitats. Others are concerned about biodiversity for ethical reasons, believing that humans have a moral obligation to conserve and protect other species. Biodiversity also has utilitarian functions; maintaining biodiversity is sometimes an important step to maintaining sustainable ecosystems that provide important services and products for humans. Much of the public interest and concern for biodiversity has focused on wildlife. Although this term is defined differently by different people, for the purposes of this book we will use a relatively narrow definition: by "wildlife" we mean vertebrates that spend part or all of their lives on land. Thus the wildlife of the Oregon Coast Range includes all of the mammals, birds, amphibians (salamanders and frogs), and reptiles (snakes, lizards, and turtles) occurring in the area (Figure 5-1).

Despite the heavy public emphasis on wildlife, it is but a small—although important—part of biodiversity. For example, of 407 species and groups of species identified in the Northwest Forest Plan as species of management concern (known as "survey and manage species"), 16 are vascular plants, 104 are lichens and bryophytes, 234 are fungi, 4 are "functional groups" of invertebrates (each group consisting of many species), and only 6 are wildlife as defined here (Del Norte salamander, Larch Mountain salamander [*Plethodon larselli*], Shasta salamander, Siskiyou Mountains salamander [*Plethodon stormi*], Van Dyke's salamander, and red tree vole) (USDA Forest Service and USDI Bureau of Land Management 1994). In this chapter, we focus almost exclusively on the wildlife of the Oregon Coast Range, their ecology, and their management. Some of the principles and considerations in this chapter apply to other taxa as well, but providing for wildlife is only one component of managing forests for biodiversity.

A key factor influencing the abundance and diversity of wildlife is the presence of suitable habitat. Habitat is the resources and conditions necessary to support presence, survival, and reproduction of a species through time. Although habitat is not synonymous with vegetation structure (Hall et al. 1997), habitat in forested environments clearly is influenced strongly by characteristics of vegetation. Other factors, especially interspecific interactions such as predation, competition, and mutualism, also affect the abundance and distribution of wildlife. Some species of wildlife have a strong influence on the ecology of an area, and this in turn can affect the ability of other species to persist there. Species that exert influences on the ecology of an area at a level that is disproportionate to their abundance (e.g., beaver) are known as keystone species (Power et al. 1996); they play critical roles in shaping the environment for other species. Human activities in the Oregon Coast Range forests, such as recreation, also directly influence wildlife populations (Knight and Cole 1991).

Habitat characteristics can be measured at very small spatial scales, such as the characteristics of an individual log, tree, branch, or cavity, to very large spatial scales, such as the characteristics of an entire watershed or ecoregion. Wildlife respond to characteristics of habitat at different scales; thus, understanding the influence of scale on wildlife and the interplay among spatial scales is critical to understanding the influence of forest management on wildlife populations. For example, at a small spatial scale, western red-backed voles (*Clethrionomys californicus*) are closely associated with large-diameter logs (Hayes and Cross 1987; Tallmon and Mills 1994; Thompson 1996) and deep duff layers of the soil (Rosenberg et al. 1994; Thompson 1996). At a larger spatial scale, western red-backed

Notes to Table 5-1

[1] Federal status – legal status designated by US Fish & Wildlife Service: '–' = no special status;

E = endangered; SoC = species of concern; T = threatened.

[2] State status – status designated by Oregon Department of Fish and Wildlife: '–' = status not assigned;

E = endangered; S = sensitive; T = threatened.

[3] TNC Rank – Natural Heritage Network Rank: 1 = Critically imperiled; 2 = Imperiled, vulnerable to extinction; 3 = Rare, uncommon, or threatened, but not immediately imperiled; 4 = Not rare but with cause for long-term concern; 5 = Demonstrably widespread, abundant, and secure; U = Unknown rank; X = Presumed extirpated or extinct.

voles generally will be absent from clearcuts or stands in very early stages of development, even if logs are abundant (Mills 1996). Characteristics of individual habitat components (such as individual snags or trees) and habitat conditions at the stand and landscape scales play fundamental roles in determining habitat suitability for wildlife in forested environments. Stand structure can be defined as "the physical and temporal distribution of trees and other plants in a stand" (Oliver and Larson 1996). Various components, including both dead and living plants, the architecture of those plants, the diversity of plant species, and their horizontal and vertical spatial arrangement, comprise stand structure. Landscape structure is the composition and spatial configuration of patches of habitat in a landscape (McGarigal and McComb 1995). In a forested environment, landscape structure is composed of the amount, type, and spatial arrangement of nonforested and forested stands and patches. Forest management practices, especially silviculture, influence stand and landscape structure and function, thereby influencing habitat and regional biodiversity.

Johnson (1980) developed a conceptual framework to help elucidate the interacting influences of habitat characteristics at different spatial scales on the selection of habitat by wildlife. Johnson described four hierarchical levels of selection: (1) selection of the geographic range of a species (first order), (2) selection of the home range of an individual or social group (second order), (3) selection of habitat components within a home range (third order), and (4) selection of individual forage items (fourth order). Selection at a higher order influences selection at lower orders; for example, selection of a home range (second order) limits the habitat components available for selection by an individual (third order). In some cases, selection at lower orders may influence higher-order selection as well (Bissonette 1997). Regardless of the direction of influence, Johnson's (1980) hierarchy is valuable in that it points out the importance of the interplay among spatial scales and its influence on habitat selection.

The task of managing Coast Range forests for a large number of wildlife species, each with its own specific habitat needs, is daunting. Considering the needs of multiple species by providing a broad range of habitat conditions is sometimes referred to as a *coarse-filter* approach, whereas managing for each species independently is a *fine-filter* approach (Hunter et al. 1988). Coarse-filter approaches are modeled on the full range of variability inherent in natural systems and can simplify management. If used appropriately, coarse-filter approaches can be effective in maintaining habitat for many species but, if used exclusively, they may not be adequate to maintain habitat for all species. Species with narrow distributions and highly specific habitat requirements may not be adequately maintained without management approaches tailored to meet their specific needs. There is a plethora of information pertinent to fine-filter approaches for some species—most notable in the Oregon Coast Range is the northern spotted owl (*Strix occidentalis*)—but an in-depth review of the management of individual species is beyond the scope of this chapter. A combination of coarse- and fine-filter approaches is probably necessary to provide for the habitat requirements of large assemblages of species in the Oregon Coast Range.

We focus on the role of habitat in Oregon Coast Range forests and its influence on wildlife populations for two reasons. First, forest management directly influences availability and quality of habitat; this interplay between forest management and habitat can strongly influence wildlife. In addition, the amount, distribution, and quality of habitat are primary factors determining the abundance, distribution, and viability of most species of wildlife in the Oregon Coast Range. We present our perspectives on management of wildlife in the Coast Range based on available scientific information. Although our focus here is almost exclusively on wildlife, some of the principles and considerations in this chapter apply to other taxa as well. Because of the importance of spatial scales in habitat selection by wildlife, we have structured this chapter to examine factors and management considerations at different spatial scales.

Habitat Components Within a Stand Influence the Presence and Abundance of Wildlife

Although many wildlife species are often thought to be associated with a particular stage or stages of forest succession, most species respond not so much to stand age per se as to structural features that tend to be correlated with stand ages under even-aged management scenarios (deMaynadier and Hunter

1995; Bunnell et al. 1997). Structural features such as dead wood and large trees, which influence the presence and abundance of many wildlife species, can be maintained in stands of all ages, using appropriate management strategies. Special features such as the presence of water, caves, cliffs, or other formations also influence habitat suitability for some species. While these features are usually not managed directly, the condition of the stand in which they occur may be an important factor determining their value to wildlife.

Dead wood

Dead wood, either standing in the form of snags or dead portions of a tree's bole or branches or fallen in the form of a log, is a critical element of wildlife habitat in the Oregon Coast Range (Figure 5-2). Dead wood follows a reasonably predictable pattern of decay over time, and is used differently by different species as it decays.

Snags provide habitat for many species of wildlife. An important function of standing dead wood in forest ecosystems is providing cavities for nesting, roosting, or denning by birds and small mammals. In the Oregon Coast Range, most cavities used by wildlife are fully or partially excavated by species referred to as cavity excavators—sometimes called primary cavity excavators or primary cavity nesters. Woodpeckers comprise the principal group of cavity excavators. Some other species of birds, such as the chestnut-backed chickadee (*Poecile rufescens*) and red-breasted nuthatch (*Sitta canadensis*), also sometimes excavate cavities. Cavity excavators excavate at least one new cavity every year and do not reuse nest sites in subsequent years. The reasons for this are not well understood, but it could be a strategy to minimize exposure to parasites. Over time, cavity excavators create numerous cavities that are subsequently available for use by other species of wildlife. The diverse group of species that use but do not excavate cavities is referred to as secondary cavity users. Because of their disproportionate influence on the ecology of the area, we consider cavity excavators in the Oregon Coast Range to be keystone species.

Selection of snags by cavity excavators is influenced by characteristics of the snag and the surrounding habitat. Key factors influencing use of snags are size (especially diameter) and decay class. There is apparently no upper size threshold for snags used by cavity users; consequently, large

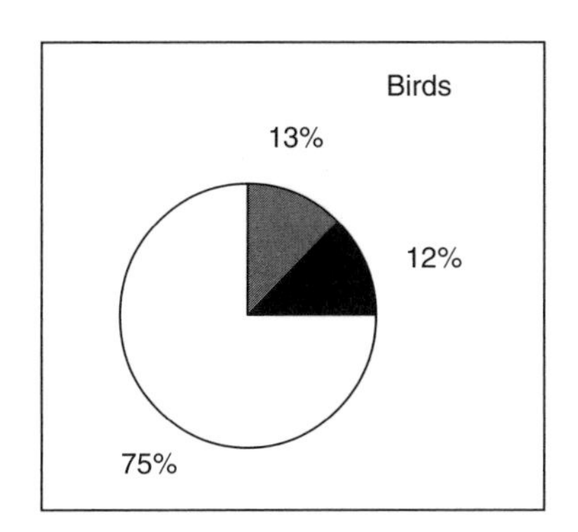

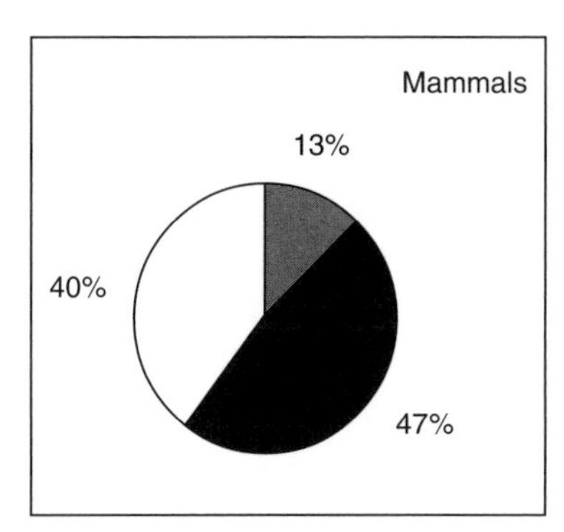

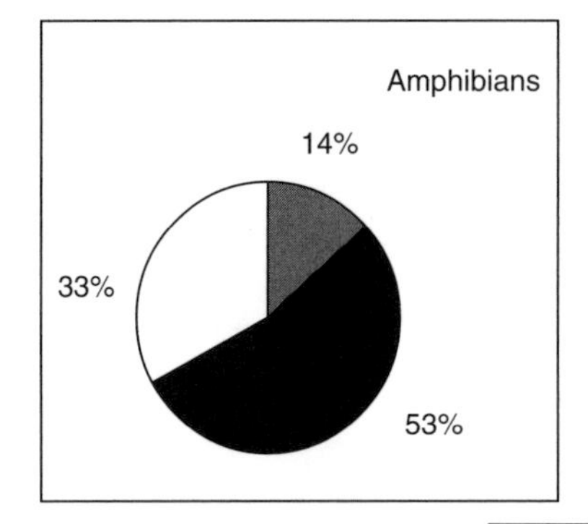

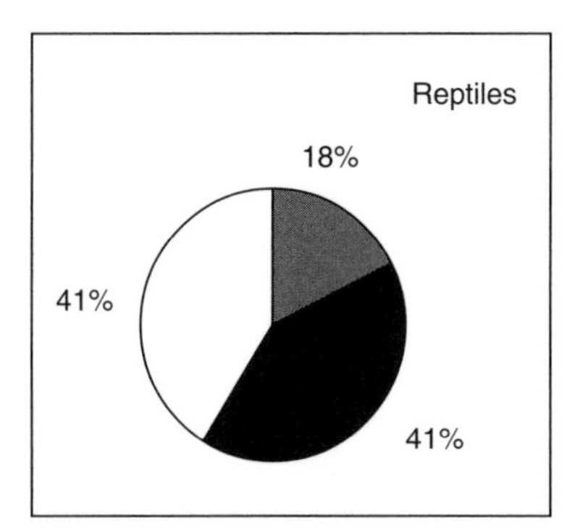

Dependent
Closely associated
Neutral/Unknown

Figure 5-2. Proportion of species in the Oregon Coast Range that use dead wood (based on McGarigal and McComb 1993).

snags provide suitable habitat for all species of wildlife that use cavities, whereas small snags provide habitat for only a subset (Figure 5-3). If snags of an optimal size are unavailable, birds sometimes will use snags that are smaller than desirable. Because only a limited number of young can physically fit in undersized cavities, reproductive output of birds using cavities in small-diameter structures can be low.

Snags decay over time as a result of the colonization of decay organisms (see Chapter 8) and physical weathering. Use of snags for nesting varies with level of decay. For example, in a study in the Oregon Cascade Range, Church (1997) found that each of the three most abundant species of woodpeckers in the region most frequently nested in snags of different decay class. Red-breasted sapsuckers (*Sphyrapicus ruber*) typically nested in hard snags, hairy woodpeckers (*Picoides villosus*) in snags of intermediate decay, and northern flickers (*Colaptes auratus*) in snags with more advanced decay.

Living trees with dead portions of sufficient size provide nesting habitat for cavity excavators and secondary cavity users (Figure 5-4 in color section following page 84). These structures often have greater longevity than snags. In the Oregon Coast Range, older bigleaf maple (*Acer macrophyllum*; especially in riparian areas) and Oregon white oak (*Quercus garryana*; predominantly at the margin of the Coast Range and the Willamette Valley) typically have numerous dead branches (Gumtow-Farrior 1991), many of which are often large enough to harbor cavities. In addition, cavities often form in these species of trees when branches break off and decay organisms invade the wood.

Management for cavity users generally has focused on providing nesting habitat for cavity excavators. This approach has assumed that if the nesting needs of cavity excavators are met, their foraging needs and the nesting, roosting, and foraging needs of secondary users would also be met (Neitro et al. 1985). However, recent evidence suggests that this assumption may not always be valid, and that foraging requirements of cavity users are an important consideration for management (Weikel and Hayes 1999). Abundance of some species of cavity-nesting birds is greater in thinned than unthinned stands, despite lower densities of snags (Hagar et al. 1996; Weikel 1997). This difference may result from changes in availability of forage following thinning. In the Tillamook Burn, hairy woodpeckers forage extensively on large-diameter snags in advanced state of decay, as well as on living trees and large logs (Weikel and Hayes 1999). Living trees with dead branches provide important foraging and nesting resources for cavity users. Dead branches are used heavily for foraging by brown creepers. Given the diverse foraging requirements of cavity-nesting birds (Figure 5-5), it is important to consider both nesting and foraging needs when managing for cavity users.

Although representatives of every order and most families of mammals in the Coast Range use snags, snags are particularly important for some mammals, such as bats, flying squirrels, and martens (*Martes americana*) (now absent or rare throughout most of the Coast Range). Patterns of use of snags by secondary cavity users vary with species, but many of the principles discussed for cavity excavators apply. Secondary cavity users generally use large-diameter snags more frequently than small-

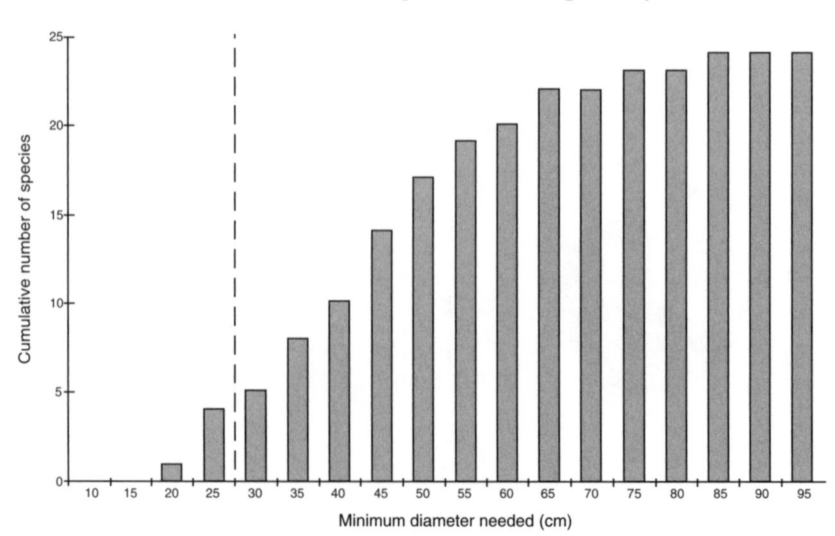

Figure 5-3. Relationship between number of species of wildlife and diameter of snags used for nesting (birds) or denning and roosting (mammals). The dotted vertical line indicates current guidelines for diameter of snags retained in harvest units according to current Oregon Forest Practices (Oregon Department of Forestry 1997). Numbers are based on minimum mean values reported minus one standard deviation.

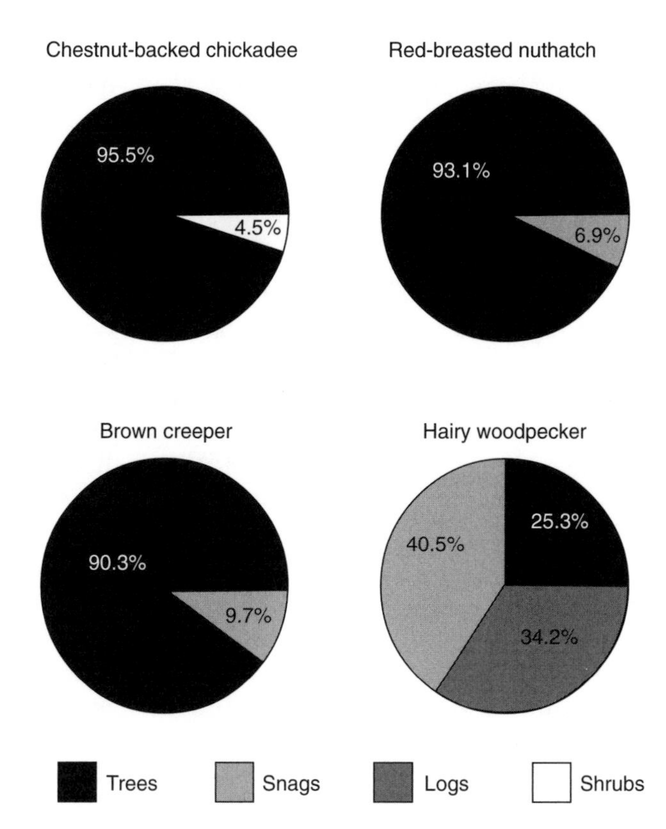

Figure 5-5. Substrates used for foraging by four species of cavity-nesting birds in the Oregon Coast Range (based on Weikel and Hayes 1999).

diameter snags. For example, in the Oregon Cascades, Ormsbee (1996) found that the average diameter of snags used by long-legged myotis (*Myotis volans*) was 97 cm, 24 cm greater than the average diameter of randomly chosen snags. These results are similar to those of several studies in diverse geographic locations, indicating that several species of bats select snags that are of large diameter, tall, or both, for roosting (Barclay et al. 1988; Campbell et al. 1996; Sasse and Pekins 1996; Vonhof 1996; Vonhof and Barclay 1996; Brigham et al. 1997; Rabe et al. 1998; Waldien et al. 2000). Snags with large diameters are also selected by northern flying squirrels (*Glaucomys sabrinus*) for both summer and winter den sites (Martin 1994; Clark 1995; Carey et al. 1997; Feen 1997).

Sloughing bark on snags and dying trees provides a special habitat feature that is valuable to several species of wildlife (Figure 5-6 in color section following page 84). Brown creepers frequently nest in the space between sloughing bark and the tree bole, and amphibians, reptiles, and bats often use this space for roosting or refuge.

Although not a replacement for taller snags, high-cut stumps also are used for foraging by birds (Morrison 1992). High-cut stumps are rarely used for nesting by most species of cavity nesters (Morrison 1992), but chestnut-backed chickadees frequently excavate nests in old, large-diameter stumps. Crevices formed from sloughing bark on stumps provide roost sites, protective cover, and rest sites for several species. In the central Oregon Cascade Range, D. L. Waldien et al. (OSU, Fisheries and Wildlife, personal communication) found northern alligator lizards (*Elgaria coerulea*), western fence lizards (*Sceloporus occidentalis*), garter snakes (*Thamnophis* spp.), skinks (*Eumeces skiltonianus*), clouded salamanders (*Aneides ferreus*), and other species using these crevices. In addition, Waldien et al. (2000) and Vonhof and Barclay (1997) documented extensive day-roosting by long-eared myotis (*Myotis evotis*) in crevices in stumps, though other species of bats appear to use stumps only rarely.

Logs are an important habitat component, especially for small mammals and amphibians (Figure 5-7 in color section following page 84). Logs and understory cover (herbs and shrubs) are the primary determinants of abundances of some species in forests of the western hemlock zone of Oregon and Washington (Carey and Johnson 1995). Wildlife use logs for foraging, roosts, rest sites, perches, cover, and resting areas, depending on the size and decay stage of the log. Logs provide sources of food for small mammals and amphibians in the form of insects and fungi, moist and protected microsites, and protected runways within, adjacent to, and under the log. Like snags, logs are used for foraging by some species of birds, such as the hairy woodpecker (Weikel and Hayes 1999). Logs also provide den sites for a variety of species, including several species of forest carnivores.

The western red-backed vole is closely associated with logs (Doyle 1987; Hayes and Cross 1987; Tallmon and Mills 1994; Thompson 1996; Martin 1998), using them heavily as runways and cover during travel. In southwestern Oregon, Tallmon and Mills (1994) found that radio-tagged western red-backed voles were located in or beside logs 98 percent of the time, although logs covered only 7 percent of the study area. Thompson (1996) found that the voles' movements were concentrated in areas with high volumes of decayed logs and deep organic soils. The close association of this species with logs is probably the result of the amenities they provide, including cover for protection from predation combined with a source of hypogeous fungal sporocarps (the fruiting bodies of fungi that fruit beneath the ground, such as truffles and false truffles), a primary source of food for the species (Hayes et al. 1986; Hayes and Cross 1987; Clarkson and Mills 1994; Tallmon and Mills 1994). As with snags, large-diameter logs are generally more valuable to wildlife than small-diameter logs. Large logs often have more well-developed overhang along their sides, which provides protective cover for small mammals. They also decay more gradually, and thus are more persistent than smaller logs, and they may retain moisture better, providing better microsite conditions for amphibians. Large logs also can provide larger physical spaces for den sites for large animals such as black bears (*Ursus americanus*) (Figure 5-8 in color section following page 84). Relatively little research has been conducted to examine specific benefits of individual logs for wildlife, but the little research that has been conducted (e.g., Hayes and Cross 1987) and anecdotal observations generally indicate that for many species, bigger is better.

As with snags, decay class also influences the amount and type of use a log receives. Hard logs provide perch sites and feeding platforms, protective cover along their sides, and nest and den sites in their interior. As a log decays, it provides habitat for amphibians, insectivores, and many species of arthropods and fungi that reside or burrow in soft and decaying wood.

Dead wood can be managed to benefit wildlife. Three approaches can be used to manage dead wood for wildlife: (1) maintaining existing dead wood during forest management operations, (2) creating new dead wood directly by killing live trees, and (3) planning for future recruitment of dead wood. In many circumstances all three approaches can and should be used.

Existing dead wood represents a critical and unique resource that is not easily replaced. In intensively managed forests, this "legacy" wood is often larger in diameter than trees immediately available to replace it, and it is sometimes present in a range of stages of decay. Providing similar habitat components or structures that will serve the same functions as these legacy structures through management, particularly in young stands and plantations, is likely to take decades. Legacy structures currently provide the primary source of nesting habitat for cavity-nesting birds in many intensively managed forests, and these structures will never be replaced in some intensively managed forests if current practices continue into the future (Ohmann et al. 1994). Past and current forest practices frequently do not accommodate the upper range of sizes and densities of woody debris that would be represented in natural systems. Maintaining a substantial component of legacy dead wood is critical to support wildlife populations. Conserving legacy logs poses a special challenge, particularly in ground-based operations, as logging equipment can pulverize logs in intermediate and advanced stages of decay. Special attention should be given to protecting these logs during harvesting and site preparation activities.

In areas where the amount of dead wood is inadequate to meet management goals or where dead wood has been removed during forest management activities, it can be created. Several techniques are available for creating snags, including girdling a tree at its base or higher, use of herbicides, use of pheromones to attract insects, injection of fungal spores into living trees, and topping trees with chainsaws, mechanical harvesters, or explosives (Bull and Partridge 1986; Ross and Niwa 1997; Lewis 1998). Methods vary with respect to cost, longevity of the resultant snag, quality of the habitat created, and safety and logistical considerations (Lewis 1998). Recent work in the Oregon Coast Range by Chambers et al. (1997) indicates that snags created by topping Douglas-fir (*Pseudotsuga menziesii*) trees with chainsaws are used relatively quickly for nesting by cavity-nesting birds. Five years after 821 snags were created, Chambers et al. (1997) observed an average of 1.3 excavated cavities and 0.6 natural cavities per snag. No difference was detected in the number of excavations in snags that were created singly versus those in clumps. Injection of fungal spores into living trees also shows promise, and research is underway to evaluate the efficacy of this approach in the Oregon Coast Range (G. Filip, OSU, Forest Science, personal communication). Less information is available on the use of logs or trees felled and left during logging operations.

Creating snags or logs is one means of compensating for an inadequate amount of dead wood and is effective in some situations (Figure 5-9 in color section following page 84). However, this approach should not be viewed as a substitute for maintaining legacy structures. As noted above, legacy dead wood has characteristics that are not replaceable in the short term. But eventually legacy dead wood will decay. To provide sustainable habitat, provision for replacement of legacy structures over time should be made. Growing large-diameter trees for future snag recruitment is possible using a variety of silvicultural approaches (see Chapter 7) and can be accomplished by allocating a portion of the trees in a stand as legacy-replacement structures. Because unpredictable events can profoundly influence longevity of trees designated as replacement structures, an adequate number should be provided to allow for a portion of them to fall over time. Placement of legacy-replacement trees should maximize potential for long-term success by minimizing the chance of blowdown or other factors that would result in loss of the structure for its intended purpose.

Despite the importance of dead wood to wildlife, precise measures of the ecological costs and benefits of maintaining different levels of dead wood remain elusive, thereby complicating attempts to establish meaningful management goals. Models based on existing information have been developed to provide guidance on the influence of quantity, size, and state of decay of snags on cavity-nesting birds (Neitro et al. 1985). These models provide a starting point for snag management, but they should be used with recognition of the assumptions on which they are based. SRS, a model based on data presented by Neitro et al. (1985), is commonly used in western Oregon to model the influence of different snag densities on cavity-nesting birds. A key assumption of this model is that cavity-nesting bird populations are limited by nesting habitat. As a first approximation, this assumption is reasonable and has been demonstrated in some situations (e.g., Schreiber and deCalesta 1992), but other factors, such as levels of predation and quality of foraging habitat, also can strongly influence bird abundance (Weikel and Hayes 1999). These factors in turn can be influenced by stand density, habitat type (e.g., upland vs. riparian), and other site-specific characteristics. Another key assumption of this model is that the needs of secondary cavity users, including several species of cavity-nesting birds, bats, flying squirrels, and other species, will be met if the needs of primary excavators are met; however, this assumption has not been rigorously tested. We suspect that existing models that focus only on the nesting needs of primary cavity users underestimate the amount of dead wood needed to support populations of wildlife associated with snags. Models relating abundance of, or viability of populations of, cavity-nesting birds to abundance and characteristics of snags in a diversity of habitat types and conditions will ultimately provide the best tool for establishing management goals. To be effective, these models will need to incorporate a variety of factors, including quality of foraging habitat and other considerations. However, the current lack of information for many species, coupled with the poor resolution of the data (e.g., the lack of understanding of the influence of foraging habitat and the absence of information concerning the relationship between abundance and population viability), make contemporary use of this approach imprecise. As a consequence, use of currently available models coupled with consideration of the historic range of variation of the abundance of snags across the landscape may provide a valuable starting point for establishing management goals.

Several researchers contend that current regulatory guidelines concerning snags are inadequate to meet the habitat needs of wildlife, with respect to both the size and number of snags required to be provided (Mannan et al. 1980; Nelson 1989; Schreiber and deCalesta 1992; Weikel 1997). At the time of this writing, Oregon forest-practices regulations require that a minimum of five snags or live trees per hectare, with minimum diameters of 27.5 cm and heights of 10 m, be retained when the area logged exceeds 10 ha in size (Oregon Department of Forestry 1997). Mannan et al. (1980) recommend that forest managers attempt to maintain "as great a density and diversity of large snags as is possible," with an emphasis on retaining or creating snags greater than 60 cm in diameter and 15 m in height. Schreiber and deCalesta (1992) recommend providing 14 snags per ha in the central Oregon Coast Range, ranging from 28 cm to greater than 128 cm in diameter and from 6.4 to 25 m in height, with at least 10 to 40 percent bark cover. Although she does not provide specific numeric guidelines, Weikel (1997) argues that whenever possible the number and diameter should be greater than those mandated by the Oregon Forest Practices Act. In addition to providing better habitat for wildlife, large snags have greater longevity than small snags (Morrison and Raphael 1993), thereby providing habitat for longer periods of time.

There are few data available to address issues concerning dispersion of snags and the costs and benefits of leaving snags in a clumped versus scattered distribution. The best information available for the Oregon Coast Range comes from work with artificially created snags by Chambers et al. (1997). They did not find differences between the number of cavities excavated in snags created in clumped distributions and those scattered throughout the stand. In Idaho, cavity-nesting birds have demonstrated a preference for snags distributed in clumps (Saab and Dudley 1998). However, given the small amount of data collected to date, the relationship between distribution and snag use remains unclear. Given this uncertainty, the prudent course of action is to leave or create snags in a variety of spatial configurations. Managers should also be cognizant of the fact that the scale of spatial configurations used should be appropriate for the species of interest. Clumps of snags distributed at very large distances may be adequate to maintain populations of species with very large home ranges, such as the pileated woodpecker (*Dryocopus pileatus*), but to provide optimal habitat for species with smaller home ranges, such as the red-breasted nuthatch, snags or clumps of snags distributed more closely to one another would be more appropriate. Operational constraints, such as logging and safety considerations, may influence optimal distributions for snags in some situations. Hope and McComb (1994) found that forest-industry workers generally consider uniformly distributed snags to be a safety hazard and to slow harvesting operations; a majority of those workers believed that maintaining snags in clumps is a more efficient approach.

Our understanding of relationships between quantity, size, and decay class of logs and abundance of wildlife remains crude. Carey and Johnson (1995) suggest that 15- to 20-percent cover of coarse wood, well distributed across the site, would provide adequate habitat for most species of small mammals, while 5- to 10-percent cover of coarse wood would limit their populations. Thompson (1996) proposes that patches with a minimum of 236 m^3 of coarse wood per hectare be distributed over 20 percent of the area to maintain habitat for western red-backed voles. These values greatly exceed current regulatory requirements, but are well within the range of volumes of dead wood found in natural systems.

The margins of the Oregon Coast Range, where oak savanna and coniferous forest intermix, present a special issue concerning management of dead wood. Because of changes in fire regimes, management, and settlement patterns, conifers have encroached on areas that were once dominated by oak-savanna habitat. As a result, conifers have overtopped oaks in many areas, killing the oaks. The loss of oaks in these areas may impact cavity users and other species. We recommend thinning and harvest of conifers in areas where conifers are encroaching into areas previously dominated by oaks to help retain this unique resource.

Within the past two decades, the importance of dead wood in ecosystem function (see Chapter 3) and for providing habitat for wildlife has been increasingly recognized. Three primary considerations impede the maintenance of adequate amounts of dead wood: safety, economics, and (in some areas) fuel considerations. Given the importance of dead wood in forest ecosystems, we advocate consideration of these factors in placement of snags retained on the landscape, but they should not be reasons for retaining inadequate amounts or types of dead wood.

Large and unique trees

Large trees have a number of characteristics that are desirable for some species of wildlife (Figure 5-10 in color section following page 84). Large trees, particularly Douglas-fir, often have deeply fissured bark that provides roost sites for bats (Christy and West 1993) and foraging sites for insectivores such as the brown creeper (Mariani and Manuwal 1990). Large-diameter branches provide stable nest platforms for a number of species, including the marbled murrelet (Hamer and Nelson 1995). The deep crowns of large (especially open-grown) trees also provide nesting sites for canopy dwellers, a source of arthropods for insectivorous birds, and nesting and foraging sites for arboreal rodents such as the red tree vole. Furthermore, large trees are the source for recruitment of large-diameter snags and logs. Individual trees with unusual structures or "defects" and open-grown trees with branches along the entire bole (often referred to as "wolf trees") also provide unique habitat for wildlife. Partially or completely hollow western redcedar provide important habitat for a variety of species, such as bats (Ormsbee and McComb 1998) and Vaux's swift

(*Chaetura vauxi*). Trees that are partially dead can combine many of the values of snags with increased longevity. Trees with forked trunks or whorls of branches provide high-quality, stable nesting sites for several species of wildlife.

Managing for large and unique trees is a two-step process. First, individual trees must be retained during forest management activities. Maintenance of these legacy structures can provide important temporal connectivity between generations of stands. As noted above, for this approach to be effective over time, retained trees need to be located to maximize their potential for survival. For example, individual trees along a ridgetop in the middle of a clearcut and susceptible to windthrow will be unlikely to contribute to the goals of maintaining large and unique trees. Second, stand management, especially density management, can be used to produce these trees (especially large-diameter, open-grown trees) over time. It is likely that open-grown trees were relatively common in the past, but will probably be relatively rare in the future unless measures are taken to grow trees with large crowns and large-diameter branches by releasing some trees through density management or wide spacing of plantations. Details on approaches to growing and managing for large trees are presented in Chapter 7.

Special features

Non-vegetative features of a stand, including presence of freshwater seeps, caves, and cliff faces, are essential habitat for some species of wildlife. Although these features are relatively uncommon elements on the landscape, they play an important role in shaping wildlife communities because of their value to some species. For example, torrent salamanders (*Rhycotriton* spp.) are closely associated with seep habitat along headwater streams; forest practices in those areas may impact the abundance and distribution of this species (Welsh and Lind 1996). Townsend's big-eared bats (*Corynorhinus townsendii*) are closely associated with caves and cave-like habitat, and use caves for maternity and colonial roosts, as well as for hibernacula (Christy and West 1993). Cliff faces and rocky outcrops provide habitat for some species, including nest sites for peregrine falcons (*Falco peregrinus*) and roost sites for several species of bats. Because of the thermal characteristics of cliff faces, they also often provide important habitat for reptiles. Cliff faces along streams provide nesting habitat for dippers (*Cinclus mexicanus*) (Loegering 1997).

Artificial structures in forest systems can also provide important habitat for wildlife. For example, bats use bridges extensively as roost sites, particularly for night roosts. In the Oregon Coast Range, at least 119 bats were observed night-roosting together under one bridge and 9 of the 10 species of bats occurring in the Oregon Coast Range have been identified roosting under bridges (Adam and Hayes 2000). Bridges provide important day-roosts and maternity sites for bats in some areas and also serve as nest sites for dippers (Loegering 1997) and other species of birds. The influence of artificial structures on overall abundance of species can only be speculated. In some cases, use of artificial structures may be a response to loss of historically used structures in an area, such as large-diameter snags or large-diameter logs that spanned a stream. In other cases, the presence of artificial structures may have resulted in increases in abundance above historic numbers. In any case, it is clear that many of these structures currently provide valuable habitat for wildlife, and they should be taken into account in forest management operations.

Although forest management activities often do not directly alter the characteristics of special features, disturbance resulting from logging can indirectly influence use of those features, for example, by altering microclimatic conditions. Removing rock from quarries for road construction can destroy special features associated with a rock face; removal of a bridge can destroy a night roost used by bats. Special features, such as caves and cliff faces, are not abundant in much of the Coast Range, but when present they provide valuable habitat for wildlife. Given the importance of special features in providing habitat for particular species of wildlife, we recommend that special attention be given to them to assess their value to wildlife and to minimize impacts of forest management activities on wildlife.

Stand-level Characteristics Influence the Presence and Abundance of Wildlife

Stand-level characteristics, such as tree density, tree species composition and richness, canopy structure, and tree diameters, directly influence the presence and abundance of many species of wildlife, as can the interplay of stand-level characteristics with stand components.

Tree density and distribution

Density, or stocking, of trees has a dramatic effect on many aspects of stand structure (Chapter 7). Over time, tree density influences tree architecture: diameter, height, canopy depth, and diameter and length of branches. Tree density also directly influences rates of tree mortality and the recruitment of snags and logs, the development of vegetation in the understory, amount of natural regeneration, species composition of the herbs, shrubs, and trees in the understory, and survival of mid-canopy trees. The cascade of effects resulting from differences in tree density has a strong influence on the presence and abundance of wildlife in a stand (McComb et al. 1993 c; Carey and Curtis 1996; Hayes et al. 1997).

Currently, our understanding of the influence of tree density on wildlife in the Oregon Coast Range is best developed for birds, and is based primarily on one observational and one experimental study of the response of birds to the thinning of young forests. Hagar et al. (1996) compared the abundance of 20 species of birds in eight unthinned stands (average density = 495 trees per hectare; Relative Density Index = 0.34) to that in eight stands that had been thinned five to 15 years previously to an average density of 360 trees per hectare (Relative Density Index = 0.24; Figure 5-11). All of the study sites examined were dominated by 40- to 55-year-old Douglas-fir; half were located in the northern and half in the central Coast Range. Hayes et al. (1998) examined the influence of thinning on birds in an experimental study being conducted in the Tillamook State Forest and adjacent forest lands (Figure 5-12 in color section following page 84). This ongoing study is examining the responses of vegetation and wildlife to two intensities of thinning. A subset of stands was thinned to a relative density of 35 (250 to 300 trees per hectare), a level typical of operational thinning in the region. Another group of stands was more heavily thinned to a relative density of 20 (150 to 175 trees per hectare). Hayes et al. (1998) reported results from the first four years of study and subsequent analyses have been conducted on the first seven years of the study (Hayes, Weikel, and Huso, personal communication).

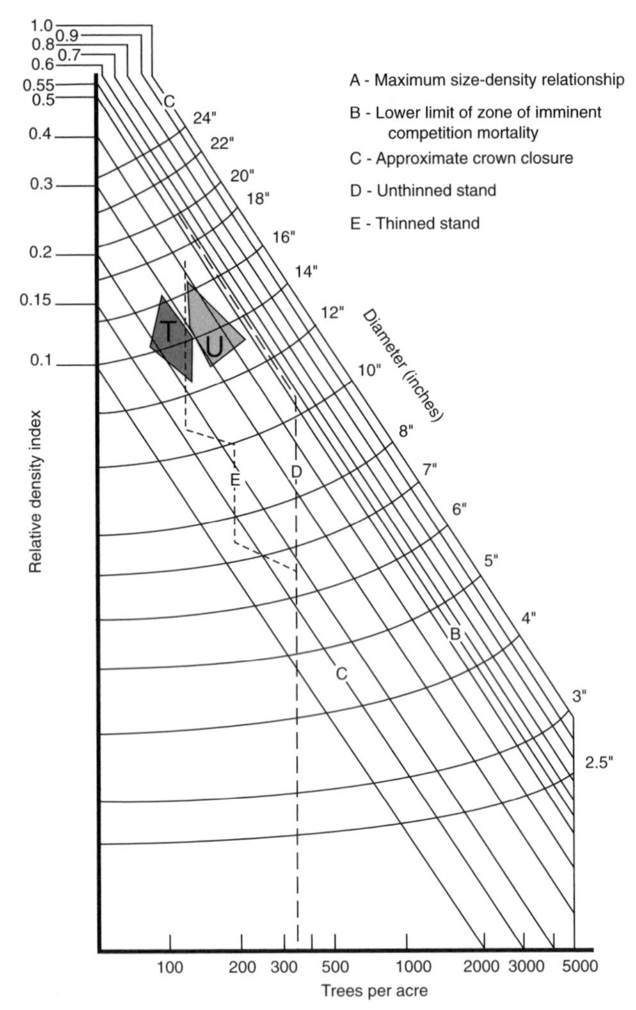

Figure 5-11. Density management diagram for Douglas-fir showing the growth trajectory of an unthinned stand (line D) and a hypothetical thinning regime (line E) with initial stocking densities of 865 trees per hectare. The thinning regime depicted in line E represents one developmental trajectory that might result in a stand with suitable habitat for Hammond's flycatchers, hairy woodpeckers, red-breasted nuthatches, and dark-eyed juncos (based on Hagar 1992). A density diagram shows the relationship of trees per acre (displayed on the X-axis) and tree size (displayed on the Y-axis) and the maximum number of trees that a site could support (a relative density of 1.0; displayed on the diagonal axis). Stands generally respond in a stereotypical fashion to tree density, with crown closure occurring at a relative density of approximately 0.15 (line C) and competition-induced mortality beginning at a relative density of approximately 0.55 (line B). Conifer densities and sizes in stands studied by Hagar et al. (1996) are depicted by the shaded areas on the diagram (T, thinned study sites; U, unthinned study sites).

Results of Hagar et al. (1996) and Hayes et al. (personal communication) are highly consistent for several species of birds (Table 5-2). Five species (hairy woodpecker, warbling vireo [*Vireo gilvus*], dark-eyed junco [*Junco hyemalis*], Hammond's flycatcher [*Empidonax hammondii*], and evening grosbeak [*Coccothraustes vespertinus*]) were more abundant in thinned stands in both studies, and evidence of higher abundance was reported for an additional four species in one or more of the studies or locations. Numbers of Pacific-slope flycatchers (*Empidonax difficilis*), golden-crowned kinglets (*Regulus satrapa*), and black-throated gray warblers (*Dendroica nigrescens*) were generally lower in thinned stands during the breeding season, although Hagar et al. (1996) found that abundance of kinglets in the winter was similar between thinned and unthinned stands. Five additional species (Hutton's vireo [*Vireo huttoni*], hermit warbler [*Dendroica occidentalis*], Swainson's thrush [*Catharus ustulatus*], varied thrush [*Ixoreus naevius*], and Steller's jay [*Cyanocitta stelleri*]) were less abundant in thinned sites in one or more areas and showed no responses or were not analyzed in other areas. Abundances of six species of birds were not demonstrably influenced by thinning or responded

Table 5-2. Evidence for response of birds to thinning based on Hayes et al. (unpublished) and Hagar et al. 1996.

Species	*Hayes et al. (unpublished)*	*Hagar et al. 1996*	
	Northern Coast Range	*Northern Coast Range*	*Central Coast Range*
Increase following thinning			
Hairy woodpecker	I	I	I
Warbling vireo	I	I	I
Dark-eyed junco	I	I	I
Hammond's flycatcher	I	I	I
Evening grosbeak	I	I	I
Townsend's solitaire	I	–	–
Western tanager	I	N	I
American robin	I	N	N
Red-breasted nuthatch	N	I	I
Decrease following thinning			
Pacific-slope flycatcher	D	D	D
Golden-crowned kinglet	D	D	D
Black-throated gray warbler	D	D	D
Hutton's vireo	D	N	N
Hermit warbler	D	N	N
Swainson's thrush	D	N	N
Varied thrush	D	–	–
Steller's jay	D	–	–
No or uncertain response to thinning			
Winter wren	N	D	I
Gray jay	N	D	I
Brown creeper	D	N	I
Chestnut-backed chickadee	N	N	N
Wilson's warbler	N	N	N
Black-headed grosbeak	–	N	N

'I' indicates evidence for an increase in abundance following thinning, 'D' indicates evidence for a decrease in abundance following thinning, 'N' indicates no significant increase or decrease in abundance was detected, and '-' indicates insufficient data to analyze trends for this species.

differently among studies or areas. Many, and possibly most, of the species that changed in abundance after thinning did not respond directly to tree density, but rather to stand characteristics that are closely correlated with tree density, such as shrub cover or availability of insects.

These findings suggest that thinning young stands in the Oregon Coast Range does not negatively impact the abundance of most species of diurnal breeding birds, and may be beneficial to several species. However, some species were negatively impacted. Furthermore, it should be recognized that abundance does not necessarily reflect habitat quality (Van Horne 1983). For habitat to contribute to maintenance of viable populations, birds using an area must be reproductively successful. Our understanding of the reproductive success of birds in the Oregon Coast Range under different management scenarios is minimal. In addition, we have little understanding of the effects of thinning on species of birds other than diurnal songbirds, including birds of prey.

Our knowledge of the response of mammals to thinning in the Oregon Coast Range is less well developed than for birds, but recent studies have shed some light on the subject. Suzuki (2000) examined the response of small mammals during the first two years after commercial thinning on the Tillamook State Forest and adjacent areas described above. Numbers of deer mice (*Peromyscus maniculatus*), creeping voles (*Microtus oregoni*), and Pacific jumping mice (*Zapus trinotatus*) increased in thinned stands. All three species began to increase after the first year, but a more substantial increase occurred during the second year. These species are typically most abundant in more open habitats, where they feed on grasses, herbaceous vegetation, or seeds (Verts and Carraway 1998). In contrast, numbers of western red-backed voles decreased after thinning. Western red-backed voles are typically associated with less-disturbed forests; they feed heavily on the truffles and other fungi occurring in these forests (Ure and Maser 1982; Hayes et al. 1986). At the same study sites, Gomez et al. (1998) examined changes in population densities of Townsend's chipmunks (*Tamias townsendii*) and northern flying squirrels. Townsend's chipmunks increased in density after stands were thinned, and abundances increased most in heavily thinned stands. Although average densities of flying squirrels were generally greater in unthinned stands, there was considerable variation among study sites and differences were not statistically significant.

Humes et al. (1999) found that, as a group, bats in the Oregon Coast Range appear to favor thinning. Although bat activity was generally greater in old-growth stands than in younger stands, levels of activity in young, thinned stands was intermediate to that in young, unthinned stands and old-growth stands.

Based on counts of elk (*Cervus elaphus*) pellet groups in the Tillamook thinning study, Adam and Hayes (1998) found that, in comparison with unthinned stands, there was evidence of slightly greater use of moderately thinned stands, and substantially more use of stands that had been heavily thinned. Although counts of pellet groups provide an imprecise index of use of an area, these data suggest that elk respond favorably to thinning in young stands, at least in situations where browse may be limiting, and is probably the result of increased forage in the understory resulting from increased solar radiation reaching the forest floor.

The distribution of trees within a stand also can influence the use of stands by wildlife. Natural disturbances on the scale of individual stands (10–100 ha), such as root-rot pockets and windthrow, leave a range of tree densities and distribution patterns, unlike most clearcuts harvested prior to the mid-1980s. Chambers et al. (1999) examined bird-community response to two alternatives to clearcut harvesting that were intended to more closely mimic stand-level structure resulting from natural disturbances: group selection and two-story harvests. Their results suggest that the density of trees retained in harvests of mature stands strongly influenced the composition and species abundance of songbird communities (Figure 5-13). Their finding that bird-community composition was similar in stands where one-third of the wood volume was removed in 0.2-hectare patches (group selection) to that in uncut control stands has profound implications for the simultaneous management of forests for wildlife and wood-fiber production. Furthermore, the retention of 20–30 trees per hectare in two-story harvests provided minimal habitat for some bird species that usually are associated with mature, closed-canopy stands (Chambers et al. 1999). Clearcuts and two-story harvests provide habitat for species associated with early-seral habitats (those provided by very young regen-

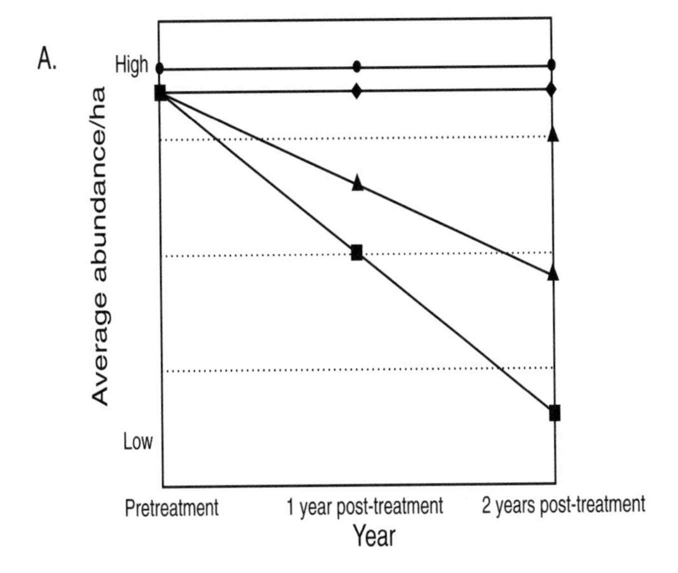

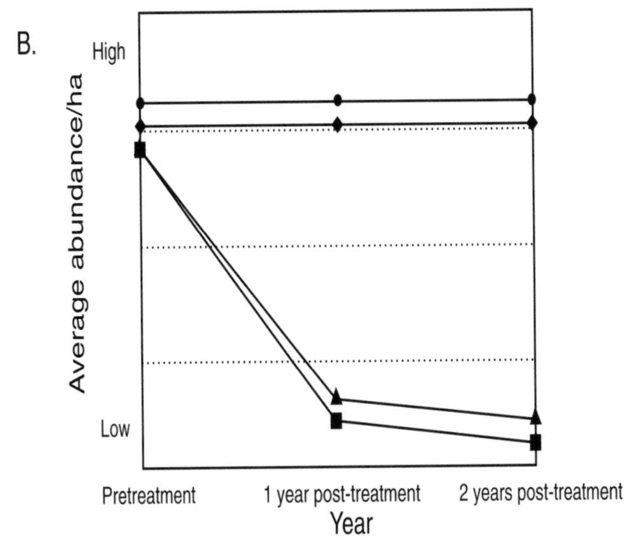

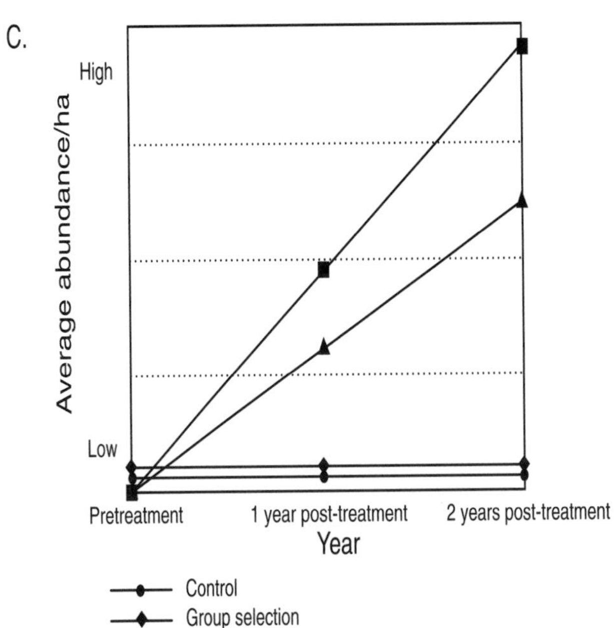

erating stands and clearcuts), and the retention of large trees also provides hiding cover, foraging habitat, and even nesting habitat for bird species associated with later seral stages (Vega 1993; Chambers et al. 1999). As a result, species richness of a breeding-bird community may be increased by 7 percent (Vega 1993) to 8 percent (Chambers et al. 1999) when trees are retained in a unit at the time of final harvest. Green trees and snags retained in clearcuts also may provide cover for other species of wildlife, such as northern flying squirrels (Feen 1997). However, although residual trees may result in occupancy of a site by some wildlife species, presence or abundance of a species does not necessarily translate to its reproductive success. Chambers (1996) found that predation rates on artificial nests were higher in two-story stands than in clearcuts, small-patch cuts, or unharvested control stands. She hypothesized that residual trees may facilitate nest predation and parasitism by providing perches from which avian predators can watch for cues indicating the location of nests.

Vertical structure of forest stands—the amount, type, and distribution of vegetation from the forest floor to the top of the canopy—can strongly influence the composition of wildlife communities (Figure 5-14). Carey and Johnson (1995) found that understory vegetation and characteristics of the forest floor (most importantly, the amount of coarse wood) profoundly influence the abundance of small mammals. Hayes et al. (1995) found that cover of salal (*Gaultheria shallon*) was highly correlated with abundance of Townsend's chipmunks in the Oregon Coast Range; salal has been found to increase in cover following thinning (Huffman et al. 1994; Bailey and Tappeiner 1998). In addition, Hayes and Gruver (2000) found that the use of canopy layers differs among species groups of bats, suggesting that stands with more complex vertical structure may

Figure 5-13. General pattern of responses of birds to three harvest treatments and unharvested control stands prior to treatment and 2 years post-treatment in the central Oregon Coast Range. A) Brown creeper and chestnut-backed chickadee. B) Steller's jay, red-breasted nuthatch, winter wren, golden-crowned kinglet, Swainson's thrush, hermit warbler, Wilson's warbler, and western tanager. C) Willow flycatcher, MacGillivray's warbler, spotted towhee, white-crowned sparrow, American goldfinch, and house wren. Adapted from Chambers (1996).

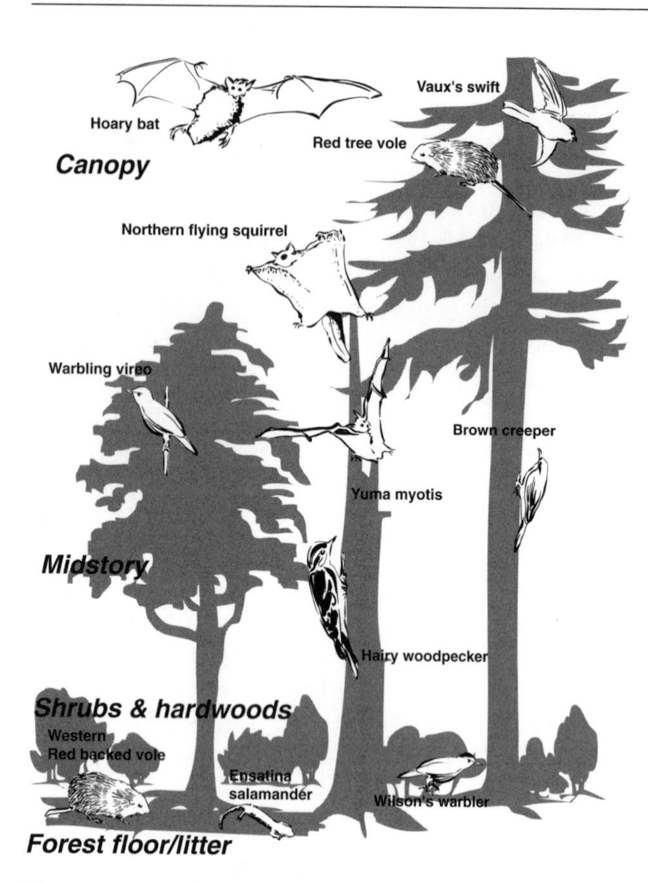

Figure 5-14. Schematic representation of vertical structure in forests of the Oregon Coast Range showing examples of species associated with different layers.

provide better habitat by providing diverse foraging opportunities.

Stands with well-developed understory vegetation provide nesting structures for shrub-nesting birds and cover for ground-nesting species. In addition, understory vegetation provides a source of forage for birds that feed on fruits, seeds, and invertebrates supported by understory vegetation.

Tree species composition interacts with vertical structure to influence wildlife communities. Because tree species differ in morphology, reproductive strategies, and phenology, a diversity of tree species contributes to local and regional diversity of wildlife by providing a variety of food and cover resources. In older conifer forests in western Oregon, hardwoods tend to occupy the lower- and mid-story canopies. These broad-leafed deciduous hardwoods substantially contribute to diversity in conifer-dominated landscapes, as does the maintenance of a diversity of coniferous tree species.

Broad-leaf deciduous hardwoods have more digestible foliage and different reproductive strategies (e.g., insect pollination, fruits, nuts) than do conifers. As a result, hardwoods and conifers provide wildlife with different food resources, both directly (through foliage and fruit) and indirectly (through invertebrates present on the trees). Some insectivorous birds, such as warbling vireos and black-capped chickadees (*Poecile atricapilla*), seem to be closely associated with deciduous trees in western Oregon (Hagar et al. 1995). Recent studies indicate that deciduous trees and shrubs support a greater diversity of moths and butterflies than do conifers (Hammond and Miller 1998); caterpillars are an important component of the diet of many insectivorous species of birds during the breeding season.

Hardwoods also provide different kinds of cover for wildlife than do conifers because of differences in branch structure and foliage. Bigleaf maple and Oregon white oak are more prone to cavities than Douglas-fir (Gumtow-Farrior 1991) and may provide a higher density of cavities for potential resting and nesting sites. Large-diameter, dead branches on living maples and oaks provide sites for cavities that can persist for long periods of time. In winter, conifers intercept snow and provide thermal cover, permitting greater mobility and foraging opportunities for some wildlife species, such as deer and elk.

Because of differences in available food and cover, wildlife communities in hardwood stands generally differ from those in conifer stands (Table 5-3). Deciduous trees thus contribute significantly to biological diversity on the stand and landscape levels in the conifer-dominated forests of western Oregon. The abundance and diversity of birds have been correlated positively with the abundance and distribution of hardwoods in conifer-dominated landscapes (Morrison and Meslow 1983; Nelson 1989; Carey et al. 1991; Gilbert and Allwine 1991; Ralph et al. 1991; McGarigal and McComb 1992; Hagar et al. 1996). Twenty-four species of mammals seem to be more abundant in hardwood than in conifer stands, whereas 11 species apparently are more abundant in coniferous than in hardwood stands (Hagar et al. 1995). Some species of arboreal mammals seem to prefer hardwoods and other conifers. Douglas-squirrels (*Tamiasciurus douglasii*), martens, and red tree voles occur only where conifers are present. In contrast, western gray squirrels (*Sciurus griseus*) occur only where hardwoods, particularly large oaks, are present.

Table 5-3. Wildlife species thought to be closely associated with deciduous trees and/or shrubs or with conifer trees in western Oregon.

Deciduous trees/shrubs	*Conifer trees*
Birds	**Birds**
Acorn woodpecker (oak)	Golden-crowned kinglet
Black-capped chickadee	Gray jay
Black-headed grosbeak	Hammond's flycatcher
Black-throated gray warbler	Hermit warbler
Cassin's vireo	Marbled murrelet
Downy woodpecker	Northern spotted owl
MacGillivray's warbler	Red-breasted nuthatch
Northern oriole	Red crossbill
Orange-crowned warbler	
White-breasted nuthatch (oak)	**Mammals**
Warbling vireo	California red-backed vole
Wilson's warbler	Douglas squirrel
Yellow warbler	Marten
Yellow-billed cuckoo	Northern flying squirrel
	Red tree vole (Douglas-fir)
Mammals	
Western gray squirrel	
White-footed vole	
White-tailed deer	

Although more species of amphibians are thought to be associated with conifers than with hardwoods (Hagar et al. 1995), deciduous forests had the highest capture rate of amphibians and reptiles of five forest types sampled in western Oregon (Gomez and Anthony 1996).

A diversity of conifers in a site also helps support diverse wildlife communities. Variations in cone production and bark characteristics among conifer species may have important implications for wildlife. For example, unlike most conifers, western hemlock produces cones every year (Packee 1990), providing a stable food source for seed-eating wildlife (Manuwal and Huff 1987). Differences in bark characteristics (e.g., roughness) among species of conifers may be important in determining use by bats for roost sites (Perkins and Cross 1988). Some bark-foraging birds, especially brown creepers, use tree species selectively because differences in bark characteristics influence the availability of invertebrate prey (Morrison et al. 1985; Mariani and Manuwal 1990). Red tree voles select Douglas-fir over other conifer species for feeding and nesting (Hayes 1996).

Managing stands to benefit wildlife

Although forest management is increasingly planned and implemented in a landscape context, most management activities on the ground occur at the stand level. Many silvicultural tools are available for stand-level management (see Chapter 7). Density management is a particularly valuable tool for managing forests to meet both wildlife and timber goals (Hagar et al. 1996; Hayes et al. 1997; Figure 5-11). Some approaches, such as uneven-aged management, have not been frequently used in the Oregon Coast Range, but may be highly conducive to meeting these goals because of their potential to produce complex stand structure (Chambers et al. 1999); we predict their use will increase in the coming years, and we advocate exploring the potential of uneven-aged management approaches.

Some species, such as the northern spotted owl and marbled murrelet, depend on older stands or stands that have old-forest structure for their survival (Forsman et al. 1984; Ralph et al. 1995; Hershey et al. 1998). Because of this and because of issues surrounding the spiritual, aesthetic, and recreational values of old-growth forests, much of the debate over forest management issues in the Pacific Northwest at both the stand and landscape levels has focused on stand age. Indeed, there is strong evidence that some species of wildlife are closely associated with particular structural conditions. Furthermore, many structural conditions are correlated with stand age under some management scenarios, but development of structural characteristics is also strongly influenced by site quality and stand density. Because of differences in developmental trajectories and structural conditions that result from different management actions, the application of the concept of ecological succession and over-reliance on stand age as a measure of habitat condition is often misguided in intensively managed stands.

The term "early-seral stage" is often used in reference to very young regenerating stands and clearcuts that generally do provide habitat for wildlife associated with early-seral stands resulting from natural disturbance; some portion of the landscape should be managed to provide habitat for these species. However, there are substantial differences between the characteristics and development of most young planted stands and early-seral habitat resulting from natural disturbance (Perry

1998). Most early-seral habitat resulting from natural disturbance has substantial amounts of standing and fallen dead wood. Schreiber and deCalesta (1992) and Chambers et al. (1997) found that snags left or created in clearcuts are heavily used by several species of cavity-nesting birds. Snags in open areas are critical for some species, such as the western bluebird (*Sialia mexicana*). To provide habitat for wildlife associated with early-seral conditions, harvesting should be done in a manner that maintains adequate levels of dead wood.

In addition, stands that develop after natural disturbances often have lower tree densities than managed plantations (Tappeiner et al. 1997). Trees grown at low densities attain large sizes rapidly and develop large limbs of forms that are used by wildlife for nesting, resting, and roosting. Furthermore, low densities of trees in naturally disturbed sites allow substantial development of shrub communities, which can then persist longer through stand development than in planted stands. Shortening the shrub-dominated stage of forest succession through planting of conifers and vegetation management likely reduces both average shrub age and shrub cover. Like trees, shrubs develop structural characteristics over time that may enhance their usefulness to wildlife. Older shrubs have thicker stems and support more epiphytes than younger shrubs (A. Rosso, OSU, Botany and Plant Pathology, personal communication). Epiphytes provide nesting material for birds and habitat for invertebrates that are food for birds and other wildlife (Gerson 1973; Pettersson et al. 1995). Populations of several species of neotropical migratory birds that are strongly associated with shrub cover have been declining over recent decades (e.g., Swainson's thrush, orange-crowned warbler [*Vermivora celata*], Wilson's warbler [*Wilsonia pusilla*], willow flycatcher [*Empidonax traillii*]) (Sauer et al. 1997). Truncation of the shrub stage of forest succession may be a contributing factor. It is also possible that some species of invertebrates or plants could require longer periods of open habitat, and we encourage additional attention to this issue.

Use of the terms "old growth" and "late-seral stage" is similarly problematic when applied to intensively managed forests. As we have discussed, some of the structures found in older forest stands that developed from natural disturbance can be created, maintained, or developed through a variety of forest management techniques. Individual components of pre-disturbance ecosystems, such as large-diameter trees, snags, and logs, can be carried forward into post-disturbance ecosystems through the careful management of legacy structures. Strategies to accomplish this include the retention of structural features at the time of final harvest and the adoption of silvicultural systems that are alternatives to the traditional even-aged system (McComb et al. 1993c; Chambers et al. 1999). In addition, stand-level characteristics typical of older forest stands can be developed in relatively short time frames through active management (Barbour et al. 1997; Hayes et al. 1997). However, these stands probably will not have all of the characteristics of late-seral stands. Hershey et al. (1998) pointed out that the ability to silviculturally manipulate stands to create the complex structural conditions required for nesting by northern spotted owls has not yet been conclusively demonstrated (although we believe it will in fact be demonstrated over time).

The initial development of most intensively managed stands in the Oregon Coast Range differs dramatically from early developmental trajectories of existing old-growth stands (Tappeiner et al. 1997). Management activities can influence the direction of young stands and can result in structural characteristics that meet the needs of many species typically associated with late-seral stage stands (Newton and Cole 1987; Barbour et al. 1997), but they will rarely be an exact replacement. Intensively managed stands can provide habitat for a wide diversity of species, and can play key roles in maintaining wildlife populations. However, given a level of uncertainty concerning the responses of some species to management and the values of old-growth stands, a strategy that combines active stand-level management with a system of reserves may be most appropriate to achieve biodiversity goals (Seymour and Hunter 1999). In addition to biodiversity considerations, old-growth stands have a variety of other values (e.g., spiritual, recreational, and aesthetic) that should be taken into account when evaluating management decisions.

When establishing stand-management guidelines, it is important to recognize that no single management prescription should be applied over vast stretches of land. The different responses of different species of wildlife to habitat structure suggest that use of multiple management prescriptions is most appropriate to maintain biodiversity (Hansen et al. 1995). For example, although research indicates that thinning provides a variety of economic and ecological benefits over multiple time frames (Hayes et al. 1997), a single thinning prescription, even variable-density thinning, repeated over an entire landscape is unlikely to meet overall objectives for the management of wildlife. Because wildlife need diverse stand conditions and landowners have varied management objectives, general recommendations for stand management are difficult. However, close attention to providing structural complexity in managed forest stands will benefit wildlife (Hansen et al. 1991). We believe that approaches suggested by Carey and Curtis (1996) for stand management to provide for biodiversity and wildlife (Table 5-4) will be effective in providing this structural complexity, and we suggest that implementation of this approach would be highly beneficial for wildlife in the Oregon Coast Range.

Providing habitat for many species of wildlife at the stand level and meeting reasonable timber objectives in the Oregon Coast Range are not mutually exclusive. Work by Chambers et al. (1999) provides evidence that as much as one-third of the volume can be removed from a stand without substantially changing songbird-community composition, at least in the short term. Special attention should be paid to highly sensitive species, and stand-management activities should be pursued in the context of the larger landscape in which they occur in order to ensure perpetuation of indigenous species in an area. We encourage land managers to strive to maintain the range of natural variability at all spatial and temporal scales of forest management.

Table 5-4. The five foundations of stand management for biodiversity, based on Carey and Curtis (1996).

1. Conserving biological legacies during management activities.
2. Planting seedlings at wide spacing, supplemented with natural or artificial regeneration of other conifers or hardwoods.
3. Minimizing the amount of area and time that stands are in the competitive exclusion stage through thinning in young stands.
4. Thinning stands and managing coarse wood to provide for a diversity of habitat niches.
5. Using long rotations on a large portion of the commercial land base.

Riparian Areas Are of Special Management and Ecological Concern

The riparian area is the region where direct interaction between terrestrial and aquatic environments occurs (Swanson et al. 1982). Natural disturbance processes originating both upland (e.g., fire, winds, and disease) and in streams (e.g., erosion and deposition) influence characteristics of riparian areas (Gregory et al. 1991). Fluvial processes create geomorphic features such as banks, benches, and terraces, providing unique microsites that result in a high diversity among riparian plant communities (Swanson et al. 1988; Gregory et al. 1991). Riparian zones are often thought to support greater biomass and diversity of plant and animal species than surrounding uplands (Thomas et al. 1979; Oakley et al. 1985; Melanson 1993) and they do so particularly in arid and nonforested regions where riparian vegetation substantially contrasts with that of the uplands. In forests of the Oregon Coast Range, however, differences in wildlife communities between riparian and upslope areas are more subtle (McGarigal and McComb 1992; McComb et al. 1993a; Kelsey and West 1998) because changes in microclimate and vegetation from streamside to ridgetop are typically not dramatic. Nonetheless, the distinct microclimate (Brosofske et al. 1997) and plant communities (Pabst and Spies 1998) of valley floors and lower slopes provide unique habitat for wildlife in western Oregon. Although wildlife are not always more abundant and diverse in riparian areas than in upslope areas, they often form a community that is distinct from that of upslope areas (McGarigal and McComb 1992; McComb et al. 1993a).

Two dominant riparian gradients influence habitat and wildlife (Johnson and Lowe 1985). The inter-riparian (or transriparian) gradient results from changes in microclimate and vegetation from streamside to ridgetop (Figure 5-15); this gradient is often viewed as subdividing landscapes into "upslope" and "riparian" habitats. A number of factors vary as one moves along the inter-riparian gradient away from the stream; for example, soil moisture and humidity decrease, and variation in temperature increases. An intra-riparian gradient results from changes in stream width, stream volume, and riparian environments that occur as one moves from headwater streams to major rivers and the ocean (Figure 5-16). Width, depth, and temperature of the stream generally increase, gradient generally decreases, and valley width typically increases as one moves down the intra-riparian gradient.

Wildlife communities along an inter-riparian gradient

Although many species of wildlife in the Oregon Coast Range are generalists with regard to the inter-riparian gradient, some species are specifically associated with either riparian or upslope habitat (Table 5-5; McGarigal and McComb 1992; McComb et al. 1993a). Species that are restricted to habitats in and adjacent to aquatic environments are considered to be riparian obligates. Roughly one-third of the wildlife species that use riparian forests in the Pacific coastal ecoregion are riparian obligates (Kelsey and West 1998). Several species of amphibians are riparian obligates because they are restricted to cool, moist habitats or have aquatic life stages. The relatively high humidity and low temperatures in riparian areas that result from cooling by evaporation along streams (Swanson et al. 1982) provide critical habitat for amphibians during the hot, dry summer months.

Among the riparian obligates, beavers (*Castor canadensis*) are a keystone species because of the magnitude of their influence on riparian habitat and the cascading effects of these influences on other species of animals and plants. Through selective herbivory of woody plants and construction of dams, beaver shape plant communities and habitat. For example, beaver avoid red elderberry (*Sambucus racemosa*; Bruner 1989), probably because it contains toxic substances (Nolet et al. 1994). This likely explains the greater abundance of red elderberry at beaver-dam sites than at unoccupied sites (Suzuki and McComb 1998). Red elderberry is an important food for band-tailed pigeons (*Columba fasciata*) (Jarvis and Passmore 1992). Patches of early-seral vegetation created by flooding near beaver dams provide habitat for several species of small mammals that are associated with meadow habitat, including Pacific jumping mice, creeping voles, Townsend's voles, and long-tailed voles (*Microtus longicaudus*) (Suzuki 1992). Furthermore, pools created by beaver dams provide habitat for coho salmon (*Oncorhynchus kisutch*) fry (Bruner 1989), rough-skinned newts (*Taricha granulosa*), and northwestern salamanders (*Ambystoma gracile*) (Suzuki 1992). Historic decreases in abundance of beaver as a result of trapping and development have likely had a substantial influence on distribution and abundance of some species of wildlife (Kelsey and West 1998).

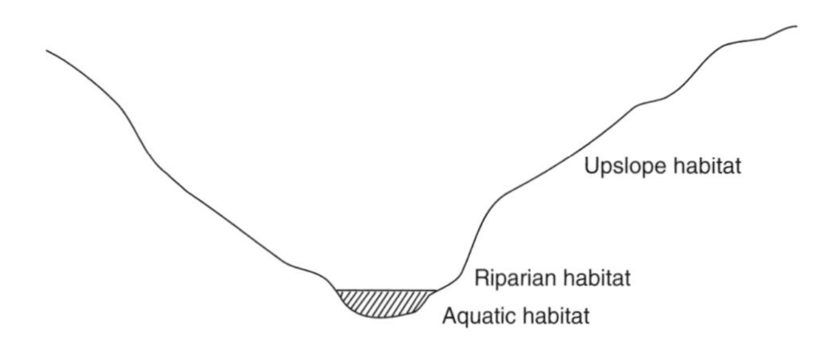

Figure 5-15. The inter-riparian gradient.

Figure 5-16. The intra-riparian gradient.

Not all reaches of stream provide suitable habitat for beaver. Suzuki and McComb (1998) found 170 beaver dams on 65 km of stream in the Drift Creek basin of the Oregon Coast Range; almost all dams (98 percent) were found on relatively small streams (first- through third-order). In addition, Suzuki and McComb (1998) compared channel geomorphology and vegetation at 40 beaver dam sites and 72 randomly selected points not occupied by beavers. They found stream gradient, stream width, and

Table 5-5. Examples of wildlife species associated with portions of the intra- or inter-riparian gradient.

Riparian
Small and headwater streams
Dunn's salamander
Pacific giant salamander
Tailed frog
Torrent salamanders
American dipper
Swainson's thrush
Winter wren
Intermediate streams or species that do not distinguish
Pacific tree frog
Rough-skinned newt
Western toad
Oregon garter snake
Beaver
Large streams and rivers
Foothill yellow-legged frog
Bald eagle
Belted kingfisher
Ducks
Osprey
Great blue heron
River otter
Pacific water shrew
Upslope
Ensatina
Western redback salamander
Northern alligator lizard
Southern alligator lizard
Western fence lizard
Western red-backed vole

valley-floor width accounted for much of the difference between occupied and unoccupied sites. Dams were found most frequently on streams with less than 3 percent gradient, that were 3 to 4 m wide, and that occurred in valleys 25 to 30 m wide. These findings are consistent with those from other areas (Howard and Larson 1985; Beier and Barrett 1987) in which physical characteristics of streams were found to be more important than vegetation in determining occupancy by beavers.

Species that are more abundant in riparian habitats but that also occur in upslope habitats are known as riparian associates. Many species of riparian associates do not respond directly to the presence of open water but rather to the plant communities and habitat features that occur in riparian zones (McComb et al. 1993b). Frequent disturbances in riparian zones, such as debris torrents and flooding caused by beaver dams and storms, create patches of early-seral habitat. As a result, species associated with early-seral habitats are sometimes more abundant in riparian than in upslope habitats (Gomez 1992; Gomez and Anthony 1996).

Several species of bats are associated with riparian areas because they require open water for drinking and they forage for aquatic insects. Hayes and Adam (1996) found four to seven times more bat activity in unlogged riparian areas dominated by alders than in adjacent riparian areas that had been clearcut. The relative proportion of use by different species also varied in response to management. Larger bats, especially silver-haired bats (*Lasionycteris noctivagans*), used logged habitat substantially more than unlogged habitat in riparian areas, but activity of smaller bats was dramatically lower in logged than in unlogged areas.

Several species of wildlife that are associated with deciduous cover (Table 5-3) may be most abundant in riparian areas because of the presence of deciduous vegetation (Figure 5-17 in color section following page 84). For example, some insectivorous birds, such as warbling vireos and Wilson's warblers, are abundant in riparian areas dominated by deciduous trees and shrubs (Hagar, personal observation). The rich arthropod community supported by deciduous hardwoods (Oboyski 1995; Hammond and Miller 1998) probably comprises an important food source for these species. Dense thickets of salmonberry (*Rubus spectabilis*), a deciduous shrub, are abundant in many Coast Range riparian areas (Hayes et al. 1996; Pabst and Spies 1998) and provide habitat for several species of wildlife. Hummingbirds feed on nectar from the flowers of salmonberry, and Swainson's thrushes, a species that is more abundant in riparian than upslope habitats (McGarigal and McComb 1992), feed on the fruit. Several species of small mammals that are more abundant in riparian areas than in coniferous uplands (Gomez 1992; McComb 1994) also may be responding to the deciduous vegetation (red alder [*Alnus rubra*], thimbleberry [*Rubus parviflorus*], and salmonberry) along streams (McComb et al. 1993a). These mammals include the white-footed vole (*Aborimus albipes*), which is rare throughout its range but appears to be most abundant in riparian areas dominated by hardwoods (Maser and Johnson 1967; Maser et al. 1981; Gomez 1992). Red alder and willow (*Salix* spp.), also common in riparian areas in the Coast Range, are abundant in the diet of white-footed voles (Voth et al. 1983).

Wildlife communities along an intra-riparian gradient

Few studies have examined changes in wildlife communities from headwater streams to marine estuaries, but some species of wildlife respond to changes in characteristics of streams and riparian areas along the intra-riparian gradient (Table 5-5). Moving downstream, riparian habitat typically becomes more distinct from that of uplands, as stream size and the influence of streams on riparian habitat increase.

Headwater streams protected by a closed canopy may be particularly important habitat for amphibians for many reasons (Vesely 1997). Small streams generally have fewer and smaller fish that prey on amphibians and the canopy closure over small streams maintains favorable shade and moisture conditions for them. Wood is an important habitat component for many species of wildlife, including amphibians. In general, smaller streams have more wood than larger ones because of the greater influence of streamside forests and the reduced transport capability of small channels (Swanson et al. 1982). Furthermore, conifers, which buffer stream microclimate year-round and are important sources of wood, tend to dominate narrow streams with steep side slopes (Minore and Weatherly 1994).

Large streams seem to be more important than small ones for birds that are riparian obligates and for large mammals that require or use riparian habitats. Mallards (*Anas platyrhynchos*), great blue herons (*Ardea herodias*), dippers, and belted kingfishers (*Ceryle alcyon*) are riparian obligates that regularly use forested riparian areas along larger streams in the Oregon Coast Range (Loegering 1997). At least one species of terrestrial bird that is associated with riparian areas, the Swainson's thrush (McGarigal and McComb 1992), is more abundant along larger streams (Hagar, personal observation). Most of the large mammals that are associated with or are dependent on riparian areas are most abundant along or limited to larger, low-elevation streams because of their body size, habitat needs, or home-range size (Raedeke et al. 1988). River otters (*Lontra canadensis*) typically are restricted to larger streams in the Coast Range. Beaver typically build dams only on smaller streams (Suzuki and McComb 1998), though they often den in banks of larger streams. The wider riparian areas that occur along larger streams encompass a greater portion of the home range of deer and elk than do the narrow riparian communities along small streams (Raedeke et al. 1988).

Managing riparian areas to benefit wildlife

Riparian areas provide a diversity of values; one of these is habitat for wildlife. The goals of current regulations concerning riparian areas in the Oregon Coast Range center on providing habitat for anadromous fish, primarily by attempting to minimize impacts of management activities on stream temperature and promoting development of large-diameter conifers to provide coarse wood in the stream (Hayes et al. 1996; see also Chapters 4 and 7). However, management of riparian areas also has important implications for terrestrial wildlife.

One approach to management of riparian areas is the designation of streamside buffers (Figure 5-18 in color section following page 148). Because of the multiple benefits provided by riparian areas and aquatic systems (ecological value, timber products, aesthetic values, clean water, and others), the large number of streams in the Oregon Coast Range, and the economic implications of streamside buffers, substantial controversy has surrounded the question of buffer size and management restrictions in riparian areas. Designation of buffers with fixed widths has been the standard approach used. Buffer width and restrictions regarding the type of management activities allowed in buffers vary depending on ownership, size, and use of the stream by anadromous fish.

Because timber harvest is sometimes restricted from riparian areas in an effort to protect aquatic habitat and water quality, protection of riparian areas sometimes provides an effective means to maintain undisturbed habitat for species of wildlife that are associated with older forests. However, reliance on riparian-area reserves to provide habitat for these species may be problematic for two reasons. First, streamside areas are subject to frequent natural disturbance. Consequently, much of the riparian habitat in the Oregon Coast Range is currently in early-seral stages (Nierenberg 1997). Because of the likelihood of disturbance, riparian reserves that are currently in late-seral stages may not provide habitat for mature-forest species over the long term—a problem common to reserve-based strategies. Riparian management strategies should anticipate disturbances such as beaver dams, blowdown, and flooding and should incorporate and accommodate these disturbances.

In addition, the habitat needs of some mature-forest species of wildlife are not met in riparian areas alone. McGarigal and McComb (1992) found that abundance, diversity, and species richness of birds and the abundance of five species of birds (brown creeper, chestnut-backed chickadee, dark-eyed junco, golden-crowned kinglet, and Hammond's flycatcher) were greater in upland than riparian habitats in the Oregon Coast Range. Hagar (personal observation) found similar patterns for bird communities in riparian and upland habitats in the Siskiyou Mountains in southwestern Oregon. Some small mammals, such as western red-backed vole, and one species of amphibian (the ensatina [*Ensatina eschscholtzii*]) also are more abundant in upslope than streamside habitats in western Oregon forests (McComb et al. 1993a; Gomez and Anthony 1996; Vesely 1997). Providing riparian reserves will not meet the habitat needs for these species unless the reserves extend well into upland habitat.

It has been suggested that riparian reserves or buffers may function as corridors for travel and dispersal by wildlife between patches of mature forest (Franklin et al. 1981; Harris 1984). The efficacy of corridors for connecting habitat patches has been questioned, although a number of studies have demonstrated that corridors can play a valuable conservation role under some conditions (Beier and Noss 1998). Riparian buffers composed of mature forest may indeed provide dispersal habitat for animals associated with both riparian and old-growth habitat, but no studies have been conducted in western Oregon to examine this question. Because several of these species are associated with mature or old-growth forests in upland habitats (McGarigal and McComb 1992; Gomez and Anthony 1996), riparian areas may not provide adequate connectivity for these species unless the area managed for corridors is wide enough to include suitable upland habitat (McGarigal and McComb 1992; McComb et al. 1993a; Hagar 1999).

Riparian obligates and associates occur throughout the intra-riparian gradient. To meet goals of providing habitat for all indigenous wildlife species, wildlife should be considered in management plans for even the smallest streams. Headwater riparian areas may be most important as habitat for amphibians and as a source of wood for higher-order streams. Management of riparian and adjacent areas on headwater streams should include maintaining large conifers whenever possible, and should provide habitat wide enough to protect amphibian habitat. Vesely (1997) found that abundance of terrestrial amphibians was greatest within 10 m of the stream, but suggested that riparian buffers >30 m wide on one side of a stream may be necessary to protect amphibians from changes in microclimate resulting from clearcutting adjacent stands. In addition to providing habitat for riparian associates, wide (>50 m) buffers along small headwater streams are likely to provide habitat for some bird species associated with mature or old-growth forests (Hagar 1999).

Riparian zones along larger streams and rivers are important to riparian obligates and associates that have large home ranges. Harvesting of timber and other human-induced disturbances have greatly changed habitat along most large streams and rivers in western Oregon. If providing habitat for wildlife is a goal, management of riparian zones along larger streams in forested regions may need to focus on reestablishing riparian vegetation across the width of the floodplain.

Riparian buffers have been shown to provide benefits for some species of wildlife, and wide buffers tend to provide benefits for more riparian-associated species and species associated with older forest than do narrow buffers (Vesely 1997; Hagar 1999). However, establishing wide, fixed buffers on *all* streams is probably ecologically unnecessary, would restrict management options, and could be economically prohibitive. A logical approach is to identify the desired amount and configuration of habitat in different conditions at the landscape scale when establishing management approaches. The width and amount of the buffers should vary depending on the ecological and geomorphological conditions of an area and on management goals. The management history of riparian areas in the Coast Range has left many streams without adequate in-stream structure, such as coarse wood, to support healthy fish habitat (Hayes et al. 1996; see also

Chapter 4). Because coarse wood in streams is important to anadromous fish populations, and because large-diameter conifers appear to provide the highest-quality wood, streamside management has emphasized reestablishment of conifers in riparian areas currently dominated by alder. However, deciduous trees are an important habitat component for several species of wildlife and riparian-zone management should provide hardwoods and early-seral patches in riparian areas. Although the amount of hardwood-dominated riparian areas currently may be more than necessary to meet most management objectives, our capacity to dramatically reduce the amount of this habitat in a relatively short time should not be underestimated.

Using a one-size-fits-all approach to management of riparian areas is probably not appropriate for managing wildlife associated with riparian areas. Designation of fixed-width buffers is administratively simple, but ecologically inefficient or ineffective because of the variability in riparian systems. Buffers may function very differently among sites with different riparian vegetation (e.g., hardwood- versus conifer-dominated stands), among areas with different management histories (e.g., upslope areas clearcut versus thinned), and within sites at different times (e.g., two years after harvest versus 30 years after harvest). We advocate an approach that considers the characteristics of individual streams and their riparian areas and that takes management goals at large spatial scales into account when developing management prescriptions. We suggest a six-step, adaptive approach to managing riparian areas for wildlife (Table 5-6).

Table 5-6. A six-step approach to managing riparian areas for wildlife.

Step 1. Set management goals. Specific riparian management goals should be established for wildlife. These goals should explicitly state species of interest and desired habitat conditions. Management planning and goals for riparian areas should be developed in the context of the larger landscapes in which they occur.

Step 2. Assess current conditions. Conditions of aquatic and riparian areas should be identified, including the amount of riparian area with different vegetative communities or stages of development, geomorphological condition, stream size, and, if possible, current status of wildlife populations.

Step 3. Develop management plans. Management prescriptions should be developed based on management goals and current conditions. Areas of specific emphasis should be identified and included in the management plan, including areas for emphasis of headwater stream amphibians, seep habitat, deciduous-dominated riparian habitat, and older forest conditions.

Step 4. Implement management plans.

Step 5. Monitor. Monitoring is a critical part of any management approach. The timing of monitoring activities will vary depending on the management activity and response system, but monitoring should be conducted at sufficiently frequent intervals to ensure that changes to the management direction can be implemented in a timely manner if goals are not being achieved.

Step 6. Evaluate and adapt. The results of the monitoring should be evaluated and management directions adapted based on those findings. Although it is not possible to anticipate all possible outcomes, likely outcomes should be identified, specific scenarios that would trigger a change in management direction should be determined, and proposed changes to management direction should be identified when the initial management prescriptions are developed.

Landscape Structure Influences the Presence and Abundance of Wildlife

In the past several years, there has been increased recognition that patterns at large spatial scales can influence the distribution and abundance of wildlife. A central concern has focused on potential negative effects of forest fragmentation and the influences of habitat edge on wildlife. The term "fragmentation" has become widely used and is sometimes (inappropriately) used synonymously with "habitat loss." Forest fragmentation is the "unnatural detaching or separation of expansive tracts [of forest] into spatially segregated fragments" (Harris and Silva-Lopez 1992). Although usually coincident, fragmentation is thus not always synonymous with habitat loss; it is possible to have habitat loss without fragmentation. Much of the concern over fragmentation and edge effects stems from observations and studies of isolated forest patches in agricultural, urban, or tropical landscapes, which document that impacts on wildlife are related to the size of habitat patches and the degree of their isolation (Forman 1997). Relatively little work has been conducted in coniferous forests of the western United States.

Landscape fragmentation is best understood in light of the needs of the species in question. For a given species, if a landscape consists of distinct patches of habitat separated by non-habitat, then that landscape is fragmented as far as that species is concerned. The response of organisms of that species to fragmentation will vary depending on factors related to the ecologyof the species, including its dispersal capability, site fidelity, and ecological niche. A landscape may be highly variable and yet not fragmented in terms of the habitat needs of the species in question. Organisms may respond to landscape structure as variegated rather than fragmented (McIntyre and Barrett 1992). Variegated landscapes consist of a series of patches that vary along a continuum with respect to suitability of habitat (Figure 5-19). The quality of the habitat patches in variegated landscapes changes through time as succession takes place. The "permeability" of patches differs along a continuum as well (Ingham and Samways 1996); some patches can be crossed by an organism relatively easily, whereas others are more hostile environments that are difficult to cross. Permeability of patches in variegated landscapes also changes through time. Although it would be convenient to be able to label landscapes unambiguously and consistently as "fragmented" or "variegated," different species can respond very differently to the same landscape. As a result, landscape structure should be considered with reference to the species of interest.

Landscape structure and species present

Landscape structure can be described by two components: landscape composition and landscape configuration. Landscape composition comprises the amount and types of habitat occurring in a given landscape. It can generally be described simply by the area or proportion of a landscape composed of

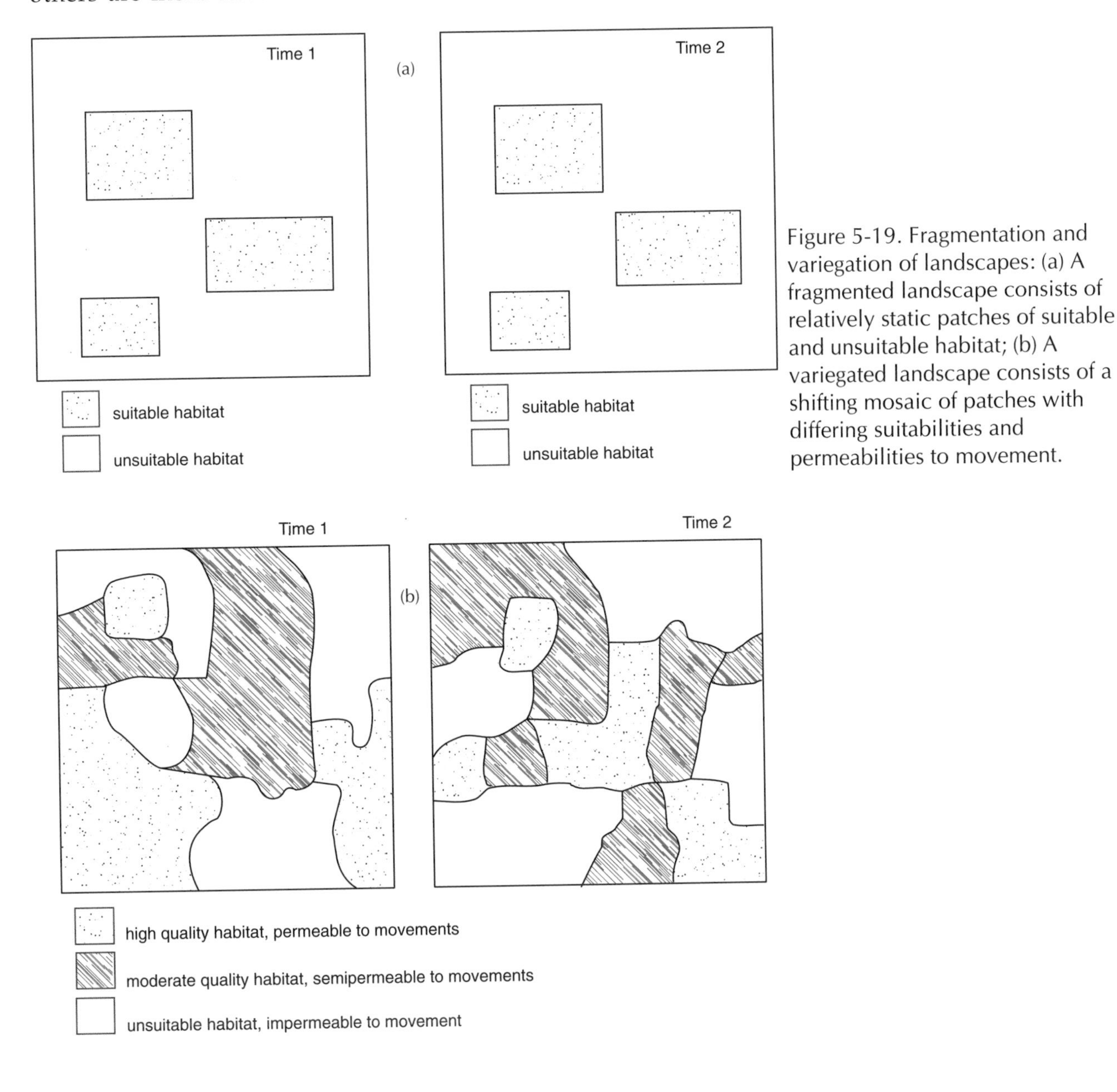

Figure 5-19. Fragmentation and variegation of landscapes: (a) A fragmented landscape consists of relatively static patches of suitable and unsuitable habitat; (b) A variegated landscape consists of a shifting mosaic of patches with differing suitabilities and permeabilities to movement.

different habitat types. Landscape configuration is the spatial patterning of those habitat types. Landscape composition has a strong influence on the distribution and abundance of wildlife species in an area. Part of the influence of landscapes on wildlife is simply the cumulation of influences occurring at smaller spatial scales. For example, species that prefer open forest habitat, such as the Townsend's solitaire (*Myadestes townsendi*), generally will be more abundant in a landscape with large amounts of open forest than in landscapes dominated by dense forest; much of this increase in abundance is simply the additive result of the increase in available habitat. In addition to the cumulative effects of small spatial scales, landscapes sometimes can have emergent properties that influence distribution and abundance of species (Bissonette 1997). Landscape-level effects most frequently result from one of three factors. First, some minimum threshold amount of a particular habitat type must be present to maintain viable populations of some species in a landscape. Second, landscape composition is also important to species requiring a mix of habitat types to thrive in an area. For example, some species of bats have distinct requirements for roosting habitat, water sources, and foraging habitat; these species will be most abundant in landscapes with a mixture of these habitats. Finally, the size and distribution of habitat patches may influence the distribution and abundance of some species.

Because of expense and logistical difficulties, relatively few studies have directly considered the influence of landscape structure on wildlife populations. As a result, knowledge of the influence of landscape structure on many species in the Oregon Coast Range is minimal. But one of the best studies conducted anywhere to date on this topic is McGarigal and McComb's (1995) study on the influence of landscape structure on breeding-bird populations in the Oregon Coast Range. They focused on the composition and configuration of landscapes with respect to mature forests in 30 subbasins in three watersheds (Figure 5-20). The researchers compared the abundances of 15 species of birds that they determined to be associated with mature forests (Table 5-7) and assessed the relationship of landscape composition and configuration to the relative abundance of birds based on nearly 25,000 detections of the 15 species of birds of interest. They concluded that landscapes with higher proportions of area in late-seral stage forest (a measure of landscape composition) had greater numbers of birds for 10 of the 15 species associated with late-seral forest (Table 5-7). More than half of the variability in numbers of gray jays (*Perisoreus canadensis*), brown creepers, winter wrens (*Troglodytes troglodytes*), and varied thrushes could be explained by the proportion of the landscape in late-seral stage forest.

In general, landscape configuration was less important than landscape composition for most of the species examined. Of nine species shown to be responsive to landscape configuration by one or more analyses, eight responded positively to edge or were most abundant in more-fragmented landscapes. The winter wren was the only species that showed evidence of a positive relationship with less-fragmented landscapes (Table 5-7). The greater importance of landscape composition than landscape configuration is consistent with other studies of northwestern coniferous forests (Rosenberg and Raphael 1986; Lehmkuhl et al. 1991).

McGarigal and McComb (1995) emphasize that landscape pattern and fragmentation may have a stronger influence on species that are rare or patchily distributed. In addition, species that are habitat specialists or have poor dispersal abilities may respond quite differently to landscape structure. Thus, it is important not to extrapolate their findings beyond the context of the study.

In a parallel study, Martin (1998) examined responses of small mammals and amphibians to landscape structure in the same sites used by McGarigal and McComb (1995). These findings suggested that landscape configuration may play a more important role for amphibians than for birds. Southern torrent salamanders (*Rhyacotriton variegatus*), Pacific giant salamanders (*Dicamptodon tenebrosus*), and tailed frogs (*Ascaphus truei*) were all most abundant in landscapes that had larger patches of mature forest with less edge habitat. Martin did not analyze landscape associations for mammals. However, the amount of mature-forest habitat in a landscape did appear to play an important role in determining distribution and abundance of 16 species of forest-floor vertebrates.

The results of Meyer et al.'s (1998) investigations of the influences of landscape structure on northern spotted owls parallel those of McGarigal and McComb (1995). Meyer et al. (1998) compared sites occupied by owls to random points in western

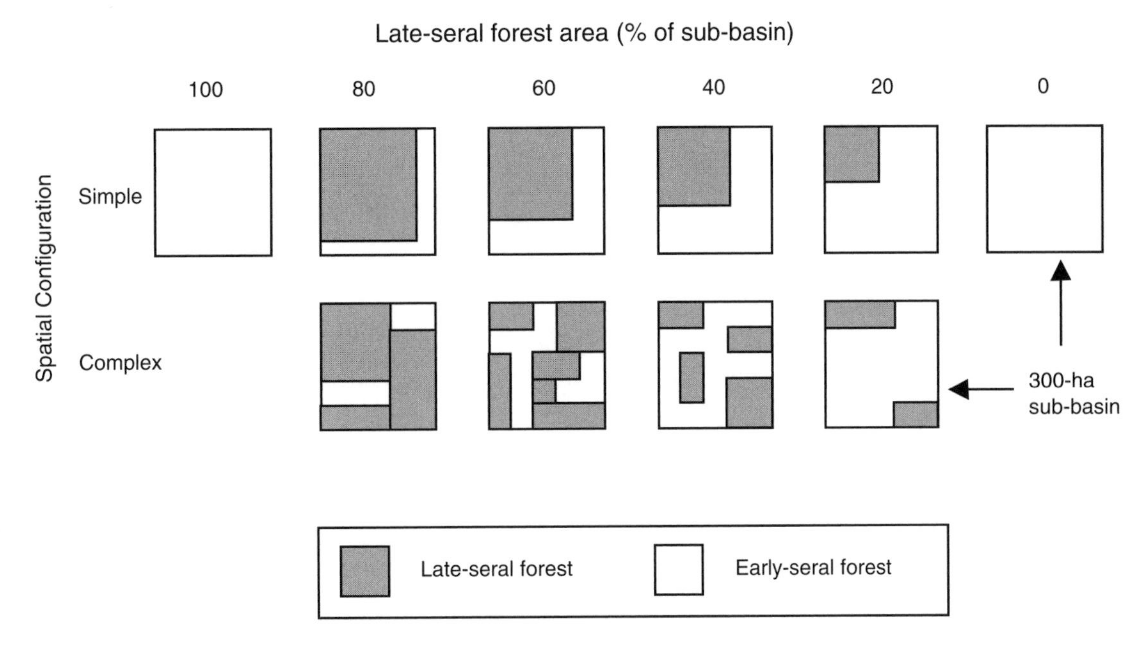

Figure 5-20. A schematic of the study design used by McGarigal and McComb (1995) to investigate the influence of landscape structure on abundance of birds associated with mature forests in the Oregon Coast Range. They selected 10 sub-basins with characteristics similar to those shown in this figure, in each of three watersheds. Each sub-basin was roughly 300 hectares in size. Using this approach, they were able to independently assess the influence of spatial configuration and composition of landscapes on the abundance of the species they studied. (Reprinted by permission of Ecological Society of America.)

Oregon to determine the influence of landscape composition and pattern on use of landscapes by owls. Landscapes used by owls had significantly more old-growth habitat than did random sites and the average size of patches of old growth was larger in landscapes used by owls than in random landscapes. However, none of the estimates for the six indices related to forest fragmentation or the four indices of patch isolation differed between used and random sites. As Meyer et al. (1998) caution, results of landscape-level analyses can vary substantially with the spatial scale chosen for analysis, and thus results should not be extrapolated to inappropriate scales. However, their findings are consistent with the notion that the amount of habitat is more important than its distribution, although patch size is likely to be important for some species.

Positive and negative effects of edge

The influence of edge on wildlife populations was recognized decades ago by Leopold (1933). Where two habitat types converge, resources found in both habitat types are available for use by wildlife. Several species are most abundant at edges. Because many game species are associated with habitat edge,

Table 5-7. Responses of 15 bird species associated with old-growth forest to landscape composition and landscape configuration.

Species	*Response to landscapes with*	
	greater proportion of late-seral forest	*increased fragmentation*
Winter wren	I	D
Varied thrush	I	N
Evening grosbeak	I	N
Chestnut-backed chickadee	I	N
Brown creeper	I	N
Hammond's flycatcher	I	N
Pileated woodpecker	I	I
Red-breasted nuthatch	I	I
Gray jay	I	I
Western wood-pewee	I	I
Red-tailed hawk	N	I
Red-breasted sapsucker	N	I
Western tanager	N	I
Olive-sided flycatcher	N	I
Red crossbill	N	N

'I' indicates increased abundance, 'D' indicates decreased abundance, and 'N' indicates no apparent influence; from McGarigal and McComb 1995.

managing areas to maximize edge was a primary focus of wildlife management for decades. In recent years, there has been increasing recognition that edges are detrimental for some species of wildlife, and that some species are associated with interior forest habitat.

In their review of the influences of edge on wildlife, Kremsater and Bunnell (1999) noted that the response of most mammals to forest edges is neutral or positive. One species that appears to be negatively influenced by edges is the western red-backed vole. Working in southwestern Oregon, Mills (1995) found that western red-backed voles exhibited a strong, negative response to forest-clearcut edges. Six times as many voles were captured in the interior of isolated forest patches than on the edge. In addition, greater densities of voles occurred in large patches of forest than in small patches. Bird species that may respond negatively to edges include varied thrushes, golden-crowned kinglets, Hammond's flycatchers, and hermit warblers. These species were not found in riparian buffers along small streams in the Oregon Coast Range, but they were found in unharvested riparian forests (Hagar 1999). Riparian buffers form narrow, linear forest tracts with high-contrast edges where they pass through recent clearcuts and may not provide suitable habitat for species associated with closed-canopy forests.

Other species of small mammals and birds differ in their response to isolation and habitat edge. Trowbridge's shrews (*Sorex trowbridgii*) were equally abundant in forest remnants and surrounding clearcuts and exhibited no edge effect, and Townsend's chipmunks were more abundant in the forest remnants than in the surrounding cut area, but were equally as abundant along the edge as in the interior of forest remnants (Mills 1996). In contrast, deer mice were most abundant in cut areas, and were seven times more likely to occur near the edge than in the interior of the forest remnants (Mills 1996). Birds that may respond positively to habitat edge include some cavity excavators in the Oregon Cascades. Nests of hairy woodpeckers, northern flickers, and red-breasted sapsuckers were most abundant near edges (Church 1997).

Managing landscapes is difficult but necessary

The task of managing landscapes for wildlife is complicated by the fact that each species responds uniquely to landscape structure. A landscape that is fragmented for one species may not be for another; an edge between two habitat types may negatively impact one species, be neutral to a second, and be favored by a third. In addition, because of differences in grain-response (the scale at which heterogeneity is perceived by a species; Forman 1997), one species may respond to a given landscape as consisting of a highly heterogeneous set of habitats, whereas a different species may respond to the same landscape as though it were homogeneous. As a result, it is difficult to make generalized statements concerning the management of landscapes. But despite the inadequacy of our current level of knowledge to fully understand all of the influences of landscape structure on wildlife in the Oregon Coast Range, our base of information is expanding and provides a point for initial recommendations. McGarigal and McComb's (1995) study points out the importance of landscape composition on the abundance and distribution of some species. Although that study examined a limited set of species, we believe its findings are likely to be generalizable to most, if not all, species of wildlife in the Oregon Coast Range. The recommendation stemming from this is obvious—maintain as much suitable habitat for as many species in a landscape as possible. The challenges resulting from this recommendation are twofold: to balance the habitat needs of different species and to balance habitat goals against other amenity and management goals.

Research results also suggest that, although landscape composition strongly influences many species of wildlife, landscape configuration is not as important for many species in the Oregon Coast Range. As McGarigal and McComb (1995) point out, the shifting mosaic of habitat patches resulting from land-management activities in the Oregon Coast Range generally does not result in isolation of species for prolonged periods of time. In addition, as they also argue, landscapes in the Oregon Coast Range are naturally heterogeneous because of their geomorphic and disturbance history, suggesting that species indigenous to the area may be adapted to survival in landscapes similar to those resulting from forest management activities. As a consequence, we suspect that landscape configuration minimally influences a wide variety of species in the Oregon Coast Range, although the empirical evidence to support this needs to be strengthened considerably.

Appropriate approaches to landscape management necessary to ensure long-term viability of populations differ depending on whether organisms respond to given landscapes as fragmented or variegated. For organisms that respond to a landscape as fragmented, land management should focus on providing specialized habitat to connect otherwise isolated populations. For organisms that respond to the landscape as variegated, management should focus on providing a mosaic of habitats between preferred habitats in a manner that does not isolate populations for prolonged periods of time.

McGarigal and McComb (1995) provide strong evidence that several species of diurnal birds perceive forested landscapes in the Oregon Coast Range as variegated, and we suspect that many other species respond similarly. As these authors point out, the nature of the disturbance regime resulting from forest management activities in the Oregon Coast Range results in a dynamic, shifting pattern of forest patches over time. Because of the dynamic nature of Coast Range forests, stands rarely are effectively isolated from one another for extensive periods for these species of birds. In addition, for many species of wildlife there is a lack of clear distinction between "habitat" and "non-habitat," as necessitated by fragmentation models, but rather a gradation of habitat quality exists. Although configuration of preferred-habitat patches is not unimportant in variegated landscapes, it is more important in fragmented landscapes. In variegated landscapes, the configuration of the intervening matrix around the preferred-habitat patches may be of greater importance than the configuration of the preferred habitat. However, the influence of configuration of patches within the matrix has not yet been examined for wildlife in the Oregon Coast Range or similar systems. Thus, although there is strong evidence suggesting that configuration of preferred-habitat patches does not strongly influence the abundance of several species in the Oregon Coast Range (McGarigal and McComb 1995; Meyer et al. 1998), it does not necessarily follow that configuration of the intervening matrix is unimportant. We suspect that it may be highly significant to some species. Percolation theory (Forman 1997; With 1999) suggests that configuration of preferred patches and the matrix may become increasingly important as the proportion of the landscape occupied by preferred habitat decreases.

Rare species may be sensitive to landscape structure; unfortunately, due to statistical constraints, it is often difficult to assess patterns for rare species. Furthermore, species with poor dispersal abilities are more likely to be sensitive to landscape configuration than are those capable of dispersing long distances.

Two basic principles of conservation biology can be used to guide general directions in landscape management. First, for a variety of genetic and demographic reasons, isolated populations are more vulnerable than non-isolated populations to local extinction. Thus, connectivity between habitat patches is important in order to maintain species on a landscape. What constitutes connectivity differs among species. A species that responds to a particular landscape as fragmented may require specialized habitat, such as designated movement corridors, to prevent isolation. For a species that responds to the same landscape as variegated, maintaining an appropriate mix of high-quality habitat in a matrix of lands that are suitable for dispersal should prevent isolation of populations. Additional research is needed to determine patterns of species response to landscape structure in the Oregon Coast Range. Second, small populations are more vulnerable to local extinction than are large populations. Because of this, to maintain viable populations of a species on a landscape, habitat patches must be of adequate size, given the scale to which a particular species is responding. For example, one hectare of high-quality habitat may be adequate to maintain populations of a number of species of invertebrates, but the patch is not likely to be adequate for many species of vertebrates if it is highly isolated.

Effective management of landscapes and perceptions of landscape-level patterns and responses has been impeded by a variety of factors acting in concert. A significant challenge to managing at large spatial scales is imposed by the lack of congruence of ecological and administrative boundaries (Knight and Landres 1998). In the Oregon Coast Range, many landscapes consist of a patchwork of ownerships, each with an independent set of management goals and directions (Garman et al. 1999). Planning at large spatial scales in landscapes with mixed ownerships is often complicated by a lack of control over the habitat conditions of a significant portion of the landscape. These factors present difficulties, although not necessarily insurmountable ones (Landres 1998), to managing at large spatial scales. In situations where cooperative decisions concerning landscape-level planning are not possible or desirable, we suggest that, at a minimum, likely management patterns for neighboring parcels of land be considered when determining and evaluating management options.

Difficulties in perceiving landscape patterns and the likely consequences of management actions at the landscape level are being reduced by the use of remotely sensed spatial data and with the development of more sophisticated computer software to (1) enable mapping and tracking of landscape conditions (through use of GIS); (2) evaluate landscape structure (through use of software such as FRAGSTATS; McGarigal and Marks 1995); (3) track spatially explicit changes in landscape patterns through time resulting from management activities (through use of software such as SNAP; Sessions et al. 1994); and (4) determine likely influences on abundance of wildlife (through use of software such as ZELIG; Garman et al. 1992; Urban et al. 1999). Large-scale modeling projects, such as CLAMS (see Chapters 3 and 10), provide a context for evaluating management decisions at the regional scale. The accessibility of these models varies, but we anticipate that managers will have greatly increased access to these types of models in the coming years. Use of computer models to evaluate influences of management at large spatial scales will never fully substitute for on-the-ground monitoring of responses to management, but they can provide an extremely valuable tool for planning. Computer models will become increasingly powerful as they are refined through incorporation of additional information, particularly as information is gathered concerning response of wildlife species to landscape structure.

Summary and Conclusions

Forests of the Oregon Coast Range provide homes for a diverse array of wildlife, each with its own unique natural history. As a result, each species responds uniquely to management activities. The diversity of species and their responses to management activities provide a special challenge to land managers who are interested in understanding the influences of management activities on wildlife or in maintaining viable populations of all indigenous species in the forests they manage.

Despite the variability in species responses, a number of generalities and themes emerge. First, almost no management activity is inherently or entirely "good" or "bad" for wildlife in general; management activities currently being implemented in Coast Range forests are generally beneficial to some species and detrimental to others. Even "doing nothing" or not "actively managing" a stand is beneficial to some species and detrimental to others. To assess the influence of management activities on wildlife and develop appropriate approaches to management, it is necessary to clearly identify management goals with respect to wildlife populations, and then assess resulting or likely influences of management activities in that context.

Second, increased structural complexity leads to increased diversity. This relationship is the result of two processes. First, the habitat needs of individual species are often multifaceted. Thus, areas that provide multiple features, such as a combination of habitat for forage, roost sites, nest sites, protective cover, thermal cover, and a water source, will provide habitat for some species that would not be able to exploit areas with simpler structure. In addition, structurally complex environments provide an array of niches that can be exploited by a variety of species. As a result of these two processes, structurally complex environments provide habitat for more species than do structurally simple environments. Finally, recognition of the importance of spatial and temporal scale to species response is key to managing for wildlife. The importance of a given spatial scale (e.g., individual habitat component, stand-level characteristics, or landscape structure) differs among species of wildlife. However, neglect of any spatial scale is likely to have negative consequences for some species. Furthermore, to fully understand the influences of management activities on wildlife, we must recognize both short-term (days to several years) and long-term (several years to decades or centuries) influences on ecological characteristics and developmental trajectories of stands.

Literature Cited

Adam, M. D., and J. P. Hayes. 1998. Elk and deer, pp. 113-118 in *Effects of Commercial Thinning on Stand Structure and Wildlife Populations: A Progress Report, 1994-1997.* Coastal Oregon Productivity Enhancement Program, College of Forestry, Oregon State University, Corvallis.

Adam, M. D., and J. P. Hayes. 2000. Use of bridges as night-roosts by bats in the Oregon Coast Range. *Journal of Mammalogy*, in press.

Bailey, J. D., and J. C. Tappeiner II. 1998. Effects of thinning on structural development in 40- to 100-year-old Douglas-fir stands in western Oregon. *Forest Ecology and Management* 108: 99-113.

Barbour, R. J., S. Johnson, J. .P Hayes, and G. F. Tucker. 1997. Simulated stand characteristics and wood product yields from Douglas-fir plantations managed for ecosystem objectives. *Forest Ecology and Management* 91: 205-219.

Barclay, R. M. R., P. A. Faure, and D. R. Farr. 1988. Roosting behavior and roost selection by migrating silver-haired bats (*Lasionycteris noctivagans*). *Journal of Mammalogy* 69: 821-825.

Beier, P., and R. H. Barrett. 1987. Beaver habitat use and impact in Truckee River Basin, California. *Journal of Wildlife Management* 51: 794-799.

Beier, P., and R. F. Noss. 1998. Do habitat corridors provide connectivity? *Conservation Biology* 12: 1241-1252.

Bissonette, J. A. 1997. Scale-sensitive ecological properties: historical context, current meaning, pp. 3-31 in *Wildlife and Landscape Ecology Effects of Pattern and Scale*, J. A. Bissonette, ed. Springer-Verlag, New York.

Brigham, R. M., M. J. Vonhof, R. M. R. Barclay, and J. C. Gwilliam. 1997. Roosting behavior and roost-site preferences of forest-dwelling California bats (*Myotis californicus*). *Journal of Mammalogy* 78: 1231-1239.

Brosofske, K. D., J. Chen, R. J. Naiman, and J. F. Franklin. 1997. Harvesting effects on microclimatic gradients from small streams to uplands in western Washington. *Ecological Applications* 7: 1188-1200.

Bruner, K. L. 1989. *Effects of Beaver on Streams, Streamside Habitat, and Coho Salmon Fry Populations in Two Coastal Oregon Streams*. MS thesis, Oregon State University, Corvallis.

Bull, E. L., and A. D. Partridge. 1986. Methods of killing trees for use by cavity nesters. *Wildlife Society Bulletin* 14: 142-146.

Bunnell, F. L., L. L. Kremsater, and R. W. Wells. 1997. *Likely Consequences of Forest Management on Terrestrial, Forest-dwelling Vertebrates in Oregon*. Report M-7, Centre for Applied Conservation Biology, University of British Columbia, Oregon Forest Resources Institute, Portland OR.

Campbell, L. A., J. G. Hallett, and M. A. O'Connell. 1996. Conservation of bats in managed forests: use of roosts by *Lasionycteris noctivagans*. *Journal of Mammalogy* 77: 976-984.

Carey, A. B., and R. O. Curtis. 1996. Conservation of biodiversity: a useful paradigm for forest ecosystem management. *Wildlife Society Bulletin* 24: 610-620.

Carey, A. B., and M. L. Johnson. 1995. Small mammals in managed, naturally young, and old-growth forests. *Ecological Applications* 5: 336-352.

Carey, A. B., M. M. Hardt, S. P. Horton, and B. L. Biswell. 1991. Spring bird communities in the Oregon Coast Range, pp. 123-142 in *Wildlife and Vegetation of Unmanaged Douglas-fir Forests*, L. F. Ruggiero, K. B. Aubry, A. B. Carey, and M. H. Huff, tech. coord. General Technical Report PNW-GTR-285, USDA Forest Service, Pacific Northwest Research Station, Portland OR.

Carey, A. B., T. M. Wilson, C. C. Maguire, and B. L. Biswell. 1997. Dens of northern flying squirrels in the Pacific Northwest. *Journal of Wildlife Management* 61: 684-699.

Chambers, C. L. 1996. *Response of Terrestrial Vertebrates to Three Silvicultural Treatments in the Central Oregon Coast Range*. PhD dissertation, College of Forestry, Oregon State University, Corvallis.

Chambers, C. L, T. Carrigan, T. E. Sabin, J. Tappeiner, and W. C. McComb. 1997. Use of artificially created Douglas-fir snags by cavity-nesting birds. *Western Journal of Applied Forestry* 12: 93-97.

Chambers, C. L., W. C. McComb, and J. Tappeiner II. 1999. Breeding bird community responses to 3 silvicultural treatments in the Oregon Coast Range. *Ecological Applications* 9: 171-185.

Christy, R. E., and S. D. West. 1993. *Biology of Bats in Douglas-fir Forests*. General Technical Report PNW-GTR-308, USDA Forest Service, Pacific Northwest Research Station, Portland OR.

Church, T. A. 1997. *Habitat Associations of Cavity Nesters at Multiple Scales in Managed Forests of the Southern Oregon Cascades*. MS thesis, Oregon State University, Corvallis.

Clark, A. B. 1995. *Northern Flying Squirrel Den-sites and Associated Habitat Characteristics in Second-growth Forests of Southwestern Washington*. MS thesis, University of Washington, Seattle.

Clarkson, D. A., and L. S. Mills. 1994. *Hypogeous sporocarps* in forest remnants and clearcuts in southwest Oregon. *Northwest Science* 68: 259-265.

deMaynadier, P. G., and M. L. Hunter, Jr. 1995. The relationship between forest management and amphibian ecology: a review of the North American literature. *Environmental Review* 3: 230-261.

Doyle, A. T. 1987. Microhabitat separation among sympatric microtines, *Clethrionomys californicus, Microtus oregoni*, and *M. richardsoni*. *American Midland Naturalist* 118: 258-265.

Feen, J. S. 1997. *Winter Den Site Habitat of Northern Flying Squirrels in Douglas-fir Forests of the South-central Oregon Cascades*. MS thesis, Oregon State University, Corvallis.

Forman, R. T. T. 1997. *Land Mosaics: the Ecology of Landscapes and Regions*. Cambridge University Press, New York.

Forsman, E. D., E. C. Meslow, and H. M. Wight. 1984. Distribution and biology of the spotted owl. *Wildlife Monographs* 87: 1-64.

Franklin, J. F., K. Cromack Jr., W. Denison, A. McKee, C. Maser, J. Sedell, F. Swanson, and G. Juday. 1981. *Ecological Characteristics of Old-growth Douglas-fir Forests.* General Technical Report PNW-118, USDA Forest Service, Pacific Northwest Forest and Range Experimental Station, Portland OR.

Garman, S. L., A. J. Hansen, D. L. Urban, and P. F. Lee. 1992. Alternative silvicultural practices and diversity of animal habitat in western Oregon: a computer simulation approach, pp. 777-781 in *Proceedings of the 1992 Summer Computer Simulation Conference, Society for Computer Simulation, Reno, Nevada,* P. Luker, ed. The Society for Computer Simulation, San Diego CA.

Garman, S. L., F. J. Swanson, and T. A. Spies. 1999. Past, present, and future landscape patterns in the Douglas-fir region of the Pacific Northwest, pp. 61-86 in *Forest Fragmentation: Wildlife and Management Implications,* J. A. Rochelle, L. A. Lehmann, and J. Wisniewski, ed. Brill, Leiden, The Netherlands.

Gerson, U. 1973. Lichen-arthropod associations. *Lichenologist* 5: 434-443.

Gilbert, F. F., and R. Allwine. 1991. Spring bird communities in the Oregon Cascade Range, pp. 145-158 in *Wildlife and Vegetation of Unmanaged Douglas-fir Forests,* L. F. Ruggiero, K. B. Aubry, A. B. Carey, and M. H. Huff, tech. coord. General Technical Report PNW-GTR-285, USDA Forest Service, Pacific Northwest Research Station, Portland OR.

Gomez, D. M. 1992. *Small Mammal and Herpetofauna Abundance in Riparian and Upslope Areas of Five Forest Conditions.* MS thesis, Oregon State University, Corvallis.

Gomez, D. M., and R. G. Anthony. 1996. Amphibian and reptile abundance in riparian and upslope areas of five forest types in western Oregon. *Northwest Science* 70: 109-119.

Gomez, D. M., R. G. Anthony, and J. P. Hayes. 1998. Influence of commercial thinning on northern flying squirrels and Townsend's chipmunks, pp. 81-102 in *Effects of Commercial Thinning on Stand Structure and Wildlife Populations: A Progress Report, 1994-1997.* Coastal Oregon Productivity Enhancement Program, College of Forestry, Oregon State University, Corvallis.

Gregory, S. V., F. J. Swanson, W. A. McKee, and K. W. Cummins. 1991. An ecosystem perspective of riparian zones. *BioScience* 41: 540-551.

Gumtow-Farrior, D. L. 1991. *Cavity Resources in Oregon White Oak and Douglas-fir Stands in the Mid-Willamette Valley, Oregon.* MS thesis, Oregon State University, Corvallis.

Hagar, J. C. 1992. *Bird Communities in Commercially Thinned and Unthinned Douglas-fir stands of Western Oregon.* MS thesis, Oregon State University, Corvallis.

Hagar, J. C. 1999. Influence of riparian buffer width on bird assemblages in western Oregon. *Journal of Wildlife Management* 63: 484-496.

Hagar, J. C., W. C. McComb, and C. L. Chambers. 1995. Effects of forest practices on wildlife, chapter 9 in *Cumulative Effects of Forest Practices in Oregon: Literature and Synthesis,* R. L. Beschta, J. R. Boyle, C. L. Chambers, W. P. Gibson, S. V. Gregory, J. Grizzel, J. C. Hagar, J. L. Li, W. C. McComb, T. W. Parzybok, M. L. Reiter, G. H. Taylor, and J. E. Warila, ed. Prepared for Oregon Department of Forestry, Oregon State University, Corvallis.

Hagar, J. C., W. C. McComb, and W. H. Emmingham. 1996. Bird communities in commercially thinned and unthinned Douglas-fir stands of western Oregon. *Wildlife Society Bulletin* 24: 353-366.

Hall, L. S., P. R. Krausman, and M. L. Morrison. 1997. The habitat concept and a plea for standard terminology. *Wildlife Society Bulletin* 25: 173-182.

Hamer, T. E., and S. K. Nelson. 1995. Characteristics of marbled murrelet nest trees and nesting stands, pp. 69-82 in *Ecology and Conservation of the Marbled Murrelet,* C. J. Ralph, G. L. Hunt, M. G. Raphael, and J. F. Piatt, ed. General Technical Report PSW-GTR-152, USDA Forest Service, Pacific Southwest Research Station, Berkeley CA.

Hammond, P. C., and J. .C Miller. 1998. Comparison of the biodiversity of Lepidoptera within three forested ecosystems. *Annals of the Entomological Society of America* 91: 323-328.

Hansen, A. J., T. A. Spies, F. J. Swanson, and J. L. Ohmann. 1991. Conserving biodiversity in managed forests. *BioScience* 41: 382-392.

Hansen, A. J., W. C. McComb, R. Vega, M. G. Raphael, and M. Hunter. 1995. Bird habitat relationships in natural and managed forests in the west Cascades of Oregon. *Ecological Applications* 5: 555-569.

Harris, L. D. 1984. *The Fragmented Forest, Island Biogeography Theory and the Preservation of Biotic Diversity.* University of Chicago Press, Chicago IL.

Harris, L. D., and G. Silva-Lopez. 1992. Forest fragmentation and the conservation of biological diversity, pp. 197-237 in *Conservation Biology: The Theory and Practice of Nature Conservation and Management,* P. L. Fiedler and S. K. Jain, ed. Chapman and Hall, New York.

Hayes, J. P. 1996. *Arborimus longicaudus. Mammalian Species* 532: 1-5.

Hayes, J. P., and M. D. Adam. 1996. The influence of logging riparian areas on habitat use by bats in western Oregon, pp. 228-237 in *Bats and Forests Symposium, October 19-21, 1995, Victoria, British Columbia,* R. M. R. Barclay and R. M. Brigham, ed. Working Paper 23, British Columbia Ministry of Forests Research Branch, Victoria BC.

Hayes, J. P., and S. P. Cross. 1987. Characteristics of logs used by western red-backed voles (*Clethrionomys californicus*), and deer mice (*Peromyscus maniculatus*). *Canadian Field-Naturalist* 101: 543-546.

Hayes, J. P., and J. Gruver. 2000. Vertical stratification in bat activity in an old-growth forest in western Washington. *Northwest Science* 74: 102-108.

Hayes, J. P., S. P. Cross, and P. W. McIntire. 1986. Seasonal variation in mycophagy by the western red-backed vole, *Clethrionomys californius*, in southwestern Oregon. *Northwest Science* 60: 250-257.

Hayes, J. P., E. G. Horvath, and P. Hounihan. 1995. Townsend's chipmunk populations in Douglas-fir plantations and mature forests in the Oregon Coast Range. *Canadian Journal of Zoology* 73: 67-73.

Hayes, J. P., M. D. Adam, D. Bateman, E. Dent, W. H. Emmingham, K. G. Maas, and A. E. Skaugset. 1996. Integrating research and forest management in riparian areas of the Oregon Coast Range. *Western Journal of Applied Forestry* 11: 1-5.

Hayes, J. P., S. S. Chan, W. H. Emmingham, J. C. Tappeiner II, L. D. Kellogg, and J. D. Bailey. 1997. Wildlife response to thinning young forests in the Pacific Northwest. *Journal of Forestry* 95(8): 28-33.

Hayes, J. P., J. M. Weikel, and M. D. Adam. 1998. Abundance of breeding birds, pp. 39-60 in *Effects of Commercial Thinning on Stand Structure and Wildlife Populations: A Progress Report, 1994-1997.* Coastal Oregon Productivity Enhancement Program, College of Forestry, Oregon State University, Corvallis.

Hershey, K. T., E. C. Meslow, and F. L. Ramsey. 1998. Characteristics of forests at spotted owl nest sites in the Pacific Northwest. *Journal of Wildlife Management* 62: 1398-1410.

Hope, S., and W. C. McComb. 1994. Perceptions of implementing and monitoring wildlife tree prescriptions on national forests in western Washington and Oregon. *Wildlife Society Bulletin* 22: 383-392.

Howard, R. J., and J. S. Larson. 1985. A stream habitat classification system for beaver. *Journal of Wildlife Management* 49: 19-25.

Huffman, D. W., J. C. Tappeiner II, and J. C. Zasada. 1994. Regeneration of salal (*Gaultheria shallon*) in the central Coast Range forests of Oregon. *Canadian Journal of Botany* 72: 39-51.

Humes, M. L., J. P. Hayes, and M. Collopy. 1999. Activity of bats in thinned, unthinned, and old-growth forests in western Oregon. *Journal of Wildlife Management* 63: 553-561.

Hunter, M. L. Jr., G. L. Jacobson Jr., and T. Webb III. 1988. Paleoecology and the coarse-filter approach to maintaining biological diversity. *Conservation Biology* 2: 375-385.

Ingham, D. S., and M. J. Samways. 1996. Application of fragmentation and variegation models to epigaeic invertebrates in South Africa. *Conservation Biology* 10: 1353-1358.

Jarvis, R. L., and M. F. Passmore. 1992. *Ecology of Band-tailed Pigeons in Oregon.* Biological Report 6, US Fish and Wildlife Service, Washington DC.

Johnson, D. H. 1980. The comparison of usage and availability measurements for evaluating resource preference. *Ecology* 61: 65-71.

Johnson, R. R., and C. H. Lowe. 1985. On the development of a riparian ecology, pp. 112-116 in *Riparian Ecosystems and Their Management: Reconciling Conflicting Uses,* R. R. Johnson, coord. General Technical Report RM-120, USDA Forest Service, Rocky Mountain Forest and Range Experiment Station, Fort Collins CO.

Kelsey, K. A., and S. D. West. 1998. Riparian wildlife, pp. 235-258 in *River Ecology and Management, Lessons from the Pacific Coastal Ecoregion,* R. J. Naiman and R. E. Bilby, ed. Springer-Verlag, New York.

Knight, R. L., and D. N. Cole. 1991. Effects of recreational activity on wildlife in wildlands. *Transactions of the 56th North American Wildlife and Natural Resources Conference* 56: 238-247.

Knight, R. L., and P. B. Landres. 1998. *Stewardship across Boundaries.* Island Press, Washington DC.

Kremsater, L., and F. L. Bunnell. 1999. Edge effects: theory, evidence, and implications to management of western North American forests, pp. 117-153 in *Forest Fragmentation: Wildlife and Management Implications,* J. A. Rochelle, L. A. Lehman, and J. Wisniewski, ed. Brill, Leiden, The Netherlands.

Landres, P. B. 1998. Integration: A beginning for landscape-scale stewardship, pp. 337-345 in *Stewardship Across Boundaries,* R. L. Knight and P. B. Landres, ed. Island Press, Washington DC.

Lehmkuhl, J. F., L. F. Ruggiero, and P. A. Hall. 1991. Landscape-scale configurations of forest fragmentation and wildlife richness and abundance in the southern Washington Cascade Range, pp. 425-442 in *Wildlife and Vegetation of Unmanaged Douglas-fir Forests,* L. F. Ruggiero, K. B. Aubry, A. B. Carey, and M. H. Huff, tech. coord. General Technical Report PNW-285, USDA Forest Service, Pacific Northwest Research Station, Portland OR.

Leopold, A. 1933. *Game Management.* Scribner's, New York.

Lewis, J. C. 1998. Creating snags and wildlife trees in commercial forests. *Western Journal of Applied Forestry* 13: 97-101.

Loegering, J. P. 1997. *Abundance, Habitat Association, and Foraging Ecology of American Dippers and Other Riparian-associated Wildlife in the Oregon Coast Range.* PhD dissertation, Oregon State University, Corvallis.

Mannan, R. W., E. C. Meslow, and H. M. Wight. 1980. Use of snags by birds in Douglas-fir forests, western Oregon. *Journal of Wildlife Management* 44: 787-797.

Manuwal, D. A., and M. H. Huff. 1987. Spring and winter bird populations in a Douglas-fir forest sere. *Journal of Wildlife Management* 51: 586-595.

Mariani, J. M., and D. A. Manuwal. 1990. Factors influencing brown creeper (*Certhia americana*) abundance patterns in the southern Washington Cascade range. *Studies in Avian Biology* 13: 53-57.

Martin, K. J. 1994. *Movements and Habitat Associations of Northern Flying Squirrels in the Central Oregon Cascades.* MS thesis, Oregon State University, Corvallis.

Martin, K. J. 1998. *Habitat Associations of Small Mammals and Amphibians in the Central Oregon Coast Range.* PhD dissertation, Oregon State University, Corvallis.

Maser, C., and M. L. Johnson. 1967. Notes on the white-footed vole (*Phenacomys albipes*). *Murrelet* 48: 24-27.

Maser, C., B. R. Mate, J. F. Franklin, and C. T. Dyrness. 1981. *Natural History of Oregon Coast Mammals.* General Technical Report PNW-133, USDA Forest Service, Pacific Northwest Forest and Range Experimental Station, Portland OR.

McComb, W. C. 1994. Red alder: interactions with wildlife, pp. 131-138 in *The Biology and Management of Red Alder,* D. E. Hibbs, D. S. DeBell, and R. F. Tarrant, ed. Oregon State University Press, Corvallis.

McComb, W. C., K. McGarigal, and R. G. Anthony. 1993a. Small mammal and amphibian abundance in streamside and upslope habitats of mature Douglas-fir stands, western Oregon. *Northwest Science* 67: 7-15.

McComb, W. C., C. L. Chambers, and M. Newton. 1993b. Small mammal and amphibian communities and habitat associations in red alder stands, central Oregon Coast Range. *Northwest Science* 67: 181-188.

McComb, W. C., T. A. Spies, and W. H. Emmingham. 1993c. Douglas-fir forests: managing for timber and mature-forest habitat. *Journal of Forestry* 91(12): 31-42.

McGarigal, K., and B. Marks. 1995. *FRAGSTATS: Spatial Pattern Analysis Program for Quantifying Landscape Structure.* General Technical Report PNW 351, USDA Forest Service, Pacific Northwest Research Station, Portland OR.

McGarigal, K., and W. C. McComb. 1992. Streamside versus upslope breeding bird communities in the central Oregon Coast Range. *Journal of Wildlife Management* 56: 10-23.

McGarigal, K., and W. C. McComb. 1993. *Research Problem Analysis on Biodiversity Conservation in Western Oregon Forests.* Unpublished. Pacific Forest and Basin Rangeland Systems Cooperative Research and Technology Unit, Bureau of Land Management, Corvallis OR.

McGarigal, K., and W. C. McComb. 1995. Relationships between landscape structure and breeding birds in the Oregon Coast Range. *Ecological Monographs* 65: 235-260.

McIntyre, S., and G. W. Barrett. 1992. Habitat variegation, an alternative to fragmentation. *Conservation Biology* 6: 146-147.

Melanson, G. P. 1993. *Riparian Landscapes.* Cambridge University Press, Cambridge UK.

Meyer, J. S., L. L. Irwin, and M. S. Boyce. 1998. Influence of habitat abundance and fragmentation on northern spotted owls in western Oregon. *Wildlife Monographs* 139: 1-51.

Mills, L. S. 1995. Edge effects and isolation: red-backed voles on forest remnants. *Conservation Biology* 9: 395-403.

Mills, L. S. 1996. Fragmentation of a natural area: dynamics of isolation for small mammals on forest remnants, pp. 199-219 in *National Parks and Protected Areas: Their Role in Environmental Protection,* R. G. Wright, ed. Blackwell Science, Cambridge MA.

Minore, D., and H. G. Weatherly. 1994. Riparian trees, shrubs, and forest regeneration in the coastal mountains of Oregon. *New Forests* 8: 249-263.

Morrison, M. L. 1992. The use of high-cut stumps by birds. *California Fish and Game* 78: 78-83.

Morrison, M. L., and E. C. Meslow. 1983. Bird community structure in early growth clearcuts in western Oregon. *American Midland Naturalist* 110: 129-137.

Morrison, M. L., and E. C. Meslow. 1994. Response of avian communities to herbicide-induced vegetation changes. *Journal of Wildlife Management* 48: 14-22.

Morrison, M. L., and M. G. Raphael. 1993. Modeling the dynamics of snags. *Ecological Applications* 3: 322-330.

Morrison, M. L., L. C. Timossi, K. A. With, and P. N. Manley. 1985. Use of tree species by forest birds during winter and summer. *Journal of Wildlife Management* 49: 1098-1102.

Neitro, W. A., V. W. Binkley, S. P. Cline, R. W. Mannan, B. G. Marcot, D. Taylor, and F. F. Wagner. 1985. Snags (wildlife trees), pp. 129-169 in *Management of Wildlife and Fish Habitats in Forests of Western Oregon and Washington,* E. R. Brown, tech. ed. Publication R6-F&WL-192-1985, USDA Forest Service, Pacific Northwest Region, Portland OR.

Nelson, K. 1989. *Habitat Use and Densities of Cavity-nesting Birds in the Oregon Coast Range.* MS thesis, Oregon State University, Corvallis.

Newton, M., and E. C. Cole. 1987. A sustained-yield scheme for old-growth Douglas-fir. *Western Journal of Applied Forestry* 2: 22-25.

Nierenberg, T. R. 1997. *A Characterization of Unmanaged Riparian Overstories in the Central Oregon Coast Range.* MS thesis, Oregon State University, Corvallis.

Nolet, B. A., A. Hoekstra, and M. M. Ottenheim. 1994. Selection foraging on woody species by the beaver, *Castor fiber,* and its impact on a riparian willow forest. *Biological Conservation* 70: 117-128.

Oakley, A. L., J. A. Collins, L. B. Everson, D. A. Heller, J. C. Howerton, and R. E. Vincent. 1985. Riparian zones and freshwater wetlands, pp. 57-80 in *Management of Wildlife and Fish Habitats in Forests of Western Oregon and Washington,* E. R. Brown, tech. ed. Publication R6-F&WL-192-1985, USDA Forest Service, Pacific Northwest Region, Portland OR.

Oboyski, P. T. 1995. *Macroarthropod Communities on Vine Maple, Red Alder, and Sitka Alder Along Riparian Zones in Central Western Cascade Range, Oregon.* MS thesis, Oregon State University, Corvallis.

Ohmann, J. L., W. C. McComb, and A. A. Zumwari. 1994. Snag abundance for primary cavity-nesting birds on nonfederal forest lands in Oregon and Washington. *Wildlife Society Bulletin* 22: 607-620.

Oliver, D. G., and B. C. Larson. 1996. *Forest Stand Dynamics.* Updated edition. John Wiley and Sons, New York.

Oregon Department of Forestry. 1997. *The Oregon Forest Practices Act Administrative Rules and Abridged Forest Practices Act, January 1997.* Oregon Department of Forestry, Forest Practices Division, Salem.

Oregon Natural Heritage Program. 1998. *Rare, Threatened, and Endangered Species of Oregon.* Oregon Natural Heritage Program, Portland.

Ormsbee, P. C. 1996. *Selection of Day Roosts by Female Long-legged Myotis (*Myotis volans*) in Forests of the Central Oregon Cascades.* MS thesis, Oregon State University, Corvallis.

Ormsbee, P. C., and W. C. McComb. 1998. Selection of day roosts by female long-legged myotis in the central Oregon Coast Range. *Journal of Wildlife Management* 62: 596-603.

Pabst, R. J., and T. A. Spies. 1998. Distribution of herbs and shrubs in relation to landform and canopy cover in riparian forests of coastal Oregon. *Canadian Journal of Botany* 76: 298-315.

Packee, E. C. 1990. Western hemlock, pp. 613-622 in *Silvics of North America, Vol. 1, Conifers.* Agriculture Handbook 654, USDA Forest Service, Washington DC.

Perkins, J. M., and S. .P Cross. 1988. Differential use of some coniferous forest habitats by hoary and silver-haired bats in Oregon. *Murrelet* 69: 21-24.

Perry, D. A. 1998. The scientific basis of forestry. *Annual Review of Ecology and Systematics* 29: 435-466.

Pettersson, R. B., J. P. Ball, K. E. Renhorn, P. A. Esseen, and K. Sjoberg. 1995. Invertebrate communities in boreal forest canopies as influenced by forestry and lichens, with implications for passerine birds. *Biological Conservation* 74: 57-63.

Power, M. E., D. Tilman, J. A. Estes, B. A. Menge, W. J. Bond, L. S. Mills, G. Daily, J. C. Castilla, J. Lubchenco, and R. T. Paine. 1996. Challenges in the quest for keystones. *BioScience* 46: 609-620.

Rabe, M. J., T. E. Morrell, H. Green, J. C. deVos, and C. R. Miller. 1998. Characteristics of ponderosa pine snag roosts used by reproductive bats in northern Arizona. *Journal of Wildlife Management* 62: 612-621.

Raedeke, K. J., R. D. Taber, and D. K. Paige. 1988. Ecology of large mammals in riparian systems of Pacific Northwest forests, pp. 113-132 in *Streamside Management: Riparian Wildlife and Forestry Interactions,* K. J. Raedeke, ed. Contribution No. 59, Institute of Forest Resources, University of Washington, Seattle.

Ralph, C. J., P. W. C. Paton, and C. A. Taylor. 1991. Habitat association patterns of breeding birds and small mammals in Douglas-fir/hardwood stands in northwestern California and southwestern Oregon, pp. 379-393 in *Wildlife and Vegetation of Unmanaged Douglas-fir Forests,* L. F. Ruggiero, K. B. Aubry, A. B. Carey, and M. H. Huff, tech. coord. General Technical Report PNW-GTR-285, USDA Forest Service, Pacific Northwest Research Station, Portland OR.

Ralph, C. J., G. L. Hunt, Jr., M. G. Raphael, and J. F. Piatt. 1995. Ecology and management of the marbled murrelet in North America: an overview, pp. 8-22 in *Ecology and Conservation of the Marbled Murrelet,* C. J. Ralph, G. L. Hunt, Jr., M. G. Raphael, and J. F. Piatt, tech. ed. General Technical Report PSW-GTR-152, USDA Forest Service, Pacific Southwest Research Station, Berkeley CA.

Rosenberg, D. K., K. A. Swindle, and R. G. Anthony. 1994. Habitat associations of California red-backed voles in young and old-growth forests in western Oregon. *Northwest Science* 68: 266-272.

Rosenberg, K. V., and M. G. Raphael. 1986. Effects of forest fragmentation on vertebrates in Douglas-fir forest, pp. 263-272 in *Wildlife 2000: Modeling Habitat Relationships of Terrestrial Vertebrates,* J. Verner, M. L. Morrison, and C. J. Ralph, ed. University of Wisconsin Press, Madison.

Ross, D. W., and C. G. Niwa. 1997. Using aggregation and anti-aggregation pheromones of the Douglas-fir beetle to produce snags for wildlife habitat. *Western Journal of Applied Forestry* 12: 52-54.

Saab, V. A., and J. D. Dudley. 1998. *Response of Cavity-nesting Birds to Stand-replacement Fire and Salvage Logging in Ponderosa Pine/Douglas-fir Forests of Southwestern Idaho.* Research Paper RMRS-RP-11, USDA Forest Service, Rocky Mountain Research Station, Fort Collins CO.

Sasse, D. B., and P. J. Pekins. 1996. Summer roosting ecology of northern long-eared bats (*Myotis septentrionalis*) in the White Mountain National Forest, pp. 91-101 in *Bats and Forests Symposium, October 19-21, 1995, Victoria, British Columbia,* R. M. R. Barclay and R. M. Brigham, ed. Working Paper 23, British Columbia Ministry of Forests Research Branch, Victoria BC.

Sauer, J. R., J. E. Hines, G. Gough, I. Thomas, and B. G. Peterjohn. 1997. *The North American Breeding Bird Survey Results and Analysis,* Version 96.4. Patuxent Wildlife Research Center, Laurel MD.

Schreiber, B., and D. S. deCalesta. 1992. The relationship between cavity-nesting birds and snags on clearcuts in western Oregon. *Forest Ecology and Management* 50: 299-316.

Sessions, J., S. Crim, and K. N. Johnson. 1994. Resource management perspective: geographic information systems and decision models in forest management planning, pp. 63-76 in *Remote Sensing and GIS in Ecosystem Management,* V. A. Sample, ed. Island Press, Washington DC.

Seymour, R. S., and M. L. Hunter, Jr. 1999. Principles of ecological forestry, pp. 22-61 in *Maintaining Biodiversity in Forest Ecosystems,* J. L. Hunter Jr., ed. Cambridge University Press, New York.

Suzuki, N. 1992. *Habitat Classification and Characteristics of Small Mammal and Amphibian Communities in Beaver-pond Habitats of the Oregon Coast Range.* MS thesis, Oregon State University, Corvallis.

Suzuki, N. 2000. Effect of thinning on forest-floor vertebrates and analysis of habitat associations along ecological gradients in Oregon coastal Douglas-fir forests. Ph.D. dissertation, Oregon State University, Corvallis.

Suzuki, N., and W. C. McComb. 1998. Habitat classification models for beaver (*Castor canadensis*) in streams of the central Oregon Coast Range. *Northwest Science* 72: 102-110.

Swanson, F. J., S. V. Gregory, J. R. Sedell, and A. G. Campbell. 1982. Land-water interactions: the riparian zone, pp. 267-291 in *Analysis of Coniferous Forest Ecosystems in the Western United States,* R. L. Edmonds, ed. US/IBP Synthesis Series, 14th edition, Hutchinson Ross Publishing Co., Stroudsburg PA.

Swanson, F. J., T. K. Kratz, N. Caine, and R. G. Woodmansee. 1988. Landform effects on ecosystem patterns and processes. *BioScience* 38: 92-98.

Tallmon, D., and L. S. Mills. 1994. Use of logs within home ranges of California red-backed voles on a remnant of forest. *Journal of Mammalogy* 75: 97-101.

Tappeiner J. C., II, D. Huffman, D. Marshall, T. A. Spies, and J. D. Bailey. 1997. Density, ages, and growth rates in old-growth and young-growth forests in coastal Oregon. *Canadian Journal of Forest Research* 27: 638-648.

Thomas, J. W., C. Maser, and J. E. Rodiek. 1979. Riparian zones, pp. 40-47 in *Wildlife Habitats in Managed Forests, the Blue Mountains of Oregon and Washington,* J. W. Thomas, tech. ed. Agricultural Handbook No 553, USDA Forest Service, Portland OR.

Thompson, R. L. 1996. *Home Range and Habitat Use of Western Red-backed Voles in Mature Coniferous Forests in the Oregon Cascades.* MS thesis, Oregon State University, Corvallis.

Urban, D. L., M. F. Acevedo, and S. L. Garman. 1999. Scaling fine-scale processes to large-scale patterns using models derived from models: meta-models, pp. 70-98 in *Spatial Modeling of Forest Landscape Change: Approaches and Applications,* W. L. Baker and D. J Mladenoff, ed. Cambridge University Press, New York.

Ure, D. C., and C. Maser. 1982. Mycophagy of red-backed voles in Oregon and Washington. *Canadian Journal of Zoology* 60: 3307-3315.

USDA Forest Service and USDI Bureau of Land Management. 1994. *Record of Decision for Amendments to Forest Service and Bureau of Land Management Planning Documents within the Range of the Northern Spotted Owl and Standards and Guidelines of Management of Habitat for Late-successional and Old-growth Forest Related Species within the Range of the Northern Spotted Owl.* USDA Forest Service and USDI Bureau of Land Management, Washington DC.

Van Horne, B. 1983. Density as a misleading indicator of habitat quality. *Journal of Wildlife Management* 47: 893-901.

Vega, R. M. S. 1993. *Bird Communities in Managed Conifer Stands in the Oregon Cascades: Habitat Associations and Nest Predation.* MS thesis, Oregon State University, Corvallis.

Verts, B. J., and L. N. Carraway. 1998. *Land Mammals of Oregon.* University of California Press, Berkeley.

Vesely, D. G. 1997. *Terrestrial Amphibian Abundance and Species Richness in Headwater Riparian Buffer Strips, Oregon Coast Range.* MS thesis, College of Forestry, Oregon State University, Corvallis.

Vonhof, M. J. 1996. Roost-site preferences of big brown bats (*Eptesicus fuscus*) and silver-haired bats (*Lasionycteris noctivagans*) in the Pend d'Oreille Valley in southern British Columbia, pp. 62-79 in *Bats and Forests Symposium, October 19-21, 1995, Victoria, British Columbia,* R. M. R. Barclay and R. M. Brigham, ed. Working Paper 23, British Columbia Ministry of Forests Research Branch, Victoria BC.

Vonhof, M. J., and R. M. R. Barclay. 1996. Roost-site selection and roosting ecology of forest-dwelling bats in southern British Columbia. *Canadian Journal of Zoology* 74: 1797-1805.

Vonhof, M. J. and R. M. R Barclay. 1997. Use of tree stumps as roosts by the western long-eared bat. *Journal of Wildlife Management* 61: 674-684.

Voth, E. H., C. Maser, and M. L. Johnson. 1983. Food habits of *Arborimus albipes,* the white-footed vole, in Oregon. *Northwest Science* 57: 1-7.

Waldien, D. L., J. P. Hayes, and E. B. Arnett. 2000. Day-roosts of female long-eared myotis in western Oregon. *Journal of Wildlife Management* 64: 785-796.

Weikel, J. 1997. *Habitat Use by Cavity-nesting Birds in Young Thinned and Unthinned Douglas-fir Forests in Western Oregon.* MS thesis, Oregon State University, Corvallis.

Weikel, J. M., and J. P. Hayes. 1999. The foraging ecology of cavity-nesting birds in young forests of the northern Coast Range of Oregon. *The Condor* 101: 58-66.

Welsh, H. H., Jr., and A. J. Lind. 1996. Habitat correlates of the southern torrent salamander, *Rhyacotriton variegatus* (Caudata: Rhyacotritonidae), in northwestern California. *Journal of Herpetology* 30: 385-398.

With, K. A. 1999. Is landscape connectivity necessary and sufficient for wildlife management? pp. 97-115 in *Forest Fragmentation: Wildlife and Management Implications,* J. A. Rochelle, L. A. Lehman, and J. Wisniewski, ed. Brill, Leiden, The Netherlands.

Timber Harvesting to Enhance Multiple Resources

Loren D. Kellogg, Ginger V. Milota, and Ben Stringham

Introduction

Along with its customary function of extracting wood from the forest for economic benefits, timber harvesting can help achieve other resource-management objectives. It is one of the few tools available to managers for making significant changes to the vegetation structure and species composition of a stand, changes which in turn can enhance wildlife habitat or other resource values and maintain forest health. Timber-harvesting systems can also be used to place large woody debris in streams to improve fish habitat. Soil and water resources should be protected by carefully selecting equipment and harvesting methods to fit each site, minimizing the environmental impacts from timber harvesting.

If timber harvesting is to be effective in achieving other resource-management objectives, forest- and stream-resource specialists and logging specialists must interact early in the planning process. Good communication and thorough cooperation throughout the project will ensure that it is feasible from a forest-operations standpoint as well as safe, cost effective, and environmentally sound. To aid in this interaction, resource specialists need a basic understanding of forest operations and the terminology used for various harvesting systems.

This chapter provides an overview of harvesting systems used in the Oregon Coast Range, including a general presentation of the capabilities, limitations, costs, and site impacts of each system. We describe the harvest-planning process, including the four requirements of all harvesting plans, along with the different levels of harvest planning and considerations of selecting a harvesting system best suited to the site and the project's objectives.

Next, we discuss timber harvesting within the context of both even-age and uneven-age silvicultural systems used in the Coast Range, addressing appropriate types of equipment and methods, harvest planning and layout considerations, and production and cost information for each silvicultural system. Because many alternative silvicultural prescriptions call for leaving some trees in the stand, a section on minimizing stand damage is also included.

We close with a section on timber harvesting for riparian-area management, with information on appropriate equipment and harvesting methods and a description of a case study in which timber harvesting was used in a riparian management project.

Review of Harvesting Systems in the Oregon Coast Range

The various timber-harvesting systems in common use in Oregon Coast Range forests fall into four categories: cable (highlead and skyline), aerial (mostly helicopters), conventional ground based (tractors, skidders, log loaders used to swing logs, and horses), and mechanized ground based (e.g., forwarders). Each of these systems has its own capabilities, limitations, costs, and effects on the site.

Figure 6-1 shows common yarding distances and slope-steepness limitations for the basic harvesting systems. The yarding distance for each system has

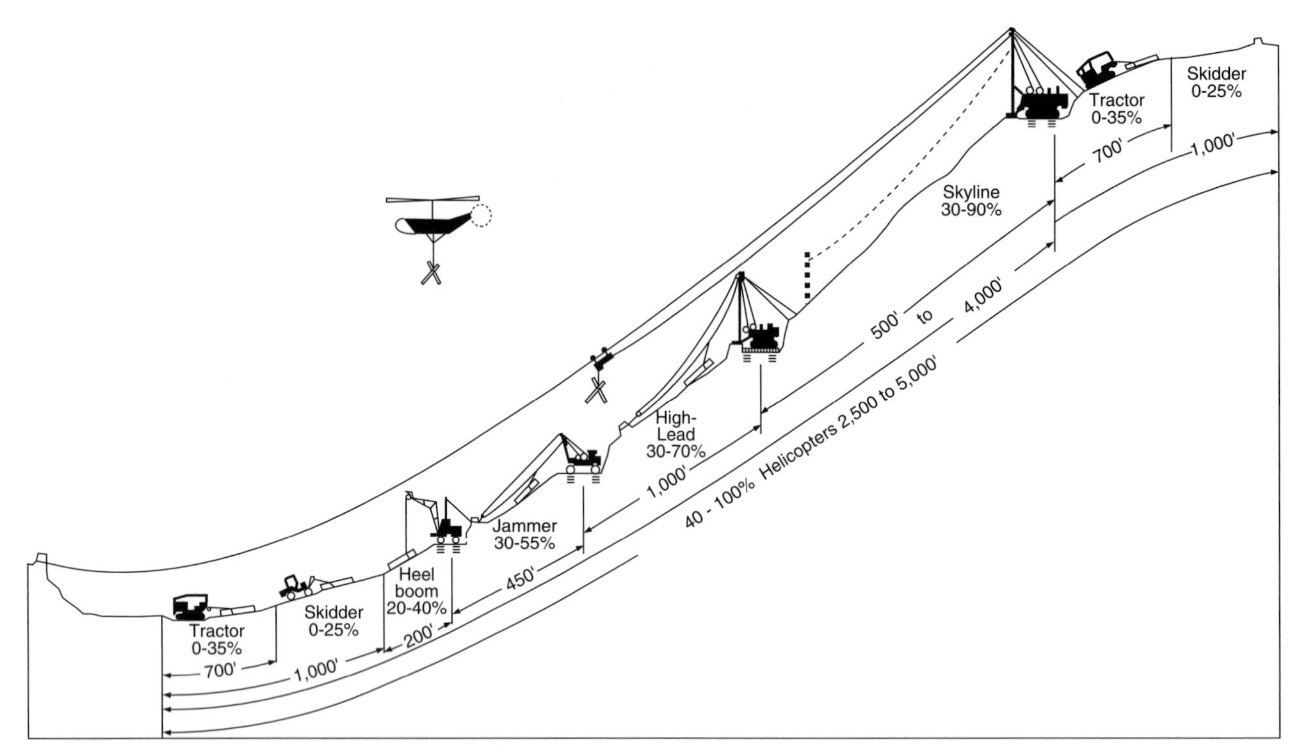

Figure 6-1. Common yarding distances and slope percent for different logging systems used in the Oregon Coast Range. From Studier and Binkley (1974).

Table 6-1. Relationship between harvesting system and soil area devoted to roads.

Harvest system	*Common maximum yarding distance (ft)*	*Area in haul roads (%)*
Conventional ground-based	1,000	4-15
Forwarder	2,000	2-7
Highlead	1,000	2-10
Short-span skyline	1,500	3-7
Long-span skyline	3,500	2-4
Helicopter	4,000	2-3

Silen (1957); Haupt (1960); Binkley (1965); Megahan and Kidd (1972); Swanston and Dyrness (1975); Kochenderfer (1977); Garland (1979); Kellogg et al. (1992).

Table 6-2. A comparison of soil disturbance from common harvesting systems.

Harvest system	*Bare soil (%)*	*Compacted soil (%)*
Tractor	35	26
Highlead	15	9
Skyline	12	3
Helicopter	4	2

Dyrness (1965, 1967, 1972); Swanston and Dyrness (1973); Clayton (1981).

Table 6-3. A comparison of characteristics for harvesting systems used in the Coast Range.

Characteristic	*Harvesting system*		
	Conventional ground-based	*Cable*	*Helicopter*
Yarding direction	Flat to downhill	Up or downhill	Up or downhill
Yarding distance (common maximum)	1,000 ft	3,500 ft	4,000 ft
Ground disturbance	Most	Minimal	Landing/service areas only
Harvest production levels	Low - medium	Low - high	High - very high
Costs	Lowest	Higher	Usually highest

an influence on the soil area devoted to roads (Table 6-1). A comparison of soil disturbance and compaction by each system is shown in Table 6-2. Recent experience with forwarders has shown that they compact and disturb the soil about as much as tractor systems, or less (Armlovich 1995; Allen 1997). Other system characteristics are compared in Table 6-3.

Felling

The felling of timber is accomplished either by a person with a chainsaw (referred to as motor-manual felling, or simply manual felling) or machines such as feller-bunchers (mechanical felling). In the past, manual felling was the main practice in the Coast Range because it was best suited to the region's large trees, steep terrain, and wet weather conditions. Today, however, mechanical felling is increasingly used, especially in thinning and in the harvest of smaller timber. Mechanical felling may have an advantage over manual felling in thinning because it keeps control of the tree during felling and allows good placement of the tree on the ground, helping to minimize stand damage during yarding. On the other hand, feller-buncher operations typically leave the whole tree (not limbed and bucked), resulting in a higher risk of stand damage during yarding. Manual or mechanical felling methods that include limbing and bucking will help reduce thinning damage.

In every harvesting operation, the felling pattern—the orientation of the felled trees on the ground—is an important factor that affects harvesting economics, the value of the wood recovered, and the amount of logging damage to the site. Three felling patterns are common: herringbone, contour, and perpendicular (Figure 6-2). The choice depends mainly on terrain, silvicultural prescription, and tree size. In young-stand thinning with cable systems, stand damage will be minimized when yarding corridors are designated with flagging prior to felling and timber is felled in a herringbone pattern (Han and Kellogg 1997). In these situations, yarding damage is reduced because the pulling lead of the logs can be easily lined up with the skyline carriage position and the angle for the logs to turn into the skyline corridor is smaller than with other felling patterns, thus reducing rubbing on residual trees. However, in partial cuttings with larger trees, a herringbone felling pattern may result in yarding

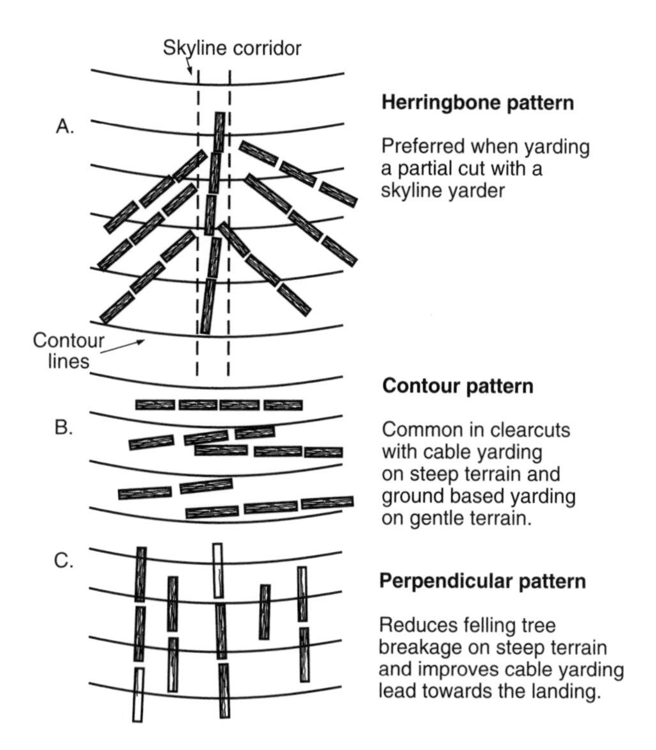

Figure 6-2. Felling patterns to facilitate yarding. From Studier and Binkley (1974).

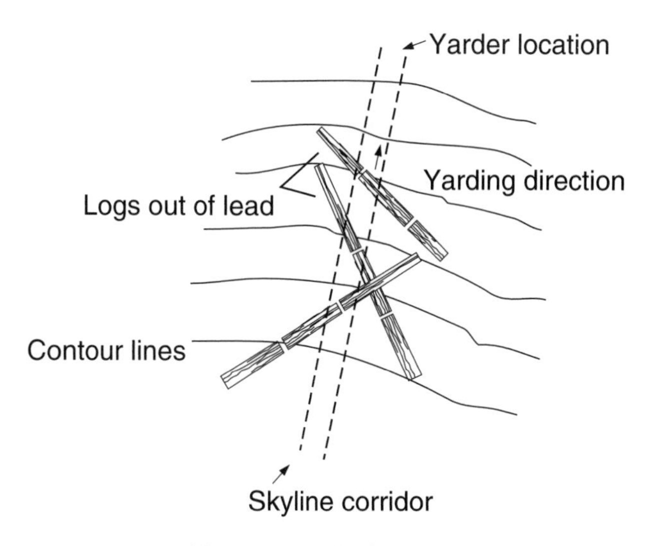

Figure 6-3. Problems created when large timber is felled in a herringbone pattern.

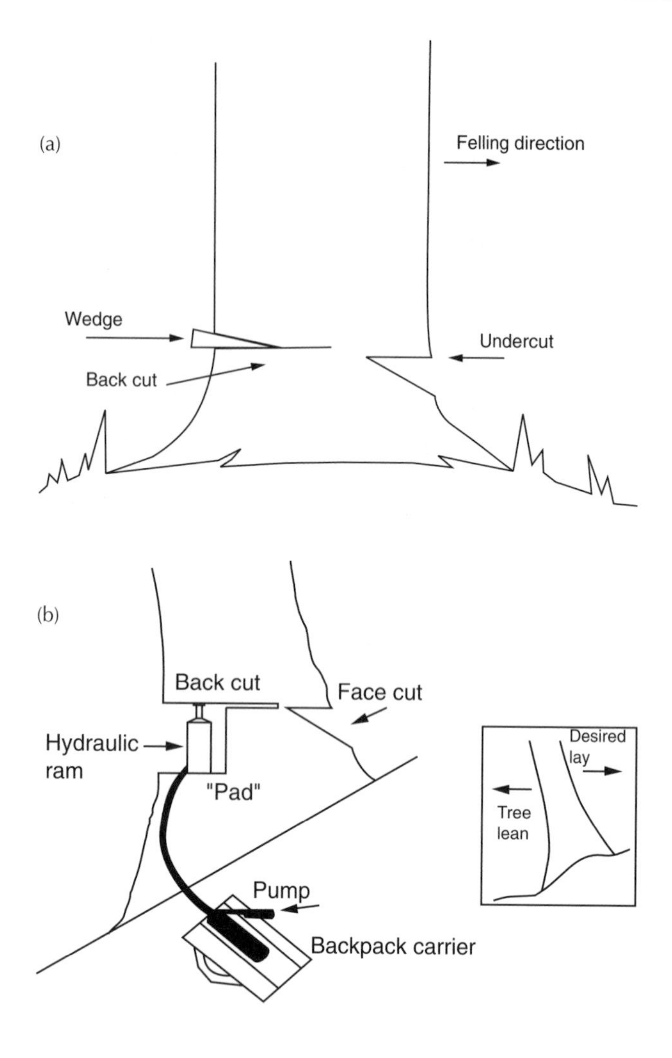

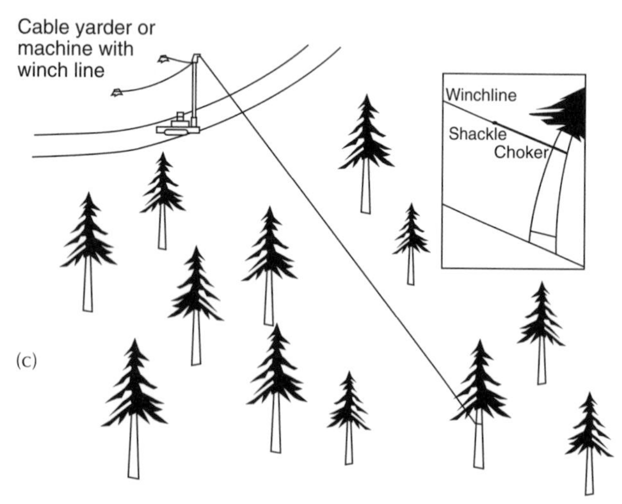

Figure 6-4. Nonmechanical and mechanically assisted directional felling techniques: (a) wedging; (b) tree jacking; (c) tree pulling.

difficulties and a high amount of stand damage because even if the butt sections of trees are in a good position for yarding, the top sections may not be (Figure 6-3). The contour felling pattern may be a better choice in these situations. Uphill timber felling in the perpendicular pattern is often used with mechanically assisted directional felling.

Special directional-felling techniques are useful in a number of situations, such as felling trees around riparian areas, clumps of smaller trees, or regeneration; felling trees toward ground-based skid trails; and felling trees on uneven terrain where breakage can occur. Directional-felling techniques may also be helpful in minimizing damage to streambanks and riparian vegetation. In addition, timber-value recovery can be greater and yarding productivity can be increased because trees are in lead for pulling to the equipment or skyline carriage position. Nonmechanical directional-felling techniques such as wedging (Figure 6-4) are limited by the weight distribution and the amount of tree lean (Garland 1983b). Mechanically assisted directional-felling techniques such as tree jacking and tree pulling (Figure 6-4) can be used for large, high-value trees; however, these techniques require more planning and are more expensive (Hunt and Henley 1981). Fuller explanation of these techniques can be found in Garland (1983b) and Hunt and Henley (1981).

Cable systems

Cable-yarding systems are commonly used on steep terrain in the Coast Range. In the past, cable systems were mainly used to log clearcuts, but these systems are well suited to thinnings and partial cuts as well. The capability of cable systems to yard relatively long distances is an advantage because fewer haul roads are needed than with ground-based systems, and the amount of soil disturbance is usually less.

Cable-yarding systems are characterized by stationary yarders equipped with multiple winches and wire rope to move logs from the felling site to a landing or roadside. Wire rope on the winches is suspended over the harvesting area and transmits the pulling forces for moving logs. The distance capabilities of these systems vary greatly depending on equipment and terrain. Logs can be yarded over distances of more than 5,000 feet, but a more realistic maximum for cable systems in the Coast Range is 1,500 to 2,000 feet. Uphill yarding is generally

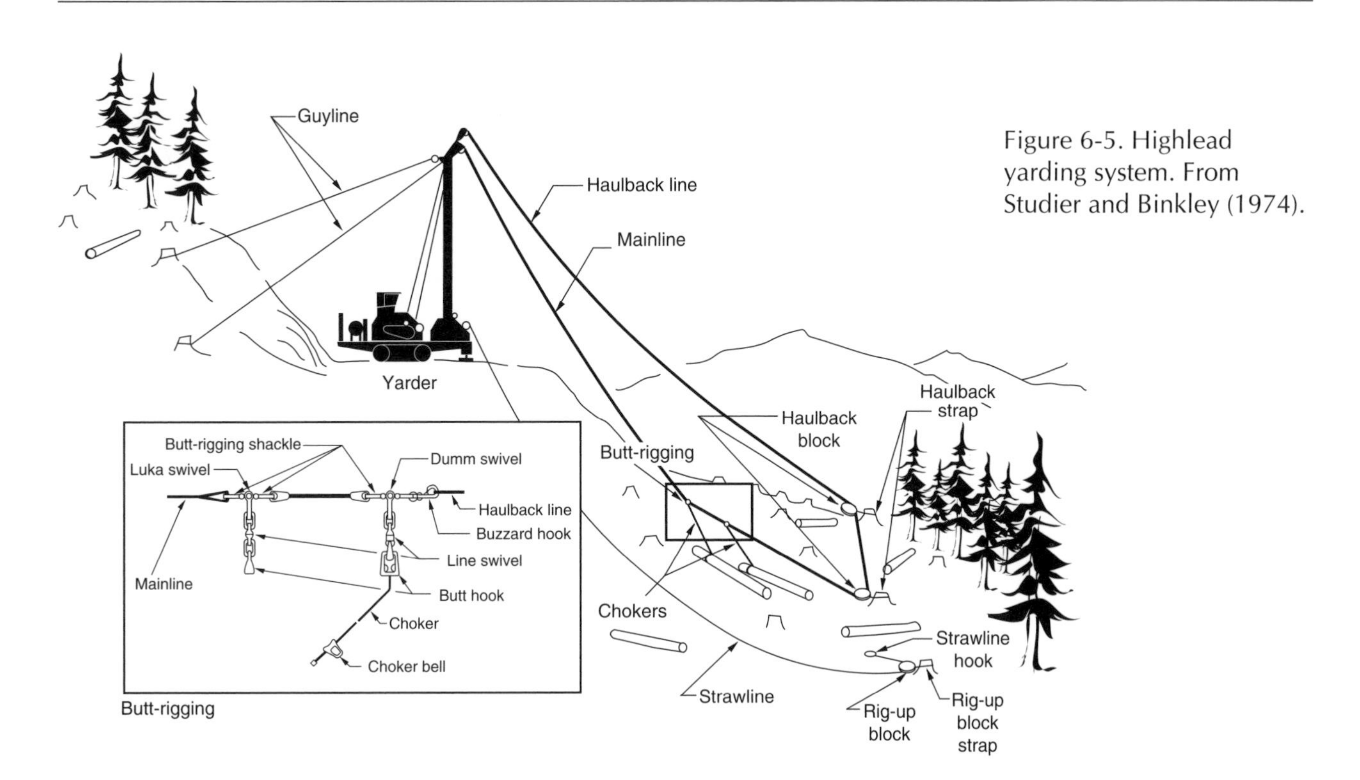

Figure 6-5. Highlead yarding system. From Studier and Binkley (1974).

preferred over downhill yarding because it is safer and minimizes potential for soil erosion. However, downhill cable yarding is sometimes used when truck roads cannot be constructed at the tops of harvest units.

We will give only a brief overview of cable systems appropriate to the Coast Range. More information can be found in Studier and Binkley (1974), Conway (1982), and Stenzel et al. (1985).

Highlead

The highlead system is the simplest type of cable logging. It does not utilize a skyline carriage and has only two operating lines, the mainline and the haulback line (Figure 6-5). These lines are connected at the butt-rigging, which consists of hardware to keep the lines from twisting and serves as the attachment point for chokers. The mainline pulls the butt-rigging and logs to the landing, and the haulback line pulls the butt-rigging and mainline back to the felling site. Lateral yarding—bringing logs from the side of the cableway into the path of the mainline—is limited to the length of the chokers.

The highlead system does not typically lift logs off the ground. Partial log suspension can be obtained by applying the brakes to the haulback drum while pulling in on the mainline drum, a procedure called "tightlining." However, this procedure is hard on yarding equipment and is usually done for only short distances to avoid obstructions on the terrain.

The highlead system is best suited to uphill yarding at maximum distances of less than 1,000 feet; downhill yarding is not usually recommended, but it may be feasible on concave terrain at shorter yarding distances. Because of its short lateral yarding range, the highlead method is best used for clearcuts; it is not well suited to thinnings and partial cuts. However, the system can be used effectively to yard portions of harvest units where skyline yarding is limited because of short yarding distances and inadequate clearance to suspend the skyline above the ground.

There are several variations to the standard highlead system. One, the jammer, is a short-distance (200-300 feet) system in which the skidding line is pulled out manually for each turn of logs. The system can be advantageous in situations where it is uneconomical to rig a cable-yarding system for short yarding distances. Also, since the equipment used is often a log-loading machine with a winch, one machine can be used for both yarding and loading.

Skyline

In skyline logging, a cableway, the skyline, is suspended between the yarding equipment and the anchor point at the end. It serves as a track for a

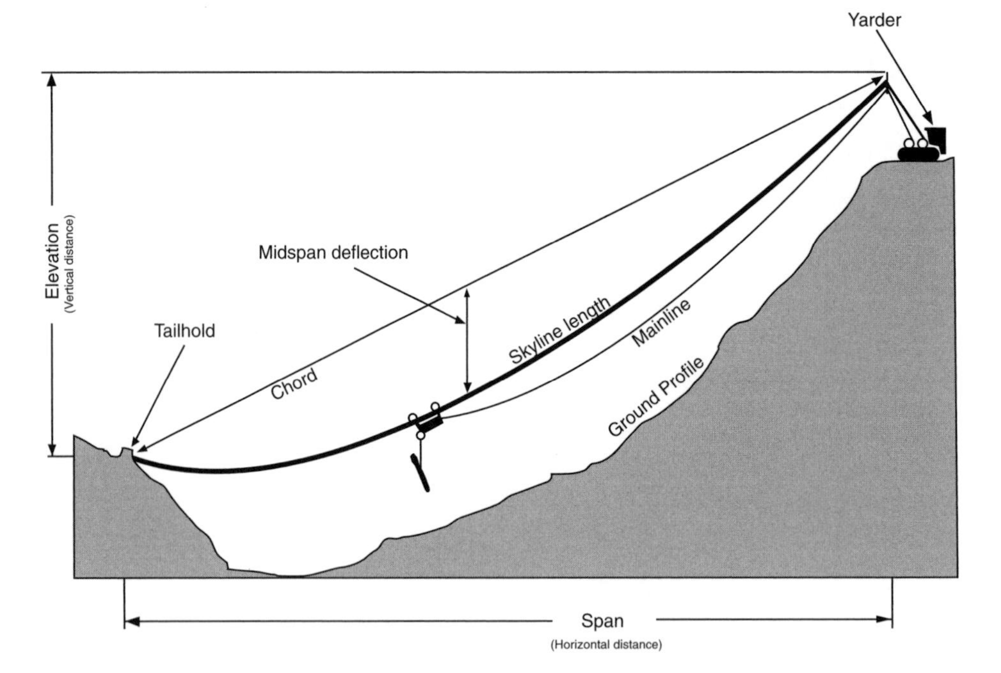

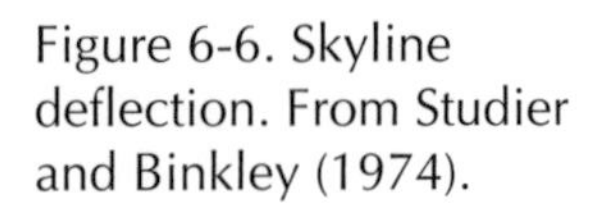
Figure 6-6. Skyline deflection. From Studier and Binkley (1974).

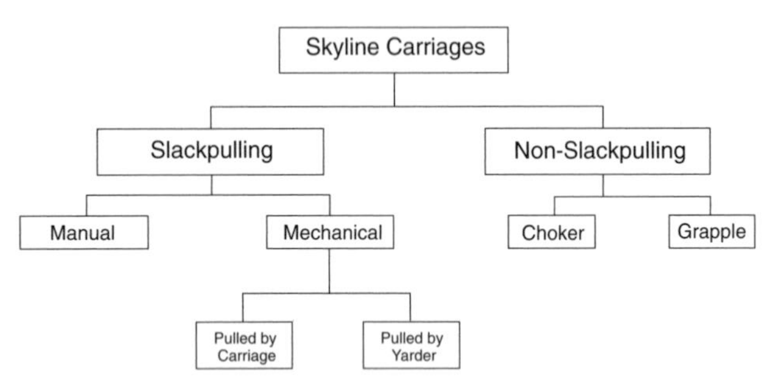

Figure 6-7. Categorization of skyline carriages. From Studier (1993).

block or skyline carriage. All skyline systems utilize a skyline carriage; the number of yarding lines varies from two to four.

Skyline systems are an important improvement over highlead because they can be used to yard timber from difficult harvesting sites with little soil disturbance. Logs are moved along the skyline either completely suspended or with one end on the ground, depending on topography and the type of equipment used. Some skyline systems can also yard logs laterally to the skyline corridor. Thus, skylines are applicable both to clearcutting and to harvest situations where some portion of the existing stand is retained.

Skyline deflection, or sag in the load-lifting line, is the key condition needed for skyline operations (Figure 6-6). The carrying capacity of skyline systems is directly related to skyline tension and deflection: the greater the deflection, the higher the load-carrying capacity. Skyline deflection concepts and calculating procedures are further explained in Brown et al. (1997), Studier and Binkley (1974), and Binkley and Sessions (1978).

Two main categories of carriages are used in skyline systems: slackpulling and non-slackpulling (Figure 6-7). Slackpulling carriages can yard logs laterally into the skyline corridor with a skidding line pulled out from the carriage; non-slackpulling carriages have no lateral yarding capability (Figure 6-8). Slackpulling carriages are further categorized as manual or mechanical. In manual slackpulling carriages, the skidding line is an extension of the yarder mainline and is manually pulled through the carriage to the logs. These carriages are typically used with relatively small yarding equipment and operated at relatively short (less than 1,000 feet) yarding distances; manual slackpulling is difficult with bigger lines and longer distances.

Mechanical slackpulling carriages use a power source—a diesel, gas, or propane engine—to pay out the skidding line from the carriage. The engine is located either in the carriage or in the yarder. The engine and skidding line in the carriage are controlled by radio signals from choker setters or landing workers. Mechanical slackpulling carriages powered from the yarder require an additional drum on the yarder, referred to as the slackpulling drum (Figure 6-8). Thus, three- or four-drum yarders are required to operate this carriage, depending on the specific skyline system.

Slackpulling carriages are necessary when the skyline is rigged as two or more spans because the intermediate support jacks prevent changes to the skyline length during yarding (required for nonslackpulling carriages). The multispan carriage must be equipped with open-sided sheaves in order to pass over an intermediate support jack (Figure 6-9). The carriage may be either a manual or a mechanical slackpulling carriage.

Non-slackpulling carriages have two variations: either chokers are attached directly to the bottom of the carriage, or the carriage is equipped with a grapple. Longer chokers are used with non-slackpulling carriages than with slackpulling carriages. Grapple carriages are commonly used with swing-boom yarders. For further information on carriages, including detailed diagrams, refer to Studier (1993).

Skyline systems are categorized into three main groups: standing, live, and running. In the standing skyline system, both ends of the skyline are secured and the skyline is not raised or lowered during the yarding cycle; the skyline length remains constant (Figure 6-10). Typically the skyline is anchored at one end (tailhold) and the skyline drum brake on the yarder holds the other end of the skyline. Standing skylines utilize slackpulling carriages, either manual or mechanical. The two main standing skyline systems are shotgun and slackline. Shotgun skyline systems rely on gravity to move the carriage to the logs and a mainline to pull it in with a load; a two-drum yarder is adequate, and yarding direction is limited to uphill. A slackline system uses a haulback line to pull the carriage out and a mainline to pull it back in; a three- or four-drum yarder is required, depending on carriage requirements, and yarding direction can be either uphill or downhill. A multispan skyline system is a standing skyline consisting of two or more spans separated by intermediate supports.

In the live skyline system, the skyline is raised and lowered during the yarding cycle for each turn of logs (Figure 6-11). The skyline is anchored to a tailhold at one end and wound on a yarder drum so that it can be easily lengthened or shortened during each cycle. The skyline length adjustment occurs at the skyline drum on the yarder. Live skylines typically utilize non-slackpulling carriages. As with standing skylines, the two main live skyline systems are shotgun (for uphill yarding) and slackline (for either uphill or downhill yarding).

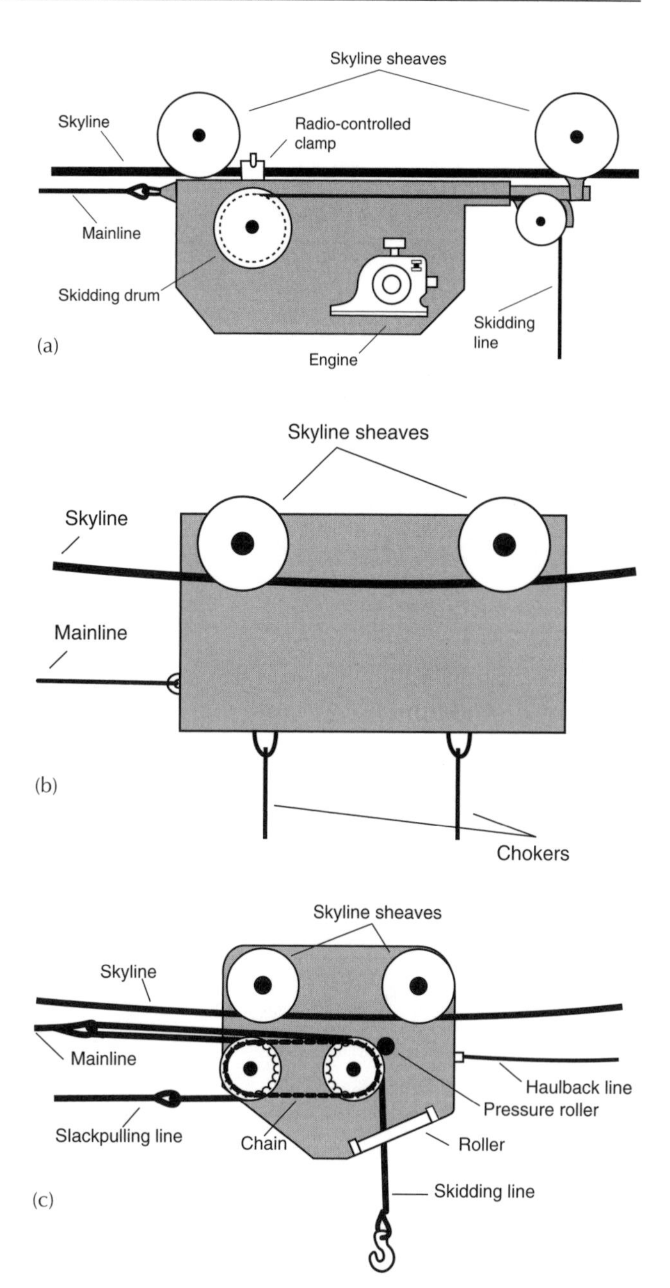

Figure 6-8. Skyline carriages: (a) mechanical slackpulling carriage with self-contained skidding line; (b) non-slackpulling carriage with chokers attached to bottom of carriage; (c) yarder-controlled mechanical slackpulling carriage with skidding line attached to mainline. From Studier (1993).

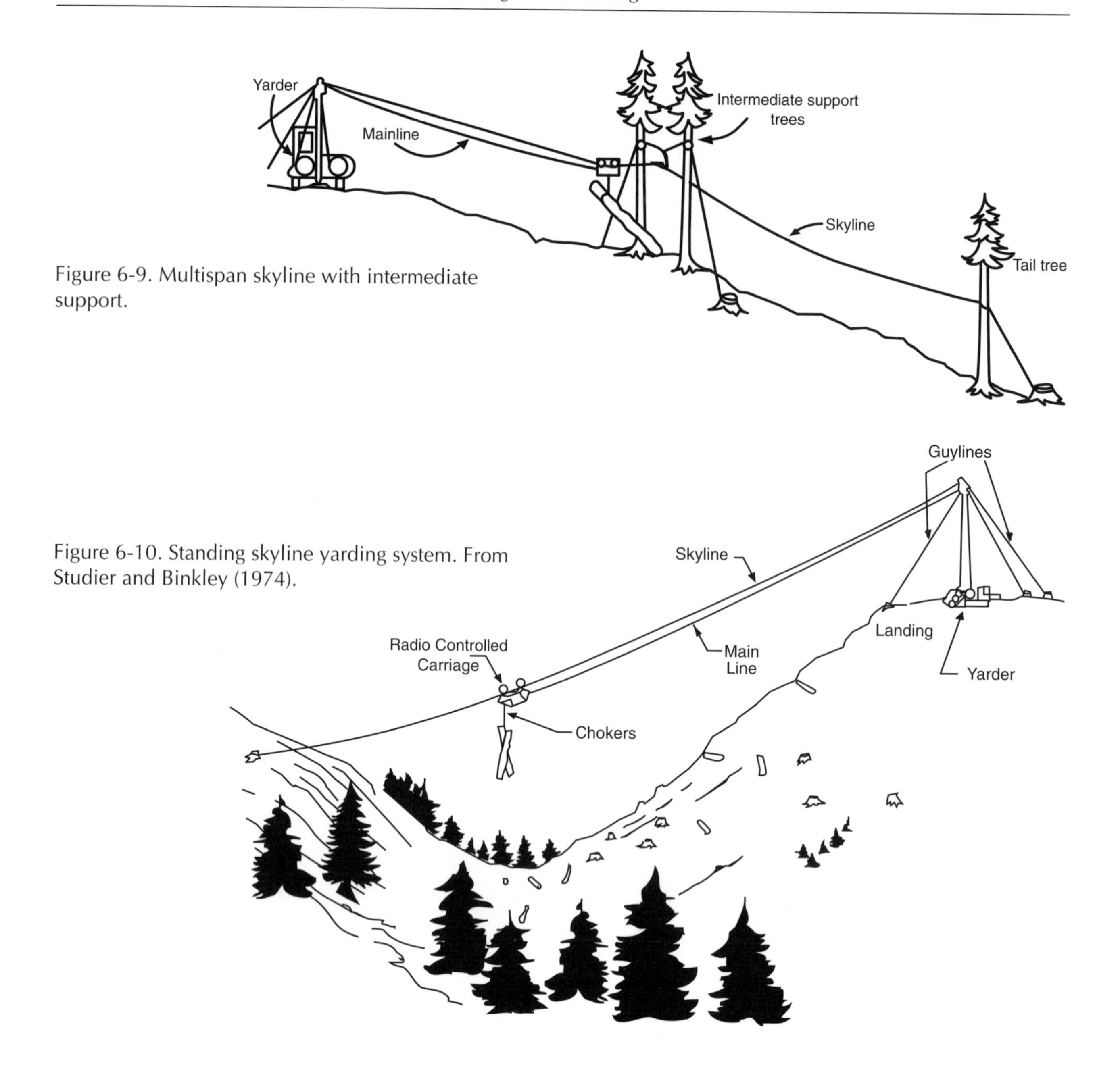

Figure 6-9. Multispan skyline with intermediate support.

Figure 6-10. Standing skyline yarding system. From Studier and Binkley (1974).

The running skyline system has two or three lines that support the log load: the haulback line, the mainline, and in some cases the slackpulling line (Figure 6-12). In the running skyline system, all lines move in and out with the carriage during each yarding cycle. Swing yarders with slackpulling carriages or grapple carriages are commonly used. Running skyline yarders are often equipped with an interlock device that allows drums to turn at different speeds. This maintains a similar tension in all running lines, which helps them support the log load. Yarding direction may be either uphill or downhill.

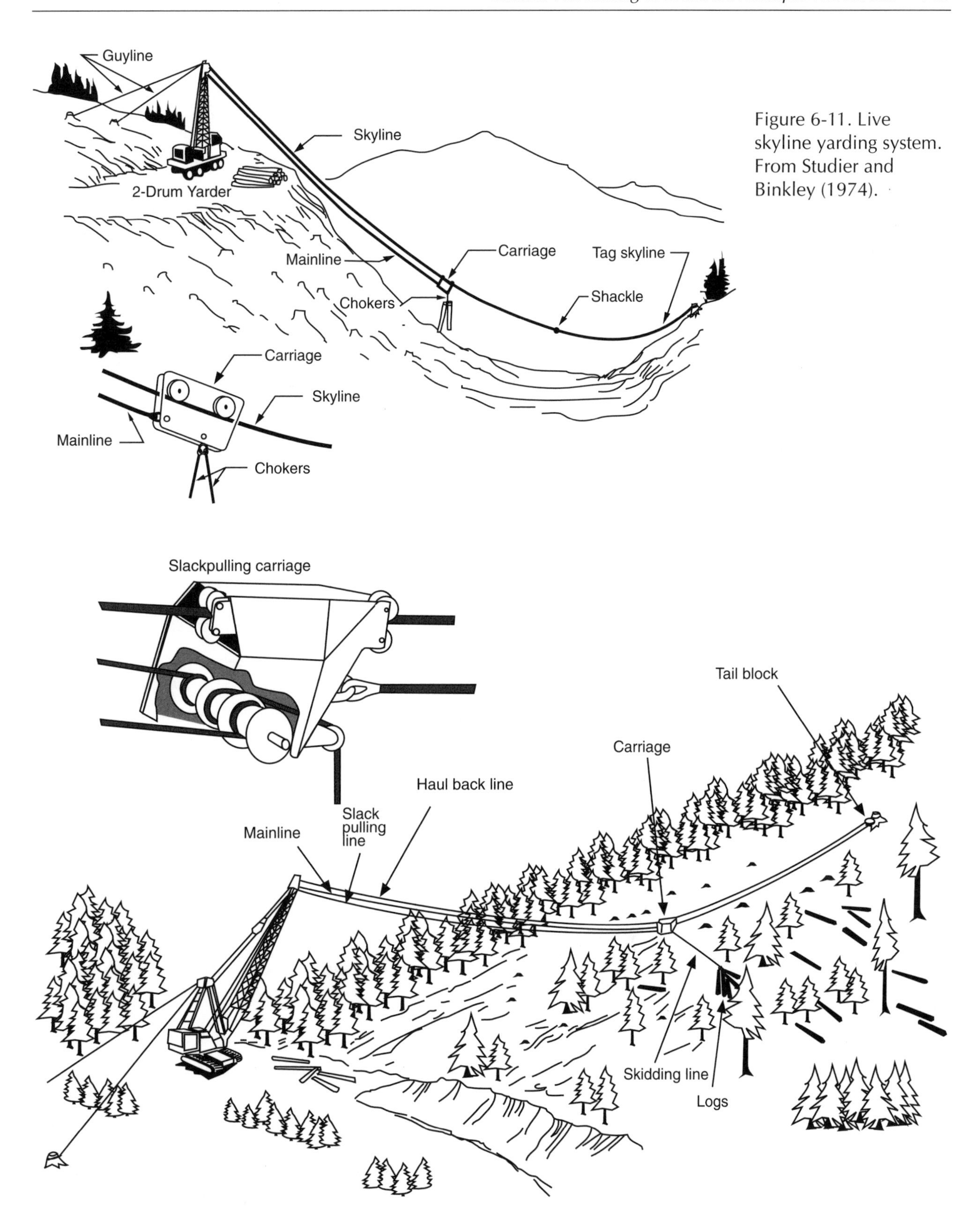

Figure 6-11. Live skyline yarding system. From Studier and Binkley (1974).

Figure 6-12. Running skyline yarding system. From Studier and Binkley (1974).

Aerial systems

Aerial harvesting systems generally use helicopters; balloons were used at times in the past. Helicopters have been used for yarding timber in the western United States since the early 1970s. They typically are used in sensitive resource areas or terrain inaccessible by road. Helicopters can move large timber volumes quickly and are thus well suited to rapid harvest of timber damaged by fire, insects, or disease. In addition, helicopter logging causes minimal ground disturbance and residual-stand damage because logs are lifted vertically off the ground and flown to the landing. Helicopters have payload capabilities ranging from about 4,000 to 28,000 pounds; lift capacities are reduced as altitude and temperature increase (Studier and Neal 1993). Helicopter logging has both higher operating costs and higher yarding production rates than ground-based yarding and cable yarding methods. Silvicultural costs after harvesting, such as site preparation and regeneration, also may be higher than in other harvesting systems.

The helicopter yarding cycle consists of the helicopter flying from the landing to the log-hooking location, hovering over the logs while a ground-crew member attaches chokers to a hook at the end of the tagline from the helicopter, vertically lifting the logs in the air, and returning to the landing, where the logs are electronically released from the hook. The yarding operation is interrupted about once every hour for refueling and twice a day for scheduled maintenance. To minimize hovering time, chokers are preset on the logs before the helicopter arrives at the hooking site; either the helicopter flies between different hooking sites to allow time for the ground crew to set chokers, or a separate choker-setting crew works ahead of the helicopter. In either situation, another light-utility helicopter may be used to return chokers to the ground crew. Because helicopters have specific payload-lifting capacities and very high operating costs, logs—especially large logs—must be bucked precisely, and the choker setters must select logs carefully so that the helicopter carries maximum payloads on each cycle.

Helicopter operations are less constrained than other harvesting systems by yarding distance and timber volumes removed per acre. A helicopter can often economically yard logs a mile or more, and it can be suitable for jobs with a low timber volume per acre if that volume is in relatively large logs or concentrated in a few areas.

Because helicopter logging goes fast, the safe, efficient movement of workers and equipment requires careful planning. When central landings are used, they must be relatively large to accommodate log storage and to segregate work activities for adequate worker safety. Two or more landings may be used, allowing the helicopter to alternate between landings. Roads can be used for landing logs; however, this may compromise safety or cause operational difficulties if logs and equipment obstruct the road.

Weather-related problems such as wind shifts and fog affect helicopter yarding more than cable yarding. Often logging crews stay in travel trailers close to the operation so they can quickly take advantage of good flying conditions. Another concern is that helicopters may not be available when and where they are needed. Helicopter logging contractors are few, and getting the right size helicopter for the most economical yarding requires planning ahead.

Ground-based systems

Ground-based skidding is usually the least expensive harvesting system when applied to the appropriate terrain—flat or gently sloping ground. Practical slope limitations are about 35 percent for downhill and 15 percent for uphill skidding. These limits are not absolute; ground-based machines can sometimes be used on steeper slopes if the steep areas are small and situated next to more gently sloping ground. Preferred skidding direction is downhill because less power is needed than for pulling logs uphill; in addition, track or tire slip may be less. Sidehill skidding requires excavating skid trails on slopes greater than about 20 percent.

Ground-based systems use tracked vehicles, including both rigid-tracked and flexible-tracked machines (with independently sprung suspension), four-wheel-drive articulated tractors such as rubber-tired skidders, hydraulic-grapple loaders, and animals such as draft horses. The machine or animal travels to the log, then moves the log over the ground to the landing. See Kellogg et al. (1993a) for definitions and drawings of this equipment. Rigid-tracked crawler-tractors typically weigh between 30,000 and 40,000 pounds. They are smaller than those used in the construction of roads and landings, although they can be used for that purpose, and they can also serve as mobile anchors for cable systems.

During skidding, an attachment on the tractor called an integral arch lifts the front ends of logs clear of the ground.

Flexible-tracked machines, sometimes described as high-speed tracked skidders or low-ground-pressure vehicles, are variants of military-type armored personnel carriers. They weigh about as much as crawler-tractors (30,000-40,000 pounds), but they are more stable on steep or irregular ground. They are usually equipped with an integral arch with a structure that is capable of supporting one end of the log load off the ground. Flexible-tracked machines typically need more maintenance and repair than rigid-tracked machines.

Rubber-tired skidders have lighter machine weights (20,000-30,000 pounds) and less pulling capacity than tracked vehicles; as a result they usually skid less volume per cycle. The front-axle section contains the engine and operator's cab; the rear-axle section contains a winch and an integral arch or grapple. The two sections are articulated at a hinge point in the middle. Chains are often used on the front wheels and sometimes on all four wheels to improve traction and protect tires from rock damage. These machines are rugged, but their turning radius is greater than that of tracked vehicles, and they are the least stable of all ground-based machines on steep or irregular ground.

A grapple or winch line and chokers are used for hooking felled timber with ground-based skidding equipment. The winch line is more common with tracked machines, while either the winch line or grapple is common on rubber-tired skidders. Grapple skidding matches well with mechanical felling and bunching, while the use of a winch-line machine matches well with manual felling.

Hydraulic-grapple loaders with track under-carriages are used in loader or shovel logging. They travel off-road over moderate terrain, moving through felled timber to cast or swing logs or felled trees from stump to roadside. Several passes through a harvest unit may be required to cover the area (Figure 6-13a). A variation of this procedure is to swing and bunch logs to skid trails, where another ground-based machine skids them to the landing (Figure 6-13b).

Horses used for logging are generally draft horses such as Belgians, Clydesdales, and Percherons. These animals weigh between 1,500 and 2,000 pounds and generally work in teams of two. On flat ground, a horse can pull a maximum of approximately its own weight, although log loads should be less than the animals' weight. Production rates are lower than for machines, but because both the initial cost and operating costs are lower, horse logging is economically competitive with machines. Production rates for a team average approximately 3,000 board feet a day, or about one logging-truck load.

With any ground-based system, a key environmental concern is soil compaction, which has been shown to reduce tree growth (Froehlich 1979; Wert and Thomas 1981), contribute to soil erosion, and impair water quality. However, mitigating measures are available. The most effective is to designate skid trails so that the area affected by skidding is kept below some portion of the total harvesting area. For example, when using directional motor-manual felling techniques and winch-line yarding with a rubber-tired skidder, a realistic objective is to keep skid-trail coverage below approximately 10 percent of the total harvest area (Garland 1983a). Properly designated skid trails can also increase the efficiency of logging operations. Mapping and keeping records of skid-trail locations with such tools as global positioning systems (GPS) and geographic information systems (GIS) can be helpful for relocating designated skid trails for future harvesting entries. Operating ground-based systems over snow or logging slash helps reduce soil compaction, and constructing water bars across skid trails helps minimize erosion. Finally, compacted soil can be tilled after logging to restore soil conditions (Andrus and Froehlich 1983).

Mechanized systems

In mechanized harvesting systems, machines operated by a few people perform the tasks formerly done by many more. The history of mechanized logging in the Coast Range is relatively short, but interest in mechanization is growing. These systems lend themselves to Coast Range conditions for several reasons. They are suitable for harvesting younger, second-growth forests, many acres of which are coming under management. The smaller-diameter trees in these forests are more uniform than the large trees in old-growth stands, making them well suited to mechanized harvesting. Management prescriptions for these forests frequently call for thinning, a job some mechanized systems can do very well. Mechanized systems tend to be safer than

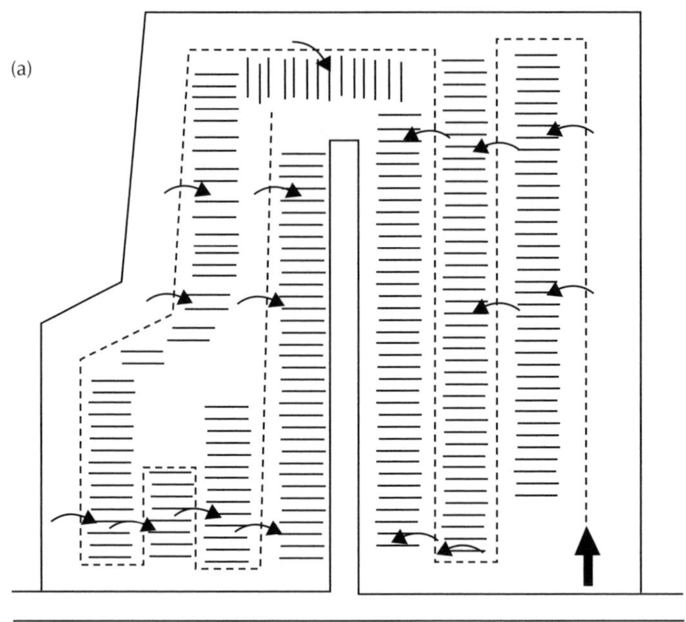

Figure 6-13. (a) Loader logging in a serpentine yarding pattern; (b) loader logging in which the loader swings bunched logs to skid trails.

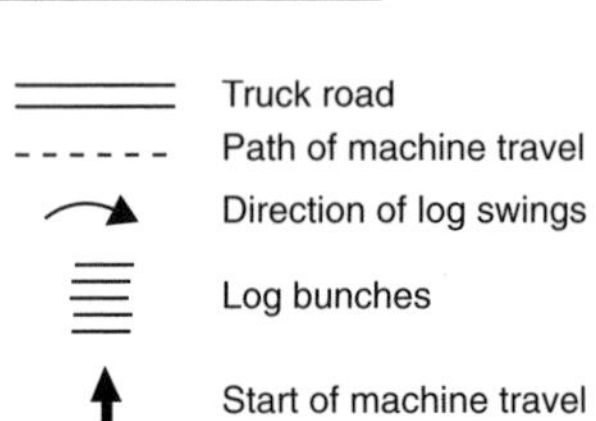

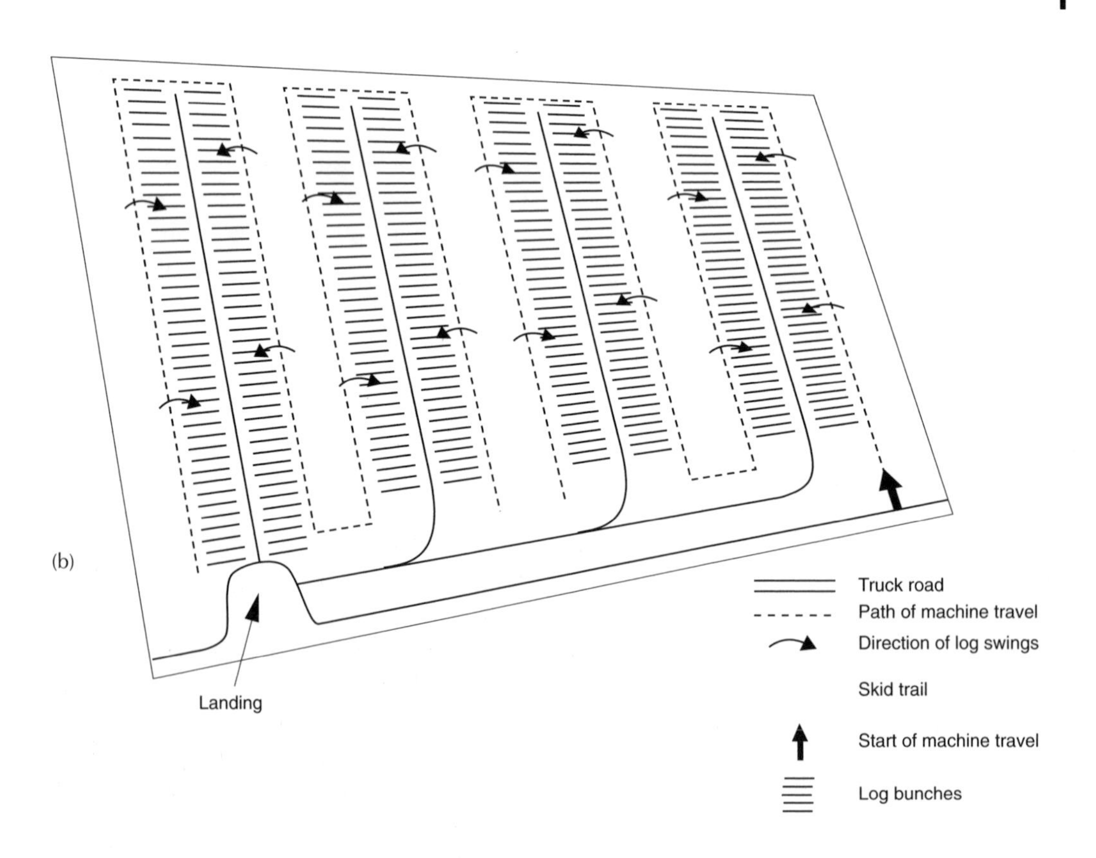

other harvesting systems because workers perform operations from inside a cab.

Although production rates are typically high, a few disadvantages associated with mechanized systems can cause problems. Repairing equipment breakdowns may require more specialized knowledge (e.g., of electronics and computer-control systems) than would be needed with conventional systems, and parts may be harder to obtain. It may be more difficult to coordinate the rate of felling with the rate of the skidding, yarding, or forwarding for an efficient operation. Operating in steep terrain or bad weather can lower production rates, and it can be costly with capital-intensive harvesting systems.

Mechanized harvesting can be categorized into three main systems: whole-tree, tree-length, and cut-to-length (Figure 6-14). These systems encompass varying combinations of people and equipment. Mechanical felling might be paired with conventional ground-based or cable systems for the yarding phase, or with a forwarder, depending on the type of system used. A more-detailed explanation of mechanized equipment can be found in Kellogg et al. (1993a).

In the whole-tree system, trees are felled and the limbs and tops are left attached. On gentle terrain, trees are cut with conventional feller-bunchers and moved to a landing with grapple skidders; on steeper terrain, self-leveling feller-bunchers and skyline systems are used. Trees are processed into logs or chips at the landing. Whole-tree logging has certain disadvantages. Large landings are typically needed, and working with full-length trees makes it difficult to control damage to residual trees. Because limbs and tops are brought to the landing, large slash piles can result, and limited slash remains in the stand for nutrient cycling or to reduce soil compaction on skid trails. On the other hand, clean logging of the site may eliminate the need for burning and aid in site preparation, and there may be opportunities for utilizing slash at the landing.

In the tree-length system, trees are felled, delimbed, and topped at the stump, but not cut into logs. Typically a single- or double-grip harvester is paired with a grapple skidder to fell and skid the trees. Trees may be transported to the mill as tree-length or cut into logs at the landing before transport. With this system, limbs and tops are left in the unit. Landings are typically smaller than those in a whole-tree system. Damage to residual trees may be difficult to control, as it is for whole-tree skidding.

The cut-to-length system solves some of the disadvantages of the other two systems. Trees are felled with a single-grip harvester, delimbed, topped, and cut into logs at the stump. The logs are transported to the landing with a forwarder, which unloads them onto roadside decks or directly onto truck trailers. The cut-to-length system leaves tree limbs and tops scattered over the site. Trees are commonly processed in front of the harvester, so that the machine travels over the slash, producing less soil compaction than would otherwise occur (Allen 1997). Designated equipment trails may be used with this system, so that the forwarder utilizes the same trails as the harvester, thus keeping the area covered in trails to approximately 20 percent of the site. The soil impacts of the forwarder are greater than those of the harvester because the forwarder makes multiple passes over equipment

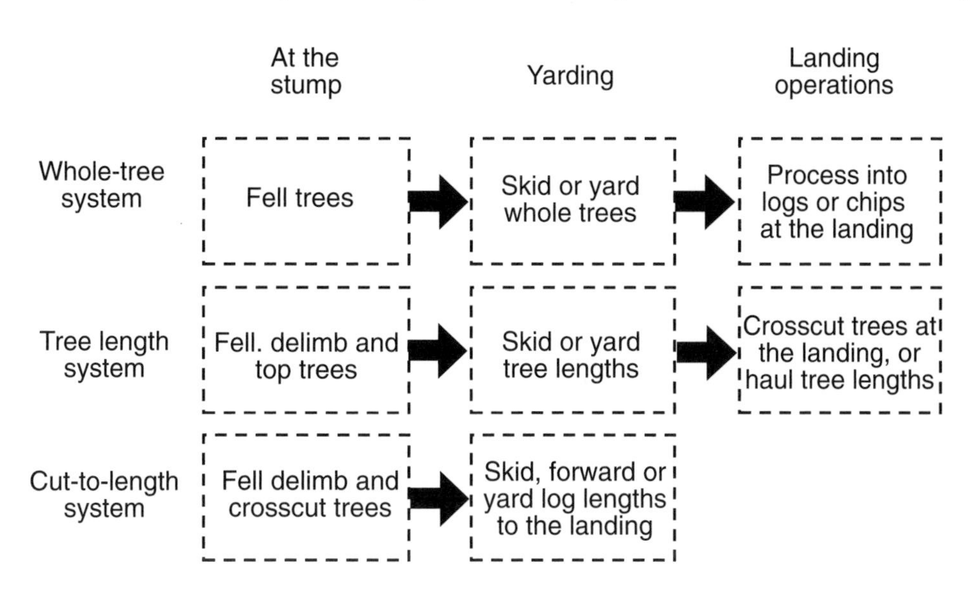

Figure 6-14. Types of mechanized harvesting systems. From Kellogg et al. (1993a).

trails carrying heavy log loads. However, specific impacts vary according to steepness of terrain, soil type, number of machine passes, and amount of slash on equipment trails (Armlovich 1995; Allen 1997). The cut-to-length system operates with smaller roadside landing areas than do the other two systems, and cut-to-length logging causes less damage to residual trees.

Harvest Planning to Achieve Forest and Stream Resource Objectives

Harvest planning is inseparable from broader silvicultural objectives for a site, because harvesting is used to accomplish much of the work that must be done to meet management objectives. Harvesting is the main tool available for making major structural changes in forest vegetation, such as converting an even-aged to an uneven-aged stand.

Many forest operations in the Coast Range—such as young-stand thinning, uneven-age stand prescriptions, logging around riparian areas, harvesting on steep dissected terrain with unstable slopes, and the use of skylines, helicopters, and new mechanized harvesting systems—can be characterized as being on the relatively high end of silvicultural and logging complexity. These complex projects are more likely to accomplish the desired management objectives (e.g., silvicultural, environmental, social, and economic) when they are carefully planned in conjunction with harvesting specialists (Figure 6-15). The harvesting specialist can help anticipate obstacles to successful project implementation and can identify workable solutions. While the specialist's expertise is important with relatively simple silvicultural systems (e.g., even-age management and clearcut harvesting) with manual logging (e.g., ground skidding on gentle terrain away from riparian areas), it is even more important with more complex silvicultural systems (e.g., uneven-age management and single-tree selection) and more challenging logging operations (e.g., multispan skyline yarding).

Figure 6-16 outlines the harvest-planning process, starting with a broad forest resource assessment and moving to a more-detailed, site-specific, project-level plan. This process contains many detailed steps, the precise nature and order of which will vary with each organization and landowner. The important thing, however, is that harvest-system planning and economic analysis be incorporated

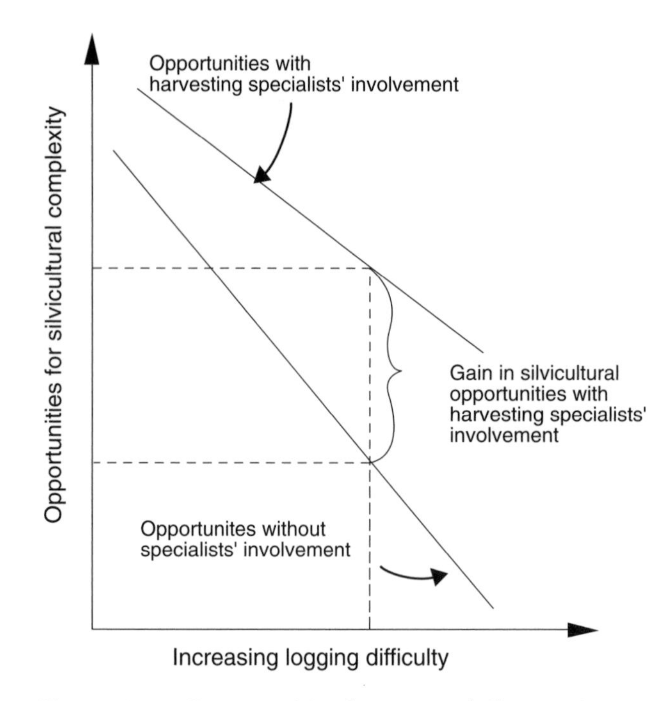

Figure 6-15. Opportunities for successfully carrying out silviculturally complex operations with interaction between forest managers and harvesting specialists, and without such interaction.

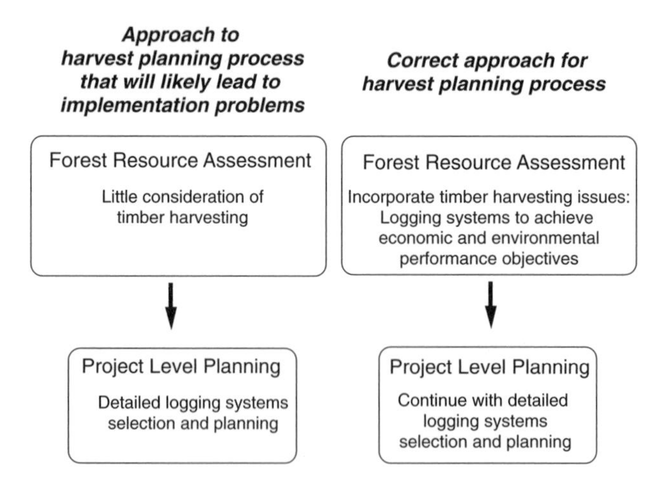

Figure 6-16. The harvest planning process: likely implementation problems contrasted with the correct approach.

Figure 5-18. A narrow riparian buffer in the Oregon Coast Range. (D. Vesely photo)

Figure 7-1. A dense young stand (12–15 years old) of over 200 trees per acre, typical of many acres in the Coast Range. This stand is a mixture of planted Douglas-fir and natural western hemlock and red alder (lower right). Stands like this can originate from planted Douglas-fir, natural regeneration, or combinations of the two.

Figure 7-2. This uncontrolled salmonberry made a dense cover that excluded all the conifers except the one in the lower left. Continual sprouting of new stems will enable salmonberry to persist and exclude other trees and shrubs for many years. Before cutting, this site had a stand of alder with an understory of salmonberry (Figure 7-8a).

Figure 7-3. A 5-year-old plantation with Douglas-fir saplings overtopping salmonberry, bracken fern, and other shrubs. Shrubs had been controlled about 3 years previously to promote Douglas-fir establishment. Note the height of the background red alder—at this age exceeding that of the Douglas-fir.

Figure 7-4. A 12-year-old Douglas-fir plantation that has just been thinned to about 180 trees per acre. This thinning was done to maintain healthy crowns and promote the growth of large trees with commercial value at about age 20–30. Thinning will also delay the onset of crown closure and maintain a diverse understory of shrubs and herbs.

Figure 7-5 (a)

Figure 7-5. Douglas-fir 8 years old growing under three levels of competition from tanoak in southwestern Oregon. The seedling growing in high competition (a) will probably not survive as the tanoak continues to overtop it. The seedling in moderate competition (b) will overtop the tanoak, but the tanoak will persist in the understory of the developing new stand. The seedlings in (c), where all the tanoak was removed, will grow the fastest. The stand shown in (c) will produce more wood in a shorter time than that in (b). The stand in (b) will produce considerable wood as well as an understory of tanoak and other plants. The stand in (a) will likely become a pure tanoak stand.

Figure 7-5 (b)

Figure 7-5 (c)

Figure 7-6. Douglas-fir stands 50 years old that have grown at various densities (Marshall et al. 1992). Stand A was thinned heavily 10–25 years previously, stand B was thinned lightly during the same period, and stand C was not thinned. See Table 7-1 for results of thinning and stand development.

Figure 7-6 (a)

Figure 7-6 (b)

Figure 7-6 (c)

Figure 7-8 (a)

Figure 7-8 (b)

Figure 7-8 (d)

Figure 7-8 (c)

Figure 7-8 (e)

Figure 7-8. A heavy cover of salmonberry has formed under alder (a). A dense network of rhizomes (b) was excavated from a 6-ft x 6-ft plot beneath a salmonberry community like the one in the background and in (a). Salmonberry communities can produce over 10 miles of rhizomes to the acre, with buds every 2-3 inches. New aerial stems sprout from these buds to replace older ones as they die or are killed by fire or other disturbance. Salal (c) also forms a dense cover and a dense network of rhizomes below ground. This bundle was excavated from a plot of about 3 x 3 ft. A dense community of salal may have over 50 miles of rhizomes to the acre. Oregongrape (d) has a much less dense and vigorous rhizome network than salal. It invades mainly by seeding, but it can produce new aerial stems from rhizomes after disturbance to its top. Vine maple (e) reproduces vegetatively. Its long, drooping stems, bending to contact the ground, take root and produce new upright stems. Sprouting is accelerated when branches are pinned to the ground by natural debris fall or logging slash from thinning.

Figure 7-10 (a)

Figure 7-10 (b)

Figure 7-10 (c)

Figure 7-10 (d)

Figure 7-10. Stand in (a), about 100 years old, has a natural opening caused by wind in which an understory of Douglas-fir has regenerated. Stands in (b) and (c), 60 years old, were heavily thinned about 15 years previously, leaving <80 trees per acre. Because there were seed-producing Douglas-fir and hemlock in the stands, a conifer understory became established and a multi-story condition is now developing in both stands. Additional thinning of the stand in (b) may be needed to encourage further development of the understory conifers. Salal and bracken fern have become established in the stand, forming a dense understory that will likely preclude more conifer establishment unless it is controlled. Stand in (d), 80 years old, was thinned at age 40 to 40-50 trees per acre and underplanted with hemlock. There were no seed-producing hemlock in the stand, so hemlock was planted along with western redcedar (right). At age 80, the Douglas-fir canopy has closed and a multi-storied stand is developing, but further thinning of the overstory may be needed.

(a)

(b)

Figure 7-12. Leaving large trees and snags when a stand is harvested (a) can provide habitat for birds as the young stand develops. Beginning at about 50 years these stands can provide some of the characteristics of old-growth forests. Topping large trees in young-growth stands (b) causes twin tops to develop, creating a potential nesting site for species that use older forests (80 or more years old).

Figure 7-13. Good-quality seedlings of red alder in a nursery bed, ready for lifting and planting.

Figure 7-14. Sprouts that originate high on a stump are easily broken. This is a madrone sprout.

Figure 7-17. A Coast Range stream surrounded by a young, mixed-species forest. Note the large boulders in the stream and the logs that will probably fall into the streambed during high flows.

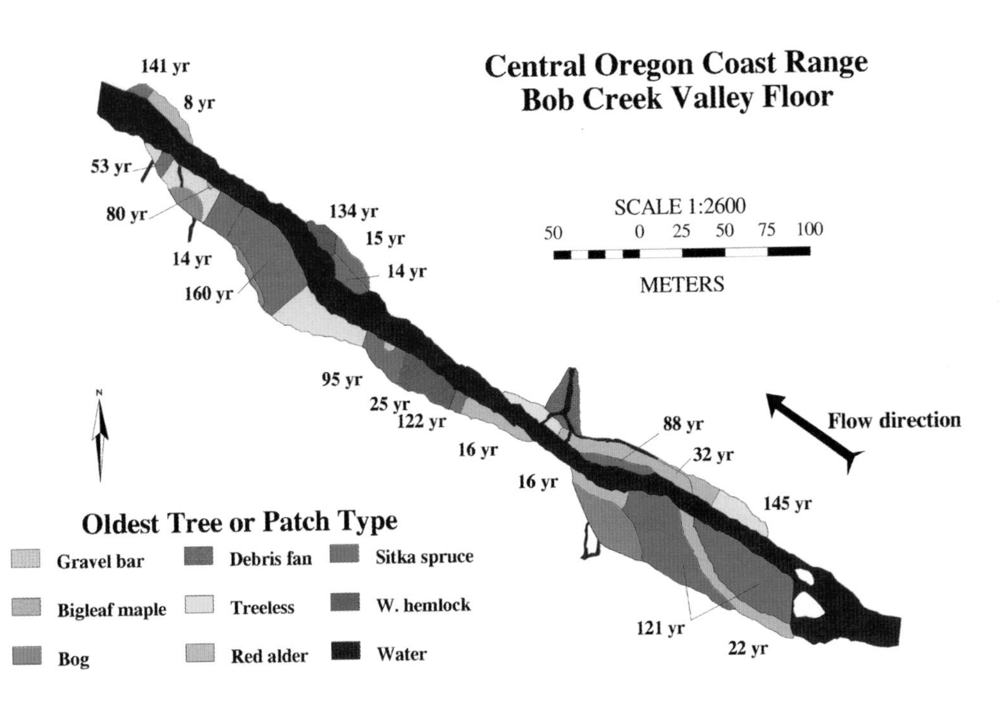

Figure 7-18. An example of the considerable variation in ages and types of vegetation and substrates along approximately 0.4 mile of stream in the coastal forests of western Oregon. These patterns of vegetation and substrate are likely the result of natural processes, since there was no evidence of human-caused disturbances (K. Avina and D. Hibbs, personal observation).

Figure 7-20. A buffer strip between two clearcuts. On the near side the forest is alder and maple; on the far side, Douglas-fir. Both are unmanaged.

Figure 8-1a. **Laminated root rot**: *Phellinus weirii* can attack and kill young trees. Seedlings may be killed within a few months of planting if their roots are placed in contact with an active inoculum source. A sheath of ectotrophic mycelia usually covers the roots and belowground portion of the stem of a diseased seedling.

Figure 8-1b. **Laminated root rot**: Pockets of dead trees may develop at an early stand age. The roots of susceptible seedlings planted near an infected stump may come into contact with the infected stump or infected roots, become infected, and be killed within a few years. Infected stumps serve as long-lived inoculum that enables the fungus to remain on the site and initiate the disease in the replacement stand.

Figure 8-1c. **Laminated root rot**: Young Douglas-fir infected by *Phellinus weirii.* Note the distress crop of smallish cones, reduced terminal growth, short branchlets, and yellowish needles that are smaller and fewer than on healthy trees. The dead tree at the right was killed by laminated root rot.

Figure 8-1d. **Laminated root rot**: Laminated root rot causes changes in host crown shape and appearance. Diseased Douglas-firs (center and right edge of the picture) at the edge of a disease center have rounded tops (resulting from stunted height growth), bushy branch ends (resulting from reduced branch growth and shortened needle), and thinning foliage. The tree at the right also has a distress crop of smallish cones typical a year or two before the tree dies from laminated root rot.

Figure 8-1e. **Laminated root rot**: Uprooted Douglas-firs with root wads, which may resemble a closed fist, are characteristic of laminated root rot. Decayed roots have broken close to the root collar, leaving only stubs. Broken-off roots left in the ground remain intact for decades as a food base for the fungus and as the source of new infections on the same site.

into the early steps of the forest resource assessment. In many cases, discussion of specific harvest systems is left until one of the last steps in the planning process. Failure to consider at an early stage the capabilities, limitations, economics, and environmental performance of harvesting can make it impossible to meet all of the management objectives. For instance, managers who plan without considering harvesting details may unwittingly close off certain harvesting options, and the ones that remain may be physically incapable of doing the job, or incapable of doing it in a cost-effective, environmentally sound manner. This problem is less likely to occur, as shown in Figure 6-16, when the capabilities and limitations of harvesting systems are considered earlier in the process, along with other critical economic and environmental factors.

Four harvest-planning requirements are common to all operations:

1. *The harvest system selected must be physically capable of accomplishing the silvicultural and other resource-management objectives.*

Physical capabilities are dependent on site-specific factors such as tree size, ground slope, and yarding distance. In addition, the physical capabilities of the equipment, such as cable diameters, tower heights, drum capacity, and engine horsepower, must be considered, along with logging safety (Oregon Occupational Safety and Health Division 1995). For example, the harvest plan may call for full suspension of logs over a stream in a riparian management area. Thus skyline deflection is a critical factor, and the harvest unit must be laid out in such a way that adequate deflection can occur to safely yard log loads and maintain stream clearance. The yarding distance or span length, shape of the topography, skyline-anchor location, and tower height must be considered together in order to achieve this goal.

2. *The harvest system selected must meet environmental and other legal requirements.*

Environmental requirements must be followed, for example, to meet forest-practices regulations as well as to minimize soil disturbance and compaction, to maintain streambank integrity, to prevent soil erosion, to maintain water quality, to protect cultural sites and wildlife habitat, to protect residual trees from logging damage, and to minimize landing size. Harvest operations conducted upslope from riparian areas must take particular care to minimize impacts to the riparian zone.

3. *The harvest system selected must be socially acceptable.*

Logging must be conducted as prescribed in the silvicultural treatment plan with consideration of the interests and concerns of the general public and special-interest groups. Local community and workforce needs should also be considered.

The approach to accommodating social issues varies depending on legal obligations and landowner ideas. Planning tools such as visual simulation can be used to aid thinking and to project into the future what the forest operations will look like. Project-level planning, such as identifying scenic buffers and screens and using signs to help explain forestry operations, can be helpful (Garland 1998).

4. *The harvest system selected must be economically efficient and feasible.*

There are two parts to this statement: efficient and feasible. First, harvesting systems should meet the other three requirements without incurring unnecessary costs; harvesting should be cost efficient. For example, appropriately planned cable logging may adequately meet environmental and social requirements, making it unnecessary to use a higher-cost method such as helicopter logging.

Secondly, the harvesting operation should be economically feasible; the value of timber removed should cover the cost. If harvesting must be done at a loss, additional financial subsidies or considerations are needed to carry the operation forward. Examples of such situations could include creating wildlife habitat by the selective removal of trees, removing dead or diseased trees to eliminate a fire hazard, and smallwood thinning. In each case a thorough economic analysis must be completed. If the value of timber removed does not cover the harvesting cost, a land manager may still decide the anticipated benefits justify a financial subsidy.

As we have stressed already, it is important to consider harvest-planning objectives and constraints at every level of forest resource management planning. Often there is some overlap between the different levels of planning. Figure 6-17 shows a simplified breakdown of the levels of planning that are closest to forest operations. For effective implementation and continuous improvement, information and communications should flow freely among those responsible for all levels of planning.

Strategic planning encompasses a wide range of forest resource issues, including wildlife habitat, visual corridors, riparian management areas, and

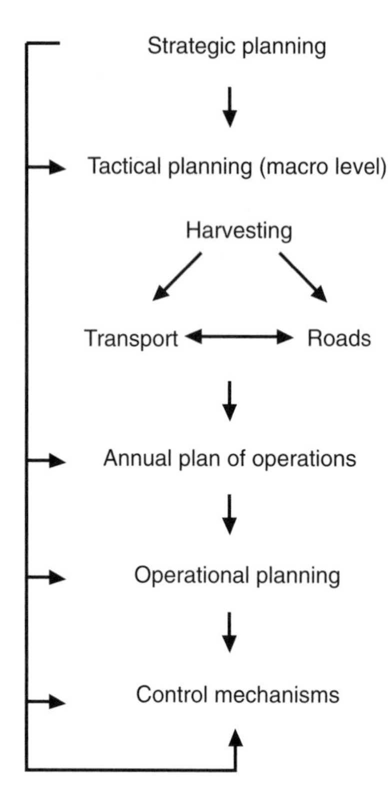

Figure 6-17. Levels of harvest planning. From Brink et al. (1995).

timber removal. In addition, issues such as market requirements and timber supply are considered. Strategic planning sets the direction for important issues that must also be addressed at other levels of planning. Tactical planning focuses on identifying and scheduling harvest, road, and transportation plans that will achieve the strategic management objectives. Strategic planning is also referred to as landscape-level planning, or large-area planning and it typically covers a time period of approximately three to five years.

Landscape-level planning is planning for an entire management area for the best realization of all objectives identified in the strategic plan. The emphasis is on "paper planning," using topographic maps, digital terrain models, aerial photos, and GIS or maps of soils and forest cover. Only limited field reconnaissance, verification, and layout are completed at this level of large-area planning, but harvest boundaries and road locations are identified that match with the requirements of the logging systems to be used. Landscape-level planning offers a good opportunity to solve problems that arise beyond the stand level. For example, a landscape-level plan may target scattered areas that pose problems for conventional logging systems—areas where conventional logging is impractical, for example, because of inadequate skyline deflection and long yarding distance, or because of resources that must be protected. The large-area plan might solve this problem by combining these areas into one logging project in order to justify a more costly logging system, such as helicopter yarding. The annual plan of operations flows out of the strategic and tactical plans.

Operational planning involves stand-level planning, or unit planning. Here there is a more thorough field reconnaissance, laying out of the harvest operations, selection of specific equipment, formulation of detailed operation plans, and economic analysis and budgeting. Stand-level planning should be completed within the context of previously developed landscape-level plans. Practicing separate unit-by-unit planning without a landscape-level blueprint leads to problems and inefficiencies in forest operations.

Control mechanisms involve implementation of the harvest plan and monitoring its effectiveness in meeting forest resource objectives. Good record keeping during earlier levels of planning and communication can be helpful.

Harvesting systems are selected on the basis of site conditions and the management objectives for the site. Factors affecting the choice include: (1) topography (slope steepness and variability); (2) soil (composition, sensitivity to disturbance); (3) timber characteristics (tree size and volume per acre); (4) potential constraints on road access and transportation; and (5) yarding distance and direction.

The selection process is often complex; many factors must be weighed. When a portion of the existing stand is to be retained, for example, the characteristics of the stand following each harvesting entry—not only the first one—should be assessed. The number, size, and distribution of trees to be retained will influence what can be done in the initial harvesting operation, and future operations will be affected by how the residual stand has been shaped by subsequent growth and development.

In selecting harvesting systems, there are several identifiable points where decisions must be made (Figure 6-18). Careful consideration of choices at these points will lead planners toward the best harvesting system or combination of systems for each situation. More examples for each harvesting system can be found in MacDonald (1999).

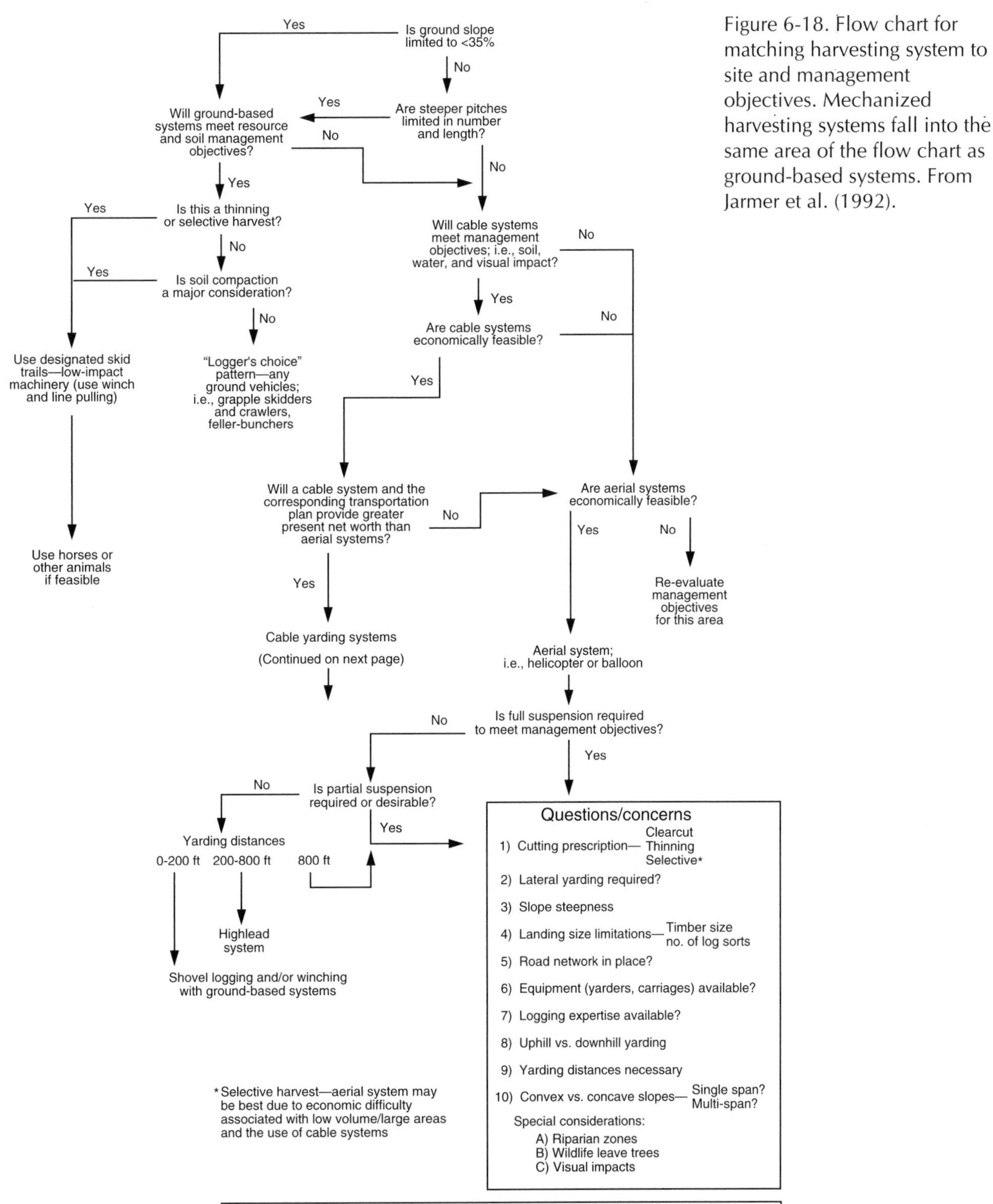

Figure 6-18. Flow chart for matching harvesting system to site and management objectives. Mechanized harvesting systems fall into the same area of the flow chart as ground-based systems. From Jarmer et al. (1992).

Skyline system types—capabilities/limitations

Standing skyline—skyline length fixed, slackpulling carriage required, long yarding distances possible, two-drum yarder required, all multi-span skylines this type.

Running skyline—uphill or downhill, clear-cut or other regeneration methods, lateral yarding with correct carriage, interlock drums desirable.

Live skyline—gravity outhaul with uphill yarding only, clear-cut or partial cut with proper carriage, lateral yarding with proper carriage.

Slackline system—uphill or downhill, long yarding distances possible, good for full suspension, three-drum yarder required, can be expensive.

Multi-span sytem—good for convex slopes, broken terrain, uphill or downhill, good for thinning, supports require additional rigging time and expertise.

Even-age Silvicultural Systems

Even-age silvicultural systems are widely used in the Oregon Coast Range. These include clearcutting; clearcutting with retention of trees such as seed-trees, shelterwood, and wildlife-trees; and commercial thinning of even-aged stands.

There has been increased interest in commercial thinning in recent years, both to enhance multiple forest resources and to harvest wood to achieve economic benefits. Through commercial thinning, a landowner captures mortality, gains an early return on investments, and maintains or increases the growth rate of residual trees. Because prices and demand have increased in recent years, marketing of smaller trees has become easier than it was in the past.

Both public and private landowners are also interested in commercial thinning as a way to increase diversity of plant species, increase the structural diversity of forest stands, and enhance wildlife habitat. For example, a commercially thinned stand may be underplanted with a variety of tree species, or the understory may be allowed to develop from natural regeneration, in order to encourage the development of a multi-storied stand. Current research indicates that if young stands are thinned to low residual densities to stimulate rapid growth of dominant trees, some of the old-growth habitat characteristics needed by certain wildlife species can develop in a matter of decades instead of centuries (Newton and Cole 1987; O'Hara 1990). Further details on silviculture in the Coast Range are presented in Chapter 7.

Appropriate harvesting systems

Virtually any harvesting system can be used successfully in clearcut harvest operations. Other even-age silvicultural practices, such as seed-tree, shelterwood, wildlife-tree-retention, and even-age-thinning prescriptions, offer less flexibility in the selection of an effective logging system. The more trees retained within a harvest area, the more selective a planner must be in order to match a logging system to silvicultural objectives to ensure safe, efficient, socially acceptable, and environmentally sound harvest operations.

Clearcutting

On steep terrain or in areas where soil disturbance must be avoided, cable-logging systems should be used. Highlead and live skyline cable systems may be used for clearcut silviculture. These systems are simple and capable of yarding substantial payloads. A live skyline system typically uses a non-slack-pulling carriage, which limits the lateral yarding capability and increases the number of cable roads required to yard a given area. Both highlead and live skyline systems can also be used in clearcuts with tree retention, as long as the individual leave trees or groups are situated where they will not inhibit logging production or create safety hazards (Figure 6-19). Standing and running skyline systems incorporating slackpulling carriages can also be used very effectively in any clearcut operation. Accurately positioning the carriage to yard laterally enables logging crews to work around leave trees safely and efficiently. Lateral yarding makes it possible to space cable roads more widely. This reduces the number of lift trees and anchors required for a given area and may reduce total rigging time.

Helicopter logging has been used successfully in all silvicultural systems. Helicopters are capable of

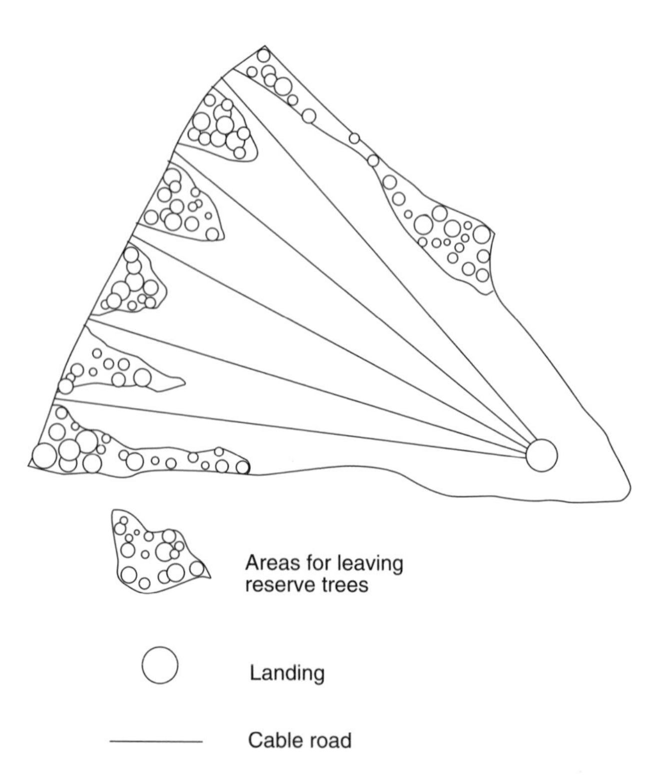

Figure 6-19. Reserve tree clumps compatible with highlead and live skyline systems. From Oregon Occupational Safety and Health Division et al. (1995).

large payloads and high production rates at relatively long yarding distances. Flying logs downhill is preferred, although uphill yarding is also feasible.

Ground-based logging systems can be configured in a variety of ways to meet clearcut harvest goals. Manual felling with winch-line skidding is commonly used in clearcut logging. Feller-bunchers can also be used with skidders or tractors. Feller-buncher operations usually result in closer skid-trail spacing than manual felling and winching operations. Loader logging in clearcuts is common. Again, in clearcuts where trees are retained, care must be taken in locating leave trees. In all ground-based logging systems, soil impacts can be reduced by using designated skid trails or tilling the soil after harvest.

Fully mechanized systems can be successfully used for clearcutting if tree size is within the operating capabilities of the equipment. Because single-grip harvesters have limited reach, equipment trails may be spaced more closely than for ground-based skidding machines with winches. Tree-length and whole-tree harvesting methods can be used if leave trees are widely spaced and skid trails are carefully laid out.

Thinning

Thinning operations can be carried out with cable systems if yarders and carriages are matched to stand and terrain conditions. Yarders should be selected for their ability to operate the most efficient cable system for a given harvest area. Standing skylines and multispan systems with slackpulling carriages are the ones most commonly used for thinning operations. A multispan system often makes it possible to use shorter spans. The shorter the span, the less the lateral deflection of the skyline, and the less the potential for damage to the remaining stand. Running skyline systems are sometimes used to thin stands, but the carriages used in these systems are currently incapable of passing over intermediate supports; thus, they are limited to single spans. Yarders equipped with swing booms have advantages over those with stationary towers when logs are landed alongside the road rather than at a central landing. Swing yarders can rotate to the side to land logs on the road, thus eliminating the need for large landing areas. Skyline corridors are commonly cut 10 to 15 feet wide and spaced 100 to 150 feet apart.

Selection of the right carriage is critical in cable thinning operations. Both manual and mechanical slackpulling carriages may be used. Carriage position control when lateral yarding will improve productivity and reduce the potential for residual-stand damage.

Aerial systems, specifically helicopters, can be used for thinning, but wood value, tree size, and thinning intensity must be carefully evaluated to determine whether the higher cost is justified. Helicopter payload capability must be matched to stand characteristics and thinning intensity. Thinning intensity will determine how much volume is available over the harvest area. Conversely, thinning intensity determines stand-canopy density. At densities of 50 percent or greater, locating the hooking crew and lifting logs takes enough time to adversely affect production. In addition, as crown densities increase, ground crews are exposed to more hazards.

Skidders and tractors are commonly used to thin stands in the Coast Range on flat or moderately steep slopes during dry periods of the year. Manual felling and winch-line skidding from designated skid trails are effective in reducing soil impacts and damage to the residual stand. Tractors have greater maneuverability than rubber-tired skidders and thus lower potential for stand damage when operated carefully. Horse logging is most efficient when removing small trees from relatively flat terrain in situations where environmental concerns preclude the use of machines or where small volumes are to be removed. Several conditions must be in place if horse logging is to be successful. Slope of the ground should be 35 percent or less, and all landings should be located for downhill skidding. Tree size should be such that the largest tree can be bucked to logs each weighing no more than one-third of the horse's or team's weight. Trees must be felled in lead (toward the skid trail) for ease of skidding, and yarding distances should be limited to 300-400 feet (Fieber and Robson 1985). Heavy amounts of logging slash and brush will impede production and should be removed before logging.

Mechanized logging systems are very effective in thinning even-age stands. Cut-to-length operations that use a single-grip harvester and forwarder are especially appropriate, and an experienced operator can thin a stand with virtually no damage to residual trees. Tree-length and whole-tree harvesting is also productive, but because logs are longer, there may be higher levels of residual damage.

Harvest planning and layout approaches

Any silvicultural or harvesting system must be planned at both landscape and unit (stand) levels. First, managers should use topographic maps, digital terrain models, and aerial photos along with limited field reconnaissance, to develop a preliminary approach for the harvest area. Initial decisions are then made regarding road locations, landing locations, and appropriate harvesting systems. Once landscape-level planning is completed, stand-level planning can begin. This section outlines decisions that arise in clearcut and thinning operations as the planning progresses from general landscape concerns to specific stand-level concerns.

Clearcutting

For cable operations, roads and unit boundaries must be located in light of the operating capabilities of specific yarding equipment. For example, a standing skyline system capable of passing over intermediate supports can be used to log areas with limited skyline deflection, eliminating the need to build roads into these areas. Intermediate support capability also permits longer yarding distances. These considerations make a difference in where harvest boundaries may be located. Equipment should be evaluated with one of the available payload-analysis computer programs, such as LOGGERPC 3.2 (Brown et al. 1997), to determine whether it is capable of yarding the harvest area. Locations and size requirements for tail trees, intermediate support trees, and guyline and skyline anchors must be decided on and then field-verified prior to final field layout. Specific requirements and recommendations for lift trees, anchors, and rigging are provided in the Oregon Occupational Safety and Health Division et al. (1995) logging-safety regulations. It should be stressed that all intermediate support trees and anchors must be marked and mapped before logging begins.

Planning and layout for aerial operations must take into consideration several concerns specific to aerial logging systems. Aerial systems require larger landings than cable or ground-based logging systems. Because helicopter operations generate high volumes of logs very quickly, they generally need to have more than one landing available for log drop. Helicopter operations may also require fueling and service areas in addition to landings. Flight distance, vertical direction, and altitude changes are critical factors in planning feasible aerial operations. Rates of ascent and descent must be considered when locating landings. Time of year must be considered when scheduling aerial operations due to their sensitivity to weather conditions. More information on unit-level planning for helicopter operations can be found in Studier and Neal (1993).

Harvest planning and layout for ground-based and mechanized harvest systems should take into consideration the sizes and value of the trees, distribution and density of leave trees, steepness of the slope, direction of skidding or forwarding (uphill or downhill), and distance the logs must be transported. Evaluation of these factors will lead to decisions on final system selection, road and landing spacing and locations, unit-boundary locations, and skid-trail and forwarding-trail design. Each system or combination of machines has specific operating characteristics that should be taken into account to design economically and environmentally sound harvest units.

Wildlife trees and snags can be retained in clearcut harvest areas if careful planning is done before the beginning of operations. Objectives for reserve trees, specific site conditions, equipment capabilities, the overall harvesting process, and above all safety must be evaluated. It is also important to consider future forest activities, such as site preparation and planting, when leaving wildlife trees and snags. Snags are generally not left in areas where there are workers on the ground because of overriding safety concerns. An option is to leave a ring of green trees around a particularly desirable snag or patch of snags to provide a protective buffer. More information on selecting wildlife trees and snags can be found in Oregon Occupational Safety and Health Division et.al (1995).

Thinning

Successful commercial thinning operations will require more-detailed planning and layout than clearcut harvests. Thinning requires careful logging around the remaining trees, and it may involve repeated entries over the life of the stand. Decisions must be made that affect all entries, not just the initial one. Regardless of the logging system used, proper planning and field layout are critical for meeting silvicultural goals, maintaining or enhancing stand value, and minimizing environmental impacts.

Cable thinning operations require the most detailed levels of planning and layout. Roads and landings should be located where they can be used for subsequent entries. A yarding pattern should be designed so as to minimize stand and site damage. Such a design might call for minimal sidehill yarding, parallel skyline corridors, and designated rub trees along skyline corridors. Tail trees and intermediate support trees should be protected from damage if they will be needed for future entries. Guylines for towers and lift trees may need to be attached to standing trees following Oregon Occupational Safety and Health Division et al. (1995) regulations, if creating stump anchors will adversely affect future operations.

In cable thinning operations, field layout—which, as in all harvest operations, should proceed only after all planning decisions have been evaluated—is critical for meeting both current and future objectives. Cable setting boundaries, skyline corridors, tail trees, intermediate support trees, and required guyline anchors should all be marked and mapped prior to felling. Flagging of skyline corridors will help fallers keep trees felled to lead with the corridor. Clearly marking lift trees and anchor stumps should eliminate any inadvertent felling of these structures. Communication among harvest planners, layout personnel, resource managers, and logging contractors is crucial for success in any thinning operation.

Commercial thinning with helicopters will require even more attention to planning and layout. Productivity of these systems is highly dependent on piece size and volume distribution. Crown density is also an important consideration because the pilot must be able to see the load hook as it is lowered. Critical to the economic feasibility of these systems is the ability to maximize each turn's payload. Volume removal should be designed to provide the rigging crew the opportunity to build adequate payloads. Productivity depends heavily on piece size; thus, it may be more cost-efficient to use these systems only for later entries when piece size is larger or use a smaller helicopter.

In ground-based thinning, the skid-trail or forwarding-trail system becomes a part of the dedicated transportation system within the unit, and it should therefore be designed to minimize site impacts. Trails should be well spaced, located to minimize soil rutting and to avoid converging, and located to facilitate current and future stand entries. Trails should be flagged prior to the beginning of operations. Rub trees should be designated in order to protect the residual trees adjacent to trails. As with all operations, communication among planners, field personnel, resource managers, and logging contractors before and during harvest activities will increase the likelihood of safe, productive, and environmentally sound operations.

Production and cost information

In this section we review recent research on harvesting production and costs applicable to the Oregon Coast Range for even-age silvicultural systems. Some studies compared different silvicultural systems or thinning intensities, some compared different logging systems, and some estimated production and cost for a specific logging system.

Studies have shown that clearcutting has lower costs per unit volume removed than commercial thinning. Logging costs for commercial thinning vary considerably among different sites, depending on the logging system used, thinning intensity, and stand conditions such as tree size (Hayes et al. 1997). In general, logging costs per unit of wood harvested are higher for light thinnings than for heavy thinnings, although light thinnings can still be economically feasible.

A study conducted by Kellogg et al. (1991) compared clearcutting with a two-story silvicultural treatment that called for leaving 12 overstory trees per acre. Two logging systems were used for each: ground based and cable. The study site was a naturally regenerated stand of Douglas-fir (*Pseudotsuga menziesii*) in the Oregon Coast Range with an average diameter at breast height (dbh) of 23 inches. For the ground-based system, total logging costs (layout, felling, and skidding) were 16 percent higher for the two-story treatment than for the clearcut. For the cable-yarding system, total logging costs were 23 percent higher for the two-story treatment than for the clearcut. For both systems, the higher costs for the two-story treatment were due mainly to higher layout and yarding costs; felling costs were only slightly higher.

Kellogg et al. (1996a) examined harvesting production rates and costs for three silvicultural prescriptions at two sites, Yachats and Hebo, in the Oregon Coast Range. The three prescriptions—three thinning densities—were chosen to investigate the effect of different commercial-thinning intensities on accelerating the development of old-growth habitat characteristics in young stands. (Other researchers are investigating ongoing aspects of the integrated project, including wildlife-habitat suitability, tree growth, and understory-vegetation development after thinning [Tucker et al. 1993].) The study sites were 33-year-old Douglas-fir stands that were thinned to residual densities of 100, 60, and 30 trees per acre (tpa). Tree size averaged 11 inches dbh. For the harvesting portion of the project, detailed time studies were conducted on manual felling and uphill skyline yarding with small yarders. The 30-tpa treatment had the highest production rate and was the least costly to harvest. Total harvesting costs of the other two treatments averaged from 6 percent more (60 tpa) to 12 percent more (100 tpa). Figure 6-20 shows total harvesting costs at the two sites for each silvicultural treatment.

In the same study, Kellogg et al. (1996a) also found that differences in logging techniques affected production rates and costs. Trees were manually felled at the first site. The tops were left attached to the last logs, and only the top sides of the trees were limbed. These measures increased the felling production rate over that at the second site, where loggers bucked the tops off the trees and limbed all three sides of the logs. In addition, road-change times were shorter at the first site, where workers rigged tail trees, than at the second site, where they used a mobile tailhold outside the unit, resulting in long rigging distances and problems in aligning the skyline with the thinning corridors. However, the parallel skyline-road layout used at the second site required fewer roads per acre and less time spent in road changes than did the fan-shaped layout at the first site. Overall, the first site was more productive and had lower total harvesting costs (Figure 6-20).

Helicopter logging production and cost were also investigated in conjunction with this study, to determine whether the use of small- to medium-lift helicopters can be an economical alternative for commercial thinning of young stands in steep terrain (Born 1995). The study area consisted of several stands of Douglas-fir that were thinned to 100 residual trees per acre. A Sikorsky S-58T helicopter with an external lift capacity of 5,000 pounds was used to yard the logs. The average yarding distance was 775 feet, and the average slope was about 30 percent. Trees removed averaged 11 inches dbh. The helicopter system yarded an average of 4.71 thousand board feet (mbf; net scale) per flight hour. Total logging costs for the helicopter system were approximately $100/mbf higher than the total cable-logging costs determined by Kellogg et al. (1996a) for a similar unit nearby that was thinned to the same intensity. This cost comparison included an allowance for the road construction that would have been necessary if the helicopter unit had

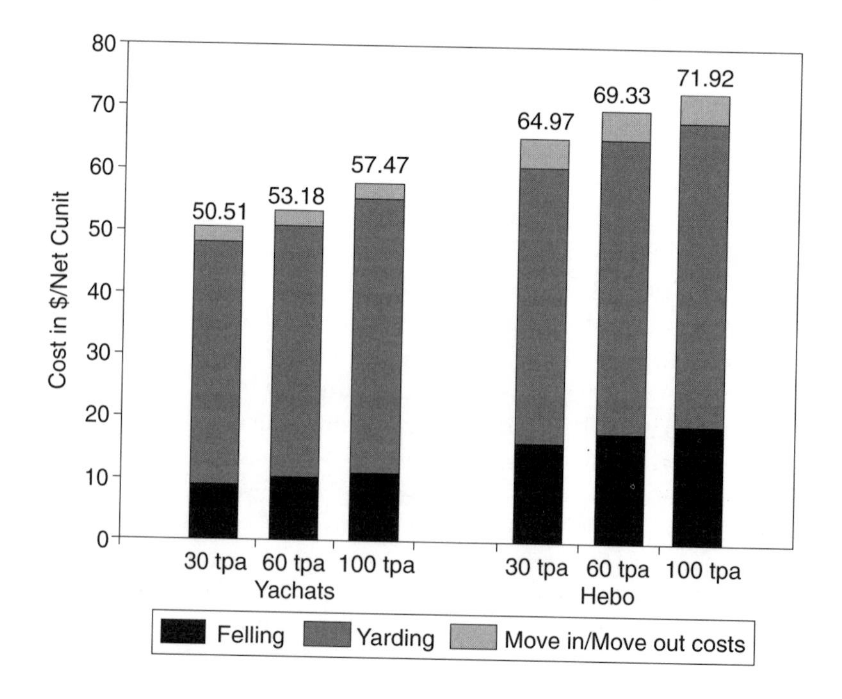

Figure 6-20. Total harvesting costs for three thinning prescriptions at two sites. Costs do not include profit and risk allowances or hauling costs. From Kellogg et al. (1996a).

been cable logged. Although the cost estimates for cable logging were much lower than for the helicopter, small-log values at the time of the study were high enough that the revenue to the landowner would have exceeded the cost of harvesting under either system. Helicopters should be considered when there are objectives other than strictly economic, such as environmental concerns, time constraints, limited access, and physical limitations of other systems (Born 1995).

Another study compared production rates and costs for two sizes of yarders at two thinning intensities (Hochrein and Kellogg 1988). The study site was a Douglas-fir stand in the western Cascades of Oregon. Trees averaged 12 inches dbh on steep slopes, and yarding distances averaged less than 1,000 feet. Costs for commercially thinning small timber were lower for a small yarder with a five-person crew than for a mid-size yarder with an eight-person crew. Yarding and loading costs for the small yarder were 11 percent lower in a light thinning (80 trees per acre removed) and 12 percent lower in a heavy thinning (125 trees per acre removed) than for the mid-size yarder. Yarding and loading costs for both yarders were 20-22 percent higher in the light thinning than in the heavy thinning.

In a western hemlock-Sitka spruce (*Tsuga mertensiana-Picea sitchensis*) stand thinned with a mid-size yarder, skyline yarding costs were 14 percent higher for a light thinning (removing 50 percent of the stand volume) than for a heavy thinning (removing 66 percent of the stand volume) (Kellogg et al. 1986). A herringbone-strip treatment that removed 51 percent of the stand volume cost 8 percent less to yard than did the heavy thinning because of the logging efficiences of felling and yarding logs in strips. The study site was in the Oregon Coast Range; trees averaged 13 inches dbh.

A study in the western Cascades of Oregon investigated the productivity and cost of using a mechanized cut-to-length system to commercially thin a second-growth stand of Douglas-fir and western hemlock (Kellogg and Bettinger 1992). Tree size before thinning averaged 10 inches dbh, and about one-third of the stand volume was removed. Delay-free production rate for the single-grip harvester averaged 11.4 cunits (1 cunit=100 cubic ft) per hour, and 7.8 cunits per hour when delays were figured in. There was no significant difference in harvester production between stands marked prior to logging and those in which the trees were selected by the operator. Delay-free production levels for the forwarder averaged 4.3 cunits per hour, and 3.3 cunits per hour including delays. When landing space was limited, a two-pass forwarding technique, in which sawlogs and pulpwood were separated by load, was more productive than a single-pass technique, with products mixed on each load and sorted at the landing. Thinning cost for the cut-to-length system was $35.37 per cunit, excluding hauling and a profit-and-risk allowance.

A series of studies conducted by the Department of Forest Engineering at Oregon State University between 1972 and 1979 evaluated several harvesting systems for thinning young stands on steep terrain (Kellogg 1980). Stands averaged 35 to 40 years old, with trees of 10-14 inches dbh. Thinning intensity varied over the different studies; an average of 40 percent of the stems were removed in the thinning operations. Several general trends were demonstrated by the studies. Felling and bucking production rate increased with thinning intensity. Tractor-yarding production rate decreased with increasing slope percent. Total cost for skyline logging ranged from 1.5 to 1.67 times that of tractor logging on slopes up to 40 percent. Prebunching logs to the skyline corridor increased yarding production for the machine on the landing. Finally, adding intermediate supports when small yarders with short towers were used extended yarding distances on convex slopes.

Uneven-age Silvicultural Systems

Uneven-age silvicultural systems, sometimes called partial-cut treatments, are increasingly being used to achieve a variety of forest resource values, such as wildlife habitat and aesthetics, in addition to wood production. Timber harvesting can be used to shift even-aged stands onto a path of development leading to an uneven-aged condition, and it is the principal means to carry out uneven-age silvicultural prescriptions. Uneven-age silvicultural systems are divided into two broad categories: group selection (patch or gap cuts) and single-tree selection.

Appropriate harvesting systems

Successful cable harvesting operations in group-selection and single-tree-selection treatments are highly dependent on the equipment and cable system chosen. Most uneven-age prescriptions call for removing trees of widely varying sizes, making equipment selection more difficult as the variation in tree size and volume distribution increases. A standing skyline with slackpulling carriage, single-span or multispan, works well for partial-cut treatments. Slackpulling carriages provide good control during lateral yarding—logs can be yarded around residual trees and through difficult terrain with little or no damage to the remaining stand. Lateral slackpulling capabilities make it possible to increase the distance between skyline corridors, reducing the amount of rigging time required. A running skyline with slackpulling carriage may also be used in partial-cut prescriptions. One drawback, however, is that a running skyline system cannot operate over intermediate supports, and thus may be unsuitable for use in areas of broken or convex terrain.

Helicopters are well suited to uneven-age silvicultural prescriptions. Helicopter logging requires fewer roads than other harvesting systems, causes minimal stand damage, and leaves virtually no impacts on the soil. As with even-age harvesting, economic feasibility depends on maximizing payload. Group and single-tree silviculture prescriptions can be designed for optimal payloads over large harvesting areas. Helicopter yarding requires an adequate visibility through the canopy to see the ground crew, lower the tag line to the ground, and minimize stand damage while lifting the logs vertically above the canopy. To make single-tree-removal silviculture prescriptions feasible and safe may require removing extra trees. Wildlife trees or snags can be left in helicopter-harvest areas if they are well marked and ground crews are aware of their locations. The use of a hydraulically controlled grapple at the end of the helicopter tag line (instead of chokers) will eliminate the risk that falling debris might injure workers on the ground.

Ground-based logging systems are also well suited to uneven-age prescriptions. Skid trails should be designed as a permanent part of the transportation system. Feller-bunchers may be used, but should be limited to patch or group removal. Skid trails used in conjunction with feller-bunchers should be designed to allow for skidding tree-length material without damage to residual trees. Manual felling and winch-line skidders or tractors working from designated skid trails usually result in the least damage to the site. Horse logging can be used when removing small trees in partial cuts; however, it is more limited by large tree sizes than other ground-based systems.

Mechanized operations may be well suited to uneven-age silviculture depending on the range of tree sizes being removed. Single-grip harvesters can handle a range of tree diameters up to approximately 24 inches; however, productivity and safety are better with trees averaging 10–14 inches in diameter. These machines can cut small stems more efficiently than manual felling and they have the capability of maintaining good directional control over trees being cut, thereby minimizing or eliminating damage to the remaining stand. If stands contain trees larger than the felling and processing head can effectively handle, manual felling will be required to cut the larger trees, and stage felling will sometimes be necessary. Cut-to-length systems may be advantageous in uneven-age silviculture prescriptions because the equipment can process trees into relatively short segments and pile them alongside the forwarding trails. Forwarders can remove the logs with little or no damage to the residual stand. Site damage is also minimized by the accumulation of slash in the trails, which cushions the impacts of the machinery on the soil.

Harvest planning and layout approaches

For uneven-aged systems to accomplish long-term management objectives, harvest planners and silviculturists must work together to design prescriptions that can be accomplished in several entries without destroying previously planted or naturally regenerated trees. Harvest planners and logging contractors need to determine the location of designated skid trails and skyline corridors for the entire unit during the initial planning phase, rather than for only the first harvesting entry (Kellogg et al. 1991).

Future entries will most likely require more directional felling due to limited options for felling timber into areas that will minimize damage to patches of small trees. With some uneven-age prescriptions, even careful planning probably cannot eliminate all damage to regeneration. For example, based on a study designed with 0.5-acre openings, Kellogg et al. (1991) indicated that some damage to small trees in replanted patches will be inevitable in future entries because tall trees, when felled, cannot be contained in openings that small.

When using cable systems in uneven-age prescriptions, it is important to leave trees in appropriate locations for future guyline and tailhold anchors (Kellogg et al. 1991), as well as for tail trees and intermediate support trees (Alarid 1993). Anchor stumps used in one entry will most likely not be sound for use in future entries and rigging trees used for one entry may sustain enough damage or stress that they are not suitable for use in future entries. Tail trees can be protected during the first entry by using rigging gear such as tree plates or nylon straps. Even with this level of planning, however, future entries may involve more-costly rigging alternatives, such as buried logs or substitute earth anchors in place of conventional stump anchors.

As with other silvicultural systems, the location of wildlife trees or trees to be converted to snags needs to be considered in light of future felling, skidding, and yarding activities. These trees can be potentially hazardous to workers in future entries (Kellogg et al. 1991).

Production and cost information

Past studies have shown that group-selection and individual-tree-selection harvests cost more than clearcutting—how much more depends on each site and situation. Appropriate harvest planning will help keep total harvesting costs from escalating greatly.

Three studies conducted recently in the Coast Range give examples of harvesting production rates and costs for uneven-age silvicultural systems. The first study compared two-story and group-selection (0.5-acre openings) treatments with clearcutting, with ground-based and skyline harvesting methods used in all three treatments (Kellogg et al. 1991). The second study compared skyline harvesting of five group-selection treatments (small patch cuts of 0.5 to 3 acres) with clearcutting (Edwards 1992; Kellogg et al. 1996b). The layout for each treatment in the second study is shown in Figure 6-21, including the skyline-road configuration (fan or parallel) as indicated by the locations of landings and designated skyline corridors. The third investigation was a case study of an individual-tree-selection harvest in a Douglas-fir stand being converted to an uneven-age condition (Alarid 1993).

In the first study, layout for the two-story and group-selection treatments was 2 to 5 times longer in hours per acre than that for the clearcut, largely because of detailed planning for locations of skid trails and skyline corridors (Kellogg et al. 1991). In the second study, planning and layout times were 4 to 7 times longer in hours per acre for the group-selection treatments than for the clearcut (Edwards 1992). Planning and layout costs for the individual-tree-selection harvest in the third study (Alarid 1993) were similar to those of the group-selection treatments in the second study (Edwards 1992). For all three studies, the cost of planning was relatively low, ranging from 1.9 percent to 8.4 percent of total harvesting costs (in $/mbf).

In the second study, Edwards (1992) found that the most efficient group-selection treatments to lay out were the 2- to 3-acre strips and wedges (units 2 and 5), compared to the 0.5-acre and 1.5-acre gap cuts (units 3, 4, and 6) (Figure 6-21). This was because the strips and wedges could be boundary-marked for felling, and skyline-road location and mapping were relatively easy. Smaller group-selection patches required additional planning time for mapping, ground layout, and marking of trees. The 0.5-acre and 1.5-acre gap cuts with fan skyline corridors (units 3 and 4) were the most difficult to plan because of the increased effort of flagging skyline roads to a central landing.

In the first study, ground-skidding costs for the group-selection cuts (0.5-acre openings) were only 2.5 percent more than for clearcutting (Kellogg et al. 1991). In the second study, costs of felling and skyline yarding showed little difference among the five group-selection treatments, and between group selection and clearcutting (Edwards 1992; Kellogg et al. 1996b). However, the cost of road and landing changes showed major differences among treatments. Three aspects of changing roads or landings were important: the time required to make the change, the number of changes required for a harvest unit, and the volume harvested between changes (Kellogg et al. 1996b).

Skyline yarding production rates and costs, including road changes, for the three studies are shown in Table 6-4. The data show general trends and examples of production and costs; the three studies cannot be directly compared because of variations in site conditions and the logging equipment used.

Total harvesting costs for the second study (Edwards 1992) are shown in Figure 6-22. Total costs for cable yarding of various group-selection cuts

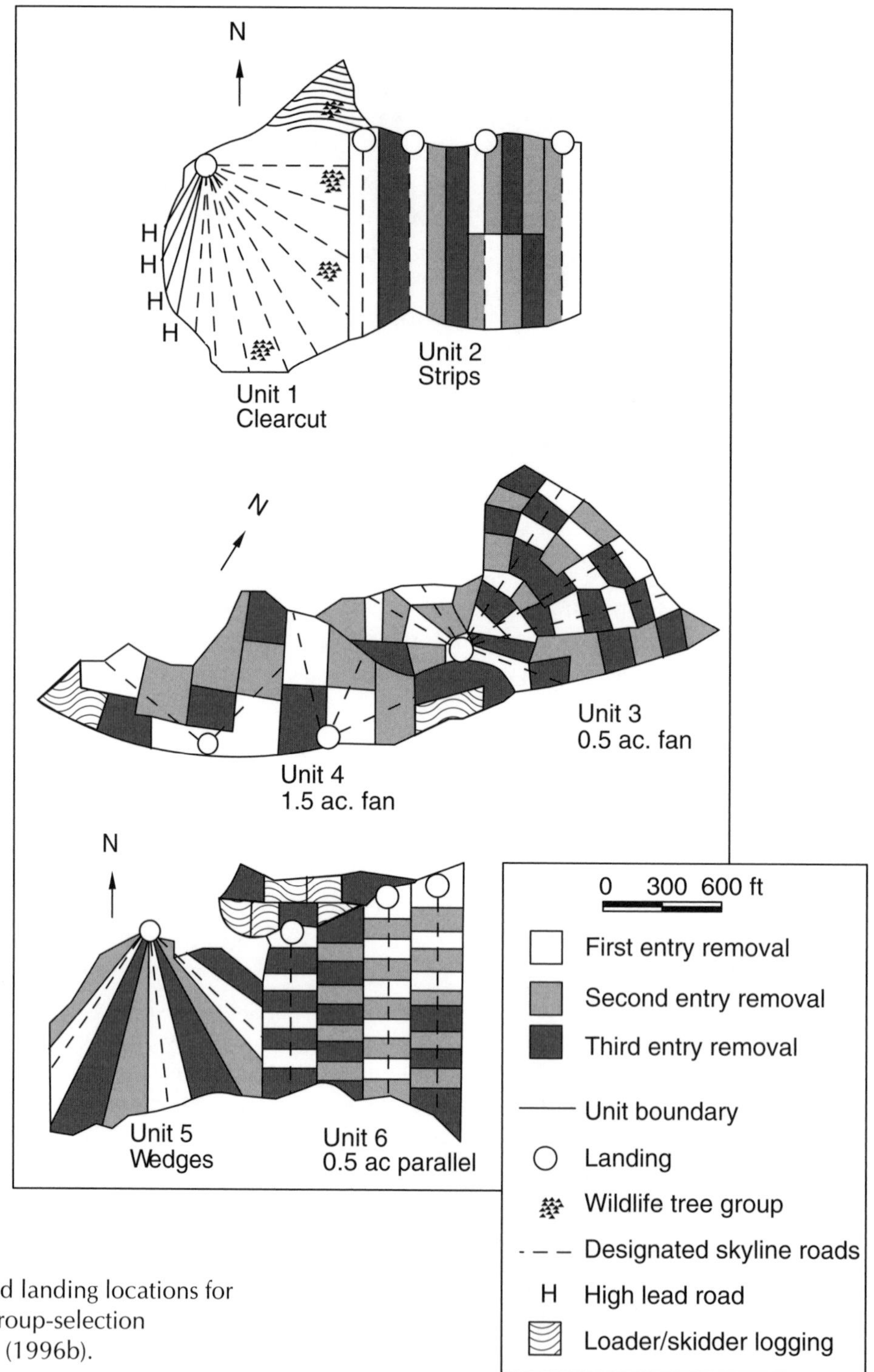

Figure 6-21. Logging layout and landing locations for a clearcut treatment and five group-selection treatments. From Kellogg et al. (1996b).

Table 6-4. Skyline yarding production rates and costs, including road and landing changes, for three harvest areas in the Coast Range.

Treatment	*$/mbf*	*Mbf/hr*	*Bf/log*	*Logs/turn*	*AYD*[2]
1) Kellogg et al. 1991: TTY-50, MSP carriage					
Clearcut	43.18	8.10	391	NA[1]	464
2-story	52.06	6.51	374	NA[1]	568
0.5-ac patch	53.36	6.38	391	NA[1]	546
2) Edwards 1992: TMY-70, S-35 drumlock carriage					
Clearcut	60.44	6.10	216	3.32	627
Strip	60.49	6.09	171	4.21	424
0.5-ac fan	68.88	5.71	173	3.80	725
1.5-ac fan	61.92	5.95	179	3.19	401
Wedge	53.90	6.83	173	3.25	621
0.5-acre parallel	60.49	6.09	169	3.50	404
3) Alarid 1993: TTY-50, MSP carriage					
Uneven-age	58.51	4.70	318	2.10	334

[1] Information not available
[2] Average yarding distance
Costs do not include move-in/move-out, profit and risk allowances, felling costs, or hauling costs; all costs in 1993 dollars; from Alarid (1993).

were between 7 and 32 percent higher than for clearcutting: 7 percent higher on wedge cuts, 16 percent higher on strip cuts, 18 percent higher in 0.5-acre patches with parallel skyline roads, 22 percent higher in 1.5-acre patches with fan skyline roads, and 32 percent higher in 0.5-acre patches with fan skyline roads (Edwards 1992; Kellogg et al. 1996b). This range reflects the increasing complexity of yarding various types of group-selection treatments, and the influence on costs of the size and arrangement of patches and the speed with which road and landing changes can be made.

Factors associated with higher harvesting costs in uneven-age stands compared to clearcuts are similar to factors affecting costs for commercial thinnings. Factors affecting all types of harvesting systems include the cost of moving equipment in and out, lower timber volumes removed than in clearcutting, variability of log sizes, and increased road-maintenance costs for several harvest entries (Kellogg 1994). When cable systems are used, additional factors include time spent making road changes and rigging tail trees and intermediate supports, and, in future entries, the need for alternative cable anchors (e.g., equipment, buried logs, rock bolts, substitute earth anchors) if live trees are not available (Kellogg 1994).

Minimizing Damage to Residual Trees

Most silvicultural prescriptions now call for leaving some of the trees in the stand, making it important to minimize damage to residual trees during harvesting operations. The most typical damage to trees is scarring of the bole. Other types of damage include partial removal of the crown, broken tops, root damage, and damage to regeneration.

Damage may result in decay and subsequent value loss in the volume and quality of the residual timber, and it can make trees susceptible to disease or insect attack. The occurrence of decay in logging scars is related to both the size of the scar and its location on the tree (Wright and Isaac 1956; Wallis et al. 1971). Small scars are less likely to decay than large scars. Hunt and Krueger (1962) found that scars with decay were on average five times larger than scars without decay (140 in^2 versus 27.4 in^2), and that the lower a scar is on the bole, the more likely it is to contain decay. Deep scars can result in higher wood loss than surface wounds (Wallis and Morrison 1975).

Several studies indicate that the amount of decay from logging wounds may be small. However, scarring is largely confined to the butt log, the most valuable portion in a tree. In studying decay from 10-year-old logging scars in 11+ year-old stands,

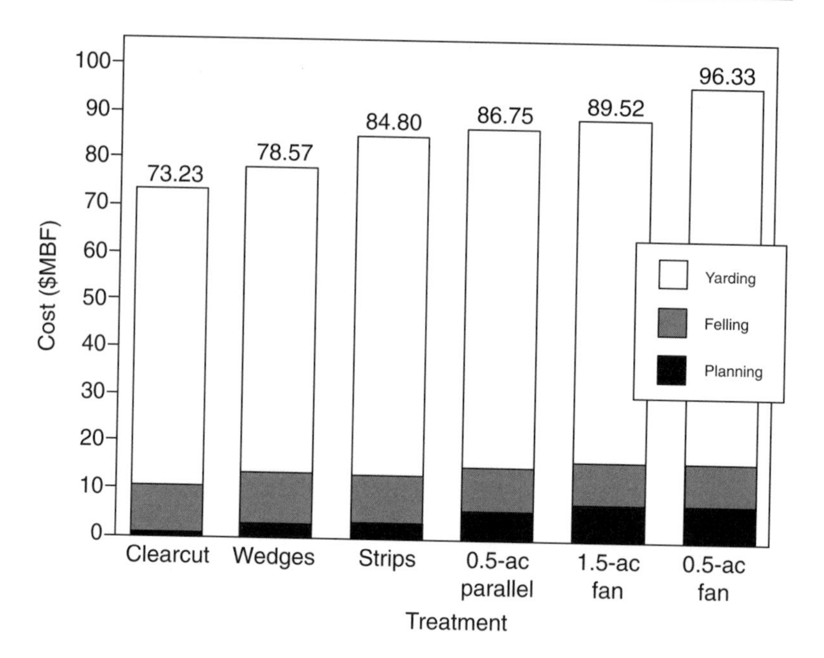

Figure 6-22. Total harvesting costs for clearcut and five group-selection treatments. Costs do not include profit and risk allowances or hauling costs. From Kellogg et al. (1996b).

Shea (1961) found that 81 percent of the scars on Douglas-fir trees were within 4.5 feet of the ground and that western hemlock was considerably less resistant to decay caused by thinning injuries than the Douglas-fir. Over a ten-year period, Douglas-fir lost 1.4 percent of its gross volume, whereas western hemlock lost 6 percent. Hunt and Krueger (1962), and Goheen et al. (1980) reported a similarly low incidence of decay and volume loss. However, scar age is an important predictor of future timber-volume loss (Goheen et al. 1980). Extended rotations can heighten the effect of stand damage from early thinning, unless the stand is subsequently thinned to remove severely damaged trees.

Han and Kellogg (2000a) compared damage to residual trees from four harvesting systems used in a commercial thinning conducted in young Douglas-fir stands: tractor, mechanized cut-to-length, skyline, and helicopter. The most common type of damage in each logging system was scarring, accounting for over 90 percent of the total damage. The ground-based and mechanized systems resulted in larger scars and more gouging and root damage than the skyline and helicopter systems, though some crown-removal and broken-top damage occurred with the skyline and helicopter systems. Between 7.5 percent and 41.3 percent of total residual trees were scarred, but only 1.4 percent to 6.9 percent had scars larger than 1 ft^2.

Minor stand damage may have a positive benefit if creating wildlife habitat is an objective (Pilkerton et al. 1996). Stand damage may create snags, as well as roosting and nesting sites in broken-topped trees.

Monitoring of stand damage

Monitoring of stand damage during harvesting operations gives managers valuable feedback with which to evaluate practices and minimize damage. If too much damage is occurring, the harvesting crew can take corrective measures. Assessing stand damage after harvesting is completed can also yield detailed information on the types and amounts of damage to residual trees. This information can be used to update growth and yield projections for the residual stand.

Variables that are typically recorded during a detailed stand-damage survey include tree species; tree diameter; width, length, and area of scar; height to scar base; orientation of scar on the bole (toward landing, away from landing, toward corridor, away from corridor); gouge damage, measured as depth penetrating sapwood; gouge area; distance from the center line of the equipment trail or skyline corridor; distance from landing; number of scars on the tree; broken top; partial crown removal; and root damage (Pilkerton et al. 1996). Scar area can be obtained by tracing the outline of the scar onto a piece of paper and measuring the tracing with a planimeter. Scars that are too high to be reached by hand can be measured using a camera equipped with a 70- to 210-mm zoom lens and a scale mounted on a level rod (Bettinger and Kellogg 1993).

Han and Kellogg (2000b) compared four sampling methods for quantifying stand damage in an area commercially thinned with manual felling and skyline yarding. Results were compared to a 100-percent survey of the study area. The four sampling methods tested were: (1) line-circular 0.1-acre plots; (2) random circular plots; (3) strip transects; and(4) blocks centered on corridors.

With a few exceptions, all the sampling methods performed within the allowable error rate of 10 percent. Ease of implementation was highest with the line-plot method and lowest with the block and random-plot methods. Statistically, the random-plot method is the most robust (least potential bias), but it is the least efficient for forest sampling. Han and Kellogg (2000b) concluded that, while all the sampling methods provided an acceptable estimate of residual damage, the line-plot method was the least cumbersome to carry out. This study also found that 50 to 80 percent of all damaged trees were located within 15 ft of the center line of the skyline corridor or equipment trail. These findings led the authors to propose a quick survey method to assess residual stand damage at the operating unit level for in-progress and post-thinning operations (Han and Kellogg 2000b).

Harvesting approaches that minimize stand damage

Several factors influence the level of stand damage from a thinning operation (Kellogg et al. 1986; Han and Kellogg 1997). Some are general and apply to any type of harvesting system; others are specific to certain harvesting systems. Stand damage caused by thinning can be effectively minimized by well-prepared thinning plans and careful logging.

General factors are:

Time of year. Logging damage is likely to be higher during early summer, when tree bark is relatively loose, than at other times of the year, when tree bark is tighter and less sap is flowing. Extra care in felling and yarding operations are needed in this season.

Tree species. Because of its thick bark, Douglas-fir tends to be less susceptible to scarring than western hemlock and other thin-barked species. Scars on Douglas-fir are also less susceptible to wood-decaying fungi than those on western hemlock (Hunt and Krueger 1962).

Tree size. Harvesting large trees is more likely to damage residual trees.

Planning and layout. A well-planned harvesting operation, including flagging of skyline corridors and skid trails, can minimize residual damage with more efficient felling and yarding.

Loggers' effort and experience. The logging crew's experience, skill, and care taken during harvesting affect the amount of damage. Skyline carriage positioning during lateral yarding and the use of tree pads are examples of logging practices that can affect levels of stand damage. When minimizing stand damage is a management objective, that must be communicated to the logging crew. Being careful takes extra time, which lowers logging productivity and increases costs (Han et al. 2000).

Monitoring of damage during harvesting. Monitoring provides immediate information on how successfully the logging crew is minimizing damage and offers an opportunity to make changes during harvesting.

Factors specific to harvesting systems are described below.

Skyline systems

Width of skyline corridors. Initial narrow corridor widths (10-15 feet depending on the logging system and volume removal) and the use of rub trees along the corridor (removed after yarding when severely damaged) will result in lower final damage levels (Kellogg et al. 1986).

Skyline height. High skylines can damage or remove crowns of adjacent crop trees, and the lateral excursion of a skyline can damage crowns or boles of remaining trees, depending on the skyline height. The use of intermediate supports or rub trees reduces the amount of skyline lateral excursion (Han and Kellogg 1997). In addition, high skylines can result in full suspension of logs, making them harder to control than when only one end is lifted above the ground (Kellogg et al. 1986).

Skyline alignment. Corridors should be straight and the skyline placed in the center. Offsetting the yarder on the landing or using tail trees other than those planned can skew the skyline to either side of the corridor, often excessively damaging corridor trees (Kellogg et al. 1986).

Sidehill yarding. Thinning corridors should be laid out to eliminate sidehill yarding, in which logs can slide downhill and damage trees on the lower side of the corridor.

Skyline corridor spacing. An average spacing of 150 feet for skyline corridors is appropriate for thinning operations. Wider spacing requires more lateral yarding distance and increases the likelihood of damage during lateral inhaul of logs to the corridor.

Felling pattern. Directional felling or a herringbone felling pattern helps reduce damage.

Tractor systems

Width of skid trail. Han and Kellogg (1997) observed far less damage in tractor-logged units where old skid trails 18-22 feet wide were used, than in those where skid trails were 14 feet wide or narrower. For all tractor-logged units in their study, the average distance of damaged trees from the center of the skid trail was 11 feet.

Condition and alignment of skid trail. When removing trees to make skid trails, it is important to cut stumps low. High stumps force the tractor to one side of the trail, increasing the chance of damaging adjacent trees (Han and Kellogg 1997). Corners and sharp curves in the skid trail will likely result in damage to adjacent trees from tires or logs passing by during logging.

Skid-trail spacing. An average spacing of 120 feet for skid trails is appropriate for thinning operations. Wider spacing between skid trails requires increased winching distance, increasing the chance that residual trees will be rubbed by the winch line or by logs being skidded.

Felling pattern. Directional felling or a herringbone felling pattern helps reduce damage.

Landing. Using a large central landing with a decking place tends to result in less logging damage to trees adjacent to the landing than using continuous roadside landings (Han and Kellogg 1997). Tree pads can be used to reduce damage when logs are stacked along residual roadside trees.

Mechanized cut-to-length systems

Sorting and bunching logs. The forwarder can control logs better if the harvester sorts logs according to saw logs, chip logs, and diameter classes, and bunches them perpendicular to the hauling direction.

Properly spaced equipment trails. The maximum spacing between equipment trails is usually 50 to 60 feet. Appropriate trail spacing eliminates the need for the harvester to move off the trail to cut trees because of the boom's reach limitations.

Timber Harvesting for Riparian-area Management

Many riparian areas along coastal Oregon streams are dominated by hardwood overstories and shrub understories as a result of past fires, floods, and timber-harvesting practices. These streams often lack the large woody debris important for creating pools and cover for fish habitat. Although streamside hardwoods contribute some woody material, this material is generally smaller and decomposes more quickly than that from conifers.

Harvesting systems can be used to carry out riparian management projects designed to address this lack of large woody debris in streams. Two types of projects fit especially well with using harvesting systems. One approach involves removing existing hardwoods in patches along the stream so that conifers can be established in riparian areas and eventually become sources of large woody debris. Another approach involves placing large woody debris directly into debris-poor streams to enhance fish habitat. For more information on other types of stream-habitat improvement projects on private industrial forest lands in Oregon, refer to Maleki and Moore (1996).

Appropriate harvesting systems

To be effective, harvesting systems for riparian-area-management projects must meet stream-habitat needs and objectives in a safe and economically efficient manner. In addition, harvesting operations upslope from riparian areas should be conducted with methods that minimize impacts to the riparian zone. Trees can be removed from riparian areas with minimal impact to streamside vegetation, streambank stability, and water quality. It is important to select the appropriate harvesting system for the site and stand characteristics and the management objectives. Directional felling techniques (wedging, jacking, or pulling) may be necessary to avoid damage to waterways, streambanks, and riparian vegetation. Directional felling will also increase yarding and skidding productivity in and near riparian areas because trees are in lead for yarding and skidding operations.

In steep coastal terrain, damage can be minimized by using standing skyline systems adjacent to and within riparian management areas. When operating slackpulling carriages, standing skyline systems can usually achieve partial or full suspension of logs. This in turn can reduce or eliminate soil damage adjacent to and upslope from riparian areas. In cases where timber is yarded from the opposite hillside, narrow corridors will have to be cut through the riparian zone; however, logs can usually be fully suspended over the stream if cross-channel yarding is required. Slackpulling carriages provide greater lateral yarding distances than non-slackpulling carriages, thus reducing the number of skyline corridors through a riparian area. The standing skyline system has the fewest moving lines, especially a system without a haulback line, and therefore has the least potential for damage to riparian vegetation. If this configuration of equipment is not available, a standing skyline with haulback or a running skyline system with slack-pulling carriage can also be effective in yarding around riparian areas and in placing logs or pulling trees into desired positions. Highlead and live skyline systems with non-slackpulling carriages are not recommended when harvesting in a riparian area or carrying out stream enhancement activities.

Helicopter logging can be conducted in riparian areas with minimal impact, but it is not always the best option. Areas with dense canopies are harder to log safely and efficiently. In addition, many Coast Range streams are located in narrow, v-shaped canyons where adverse wind and fog conditions and the proximity of sidehills can make aerial operations unsafe. In these instances, cable systems may be the best option. However, helicopters are efficient and economical in removing scattered large trees, a common harvest prescription in riparian management areas.

Ground-based operations can also be used to harvest timber from riparian areas. Careful planning and layout are required to minimize soil disturbance caused by road and skid-trail construction. Uphill skidding, away from the stream, will most likely be necessary. Machine capabilities should be evaluated with respect to steepness of slopes and sizes of logs to be moved. Forest-practice rules require maintaining equipment-exclusion zones in most riparian areas. Using winch-line skidders or tractors and directional felling will allow equipment to stay outside the exclusion zone by reducing the number of trails required to access an area. Feller-bunchers can provide good tree placement away from the stream, but their reach is limited, and therefore they may require more trails. Loader logging may be possible within riparian areas, especially when the loader is equipped with a winch line for jammer-style yarding. Loaders can pull logs away from the stream and swing logs to skid trails. These machines can also be used to install large woody debris in streams.

Horses can remove trees from riparian areas with very little impact. Log size, distance to a landing or road, and direction of travel will determine the feasibility of this method. Blocks and tackle can be used with horses to help lever large logs into desired positions for debris placement.

Mechanized operations can remove trees from a riparian area with few if any adverse impacts. Single-grip harvesters can fell and then process trees away from the riparian zone, reducing the need for equipment to operate close to or within the riparian area. However, because their reach is limited, and because tree sizes may be larger than the harvester can handle, some manual directional felling may be required.

Harvest planning and layout approaches

As with any harvesting operation, the key to success in integrating harvesting with riparian-area management is adequate planning. This begins with long-range harvest planning to identify stream-habitat improvement projects with the greatest chance of success. It is important to include a harvesting specialist early in the planning process for those stream enhancement projects that involve a harvesting prescription.

Stream enhancement projects may be conducted by themselves or as part of a larger upslope harvest operation. Stand-alone projects—those in which the sole objective is to improve stream habitat—may be the simplest to coordinate because the work does not have to mesh with that of a harvest operation. These generally call for ground-based equipment, which is more economical and safer than cable systems or helicopters, and are best suited to areas with easy road access. A disadvantage of stand-alone projects is that key woody-debris structural components, such as large logs, may not be readily available on site.

Stream enhancement activities may also use the equipment and crew from a logging operation but be conducted during weekends or other slack times. This option takes advantage of available equipment and crew, but does not interfere with logging production. The harvest crew can stockpile logs in locations where they will be needed for placement in the stream. A project of this type can enhance streams that do not have roads nearby but that are accessible to cable harvesting systems. Long-range scheduling and planning are essential when choosing this option.

A third option involves using harvesting equipment and crew for stream enhancement activities during an upslope logging operation. For example, during a hardwood-conversion project in a riparian area, the harvesting equipment and crew could also place debris structures in the stream. In an operation that calls for cable yarding a unit upslope from a riparian area, trees removed for skyline corridors that cross the riparian management area could be used to create debris structures in the stream. With good planning, such coordinated efforts can be efficient and economical.

Safety must always be the first consideration when cable-yarding equipment is used for placing woody debris in streams. Operational and cost efficiencies are also important. One way to improve efficiency of woody-debris placement with cable systems is to use blocks to better guide the angle and direction of the pulling. For example, a block can be used to pull trees down toward the ground, instead of up toward the skyline (Figure 6-23). Blocks can also be used to achieve a favorable placement of logs and trees in the stream. It takes a few extra minutes to rig a block, but often the block can be prerigged by one person while the rest of the crew continues logging. Even if the whole crew is working on debris placement, using blocks can help crews achieve the desired placement faster.

Time efficiency can also be gained by pulling a log or tree from more than one skyline road for better placement in the stream, instead of spending a long time getting proper placement from the first road. For example, during the case study described below, five attempts were made to place one conifer log (as a key piece for a debris structure) from three different skyline roads, but the total time involved was only 19 minutes (Kellogg et al. 1997). Other techniques that work well include rehooking chokers for better leads and hooking logs with more than one choker for better control during placement in the stream from a skyline span.

Whenever stream enhancement activities are integrated with timber harvesting, crews that understand the riparian management objectives will usually find a way to meet them. Professional programs can train logging crews to carry out operations successfully. In addition, loggers must be compensated for the extra work involved in completing stream enhancements, because it does reduce harvesting productivity.

A Case Study of Timber Harvesting for Active Riparian Management

In this section, a COPE case study is described to give an example of integrating stream enhancement activities with harvesting, and to show the production rates and costs that resulted. For a detailed description of this study, refer to Kellogg et al. (1993b, 1997) and Skaugset et al. (1994).

The study was conducted on three sites in the central Oregon Coast Range, all located on fish-bearing streams on private industrial forest land. The sites were Bark Creek, Buttermilk Creek, and Hudson Creek. Each site was dominated by red alder (*Alnus rubra*) and lacked enough large woody debris for good fish habitat. The study was conducted between 1992 and 1994.

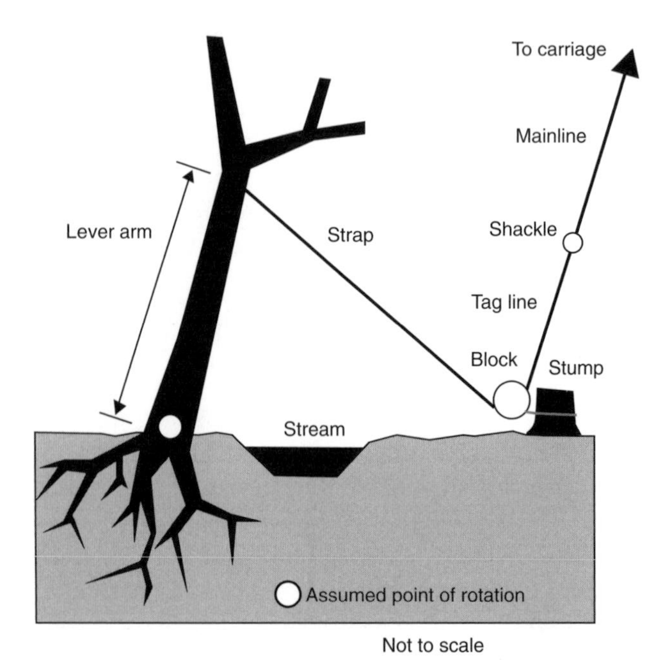

Figure 6-23. Use of a block for tree pulling with a skyline system. From Kellogg et al. (1997).

At each site, a harvest unit was established that included about 2,000 feet of the main stream reach. Hardwoods were harvested from one 600-foot opening and two 300-foot openings along the stream at each site, and the cleared area was replanted with conifer seedlings. Between the harvested openings, 100-foot buffer strips were left along each side of the stream. Adjacent upslope areas were clearcut. Skyline logging systems were used to harvest the riparian openings and upslope areas.

Woody debris was installed in the 300-foot openings. Two types of large woody debris were used: large conifer logs yarded from the upslope harvest area, and large streamside alders pulled over with rootwads intact. At Bark Creek and Hudson Creek, debris was placed by the logging crew with the skyline yarding equipment during harvesting operations. At Buttermilk Creek, logging was completed in spring, but debris-placement activities were deferred until summer. Manual felling and a hydraulic excavator were used for debris placement there.

Debris-placement activities were timed and observations recorded at each site. Costs were calculated for the amount of time the equipment and crew were involved with these activities, and for the value of timber placed in the streams.

The resulting costs for constructing large-woody-debris structures at the three study sites are shown in Table 6-5. On the sites where skyline yarding equipment was used for stream enhancement, costs of the equipment and crew ranged from 25 percent of the total debris-placement cost at Bark Creek to 30 percent at Hudson Creek. At Buttermilk Creek, equipment and labor costs were 46 percent of the total. The higher cost at this site was primarily due to the additional costs of transporting the excavator to and from the site, which did not apply to the other sites because the equipment and crew were already there for harvesting activities.

At all three sites, the greatest portion of the total cost came from the value of the timber used for stream enhancement. The amount of time needed to place logs or trees in the stream was relatively short, averaging 11.4 to 13.8 minutes per installation.

This case study suggests that stream improvement can be successfully integrated with harvesting operations. If stream enhancement activities are well planned and executed, the additional cost of using the logging equipment and crew while they are on site for harvesting is minimal. Using a hydraulic excavator and faller after harvesting is also feasible, although additional transportation costs are incurred to move the excavator to and from the site.

Table 6-5. Summary of debris-placement costs.

Cost Items	*Bark Creek*		*Buttermilk Creek*		*Hudson Creek*	
	No.	*Value ($)*	*No.*	*Value ($)*	*No.*	*Value ($)*
Timber Costs[1]						
Douglas-fir logs	2	495	6	580	2	445
Alder logs	3	161	15	716	8	348
Cedar logs					2	636
Timber subtotal		656		1,296		1,429
Operations Costs						
Skyline yarding equipment and crew[2]		162				477
Faller[3]				134		
Hydraulic excavator and operator[4]				590		
Move in/out				400		
Operations subtotal		162		1,124		477
Total Costs		**$818**		**$2,420**		**$1,906**

[1] Timber costs (stumpage) based on: Douglas-fir: $500/mbf; alder: $310/mbf; cedar: $400/mbf.

[2] Yarding cost based on $172/hour for ownership, operating, and labor costs for yarder, carriage, and six-person crew.

[3] Faller cost based on $30/hour.

[4] Hydraulic excavator and operator cost based on $135/hour.

Summary

Today's forest operations in the Coast Range—many of which call for young-stand thinning, uneven-age management, logging in and near riparian areas, operating in steep terrain dissected with unstable slopes, and the use of skylines, helicopters, and new mechanical harvesting systems—are on the relatively high end of silvicultural and logging complexity. As project complexity increases, the interaction of different resource specialists becomes even more important.

A harvesting system must meet four criteria to be successful. It must (1) be physically capable of accomplishing the management objectives; (2) meet environmental requirements and appropriate regulations; (3) be socially acceptable; and (4) be economically efficient and feasible.

Ground-based logging systems are usually the least expensive of all the harvesting systems when they are used on the appropriate terrain—flat to moderately sloping ground. These systems generally result in more soil compaction and disturbance than other harvesting systems. However, their impacts can be reduced by using designated skid trails, or mitigated by soil tillage.

Cable systems are well suited to the steep slopes of the Coast Range. They are more expensive to operate than ground-based systems, but are less damaging to the soil. A wide variety of cable systems and yarder-carriage combinations are available to meet different silvicultural prescriptions, resource objectives, and site situations.

Helicopter systems are the most expensive to operate, but result in the fewest environmental impacts. They are useful in sensitive resource areas or areas inaccessible to other harvesting systems. A helicopter can often economically yard logs a mile or more. Helicopters are capable of moving large volumes of timber quickly, making them suited to the rapid harvest of timber damaged by fire, insects, and disease. Use of helicopters in areas loggable by other means is increasing; increasingly stringent water-quality rules (often related to various protective listings of fish species) are making helicopter operations more attractive for winter operations in these areas. Helicopters are, however, more susceptible than other harvesting systems to weather-related problems such as fog and wind shifts.

Mechanized harvesting systems are relatively new to the Coast Range. They are well suited to second-growth forests, which tend to have more uniform, smaller-diameter trees than old-growth forests. Mechanized systems disturb the soil no more than ground-based systems, and frequently less. Their impacts can be reduced by using designated equipment trails and by accumulating a mat of slash on the trails to cushion the weight of the harvester and forwarder.

Timber harvesting can achieve alternative silvicultural prescriptions by creating openings and promoting structural diversity in stands, which in turn can enhance wildlife habitat and other resource values. Although harvesting costs for light thinnings and partial cuts are higher than for heavy thinnings or clearcuts, they can still be economically feasible with proper harvest planning.

Alternative silvicultural prescriptions for thinnings or partial cuts require attention to minimizing damage to residual trees. Perhaps the most important factor affecting stand damage is the loggers' effort and experience, reflected in the skills and decisions made by the logging crew during harvesting. Monitoring stand damage during harvesting can provide important feedback on how well the crew is minimizing damage, and provides an opportunity to make changes during harvesting if excessive damage is occurring.

Riparian-area enhancement can be successfully integrated with traditional upslope timber harvesting. Stream enhancement activities can be cost-effective if they are well planned and executed. Additional cost for stream enhancement is minimal if cable-logging equipment and crew are used in conjunction with an upslope harvesting operation.

A wide range of machines and harvesting systems are available to meet different resource objectives, silvicultural prescriptions, and site-specific conditions. Operating the machine or harvesting system within its appropriate range minimizes the risk of exceeding operational and environmental limits. Operating outside the range will increase the risk of generating unacceptable results. Therefore, selecting the appropriate equipment and system for each site is a matter of matching the features of the equipment with site conditions and resource management requirements.

Harvesting operations in the Oregon Coast Range can be conducted in ways that enhance fish and wildlife habitat, minimize impacts to soil and water, and remove wood from the forest for economic and social benefits. However, these objectives can be accomplished only through careful project planning and effective interaction of resource specialists and harvesting specialists early in the planning process.

Literature Cited

Alarid, S. 1993. *Production and Cost Analysis of a Skyline Cable Logging System Operating in an Uneven-age Management Prescription.* MF paper, Department of Forest Engineering, Oregon State University, Corvallis.

Allen, M. M. 1997. *Soil Compaction and Disturbance Following a Thinning of Second-growth Douglas-fir with a Cut-to-length and a Skyline System in the Oregon Cascades.* MF paper, Department of Forest Engineering, Oregon State University, Corvallis.

Andrus, C. W., and H. A. Froehlich. 1983. *An Evaluation of Four Implements Used to Till Compacted Forest Soils in the Pacific Northwest.* Research Bulletin 45, Forest Research Laboratory, Oregon State University, Corvallis.

Armlovich, D. 1995. *Soil Compaction Study on a Cut-to-length Mechanized Harvesting System.* MF paper, Department of Forest Engineering, Oregon State University, Corvallis.

Bettinger, P., and L. D. Kellogg. 1993. Residual stand damage from cut-to-length thinning of second-growth timber in the Cascade Range of western Oregon. *Forest Products Journal* 43(11/12): 59-64.

Binkley, V. W. 1965. *Economics and Design of a Radio-controlled Skyline Yarding System.* Research Paper PNW-25, USDA Forest Service, Pacific Northwest Forest and Range Experiment Station, Portland OR.

Binkley, V. W., and J. Sessions. 1978. *Chain and Board Handbook for Skyline Tension and Deflection.* Miscellaneous Publication, USDA Forest Service, Pacific Northwest Region, Portland OR.

Born, R. G. 1995. *Production and Cost Analysis of a Helicopter Thinning Operation in the Oregon Coast Range and Comparison to HELIPACE Production Estimates.* MF paper, Department of Forest Engineering, Oregon State University, Corvallis.

Brink, M. P., L. D. Kellogg, and P. W. Warkotsch. 1995. Harvesting and transport planning—a holistic approach. *South African Forestry Journal* 172: 41-47.

Brown, C., L. Kellogg, J. Sessions, C. Jarmer, and G. Milota. 1997. *LOGGERPC 3.2 User's Manual.* Department of Forest Engineering, Oregon State University, Corvallis.

Clayton, J. L. 1981. *Soil Disturbance Caused by Clearcutting and Helicopter Yarding in the Idaho Batholith.* Research Note INT-305, USDA Forest Service, Intermountain Forest and Range Experiment Station, Ogden UT.

Conway, S. 1982. *Logging Practices: Principles of Timber Harvesting Systems.* Miller Freeman Publications, Inc., San Francisco CA.

Dyrness, C. T. 1965. *Soil Surface Conditions following Skyline Logging.* Research Note PNW-55, USDA Forest Service, Pacific Northwest Forest and Range Experiment Station, Portland OR.

Dyrness, C. T. 1967. *Soil Surface Conditions following Skyline Logging.* Research Note PNW-55, USDA Forest Service, Pacific Northwest Forest and Range Experiment Station, Portland OR.

Dyrness, C. T. 1972. *Soil Surface Conditions following Balloon Logging.* Research Note PNW-182, USDA Forest Service, Pacific Northwest Forest and Range Experiment Station, Portland OR.

Edwards, R. M. 1992. *Logging Planning, Felling, and Yarding Costs in Five Alternative Skyline Group Selection Harvests.* MF paper, Department of Forest Engineering, Oregon State University, Corvallis.

Fieber, W., and T. Robson. 1985. *Jennie Springs–Horse Logging as a New Enterprise.* USDA Forest Service, Pacific Southwest Region, San Francisco CA.

Froehlich, H. A. 1979. Soil compaction from logging equipment: effects on growth of young ponderosa pine. *Journal of Soil and Water Conservation* 34: 276-278.

Garland, J. J. 1979. *Yarding Systems on Unstable Land.* Unpublished summary table, Department of Forest Engineering, Oregon State University, Corvallis.

Garland, J. J. 1983a. *Designated Skidtrails Minimize Soil Compaction.* Extension Circular EC-1110, Oregon State University Extension Service, Corvallis.

Garland, J. J. 1983b. *Felling and Bucking Techniques for Woodland Owners.* Extension Circular EC-1124, Oregon State University Extension Service, Corvallis.

Garland, J. J. 1998. *Timber Harvesting: How We See It!* Produced under an interagency agreement with Forest Engineering Department, Oregon Forest Resources Institute, and Oregon State University, Corvallis.

Goheen, D. J., G. M. Filip, C. L. Schmitt, and T. F. Gregg. 1980. *Losses from Decay in 40- to 120-year-old Oregon and Washington Western Hemlock Stands.* R-6-FPM-045-1980, USDA Forest Service, Pacific Northwest Region, Forest Pest Management, Portland OR.

Han, H. S., and L. D. Kellogg. 1997. Comparison of damage characteristics to young Douglas-fir stands from commercial thinning using four timber harvesting systems, pp. 76-85 in *Proceedings of the 20th Annual Meeting of the Council on Forest Engineering, Rapid City, South Dakota.* Council on Forest Engineering, Corvallis OR.

Han, H.S., and L. D. Kellogg. 2000a. Damage characteristics in young Douglas-fir stands from commercial thinning with four timber harvesting systems. *Western Journal of Applied Forestry* 15: 27-33.

Han, H. S., and L. D. Kellogg. 2000b. A comparison of sampling methods for measuring residual stand damage from commercial thinning. *Journal of Forest Engineering* 11: 63-71.

Han, H. S., L. D. Kellogg, G. M. Filip, and T. D. Brown. 2000. Scar closure and future timber value losses from thinning damage in western Oregon. *Forest Products Journal* 50: 36-42.

Haupt, H. F. 1960. Variation in aerial disturbance produced by harvesting methods in ponderosa pine. *Journal of Forestry* 58: 634-639.

Hayes, J. P., S. S. Chan, W. H. Emmingham, J. C. Tappeiner II, L. D. Kellogg, and J. D. Bailey. 1997. Wildlife response to thinning young forests in the Pacific Northwest. *Journal of Forestry* 95: 28-33.

Hochrein, P. H., and L. D. Kellogg. 1988. Production and cost comparison of three skyline thinning systems. *Western Journal of Applied Forestry* 3: 120-123.

Hunt, D. L., and J. W. Henley. 1981. *Uphill Falling of Old-growth Douglas-fir*. General Technical Report PNW-122, USDA Forest Service, Pacific Northwest Forest and Range Experiment Station, Portland OR.

Hunt, J., and K. W. Krueger. 1962. Decay associated with thinning wounds in young-growth western hemlock and Douglas-fir. *Journal of Forestry* 60: 336-340.

Jarmer, C. B., J. W. Mann, and W. A. Atkinson. 1992. Harvesting timber to achieve reforestation objectives, pp. 202-230 in *Reforestation Practices in Southwestern Oregon and Northern California*, S. D. Hobbs, S. D. Tesch, P. W. Owston, R. E. Stewart, J. C. Tappeiner II, and G. E. Wells, ed. Forest Research Laboratory, Oregon State University, Corvallis.

Kellogg, L. D. 1980. *Thinning Young Timber Stands in Mountainous Terrain*. Research Bulletin 34, Forest Research Laboratory, Oregon State University, Corvallis.

Kellogg, L. D. 1994. Harvest planning for high quality forestry on steep ground, pp. 99-104 in *High Quality Forestry Workshop: The Idea of Long Rotations*, J. F. Weigand, R. W. Haynes, and J. L. Mikowski, ed. Special Paper 15, CINTRAFOR, the Center for International Trade in Forest Products, University of Washington, Seattle.

Kellogg, L. D., and P. Bettinger. 1992. Thinning productivity and cost for a mechanized cut-to-length system in the Northwest Pacific Coast region of the U.S.A. *Journal of Forest Engineering* 5(2): 43-53.

Kellogg, L. D., E. D. Olsen, and M. A. Hargrave. 1986. *Skyline Thinning a Western Hemlock-Sitka Spruce Stand: Harvesting Costs and Stand Damage*. Research Bulletin 53, Forest Research Laboratory, Oregon State University, Corvallis.

Kellogg, L. D., S. J. Pilkerton, and R. M. Edwards. 1991. Logging requirements to meet new forestry prescriptions, pp. 43-49 in *Forest Operations in the 1990s: Challenges and Solutions. Proceedings of the 14th Annual Meeting of the Council on Forest Engineering, Nanaimo, British Columbia.* Council on Forest Engineering, Corvallis OR.

Kellogg, L. D., P. Bettinger, S. Robe, and A. Steffert. 1992. *Mechanized Harvesting: A Compendium of Research*. Forest Research Laboratory, College of Forestry, Oregon State University, Corvallis.

Kellogg, L. D., P. Bettinger, and D. Studier. 1993a. *Terminology of Ground-based Mechanized Logging in the Pacific Northwest*. Research Contribution 1, Forest Research Laboratory, Oregon State University, Corvallis.

Kellogg, L. D., S. J. Pilkerton, and A. E. Skaugset. 1993b. Harvesting for active riparian zone management and the effects on multiple forest resources, in *Environmentally Sensitive Forest Engineering, Proceedings of the 16th Annual Meeting of the Council on Forest Engineering, Savannah, Georgia*. Council on Forest Engineering, Corvallis OR.

Kellogg, L. D., G. V. Milota, and M. Miller Jr. 1996a. A comparison of skyline harvesting costs for alternative commercial thinning prescriptions. *Journal of Forest Engineering* 7(3): 7-23.

Kellogg, L. D., P. Bettinger, and R. M. Edwards. 1996b. A comparison of logging planning, felling, and skyline yarding costs between clearcutting and five group-selection harvesting methods. *Western Journal of Applied Forestry* 11: 90-96.

Kellogg, L. D., G. Milota, M. Miller, and S. Pilkerton. 1997. *Integrating Stream Habitat Improvement with Harvesting*. COPE Report 10(1,2): 27-31, Oregon State University, Corvallis.

Kochenderfer, J. N. 1977. Area in skid roads, truck roads, and landings in the central Appalachians. *Journal of Forestry* 78: 507-508.

MacDonald, A. J. 1999. *Harvesting Systems and Equipment in British Columbia.* FERIC Handbook No. HB-12, British Columbia Ministry of Forests, Forest Practices Branch, Victoria BC.

Maleki, S., and K. Moore. 1996. *Stream Habitat Improvement Projects on Private Industrial Forest Lands: Results of an Inventory Conducted by the Oregon Department of Fish and Wildlife for the Oregon Forest Resources Institute.* Oregon Department of Fish and Wildlife, Fish Research and Development Section, Salem.

Megahan, W. F., and W. J. Kidd. 1972. Effects of logging and logging roads on erosion and sediment deposition from steep terrain. *Journal of Forestry* 70: 136-141.

Newton, M., and E. C. Cole. 1987. A sustained-yield scheme for old-growth Douglas-fir. *Western Journal of Applied Forestry* 2: 22-25.

O'Hara, K. L. 1990. Twenty-eight years of thinning at several intensities in a high-site Douglas-fir stand in western Washington. *Western Journal of Applied Forestry* 5: 37-40.

Oregon Occupational Safety and Health Division. 1995. *Oregon Occupational Safety and Health Code*. Oregon Administrative Rules, Chapter 437, Division 6, Forest Activities. Department of Consumer and Business Services, Salem.

Oregon Occupational Safety and Health Division, Associated Oregon Loggers, Oregon Department of Forestry, Oregon Department of Fish and Wildlife, USDA Forest Service, and Bureau of Land Management. 1995. *Oregon Guidelines for Selecting Reserve Trees.* USDA Forest Service, Portland.

Pilkerton, S. J., H. S. Han, and L. D. Kellogg. 1996. Quantifying residual stand damage in partial harvest operations, pp. 62-72 in *Proceedings of the 19th Annual Meeting of the Council on Forest Engineering, Marquette, Michigan.* Council on Forest Engineering, Corvallis OR.

Shea, K. R. 1961. *Deterioration Resulting from Logging Injury in Douglas-fir and Western Hemlock.* Forestry Research Note No. 36, Weyerhaeuser Company, Forestry Research Center, Centralia WA.

Silen, R. 1957. More efficient road patterns for a Douglas-fir drainage. *The Timberman* 56(6): 82-86, 88.

Skaugset, A., L. Kellogg, S. Pilkerton, and M. Miller. 1994. *Active Riparian Zone Management Project: Preliminary Logging Operations Research Results.* COPE Report 7(1): 2-5, Oregon State University, Corvallis.

Stenzel, G., T. A. Walbridge, Jr, and J. K. Pearce. 1985. *Logging and Pulpwood Production.* 2nd Edition, John Wiley & Sons, New York.

Studier, D. D. 1993. *Carriages for Skylines.* Research Contribution 3, Forest Research Laboratory, Oregon State University, Corvallis.

Studier, D. D., and V. W. Binkley. 1974. *Cable Logging Systems.* Miscellaneous Publication, USDA Forest Service, Pacific Northwest Region, Portland OR.

Studier, D., and J. Neal. 1993. *Helicopter Logging: A Guide for Timber Sale Preparation.* Miscellaneous Publication, USDA Forest Service, Pacific Northwest Region, Portland OR.

Swanston, D. N. and C. T. Dyrness. 1973. Managing steep land: I. Stability of steep land. *Journal of Forestry* 71: 264-269.

Swanston, D. N., and C. T. Dyrness. 1975. Impact of clearcutting and road construction on soil erosion by landslides in the western Cascade Range, Oregon. *Geology* 3: 393-396.

Tucker, G., B. Emmingham, and S. Johnston. 1993. *Commercial Thinning and Underplanting to Enhance Structural Diversity of Young Douglas-fir Stands in the Oregon Coast Range.* COPE Report 6(2): 2-4, Oregon State University, Corvallis.

Wallis, G. W. and D. J. Morrison. 1975. Root rot and stem decay following commercial thinning in western hemlock and guidelines for reducing losses. *Forestry Chronicle* 51: 203-207.

Wallis, G. W., G. Reynolds, and H. M. Craig. 1971. *Decay Associated with Logging Scars on Immature Western Hemlock in Coastal British Columbia.* Information Report BC-X-54, Forest Research Laboratory, Department of Fisheries and Forestry, Canadian Forestry Service, Victoria BC.

Wert, S., and B. R. Thomas. 1981. Effects of skid roads on diameter, height and volume growth in Douglas-fir. *Soil Science Society of America Journal* 45: 629-632.

Wright, E., and L. Isaac. 1956. *Decay Following Logging Injury to Western Hemlock, Sitka Spruce, and True Firs.* Technical Bulletin No. 1148, USDA, Washington DC.

Silviculture of Oregon Coast Range Forests

J.C. Tappeiner II, W.H. Emmingham, and D.E. Hibbs

What Is Silviculture and Why Is it Important?

Silviculture is the science and art of tending forest vegetation to meet one or more objectives (Daniel et al. 1979; Mathews 1989; Smith et al. 1996). In the Oregon Coast Range, silviculture is often equated with reforestation after disturbances like timber harvest and fire, but it also encompasses thinning of conifer and hardwood stands to make them more vigorous or more diverse. The objective of silviculture increasingly is to produce wood and grow large trees, to enhance habitat for plants and animals, and to do both simultaneously.

The principles and methods for reforestation, thinning, and evaluation of stand development and yield have been well established through research and practical experience. However, considerable site-specific evaluation and judgment are needed when putting silvicultural methods into practice. Coast Range forests are variable in structure and species composition, and landowners may have widely differing goals. Silvicultural practices that are appropriate for some ownerships and circumstances may not be suitable for other owners in different circumstances. Science and art must be blended in the application of silvicultural principles to actual forest situations.

In this chapter we review the biological and ecological basis for the practice of silviculture in the forests of coastal Oregon. We discuss how silvicultural practices can be used to ensure that forests are regenerated after major disturbances like timber harvest and fire. We also show how silviculture can be used to manage forests not only for wood products but for other values. We discuss how silvicultural practices can be used to develop understories of trees and shrubs, grow big trees, and create or retain snags and large logs to provide some of the complexity of structure characteristic of older forests. Finally, we present recent findings on using silviculture to actively manage riparian zones, and we discuss the silviculture of alder (*Alnus* spp.) and cottonwood (*Populus balsamifera*) plantations.

During the past several decades, the main emphasis of silviculture in Coast Range forests has been on reforestation after timber harvest. As a result, these forests now have many thousands of acres of young, dense conifer and hardwood stands (Figure 7-1 in color section following page 148). Managing these stands, which were generally reforested with the goal of producing timber, presents new challenges for practitioners of silviculture. While timber production will likely remain the primary goal for many ownerships, wildlife habitat, riparian values, and other objectives will be increasingly emphasized. Silviculture provides the concepts and tools to achieve multiple objectives in these young stands.

This chapter is intended to be an overview and synthesis, not an in-depth silvicultural text. Our purpose is to summarize information basic to Coast Range silviculture and to provide its sources. We present numerous references and we encourage the reader to use them to obtain further knowledge of this subject. A recent summary of silviculture in the Douglas-fir (*Pseudotsuga menziesii*) region is a good place to begin (Curtis et al. 1998).

Regeneration of Oregon Coast Range Forests

Natural regeneration

Timber harvest in the Oregon Coast Range began in the late 1800s, but reforestation did not become a major concern until the 1930s and 1940s. Until that time regeneration was accomplished mainly by natural seeding from seed trees or from the forest edge. The results were highly variable. Very dense Douglas-fir stands, pure hardwood stands, conifer-hardwood mixes, and shrub communities with a few hardwoods and conifers all occupied different Coast Range sites depending on climatic conditions and other site variables. Successful natural regeneration depends on several requirements, such as an ample supply of seed, low predation of seed and seedlings, a seedbed that promotes germination of seed and survival of newly germinated seedlings (bare mineral soil for Douglas-fir and alder), protection from temperature extremes, and enough water and light for the new trees to grow (Lavender 1958; Hermann and Chilcote 1965). When these requirements coincide, natural regeneration results in well-stocked stands of conifers (Figure 7-1) or hardwoods. In the past, heavy disturbance from logging or fire often left an ideal seedbed, and frequently seed was available from uncut trees and adjoining stands. In some cases well-stocked conifer stands resulted; in others, pure stands of alder or varying mixes of conifers and hardwoods occupied the site.

However, if any of these requirements is lacking, a naturally regenerated stand may have too few tree seedlings or poor seedling distribution, or it may be occupied by an undesired species. If no conifer seed is available for several years after disturbance—a likely circumstance after an extensive fire or series of fires—a relatively stable cover of shrubs and sprouting hardwoods, not a conifer stand, will result (Figure 7-2 in color section following page 148). In sum, while natural regeneration has been successful in many cases, it is also quite unpredictable and is thus not always a reliable technique for regenerating either conifer or hardwood trees. Relying on natural regeneration may not meet Oregon State Forest Practices standards in many situations (Oregon Department of Forestry 1997).

Artificial regeneration

To ensure prompt, successful regeneration of Douglas-fir and other trees, researchers developed methods of artificial regeneration, perfecting techniques for collection of seed, nursery production of container and bareroot planting stock, storage and handling of seedlings, site preparation, and planting (Cafferata 1986). The principles for successfully establishing conifers by planting have been well established and are reviewed in more detail by Cleary et al. (1978), DeYoe (1986), Margolis and Brand (1990), Duryea and Dougherty (1991), Hobbs et al. (1992), Hibbs et al. (1994), and Curtis et al. (1998). Generally, they are: (1) proper seed source selection and seed handling and storage procedures; (2) careful nursery practices to produce vigorous, physiologically conditioned seedlings; (3) site preparation to control species like vine maple, salmonberry, and other vegetation (Figures 7-2 and 7-3 in color section following page 148), and to maintain safe working conditions for tree planters; (4) careful handling and storage between the nursery and planting site; (5) prompt and careful planting; (6) monitoring of planted seedlings to evaluate early survival and reduce effects of drought, browsing, and competing vegetation; and (7) thinning to influence species composition and tree density (Figure 7-4 in color section following page 148).

Control of shrubs, herbs, and hardwoods is often crucial to successful stand establishment (Figure 7-5 in color section following page 148). It is done to ensure survival of planted seedlings, to enhance their rate of growth, or to do both. The flora of coastal Oregon includes many grasses and herbs (often exotic) as well as persistent native species like tanoak (*Lithocarpus densiflorus*), vine maple (*Acer circinatum*), bigleaf maple (*Acer macrophyllum*), salmonberry (*Rubus spectabilis*), and salal (*Gaultheria shallon*). These hardwoods and shrubs sprout vigorously after their tops are killed. They may be controlled with fire or herbicides before seedlings are planted, or with herbicides or manual cutting afterward. The ecological basis for controlling shrubs and hardwoods and methods of control have been thoroughly summarized by Conard and Emmingham (1984), Walstad and Kuch (1987), and Hobbs et al. (1992).

Oregon's Forest Practices Act requires that reforestation begin on state and private lands no later than one year after timber harvest (State of Oregon 1971). By the end of the third year, a "free-to-grow" stand— one in which tree growth will not be severely impeded by competition or animal browsing—must be established. This stand must consist of at least 200 trees per acre for most Coast Range sites. Compliance with these standards ranged from 96 to 98 percent of acres harvested between 1989 and 1994 (Oregon Department of Forestry 1997). The Bureau of Land Management (BLM) and the USDA Forest Service have similar reforestation standards.

Natural diversity in young managed stands

Today virtually all harvested sites in coastal Oregon are planted to ensure compliance with state and national reforestation requirements. On most of them, because of the high productivity of Coast Range forest land, natural regeneration fills in among planted seedlings. "Pure" Douglas-fir stands are often stocked with mixtures of western hemlock (*Tsuga heterophylla*), western redcedar (*Thuja plicata*), red alder (*Alnus rubra*), and sprouting hardwoods such as bigleaf maple, tanoak, chinkapin (*Chrysolepis chrysophylla*), and myrtle (*Myrica* spp.). In older stands in the Coast Range, advance regeneration—young conifers and hardwoods in the understory of a stand—is often present, especially on drier sites in southwestern Oregon. Establishment of regeneration in the understory is often favored by thinning stands at about 30–60 years (Bailey 1996; Bailey and Tappeiner 1998). This regeneration has the potential for growing well after the overstory is removed (Tesch and Korpella 1993; Tesch et al. 1993). With the combination of advance regeneration, natural seeding, planting of several species, and retention of sprouting hardwoods, it is possible to achieve substantial tree-species diversity in 5- to 10-year-old stands.

Patterns of Stand Development

The conditions for forest growth in the Oregon Coast Range are among the very best in temperate forests worldwide (Waring and Franklin 1979). The mild and wet winters and cool summers, especially west of the summit, are atypical of most temperate forests, which usually have greater temperature extremes and less precipitation. Soils, derived from basalt and sandstone, are generally nutrient-rich clay loams or loams weathered to depths of 3 feet or more. Because of the moist, mild climate, Coast Range conifers and evergreen hardwoods can photosynthesize during most of the year, including winter, spring, and early fall, although not at the end of the summer drought or during periods of cold weather (Emmingham and Waring 1977; Harrington et al. 1995). Consequently, these forests have very large trees (200 or more feet tall and 30-40 or more inches in diameter) in very productive stands that yield large amounts of merchantable wood (Figures 7-6a and 7-6b in color section following page 148).

The development of young planted stands and stands naturally regenerated after fire or timber harvest generally follows the pattern described in detail by Oliver and Larson (1990, 1996). A stand typically begins after a disturbance with the initiation of 200 or more new trees per acre. These trees may be planted or may start naturally from seed, as with Douglas-fir, western hemlock, and red alder, or they may emerge from sprouts that were present before disturbance, as with bigleaf maple, madrone (*Arbutus menziesii*), and tanoak. As trees reach about 10–15 feet in height—typically in 5–10 years—their height, diameter, and crown development accelerate. They begin to compete with one another, and by about 20–30 years of age the smaller trees begin to die. This stem exclusion or self-thinning process occurs most rapidly in the first 50 years of the life of a stand (Figure 7-6c in color section following page 148), but it may continue for many years. Crown closure may limit or exclude understory vegetation at this point. In older stands, root disease, windthrow, and insects also may cause mortality. Following intensive self-thinning or operational thinning, an understory of various combinations of trees, shrubs, and herbs develops. Thus over many decades, stand structure (Helms 1998) changes from one of uniform-sized trees and little understory to one with large live and dead trees and an understory of shade-tolerant shrubs, principally vine maple, and sometimes salal, hemlock, western redcedar, tanoak, and bigleaf maple (Bailey and Tappeiner 1998; Bailey et al. 1998; Schrader 1998).

The rate of development and the composition of a stand's tree and shrub species are influenced by such factors as site productivity, the species composition of the previous stand, seed availability, and stand density. Stand development will be most rapid on productive sites where trees grow fast, and self-thinning will happen sooner and be most intense where there are many trees. If the previous stand had many sprouting hardwoods and shrubs, these species will be abundant in the new stand. If many seed-producing red alder were nearby when the stand was initiated, then alder will likely dominate for many decades if it is not controlled.

There is recent evidence that on many sites old-growth stands in the Oregon Coast Range followed a pattern of development different from the one described above (Tappeiner et al. 1997; Poage 2001; Sensenig 2002). This evidence indicates that after a natural disturbance such as a fire some large, old trees remained and a few young trees that were established from seed became dominant. These young trees grew very rapidly, judging by the wide rings produced at their centers. After a disturbance-free period of perhaps 100 years or more, another disturbance such as a fire occurred and a few more rapidly growing trees became established (Figure 7-7). Thus, many of today's old-growth forests appear to have developed from relatively widely spaced trees that were established in the wake of a series of infrequent disturbances. They quickly developed large, tapered stems, big crowns, and general vigor, all of which contributed to their longevity and stature in old-growth forests. In contrast, the young stands in the Coast Range today are developing from dense, single-aged, uniformly planted stands or naturally established conifers. As we will discuss later, these different patterns of stand development have ecological and silvicultural implications for stand management.

The Importance of Shrubs and Hardwoods

Coastal Oregon forests have diverse, vigorous understory communities of shrubs and hardwoods. These plants are an integral part of the forest and are important from an ecological and silvicultural perspective. They provide cover and food for many wildlife species. They stabilize soil, fix nitrogen from the atmosphere, and affect the rate and amount of nutrient cycling (Fried et al. 1988). But they also

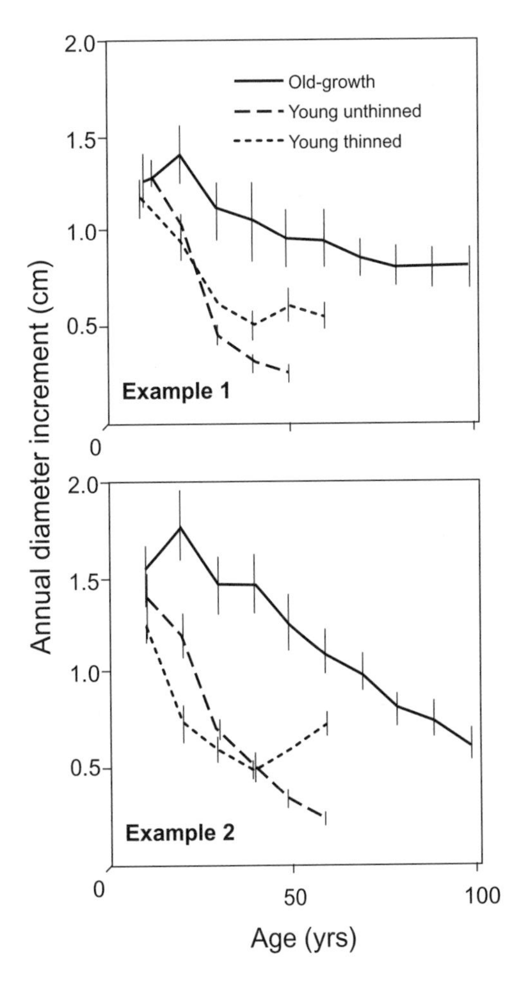

Figure 7-7. Examples of 10-year radial increment in old- and young-growth trees from two sites and from the stumps of two trees. Vertical lines represent standard errors on the graphs, which are adapted from Tappeiner et al. (1997). Dark dots mark distances between 10 annual rings.

compete with conifer and hardwood regeneration after a fire or timber harvest, and they invade the understory after natural disturbance or thinning, forming dense covers that may persist for many years and inhibit the establishment of additional trees, herbs, and other shrubs (Figure 7-2).

Shrubs and hardwoods regenerate by various methods, including seed, sprouting buds or burls, and underground stems. Bigleaf maple, vine maple, Pacific madrone, tanoak, alder, hazel (*Corylus cornuta*), ceanothus (*Ceanothus* spp.), and oceanspray (*Holodiscus discolor*) sprout from buds at the bases of their stems or from large burls at the bases of aerial stems (Tappeiner et al. 1984; Harrington et al. 1992). Rhizomatous species like salal, Oregon-grape (*Berberis* spp.), salmonberry, thimbleberry (*Rubus parviflorus*), and snowberry (*Symphoricarpos* spp.) produce a dense network of underground stems (Figure 7-8 in color section following page 148) that contain millions of buds per acre. These species continuously replace above-ground stems that die (Huffman et al. 1994; Zasada et al. 1994). Ceanothus germinates from seed stored in the forest floor (Quick 1956, 1959; Ruth 1970; Conard et al. 1985). Stored seed may be viable for hundreds of years. It quickly germinates after disturbance and produces a dense cover of shrubs within five to six years. Alder and oceanspray produce light, wind-disseminated seeds that travel many thousands of feet and germinate on mineral soil after fire or logging (Hibbs et al. 1994; Haeussler and Tappeiner 1995). Salal, Oregon-grape, bigleaf maple, salmonberry, tanoak, and vine maple reproduce from seed in the understory of conifer or alder stands. Seedlings of these species grow very slowly, but once established they persist when the overstory trees are harvested or killed by fire, wind, disease, or insects (Tappeiner et al. 1991; Tappeiner and Zasada 1993; Huffman et al. 1994; O'Dea et al. 1995; Huffman and Tappeiner 1997).

Thus, shrubs and hardwoods, with their varied mechanisms for reproduction, growth, and replacement of killed tops, are well adapted to maintaining themselves in Coast Range forests. With their potential to produce dense, persistent covers, they may hinder the establishment of desired tree species, and it may be necessary to control them. However, conifers, once established, will overtop these species and shade them out, substantially reducing their vigor during the stem exclusion stage of stand development. At that point, the reestablishment of the understory is usually slow unless the overstory density is reduced by thinning or natural disturbance (Bailey 1996; Bailey and Tappeiner 1998).

Influence of Stand Density on Tree and Stand Characteristics

The density of young stands (number of trees per acre or basal area per acre) has a major influence on the productivity of the overstory and understory. It affects not only the characteristics of individual trees but the character of the entire stand (Figure 7-9). The effects of stand density and competition on

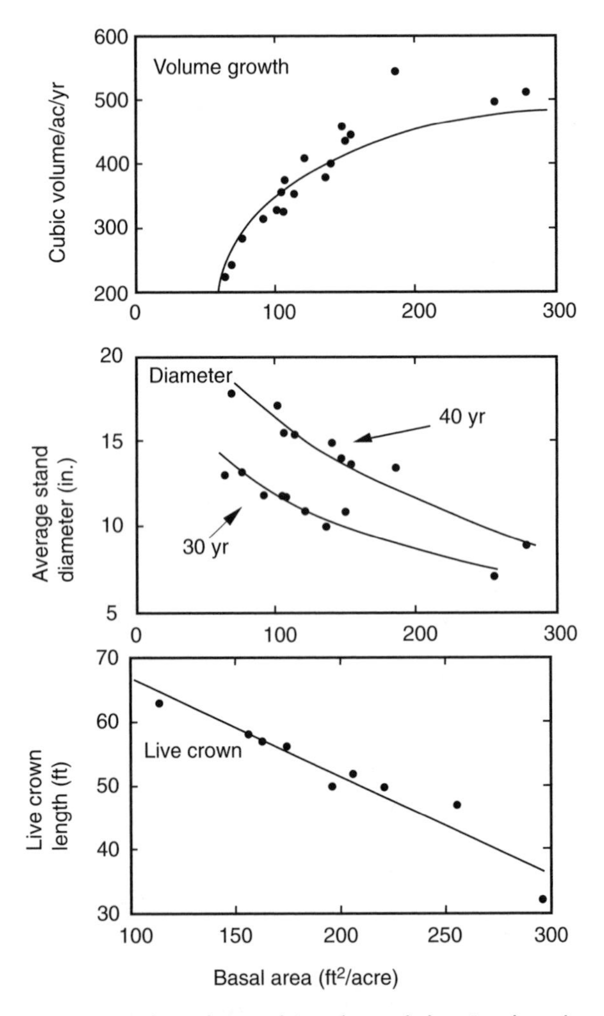

Figure 7-9. The relationship of stand density (basal area) to stand growth, average tree diameter, and length of live crown. These graphs illustrate the tradeoff between producing trees with large crowns and stems and producing high rates of volume growth. (Data from D. Marshall, personal communication.)

Table 7-1. Summary of stand and average tree characteristics of the stands in Figure 7-6 at age 50.

Stand	*A*	*B*	*C*
Trees per acre	52	207	377
Basal area (ft²/ac)	136	286	297
Relative density[1]	29	72	86
Average diameter (in.)	22	16	12
Live crown ratio (%)[2]	59	41	21
Volume removed in thinning (ft³/ac)	2,971	1,205	0
Volume in mortality (ft³/ac)	1	351	4,232
Current volume in stand (ft³/ac)	5,836	13,385	13,301
Total volume produced (ft³/ac)	8,808	14,941	17,533
Current volume growth (ft³/ac/yr)	272	507	204

[1] Curtis (1982). Maximum for Douglas-fir = 100.

[2] Percent of stem covered by crown at 45 years of age.

Thinnings were begun at age 20 (Marshall et al. 1992). All stands had over 1,000 trees per acre with an average diameter of 3.8 inches before thinning. Site index is 135 feet at 50 years. Stands A and B were heavily and lightly thinned, respectively; stand C was not thinned.

individual trees and stands have been widely documented (Assman 1970; Curtis and Marshall 1986; Marshall et al. 1992; Smith et al. 1996). They include:

Stem size. At low density, trees develop thick, tapered stems (Figure 7-6a).

Crown characteristics. At low density, trees have large branches and wide, long crowns, and much of the stem is covered with foliage and branches.

Tree vigor and stability. At low density, individual trees are vigorous and better able to resist windthrow, insects, and diseases, and they may produce more seed.

Effects on stand characteristics include:

Total stand growth and yield. Growth and yield are often low in very low-density stands. There is a tradeoff between tree size and the total volume of wood production (Table 7-1). In addition, trees growing at low density develop large branches low on the stem, resulting in knots that might reduce wood quality. However, management objectives may include leaving some trees with large branches for wildlife habitat.

Mortality from inter-tree competition. Mortality is less in low-density stands, and the trees that do die are generally considerably larger than dead trees in high-density stands.

Understory vegetation. The less dense the stand, the more developed the understory vegetation. Regeneration of herbs, understory shrubs, hardwoods, and conifers is much higher under less dense stands because there are fewer overstory trees to compete for light and water.

By manipulating stand density,, forest managers can exert considerable influence on the growth, species composition, and structural characteristics of forest stands. Stand density can be managed by controlling planting density and by thinning both young stands (10–20 years old) of planted or natural seedlings and older stands (30–100 years old or more). Reducing overstory density and shrub cover frequently enables conifers, hardwoods, and shrubs to become established in the understory (Bailey and Tappeiner 1998) (Figure 7-10 in color section following page 148).

The levels-of-growing-stock studies (Curtis and Marshall 1986; Marshall et al. 1992) demonstrate the difference in the development of young stands growing at different densities. An installation of these studies, located in the Coast Range of Oregon, is shown in Figure 7-6. From stand A, with the fewest trees, 2,971 cubic feet per acre was removed in thinning; from stand B, 1,205 cubic feet per acre was removed. In stand C, mortality of small trees (self-thinning) amounted to 4,532 cubic feet per acre; mortality is negligible in A and B (Table 7-1). Stand A has trees with the largest stems (some are 30 inches in diameter) and crowns because inter-tree competition was reduced by thinning. Stand C has produced the most total volume (including mortality), because it was stocked with over 1,500 trees per acre, most of which have died. Until age 45, stand C had the greatest current volume growth, but now stand B is growing more rapidly than stand C (507 cubic feet per acre per year compared to 204 cubic feet per acre per year) (Table 7-1). Apparently the high tree density in stand C reduced overall stand vigor, or recent mortality has reduced stand density below the level for maximum growth. There is very little shrub or herb understory in this stand.

In addition to the dramatic difference in tree and stand characteristics, the potential for the future management of these stands is quite different. Stems in stand A are larger than those in stands B or C, and trees should be resistant to damage from wind, ice, and heavy snow. The density of stand B is rapidly increasing; it could be thinned to produce a yield of wood and to increase diameter growth of the larger trees, and also to prevent loss of crowns as trees grow taller. If stand C is to be thinned, thinning must be done very carefully; only some of the small to mid-size trees should be removed. Too heavy a thinning at first would likely result in stem breakage from wind or snow. If stand B is left unthinned it will tend to become less resistant to wind. There is no evidence of pathogens in these stands, but root disease could cause mortality and canopy openings in the future.

Options for Managing Young Stands

In this section we discuss ways in which young stands (20–80+ years old) can be managed for a variety of objectives. As we will point out, new information on stand growth, tree regeneration, and understory development suggests that a variety of management strategies are possible. In addition, because of innovations in utilization of small logs and a decrease in the supply of chips and large sawlogs, log values have recently increased enough to make thinning of young stands (around age 20) profitable. Improved efficiency and greater mechanization of harvesting methods also make it more economically feasible to manage young stands.

Controlling stand density when trees are young

It is important to control stand density at an early stage of development. Height growth of Douglas-fir and other conifers is most rapid from 10 years to about 60 years. Over 70 percent of the height growth and crown development occurs during this period (Figure 7-11), as trees grow taller and add lateral branches. Overcrowding during this period retards crown development, reducing the trees' vigor, windfirmness, and ability to resume rapid growth after natural disturbance or commercial thinning. To retain large crowns, it is important to keep the lower branches from being shaded. New branches cannot replace those killed by crowding from adjacent trees. Some epicormic branches, those that originate on the stem or branches following exposure to increased light levels or fire (Helms 1998), often occur on Douglas-fir, white fir (*Abies concolor*), grand fir (*Abies grandis*), and Sitka spruce (*Picea sitchensis*) when stands are thinned. However, in young stands these branches do not produce the same well-developed crown that results from branches originating in the terminal meristem. Some prescriptions for intensive management of young stands call for pruning of lower branches to promote the growth of clear wood. Pruning should leave at least 40 percent of the stem with live crown to maintain tree vigor (Smith et al. 1996).

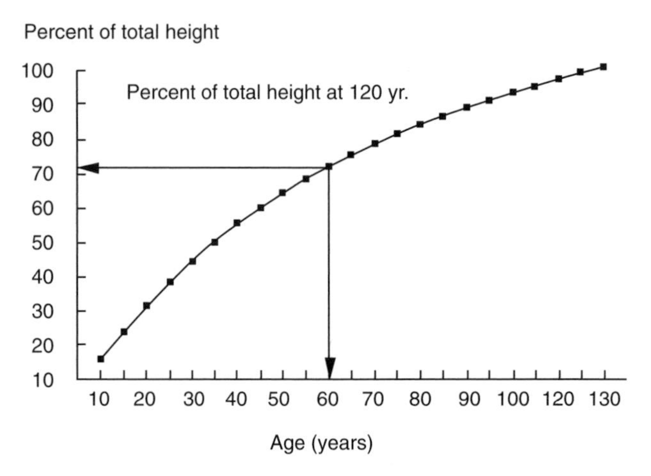

Figure 7-11. Percent of total height at 120 years for Douglas-fir on King's (1966) site class 11 (130 ft/50 yr), which is comparable to McArdle's (180 ft/100 yr). Height at 120 years of age is 200 ft. Height at 200 years is about 210 ft (McArdle et al. 1961). Over 70 percent of the potential height growth of the 40 largest trees is completed by age 60.

Options for managing young forest stands decrease as stands age, especially if trees have been growing at high densities for several decades. By that time live crowns are short, trees have a large height-to-diameter ratio (that is, they are very tall and thin), and little can be done to improve tree vigor. Trees in such stands are more likely to be damaged by snow, ice, and windstorms than trees growing at lower densities.

Thinning to achieve a variety of objectives

Recent research shows that thinning is a versatile silvicultural tool for creating a variety of stand conditions and achieving a range of resource objectives. There are several ways to thin forest stands (Smith et al. 1996). The most common methods are thinning from below and crown thinning. Thinning from below removes the smallest trees, which are most likely to die from competition, as well as some codominants and a few of the smaller dominants. Crown thinning removes some of the codominant or dominant trees to release the larger codominant and dominant trees. In practice, because of the normal heterogeneity in spacing and tree development within a stand, a prescription for thinning is likely to combine both these types.

Thinning prescriptions are developed on a stand-specific basis after accounting for current stand conditions, landowner objectives for wildlife habitat and wood production, and such considerations as logging system and log markets. Thinning strategy has been the focus of much research. For example, the levels-of-growing-stock studies (Curtis and Marshall 1986; Marshall et al. 1992) document the effects of different stand densities on tree size, crown development, and volume growth (Figure 7-6, Table 7-1). Based on these and other studies, thinning to maintain tree vigor, produce wood, and begin understory development might leave young stands (12–14 inches in diameter at breast height) well stocked, with 80–120 trees (100–150 ft^2 of basal area) per acre. Thinning young stands to these or somewhat higher densities would be a strategy for optimizing commercial wood production. Thinning to fewer than 50 trees per acre is more likely to mimic the density under which many Coast Range old-growth stands developed. This would produce large trees, but at the expense of some wood production.

Recent experience suggests that selection, or thinning from above (Smith et al. 1996), also may be a way to blend commercial wood production with development of a multi-storied stand. This approach is used by private landowners with small acreages to allow profitable harvesting in heavily stocked young stands earlier than would be possible with thinning from below (Miller and Emmingham 2001). In thinning from above, some of the dominant trees are removed to release better-formed dominant and codominant trees, as well as intermediate and suppressed trees. Although these intermediate and suppressed trees may survive and eventually be released, the remaining dominant and codominant trees provide most of the growth in stand volume (Emmingham, personal observation). For these landowners, thinning from above produces financial returns early in stand life. It may help the owner avoid high inheritance taxes by keeping the wood volume of the stand low. Stands thinned repeatedly in this manner often develop patches of advance regeneration because removal of dominant trees allows more light to reach the forest floor. Early and frequent thinning may also stimulate the development of understory herb and shrub layers, which combine with patches of regeneration to increase structural and species diversity. Stand growth and yield may, however, be as much as 50 percent lower than what is theoretically possible with thinning from below (Miller and Emmingham 2001).

Producing stands with old-growth characteristics

Recent evaluations of stand growth suggest that a regime of long rotations with thinning is viable for managing coastal Oregon forests for multiple objectives including wood production (Curtis 1992; Curtis and Marshall 1993). Rotations in forests managed only for wood production have commonly ranged from 40 to 80 years—less in recent decades as increased demand for wood has had to be met from fewer acres. However, growth of coastal forests appears to remain high well beyond 100 years. Culmination of volume growth (mean annual increment) may not occur until 120 years, and even beyond that age the mean and periodic annual growth decrease slowly. This suggests that growing stands to old ages (long rotations) with several thinnings will not reduce wood production (Staebler 1960). Long rotations are likely to lead to the development of reasonably complex multi-storied stands with at least some of the characteristics of old-growth forests, from which considerable amounts of wood could be produced by thinning. Trees over 40 inches in diameter could be produced in less than 120 years on the productive sites in the Coast Range if stands were thinned (Staebler 1960; Newton and Cole 1987; Tappeiner et al. 1997; Poage

2001). Snags and large logs on the forest floor would add habitat features as trees blow down or are weakened by root disease and other causes, or trees could be deliberately killed to provide these features where desired (Figure 7-12 in color section following page 148). Long rotations are not suitable for all ownerships, but they may be a workable way to meet multiple objectives on public lands or other ownerships where management goals encompass habitat for old-forest species, biodiversity, and wood production.

Regeneration of understory tree species

Considerable research has been conducted over the past decade on developing multi-storied stands by encouraging the regeneration of conifers and hardwoods in the understory of young stands. However, few attempts have been made to grow multi-storied stands on an operational basis. The next few paragraphs summarize recent findings on conifer establishment and development in the understory.

Thinning favors the establishment of tree seedlings and saplings in the understory of coastal forests (Del Rio and Berg 1979; Fried et al. 1988). Bailey and Tappeiner (1998) found, in a comparison of paired thinned and unthinned stands on BLM land, that seedling frequency ranged from 0.3 to 1.0 in thinned stands and 0.0 to 0.2 in unthinned stands. Douglas-fir occurred in 91 percent of the thinned stands but only 22 percent of unthinned stands. Western hemlock and western redcedar, shade-tolerant species, were the most common conifers in the understory in all stands; their densities and frequencies were highest in thinned stands. While Douglas-fir seedlings often become established following a heavy thinning, additional thinning (to ≤40 trees per acre) is likely to be needed if they are to produce a second layer (Bailey 1996), because of competition from an increasing density in the overstory with time.

Understory conifers can also be established by planting under thinned conifer stands. About 40 years ago, Alan Berg of the College of Forestry at Oregon State University thinned a 40-year-old Douglas-fir stand to 40–50 trees per acre and underplanted it with western hemlock. Today this 80-year-old stand has overstory trees averaging 30 or more inches in diameter and a well-established hemlock understory (Figure 7-10d). In nearby unthinned stands there is no multi-story development, only some Oregon-grape and vine maple in the understory, and overstory trees average just 15 inches in diameter.

Survival of five conifer species (Douglas-fir, western hemlock, grand fir, western redcedar, and Sitka spruce) planted beneath thinned, 30-year-old Douglas-fir plantations on the west slopes of the Coast Range exceeded 80 percent after three years at three thinning intensities (30, 60, and 100 trees per acre) (Emmingham 1996). Red alder and bigleaf maple also survived well, but all the maple seedlings were browsed by deer, severely restricting their height growth. Even though it is considered a shade-intolerant species, the planted red alder were twice the height of the tallest planted conifers after three years. Although Sitka spruce and western hemlock grew best, height growth of Douglas-fir averaged at least 80 percent of that of the more shade-tolerant species. Early growth patterns were affected by thinning levels only in the alder; however, future growth of underplanted conifers and hardwoods is expected to be best in the most widely thinned stands. The underplanted conifers and hardwoods probably performed well at least partly because the stand was in the stem exclusion stage, with little vegetation in the understory, when it was thinned.

Recent experience in planting Douglas-fir, hemlock, and western redcedar in the understory of 50-year-old, thinned Douglas-fir plantations at the east edge of the Coast Range indicates that control of understory shrubs at the time of planting aids seedling establishment and early growth of all species. After five years, western redcedar has higher rates of survival than Douglas-fir, grand fir, or western hemlock (Brandeis 1999), and the redcedars are larger than trees of the other species. Unlike most redcedar that is part of the natural species mix, the redcedar planted in these stands escaped browsing by deer and elk. Also, planted Douglas-fir generally survived and grew well in 0.5-acre openings in 80- to 100-year-old Douglas-fir stands (Ketchum 1994). However, dense cover of grass and shrubs, shady sites on north slopes, and deer browsing combined to reduce regeneration success in some openings.

Underplanting of conifers in thinned hardwood stands may have application in intensive management for timber in some cases, and it may also be useful in riparian areas, which we will discuss at greater length below. Underplanting of western

Figure 8-1f. **Laminated root rot**: Fallen trees in disease centers tend to occur in a random pattern of crossed stems or leaning trees—unlike storm blowdown, whch usually causes trees to fall in one direction, all at about the same time.

Figure 8-1g. **Laminated root rot**: Laminated root rot infection centers cause openings in stands. These openings characteristically have dead or declining host trees at their margins. Large centers may be occupied by tolerant conifers, shrub species, and hardwoods.

Figure 8-1h. **Laminated root rot**: Typical laminated decay caused by *Phellinus weirii*. This piece of root delaminated at the springwood into sheets, each the thickness of an annual ring. At this stage of decay, the wood often appears yellowish to light brown.

Figure 8-1i. **Laminated root rot**: Pits and zone lines are typical in root wood that has reached an advanced stage of decay. Longevity of the pathogen inside roots may be attributable to zone lines formed by *Phellinus weirii* and other fungi in colonized wood. These zone lines are protective barriers against unfavorable conditions and other organisms; they are visible as black lines when decayed wood is cut or broken.

Figure 8-1j. **Laminated root rot**: Masses of brown setal hyphae that form between layers of wood in late stages of decay by *Phellinus weirii* may look like brown felt.

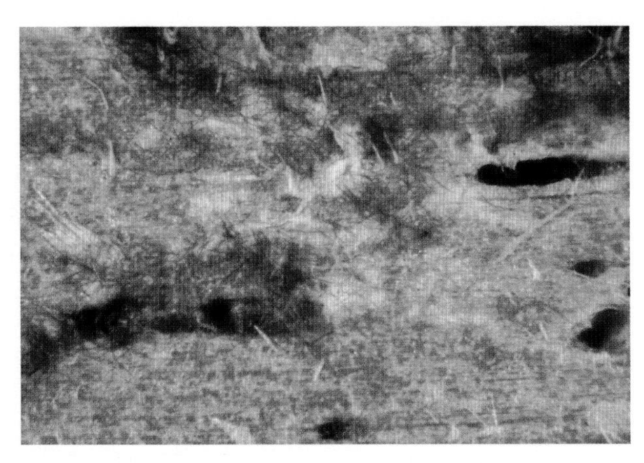

Figure 8-1k. **Laminated root rot**: Reddish-brown setal hyphae can be seen on decayed wood viewed through a hand lens. Individual setal hyphae can also be seen scattered in the ectotrophic mycelium on the surface of a root or in pockets in the bark.

Figure 8-1m. **Laminated root rot**: Light-colored ectotrophic mycelia often form a sheath over the surface of an infected root. This mycelial sheath can be seen by carefully brushing soil away from an infected root. Other fungi that might be found on the roots usually form patches of mycelia, not a continuous sheath. Viewed through a hand lens, reddish-brown, wiry setal hyphae 0.04 inch long can be seen scattered in the superficial mycelia or in pockets within the bark.

Figure 8-1n. **Laminated root rot**: Characteristic incipient decay (stain) caused by *Phellinus weirii*, on fresh Douglas-fir stump tops. The stain, typically reddish brown to chocolate brown, appears first as spots that grow as more roots are involved in the disease. These spots, or arcs, of stain finally coalesce into a continuous ring just before the tree dies. The stain usually has distinct margins defined roughly by radii and annual rings. The stain will fade when the stump top is exposed to the sun for about 30 days.

Figure 8-1o. **Laminated root rot**: Stain seen on a stump top is continuous along the grain from an infected root to the stump top. In advanced stages of the disease, the stump wood often turns to a stringy mass with a hollow at the root collar, as seen in this stump.

Figure 8-1p. **Laminated root rot**: A *Phellinus weirii*-infected Douglas-fir was decayed by the pathogen in the lower bole and broke off at stump height. Douglas-fir in coastal stands rarely breaks above the root collar.

Figure 8-2a. **Port-Orford-cedar root disease**: Extensive Port-Orford-cedar mortality on a site favorable for *Phytophthora lateralis* spread. Large, mature trees that play especially important ecological roles are being killed in large numbers.

Figure 8-2b. **Port-Orford-cedar root disease**: Port-Orford-cedar root disease-caused mortality often occurs along roads. The causal pathogen is carried in mud and organic debris that is picked up and transported along roads (or into the forest) on vehicles. Water then carries the pathogen from the road into the forest.

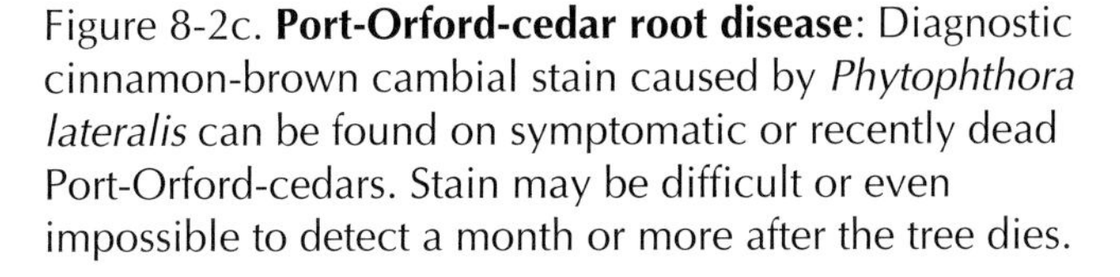

Figure 8-2c. **Port-Orford-cedar root disease**: Diagnostic cinnamon-brown cambial stain caused by *Phytophthora lateralis* can be found on symptomatic or recently dead Port-Orford-cedars. Stain may be difficult or even impossible to detect a month or more after the tree dies.

Figure 8-2d. **Port-Orford-cedar root disease**: Mortality can be particularly extensive along watercourses, such as the one shown here.

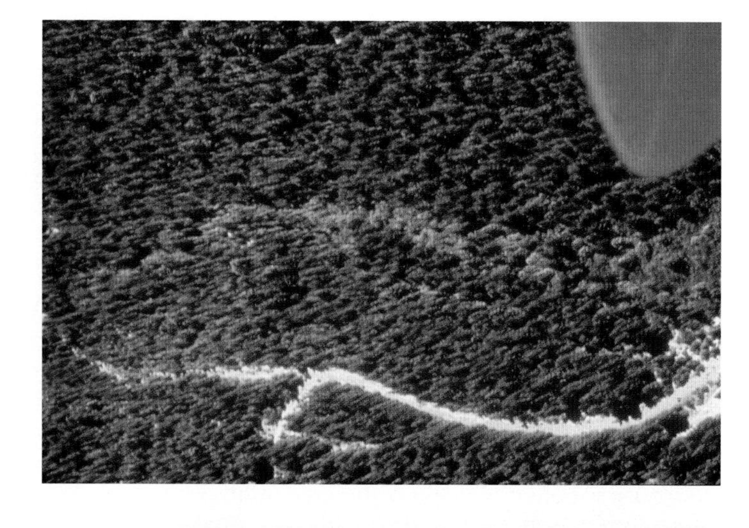

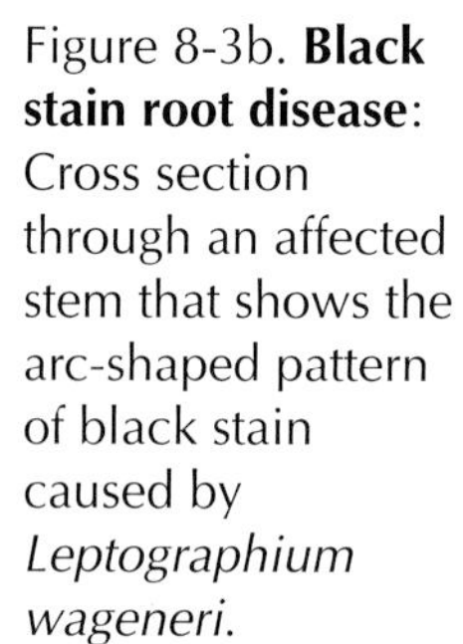

Figure 8-3a. **Black stain root disease**: Dark purple to black stain associated with black stain root disease. Stain occurs in the outer portion of the sapwood of affected roots and stems. It is most visible on symptomatic or recently killed trees. After the tree dies the stain quickly becomes difficult to distinguish from other discolorations that develop in dead trees.

Figure 8-3b. **Black stain root disease**: Cross section through an affected stem that shows the arc-shaped pattern of black stain caused by *Leptographium wageneri*.

Figure 8-3c. **Black stain root disease**: Black stain root disease frequently affects groups of trees in distinct infection centers. Typical infection centers have trees in various stages of decline near the perimeter and dead trees in the interior.

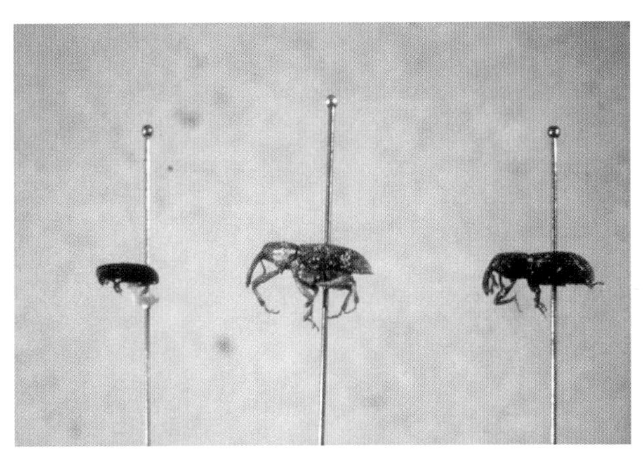

Figure 8-3d. **Black stain root disease**: Insect vectors of the black stain root disease fungus on Douglas-fir: (left to right) a root-feeding bark beetle, *Hylastes nigrinus*, and two root-feeding weevils, *Steremnius carinatus* and *Pissodes fasciatus*.

Figure 8-4a. **Annosus root disease**: Woody to leathery, shelf-like fruiting bodies of *Heterobasidion annosum* are frequently found inside hollows in large infected stumps. These conks have a dark to chestnut-brown upper surface, a white to cream pore layer with small, round, regular pores, and a white or cream sterile margin.

Figure 8-4b. **Annosus root disease**: Fruiting bodies of *Heterobasidion annosum* can be found on trees at the root collar just below the duff line, or in the root crotches. Small white to cream pustules with sterile margins and distinct pore layers that are called "button" or "popcorn" conks are the most common kind of fruiting body in these situations.

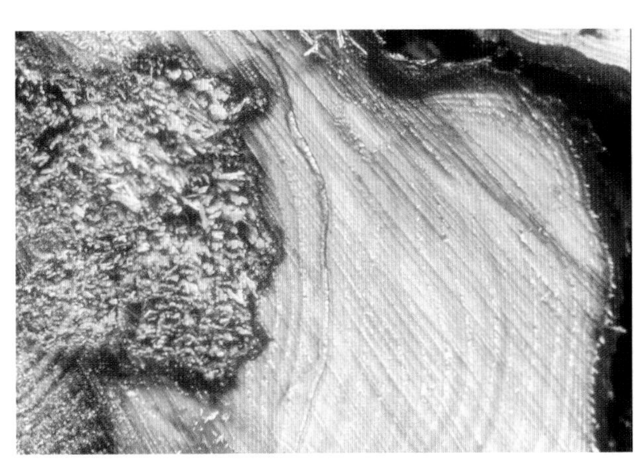

Figure 8-4c. **Annosus root disease**: Cross section through *Heterobasidion annosum*-infected western hemlock showing advanced decay caused by the fungus. Decay is a white rot.

Figure 8-4d. **Annosus root disease**: White rot with black flecking typical of *Heterobasidion annosum* in hemlocks and true firs.

Figure 8-4e. **Annosus root disease**: Longitudinal section through *Heterobasidion annosum*-infected western hemlock showing the progression from advanced decay near the stump to incipient decay.

Figure 8-5a. **Armillaria root disease**: As a result of poor planting technique, the roots of some seedlings either do not grow beyond the planting hole or have difficulty. Those seedlings become stressed and are likely candidates for Armillaria root disease. *Armillaria ostoyae* mycelial fans (A) under the bark at the base of a Douglas-fir seedling. Note the dark shoestring-like rhizomorphs (B). Rhizomorphs are dark with a white or light-colored nonwoody core.

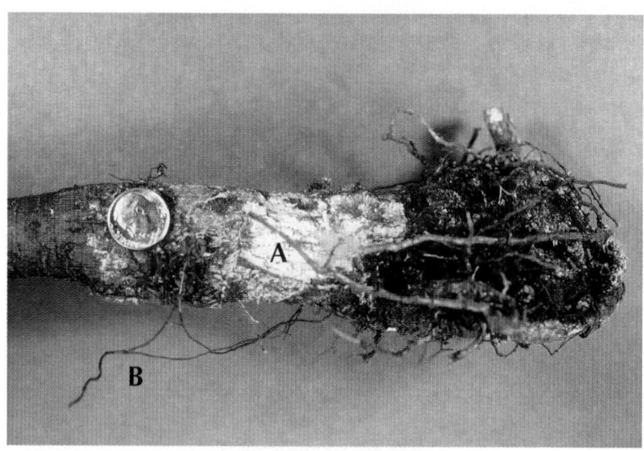

Figure 8-5b. **Armillaria root disease**: *Armillaria ostoyae* center in Douglas-fir plantation. In the Oregon Coast Range, Armillaria root disease is usually encountered in stands less than 30 years old on sites where trees are under stress. Stress may result from poor planting, use of off-site stock, soil compaction, or physical injury.

Figure 8-5c. **Armillaria root disease**: Basal resinosis is a common symptom of infection by *Armillaria ostoyae*.

Figure 8-5d. **Armillaria root disease**: Honey-colored mushrooms of *Armillaria ostoyae* appear in the fall around the bases of infected trees and stumps.

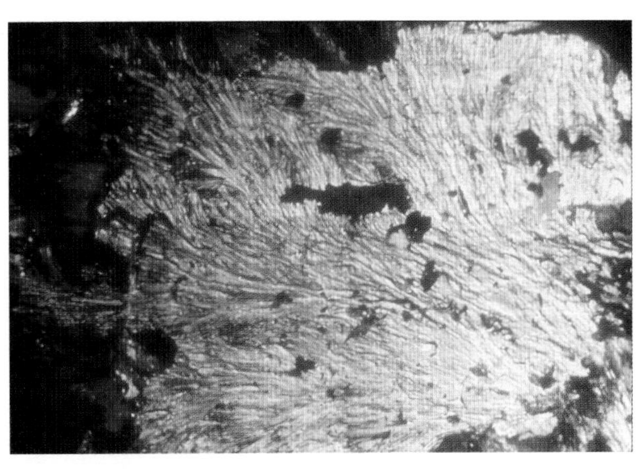

Figure 8-5e. **Armillaria root disease**: Close-up of *Armillaria ostoyae* mycelial fan found under the bark at the base of a tree. Occurrence of thick, white mycelial fans of *Armillaria ostoyae* under bark of roots and root collars is diagnostic of infection by this fungus.

Figure 8-6a. **Western hemlock dwarf mistletoe**: *Arceuthobium tsugense* var. *tsugense* plants on an infected western hemlock branch. The dwarf mistletoe obtains all of its water and most of its nutrients from its host.

Figure 8-6b. **Western hemlock dwarf mistletoe**: *Arceuthobium tsugense* var. *tsugense* plants on an infected branch of western hemlock. Fruit are nearly mature and ready for release. When a fruit is released, it falls away from the plant; the sticky seed is then forcibly ejected and may be carried 100 feet by the wind.

Figure 8-6c. **Western hemlock dwarf mistletoe**: Small, tight "witches' brooms" are induced by hemlock dwarf mistletoe infection on host branches.

Figure 8-6d. **Western hemlock dwarf mistletoe**: The bottom of an old broom with very large branches.

Figure 8-6e. **Western hemlock dwarf mistletoe**: Upper crowns of an old-growth western hemlock stand heavily infected by hemlock dwarf mistletoe. Large numbers of infections, especially in the upper part of the host's crown, can reduce growth significantly.

Figure 8-6f. **Western hemlock dwarf mistletoe**: Lower crown of a mature western hemlock stand heavily infected by hemlock dwarf mistletoe.

Figure 8-7a. **White pine blister rust**: Dead branches are seen as "flags" on infected trees. Branches die distal to the branch cankers.

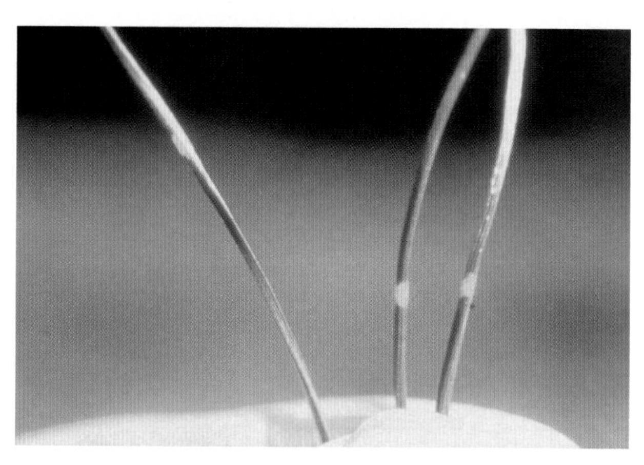

Figure 8-7b. **White pine blister rust**: Chlorotic spots on needles of white pine are a result of infection by *Cronartium ribicola*. Spores carried to the pine by air currents from *Ribes* spp. land on the needles and infect them in late summer to early fall. Spots appear 4 to 10 weeks after infection.

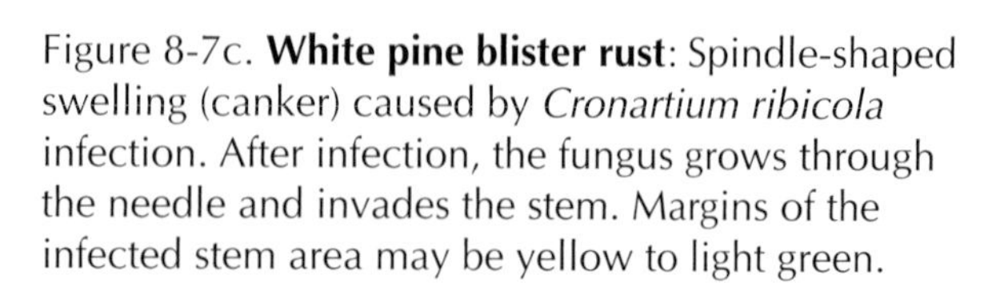

Figure 8-7c. **White pine blister rust**: Spindle-shaped swelling (canker) caused by *Cronartium ribicola* infection. After infection, the fungus grows through the needle and invades the stem. Margins of the infected stem area may be yellow to light green.

hemlock in a precommercially thinned 14-year-old red-alder stand resulted in good early survival, and height growth rates exceeded 50 cm per year (Emmingham et al. 1989). Treatment of the competing salmonberry-shrub layer with herbicides produced a positive growth response for up to three years. However, rapid crown expansion and growth of overstory red alder produced crown closure and slowed the growth of underplanted hemlock even in plots thinned to about 100 trees per acre (Emmingham et al. 1989). Additional thinning, at age 25, of the then-commercial alder stand released the hemlocks.

Recent experience on the OSU Research Forest with treatments designed to convert mature, even-aged Douglas-fir stands to uneven-aged forests confirmed that advance regeneration will respond to release. Seedlings and saplings of both Douglas-fir and grand fir responded within two years as the overstory volume was reduced from 34 to 20 million board feet (mbf) per acre. Vigorous advance regeneration was triggered by blowdown during the Columbus Day winds of 1962 and subsequent salvage operations. Skyline-cable yarding of 14 mbf per acre (average log = 23 inches in diameter) on slopes of 30-45 percent did not excessively damage advance regeneration, and there are adequate numbers of understory and midstory trees. The residual volume is now over 20 mbf per acre, and subsequent commercial harvests will probably be needed to achieve an uneven-aged structure.

Uneven-age silvicultural systems in Oregon Coast Range forest management

Early in the development of silvicultural practices in the Pacific Northwest, much harvesting was carried out by selection cutting (Kirkland and Brandstrom 1936). The goal was to remove trees likely to die and thereby, in theory, produce uneven-aged stands (Munger 1950; Curtis 1998; Curtis et al. 1998). This management proceeded largely without the help of a research program and without much formal evaluation of regeneration, tree and stand growth, impacts of disease and insects, windthrow, and other important factors. Thus there was little basis for improving the method. After a brief evaluation of the results (Isaac 1956), this partial cutting was dropped in favor of clearcutting and even-age management. This shift coincided with the beginning of new research programs that developed methods for planting, thinning, and harvesting that were compatible with even-age management. If the development of both even- and uneven-age systems had proceeded simultaneously and with the same emphasis on developing the supporting technology, a more diverse range of timber management options might have been developed for the forests of the Coast Range. This is not to say that uneven-age management would be the practice of choice throughout these forests today, but we would likely have a better understanding of sites and conditions conducive to it, and there would probably be a better-developed logging and regeneration technology to support it.

The options we have presented in this chapter for managing young stands in the Coast Range lend themselves to a mixed-species, multiple-story, or uneven-age approach. We believe mixed-species, multiple-layered stands have a role in the silviculture of Coast Range forests. Where management objectives are not heavily weighted by economic considerations, management of forests for a mix of ages, sizes, and tree species may be the best option. Indeed, on many sites it may be the only socially acceptable option that includes timber harvest, and the only alternative to forest reserves in situations where objectives are driven by aesthetic, wildlife, or other concerns.

Managing Hardwoods in the Oregon Coast Range

The Coast Range contains thousands of miles of riparian areas that contain hardwoods or a hardwood-conifer mix. The basic silviculture treatments we have discussed in the context of conifer stands may also be applied to hardwoods, although specific applications may vary because of differences in management objectives. Here we will address aspects of hardwood silviculture only where they differ from silviculture of conifers.

Regeneration

Hardwoods regenerate most commonly from seed or sprouting stumps. Managing for regeneration from seed requires an understanding of the kind of seedbed and the amount of light required by each species (Niemiec et al. 1995). Small-seeded species like alder and Pacific madrone need a mineral seedbed, high light levels, and protection from temperature extremes (Tappeiner et al. 1986; Haeussler and Tappeiner 1995). Large-seeded species like chinkapin and the oaks can be established in an intact litter layer, but success typically hinges on some control of competing vegetation. Tanoak and bigleaf maple may be established in the understory of intact, older conifer stands, but they will remain small until a disturbance opens the overstory.

Techniques for establishing alder with nursery-grown seedlings have been successfully developed (Hibbs and Ager 1989; Ahrens et al. 1992) (Figure 7-13 in color section following page 148). However, most nurseries are not familiar with the methods used to produce a high-quality seedling. Providing them with technical help would increase the success rate of hardwood regeneration in the field.

Alder requires a good supply of summer moisture but does not grow well on wet sites. Ridgetops and convex outer edges of mid-slope benches are too dry for good growth. Seedlings should be planted late by Northwest conifer standards—mid-March to mid-April—to minimize the chances of damage from a late frost. For successful planting, bareroot seedlings should have nodules with nitrogen-fixing bacteria, and they should have a height of at least 30 cm and a caliper of 4 mm.

Many hardwoods will sprout from cut stumps (Figure 7-14 in color section following page 148). If these sprouts are to be managed as reproduction, the stump should be cut low, and thinning should favor sprouts from low on the stump. By age three, sprouts can be thinned to one per foot of stump circumference.

Thinning

The strategy for managing a hardwood stand for timber production differs somewhat from methods used for conifers (Hibbs 1995). Generally, trees are kept at a relatively high density during their first years of rapid height growth to promote straight stems and natural branch pruning. Then, as tree growth begins to slow, the stand is thinned, enabling trees to develop the large crowns that are necessary for good diameter growth. In alder, the best response to thinning occurs between the ages of 10 and 18, depending on site quality. At this time the height to the base of the live crown in a dense stand is about 24 feet. Tanoak should be thinned later, between 25 and 40 years of age. In both species, early thinning produces greater crown and diameter growth than later thinning (Figure 7-15). Some stands may also need a very early thinning (before 10 years of age) if initial density is much more than 500 stems to the acre.

Experience with thinning of hardwoods is limited; most of our knowledge comes from observations of alder (Hibbs et al. 1994) and tanoak (McDonald and Tappeiner 1987). Unpublished studies of Pacific madrone, Oregon ash (*Fraxinus latifolia*), and bigleaf maple all show good responses to thinning. On dry sites, thinning Pacific madrone greatly increased total stand growth, according to preliminary study results (D. E. Hibbs, unpublished data).

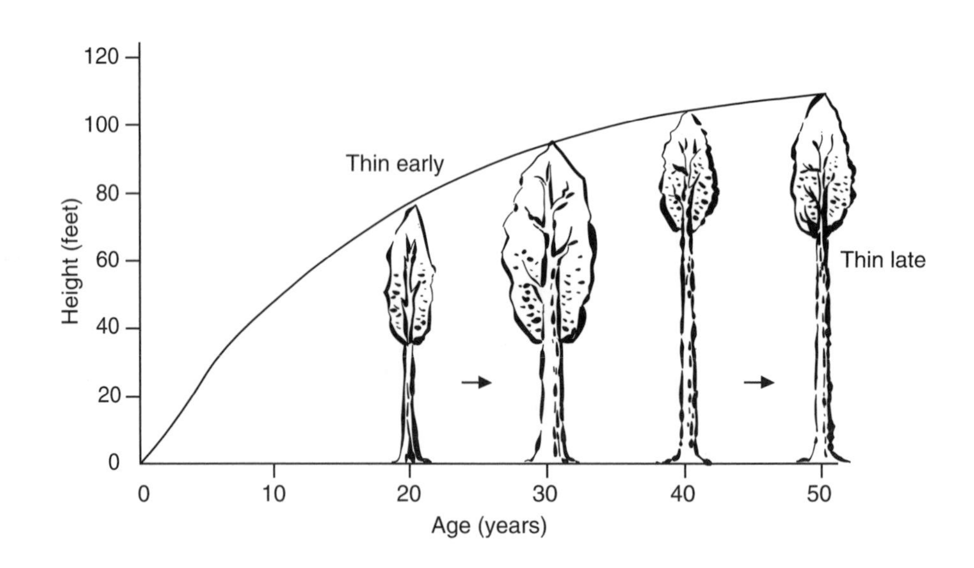

Figure 7-15. Red alder thinning response, showing that early thinnings at 20 years produce greater crown and diameter growth effects than later thinnings at 40 years.

Species mixes

Managing for a mix of conifers and hardwoods can promote structural diversity, enhance nutrient cycling, and, in the case of red alder, add significant increments of nitrogen. Madrone, bigleaf maple, and tanoak are moderately to highly shade tolerant, and all play important roles in species and structural diversity. Because it is a nitrogen-fixing species, red alder makes an important contribution to long-term site productivity. Bigleaf-maple foliage contributes to increased decomposition rates of conifer foliage in mixed forests. Silvicultural strategies can help promote these species mixes throughout the forest.

The key to successful management of species mixes is understanding the differing growth rates and shade tolerances of each species in the mix (Hibbs et al. 1994). Generally, those species that begin life with a burst of growth, like red alder, can maintain this rapid growth for only a short time (Figure 7-16). Thus, as long as the density and spatial arrangement of alder and an associated conifer, for example, are managed such that the alder cannot close canopy over the conifer, the conifer will eventually dominate. The shade tolerance of the species will determine how much conifer growth is slowed by partial shade from the hardwoods. To maintain the hardwood component over the long term, the conifer overstory will have to be thinned to provide enough light for the hardwoods.

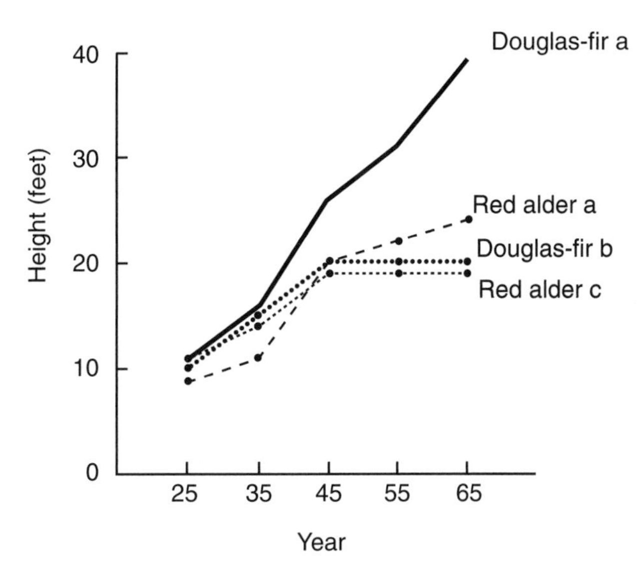

Figure 7-16. Comparison of Douglas-fir and red alder height growth from about 25–75 years of age. In pure stands (a) both species increase in height up to about 65 years of age. Douglas-fir typically grows at a faster rate than red alder after about 25 years. However, if either species is overtopped by the other (b, c), height growth is substantially reduced.

Poplar plantations

Hybrid poplar (*Populus* spp.) plantations are beginning to appear in Coast Range valleys. These plantations are established from unrooted cuttings and require a high level of weed control during establishment (Heilman et al. 1995). Plantations occupy deep, moist soils that are most often found in the valley bottom. On slopes, summer moisture is too limited for successful poplar culture.

Most plantations are grown for six to eight years at high density, about 600 stems to the acre, to produce chips for paper pulp. Cultural practices and yields for these short rotations are well known (Heilman et al. 1995; Withrow-Robinson et al. 1995). Poplar can also be grown at longer rotations and wider spacings for veneer and solid-wood products. While it is expected that veneer will be more profitable than pulp chips, cultural practices are still being developed for growing poplar for veneer.

The regulatory situation for poplar is complex. Poplar grown for rotations of less than 12 years are regulated as an agricultural crop. If the rotation is 12 years or more, forestry regulations apply. Wetland regulations will also apply; poplar is often planted on poorly drained soils or in areas where winter flooding is common.

Silviculture of Riparian Stands in the Coast Range

Silviculture is becoming increasingly important in managing Coast Range riparian stands. Management goals should be based on desired riparian conditions and functions. The conditions found today in the riparian areas of the Oregon Coast Range are highly variable owing to climate, natural-disturbance history, and human impacts. The Coast Range is climatically diverse, with strong north-south and west-east precipitation and temperature gradients that determine the riparian species and habitat conditions found at a given location. Throughout the region's history, fire, flood, and debris flows have created a variety of regeneration opportunities, resulting in riparian forest that is a diverse mix of patches of different ages and species (Figure 7-17 in color section following page 148) growing on complex landforms (Figure 7-18 in color section following page 148). The human activities of farming, logging, burning, trapping beavers, and dam building have also changed the structure, composition, and dynamics of riparian forest communities.

In all discussions of stream and streamside management, it is critical to keep in mind the dynamic nature of riparian systems. A study of central Coast Range streams that have been undisturbed by humans for at least 150 years provides one image of what some presettlement streams might have been like (Nierenberg and Hibbs 2000). This study indicated that the riparian community was composed more often of shrub-dominated patches than of patches of either hardwoods or conifers, and that hardwoods and conifers occurred at about equal frequency. Conifers seem to have been established infrequently. Patches of riparian vegetation occurred in a variety of arrangements and sizes that resulted in complex patterns (Figure 7-18).

Most people are aware of the short-term changes in riparian systems; high water one winter, for example, will knock down a few trees or shift a short section of channel. There is in addition a very long-term cycle as well (Reeves et al. 1995). In this cycle, a major fire initiates a number of debris flows that deposit gravels and wood into the stream channel, damaging vegetation and sometimes creating new surfaces for plant colonization. Initially, the stream is poor fish habitat, but it soon develops into prime habitat as the gravel and woody debris make pools, hiding cover, and gravel beds. Through a century or two, these materials are moved downstream and broken up and they eventually disappear, leaving a stream that is both wood- and gravel-poor. Another disturbance would probably be necessary to restore large wood and gravel.

Basing management strategies for riparian areas on ecological processes

Many management goals have been proposed for Coast Range riparian areas, including restoring healthy fish populations, maintaining water quality, controlling water temperature, restoring historic in-stream and vegetation conditions, and minimizing landslides. While these goals are well intentioned, they are sometimes contradictory and often based on mistaken ecological assumptions. As the study by Reeves et al. (1995) suggests, natural landslides through time have provided streams with large logs, rocks, and gravel that disrupt the system in the short term but eventually provide pools and habitat for fish and amphibians. Fire and other disturbances immediately adjacent to streams may have caused immediate damage to fish runs, but they resulted in establishment of conifer seedlings that eventually produced large trees and detritus for riparian habitat. Thus, preventing disturbance is often not the best way to manage riparian zones. Before advocating this or any other management goal, managers must have an understanding of riparian-zone ecology and basic processes. Sometimes long-term objectives, such as pristine streams, appear to conflict with short-term objectives, such as providing structure to a stream or controlling shrubs and hardwoods to start new conifer trees. It is our view that this apparent conflict can be eased, if not altogether resolved, by centering the discussion around ecosystem processes. Thus we suggest four objectives that focus on desired riparian processes: (1) to provide structural diversity in streams and flood plains, (2) to provide wildlife habitat, (3) to maintain stream productivity, and (4) to produce wood. Frequently these objectives are interrelated, and multiple objectives can be achieved simultaneously. Taken together, these objectives form a framework for clarifying and promoting management goals.

"Stream structure" commonly refers to the presence of large conifer logs in the stream, although in some streams large rock also fills this role. Current federal and state riparian management rules call for growing this wood on the streambank. However, recent research is showing that much of the large wood in Coast Range streams entered the system in debris flows from headwalls and steep side slopes (Reeves et al. 1995). These debris flows are also the main source of bed gravel. Thus, silvicultural activities focused on stream structure might consider options for managing both the riparian area adjacent to the stream and the steep upper slopes. One option might be to identify unstable slopes on which to leave or grow large trees that will eventually slide into the stream. Stands on stable slopes have a low probability of falling into the stream.

Some riparian wildlife species appear to respond to habitat associated with the high-moisture conditions on the terraces and wet depressions

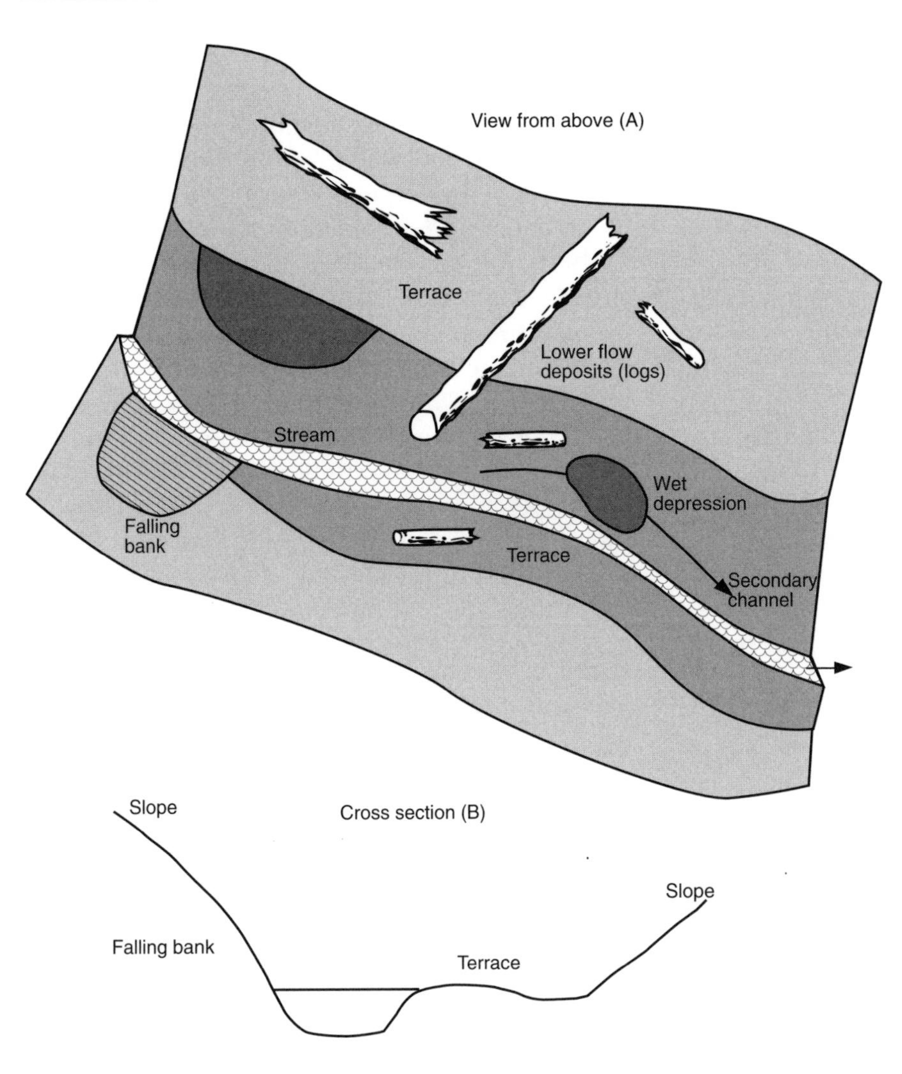

Figure 7-19. The structure of a typical riparian area around a small stream, viewed from above (a) and from the side (b). Features to note are the meandering channel, the uneven topography of the terrace (including depressions, channels, and elevated areas), and wood coming in from upslope.

found in riparian zones (McComb et al. 1993) (Figure 7-19). These species are often found at mid-slope seeps as well. Some riparian wildlife species appear to be responding to vegetation characteristics like hardwood or conifer cover or rotting logs. Many of them are also found elsewhere in the forest where these vegetation conditions occur. Because these species are so diverse, no simple or single silvicultural approach will meet their needs. Traditional silviculture has focused at the stand level, managing for the needs of a small, related group of species. Now we must also manage the riparian area at a larger scale of stream reaches, for a diversity of vegetation types and seral stages, trusting that this diversity will support the varied habitat needs of wildlife. To maximize the variety of habitats, a length of stream should probably be managed for vegetation patches of varied age and composition.

The in-stream food chain is fed by litterfall and sunlight (Gregory et al. 1991) and depends on litterfall from riparian vegetation. Conifer litter falls more evenly throughout the year than hardwood litter, has a lower nutrient content, and breaks down more slowly. A deciduous canopy provides summer shade and passes light for winter photosynthesis. Productivity of streams is often nitrogen-limited, so the nitrogen fixed by alder and released into the groundwater of riparian areas can be important.

Beyond these known principles, it is difficult to suggest how to manage riparian vegetation to aid stream productivity. Little information exists on which streams are limited by temperature, energy, or nutrients. There has been no systematic examination of how different mixes of hardwoods and conifers along a stream affect in-stream processes; all studies have been of a single location. At this time, given that both coniferous and deciduous species play different but essential roles, it appears advisable to maintain an abundance of both tree groups along a stream (Figure 7-18).

Site conditions such as drainage, soil texture, and frequency of flooding vary greatly across riparian areas. These factors affect growth rates and species of trees found or cultivated in the riparian zone. Most riparian soils are highly productive of trees, although historic disturbance and regeneration patterns have often led to incomplete and irregular stocking (Means et al. 1996). Many techniques exist to manage riparian areas intensively while minimizing environmental impacts. For best results, the basic regeneration requirements for hardwoods and conifers in the Coast Range —actively reducing competing vegetation and protecting the stand from browsing animals—should be followed through the 5- to 10-year regeneration period (Emmingham et al. 1989; Stein 1995). Competing vegetation can be controlled effectively by cutting or spot-spraying of herbicides. These hand-applied treatments can be precisely targeted, but they are more expensive than broad-spectrum treatments like prescribed fire or aerial spraying.

Harvesting in the riparian zone is best combined with harvesting in the adjacent stands because a combined harvest minimizes both cost and disturbance to the site. For example, a cable system set up for the adjacent stand can lift logs from the riparian area with minimal soil disturbance. Using ground-based equipment is more likely to result in extensive soil disturbance.

Silviculture practices to meet riparian-area management objectives

The four management objectives outlined above can provide the goals for silvicultural prescriptions. The silvicultural principles of riparian areas are the same as those used elsewhere; an important difference is that silvicultural choices are often constrained by such goals as maintaining shade to the stream or providing logs near the stream channel. Following are some ways in which conventional silvicultural principles may be used to meet riparian management objectives.

Buffers. Forested buffers (Figure 7-20 in color section following page 148) have been left along streams following clearcutting to protect in-stream and terrestrial riparian functions. Studies of the dynamics of these buffers show that they can suffer blowdown at predictable locations (Steinblums et al. 1984), but that most are, biologically speaking, surprisingly stable; that is, they develop largely as if they were still part of a continuous forest (Hibbs and Giordano 1996; D Hibbs and A Bower, personal observation). Thus buffers do generally provide protection of riparian values for the life span of the overstory trees. These studies also found that over- and understory conditions were similar to those in uncut stands. The tree, shrub, and herb layers remained intact for at least 30 years in the buffer strips (Table 7-2). The composition, abundance, species, and successional dynamics in the buffer forest are essentially the same as in riparian forest with no adjacent harvested areas. In the longer term, the lack of tree regeneration under alder-dominated riparian forests, whether in streamside buffers or in upslope stands, will result in succession to a mostly shrub-dominated community.

Regeneration. Studies of tree regeneration in Coast Range riparian areas (buffers and otherwise-undisturbed areas) indicate that conifers and hardwoods are regenerating naturally in about the southern third of the Coast Range (Minore and Weatherly 1994). However, in the middle and northern two-thirds of the Coast Range (except in the coastal fog belt and the Willamette Valley margin), regeneration of shade-tolerant tree species is limited both by competition from understory shrubs and by a lack of western hemlock, grand fir, and western redcedar seed trees (Schrader 1998). In this part of the range, natural regeneration is inadequate to replace the current riparian forest, and so active management for regeneration is needed if conifers are to be established in riparian zones.

Table 7-2. Comparison of riparian conditions (mean values) in buffer strips and undisturbed riparian stands.

Riparian condition	*Overstory basal area (ft²/acre)*	*Overstory cover (%)*	*Shrub cover (%)*
Buffer terrace	236.5 a	82.6 a	102.1 ab
Buffer slope	112.0 b	54.7 b	121.7 a
Undisturbed terrace	186.2 ab	67.6 ab	79.8 b
Undisturbed slope	122.3 ab	68.3 ab	96.2 ab

Within a column, a mean followed by the same letter is not significantly different ($p \leq 0.05$; Means et al. 1996).

The many ongoing studies of active management for tree regeneration in Coast Range riparian areas clearly show that regeneration is successful as long as adequate light is provided and browsing animals are controlled. Studies of all the native tree species show that heavy thinning of the overstory and removal of the salmonberry understory provide adequate light for shade-tolerant species (Chan et al. 1996). Small gaps (diameter > 1 tree height) are necessary for good regeneration of Douglas-fir and red alder. These species will survive and grow under lower light, but their growth will be slower than in full sunlight. Slower growth makes regeneration more vulnerable to browsing animals for a prolonged period, eventually jeopardizing survival. Western hemlock, western redcedar, and spruce grow well in partial shade but best in a high-light environment. Protection from browsing animals increases their growth and survival on most sites.

Beaver populations are thought to have important effects in Coast Range riparian areas. Many regeneration-study sites suffered heavy mortality from beavers until streamside fences were installed, and even with fences, winter floods made it possible for beavers to get into some young plantations. The long-term effect of beaver damage on riparian regeneration, stand structure, and composition is unknown.

Release. Emmingham and Maas (1994) succeeded in releasing conifer regeneration and codominant conifers, usually western hemlock and western redcedar, from competing alder. Control of alder and shrubs in the understory increased both conifer survival and height growth (Figure 7-21). Alder may be cut, girdled, or injected with herbicide. Girdling and injection work best for maintaining easy access to a site for follow-up understory treatments because the trees remain standing for several years after they die. To be successful, girdling must be done carefully and thoroughly.

Thinning. Heavy thinning that begins early in stand development has been shown to produce very large conifers and hardwoods in a relatively short time (Newton and Cole 1987). This is an important finding with respect to creating both wildlife habitat and in-stream structural diversity. Again, it is critical to manage density while the stand is young in order to maximize management options later on.

Alder and Douglas-fir dominate most riparian areas today. If managers move toward more planting and thinning in riparian zones with the goal of producing large trees, western hemlock and western redcedar will likely become much more prevalent in these areas. Because these two species cast dense shade, their increase will result in a decrease in understory cover and diversity. Overstory thinning is one way to maintain more diversity in the understory.

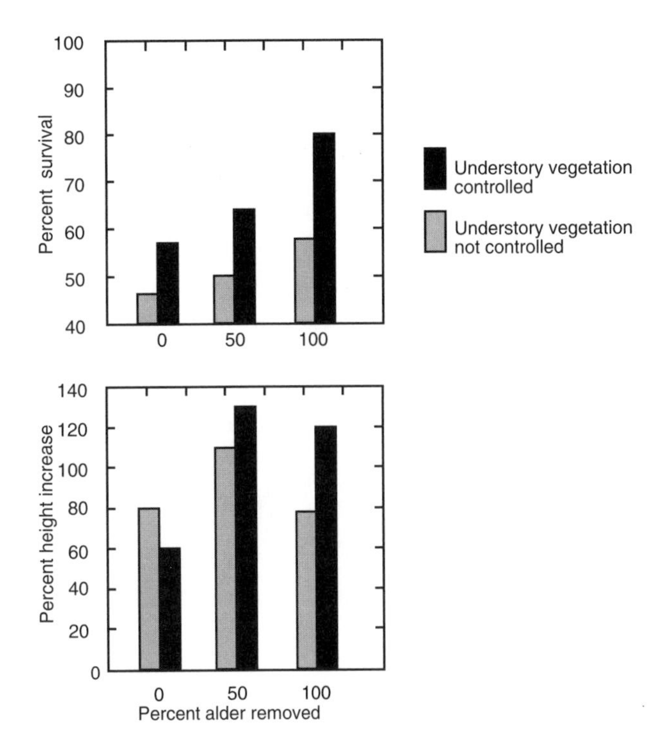

Figure 7-21. Third-year survival and growth of hemlock seedlings planted under alder with different levels of girdling of the alder.

Conclusions

The practice of silviculture is changing, not only in the Oregon Coast Range but throughout North America, with the advent of new information on forests and forest ecosystems. Primarily these changes stem from recognition of the rich flora and fauna of forests (Ruggiero et al. 1991). However, the potential for maintaining biodiversity in managed forests is just now beginning to be evaluated.

We believe there is sufficient information on forest ecology and silviculture to support new systematic, long-term, landscape-scale, well-executed forest management regimes (Kohm and Franklin 1997). In this chapter we have pointed out the wealth of new information on forest regeneration of trees and shrubs. We have the knowledge to grow diverse, multi-layered stands that are similar to older natural forests. The effects of stand density on tree and stand development are well understood. Fortunately, Coast Range conifers and hardwoods are quite long-lived, and they can be managed at ages well over 100 years for multiple objectives. Current information can be put into practice to connect silviculture and the principles of riparian and upland ecosystems presented in this book. In the coming decades, silvicultural practices will evolve, as they have in the past, into forest management policies that reflect society's current values.

Literature Cited

Ahrens, G. R., A. Dobkowski, and D. E. Hibbs. 1992. *Red Alder: Guidelines for Successful Regeneration.* Special Contribution 24, Forest Research Laboratory, Oregon State University, Corvallis.

Assman, E. 1970. *The Principles of Forest Yield Study.* Pergamon Press, New York.

Bailey, J. D. 1996. *Effects of Stand Density Reduction on Structural Development in Western Oregon Douglas-fir Forests; A Reconstruction Study.* PhD dissertation, Oregon State University, Corvallis.

Bailey, J. D., and J. C. Tappeiner II. 1998. Effects of thinning on structural development in 40- to 100-year-old Douglas-fir stands in western Oregon. *Forest Ecology and Management* 108: 99-113.

Bailey, J. D., C. Marysohn, P. S. Doescher, E. St. Pierre, and J. C. Tappeiner II. 1998. Understory vegetation in old and young Douglas-fir forests of western Oregon. *Forest Ecology and Management* 112: 289-302.

Brandeis, T. J. 1999. *Underplanting and Competition in Unthinned Douglas-fir Stands.* PhD dissertation, Oregon State University, Corvallis.

Cafferata, S. L. 1986. Douglas-fir stand establishment overview: western Oregon, and Washington, pp. 211-218 in *Douglas-fir Stand Management for the Future,* C. D. Oliver, D. P. Hanley, and J. A. Johnson, ed. College of Forest Resources, University of Washington, Seattle.

Chan, S., K. Maas-Hebner, and W. H. Emmingham. 1996. *Thinning Hardwood and Conifer Stands to Increase Light Levels: Have you thinned enough?* COPE Report 9(4): 2-6, Oregon State University, Corvallis.

Cleary, B. D., R. D. Greaves, and R. Hermann. 1978. *Regenerating Oregon's Forests: A Guide for the Regeneration Forester.* Oregon State University Extension Service, Corvallis.

Conard, S. G., and W. H. Emmingham. 1984. *Herbicides for Clump and Stem Treatment of Weed Trees and Shrubs in Oregon and Washington.* Special Publication 9, Forest Research Laboratory, Oregon State University, Corvallis.

Conard, S. G., A. E. Jaramillo, and S. Rose. 1985. *The Role of the Genus Ceanothus in Western Forest Ecosystems.* General Technical Report PNW-182, USDA Forest Service, Pacific Northwest Research Station, Portland OR.

Curtis, R. O. 1982. A simple index of stand density for Douglas-fir. *Forest Science* 28: 92-94.

Curtis, R. O. 1992. A new look at an old question: Douglas-fir culmination age. *Western Journal of Applied Forestry* 7: 97-99.

Curtis, R. O. 1998. Selective cutting in Douglas-fir: history revisited. *Journal of Forestry* 96: 40-46.

Curtis, R. O., and D. D. Marshall. 1986. *Levels of Growing Stock Cooperative Study in Douglas-fir. Report 8, The LOGS Study; Twenty Year Results.* Research Paper PNW-336, USDA Forest Service, Pacific Northwest Research Station, Portland OR.

Curtis, R. O., and D. D. Marshall. 1993. Douglas-fir rotations . . . time for reappraisal. *Western Journal of Applied Forestry* 8: 81-85.

Curtis, R. O., D. S. DeBell, C. A. Harrington, D. R. Lavender, J. B. St. Clair, J. C. Tappeiner II, and J. D. Walstad. 1998. *Silviculture for Multiple Objectives in the Douglas-fir Region.* General Technical Report PNW-GTR-435, USDA Forest Service, Pacific Northwest Research Station, Portland OR.

Daniel, T. W., J. A. Helms, and F. S. Baker. 1979. *Principles of Silviculture.* McGraw-Hill, New York.

Del Rio, E., and A. Berg. 1979. *A Growth of Douglas-fir Reproduction in the Shade of a Managed Forest.* Research Paper 40, Forest Research Laboratory, Oregon State University, Corvallis.

DeYoe, D. R. 1986. *Guidelines for Handling Seeds and Seedlings to Ensure Vigorous Stock.* Special Publication 13, Oregon State University, Corvallis.

Duryea, M. L., and P. M. Dougherty. 1991. *Forest Regeneration Manual.* Kluwer Academic Publishers, Hingham MA.

Emmingham, W. H. 1996. *Commercial Thinning and Underplanting to Enhance Structural Diversity of Young Douglas-fir Stands in the Oregon Coast Range: An Establishment Report and Update on Preliminary Results.* COPE Report 9: 2-3, Oregon State University, Corvallis.

Emmingham, W. H., and K. Maas. 1994. *Survival and Growth of Conifers Released in Alder-dominated Coastal Riparian Zones.* COPE Report 7: 13-15, Oregon State University, Corvallis.

Emmingham, W. H., and R. H. Waring. 1977. An index of forest sites for comparing photosynthesis in western Oregon. *Canadian Journal of Forest Research* 7: 165-174.

Emmingham, W. H., M. Bondi, and D. E. Hibbs. 1989. Underplanting western hemlock in a red alder thinning: early survival, growth, and damage. *New Forests* 3: 31-43.

Fried, J. S., J. C. Tappeiner II, and D. E. Hibbs. 1988. Bigleaf maple seedling establishment and early growth in Douglas-fir forests. *Canadian Journal of Forest Research* 18: 1226-1233.

Gregory, S. V., F. J. Swanson, and W. A. McKee. 1991. An ecosystem perspective of riparian zones. *BioScience* 41: 540-551.

Haeussler, S., and J. C. Tappeiner II. 1995. Germination and first-year survival of red alder seedlings in the central Oregon Coast Range. *Canadian Journal of Forest Research* 25: 1639-1651.

Harrington, T. B., J. C. Tappeiner II, and R. Warbington. 1992. Predicting crown sizes and diameter distributions of tanoak, Pacific madrone, and giant chinkapin sprout clumps. *Western Journal of Applied Forestry* 7: 103-108.

Harrington, T. B., R. J. Pabst, and J. C. Tappeiner II. 1995. Seasonal physiology of Douglas-fir saplings: response to microclimate in stands of tanoak or Pacific madrone. *Forest Science* 40: 59-82.

Heilman, P. E., R. F. Stetler, D. P. Hanley, and R. W. Carkner. 1995. *High Yield Hybrid Poplar Plantations in the Pacific Northwest.* Pacific Northwest Cooperative Extension Bulletin PNW-356, Washington State University Cooperative Extension [Pullman]; Oregon State University Extension Service [Corvallis]; University of Idaho Cooperative Extension System [Moscow]; USDA [Washington DC].

Helms, J. A. 1998. *The Dictionary of Forestry.* Society of American Foresters, Bethesda MD.

Hermann, R. K., and W. W. Chilcote. 1965. *Effects of Seedbeds on Germination and Survival of Douglas-fir.* Research Paper 4, Forest Research Laboratory, Oregon State University, Corvallis.

Hibbs, D. E. 1995. *Managing Hardwood Stands for Timber Production.* Extension Circular EC-1183, Oregon State University Extension Service, Corvallis.

Hibbs, D. E., and A. A. Ager. 1989. *Red Alder: Guidelines for Seed Collection, Handling and Storage.* Special Publication 18, Forest Research Laboratory, Oregon State University, Corvallis.

Hibbs, D. E., and P. A. Giordano. 1996. Vegetation characteristics of alder-dominated riparian buffer strips in the Oregon Coast Range. *Northwest Science* 70: 213-222.

Hibbs, D. E., D. S. DeBell, and R. F. Tarrant, ed. 1994. *Biology and Management of Red Alder.* Oregon State University Press, Corvallis.

Hobbs, S. D., S. D. Tesch, P. W. Owston, R. E. Stewart, J. C. Tappeiner II, and G. E. Wells. 1992. *Reforestation Practices in Southwestern Oregon and Northern California.* Forest Research Laboratory, Oregon State University, Corvallis.

Huffman, D. W., and J. C. Tappeiner II. 1997. Clonal expansion and seedling recruitment of Oregon-grape in Douglas-fir forests: comparisons with salal. *Canadian Journal of Forest Research* 27: 1788-1793.

Huffman, D. W., J. C. Tappeiner II, and J. C. Zasada. 1994. Regeneration of salal in the central Coast Range forests of Oregon. *Canadian Journal of Botany* 72: 39-51.

Isaac, L. A. 1956. Place of partial cutting in old-growth stands of the Douglas-fir region. Research Paper 16, USDA Forest Service Pacific Northwest Forest and Range Experiment Station, Portland OR.

Ketchum, S. J. 1994. *Douglas-fir, Grand Fir and Plant Community Regeneration in Three Silvicultural Systems in Western Oregon.* MS thesis, College of Forestry, Oregon State University, Corvallis.

King, J. E. 1966. *Site Index Curves for Douglas-fir in the Pacific Northwest.* Weyerhaeuser Forest Paper No. 8, Weyerhaeuser Forest Research Center, Centralia WA.

Kirkland, B. P., and A. J. F. Brandstrom. 1936. *Selective Timber Management in the Douglas-fir Region.* Division of Forest Economics, USDA Forest Service, Washington DC.

Kohm, K. A., and J. F. Franklin. 1997. *Creating a Forestry for the 21st Century: The Science of Ecosystem Management.* Island Press, Washington DC.

Lavender, D. P. 1958. *Effects of Ground Cover on Seedling Germination and Survival.* Research Note 34, Forest Research Laboratory, Oregon State University, Corvallis.

Margolis, H. A., and D. G. Brand. 1990. An ecophysiological basis for understanding plantation establishment. *Canadian Journal of Forest Research* 20: 375-390.

Marshall, D. D., J. F. Bell, and J. C. Tappeiner II. 1992. *Levels of Growing Stock Cooperative Study: Douglas-fir, Report 10, The Hopkins Study 1963-1983.* Research Paper PNW-448, USDA Forest Service, Pacific Northwest Research Station, Portland OR.

Mathews, J. D. 1989. *Silvicultural Systems.* Oxford University Press, Oxford.

McArdle, R. E., W. H. Meyer, and D. Bruce. 1961. *The Yield of Douglas Fir in the Pacific Northwest.* 2nd rev., Technical Bulletin 201, USDA, Washington DC.

McComb, W. C., K. McGarigal, and R. G. Anthony. 1993. Small mammal and amphibian abundance in streamside and upslope habitats of mature Douglas-fir stands in western Oregon. *Northwest Science* 67: 7-15.

McDonald, P. M., and J. C. Tappeiner II. 1987. Silviculture, ecology and management of tanoak in northern California, pp. 62-70 in *Proceedings, Symposium on Multiple-use Management of California's Hardwood Resources,* T. R. Plumb and N. H. Pillsbury, tech. coord. General Technical Report PSW-100, USDA Forest Service, Pacific Southwest Experiment Station, Berkeley CA.

Means, J. E., R. R .Harris, T. E. Sabin, and C. N. McCain. 1996. Spatial variation in productivity of Douglas-fir stands on a valley floor in the western Cascade Range, Oregon. *New Forests* 8: 201-212.

Miller, M., and B. Emmingham. 2001. Can selection thinning convert even-aged Douglas-fir to uneven-aged structure? *Western Journal of Applied Forestry* 16(1).

Minore, D., and H. G. Weatherly. 1994. Riparian trees, shrubs, and forest regeneration in the coastal mountains of Oregon. *New Forests* 8: 249-263.

Munger, T. T. 1950. A look at selective cutting in Douglas-fir. *Journal of Forestry* 48: 97-99.

Newton, M., and E. Cole. 1987. A sustained yield scheme for old-growth Douglas-fir. *Western Journal of Applied Forestry* 2: 22-25.

Niemiec, S. S., G. R. Ahrens, S. Willits, and D. E. Hibbs. 1995. *Hardwoods of the Pacific Northwest.* Research Contribution 8, Forest Research Laboratory, Oregon State University, Corvallis.

Nierenberg, T., and D. E. Hibbs. 2000. A characterization of unmanaged riparian areas in the central Oregon Coast Range of western Oregon. *Forest Ecology and Management.*

O'Dea, M., J. C. Zasada, and J. C. Tappeiner II. 1995. Vine maple clonal development in coastal Douglas-fir forests. *Ecological Applications* 5: 63-73.

Oliver, C. D., and B. C. Larson. 1990. *Forest Stand Dynamics.* McGraw-Hill, New York.

Oliver, C. D., and B. C. Larson. 1996. *Forest Stand Dynamics.* Updated edition. McGraw-Hill, New York.

Oregon Department of Forestry. 1997. *Oregon Forests (Annual Report).* Oregon Department of Forestry, Salem.

Poage, N. 2001. *Structure of Old-Growth Forests in Central Oregon Cascades and Coastal Forests.* PhD dissertation. Oregon State University, Corvallis.

Quick, C. R. 1956. Viable seeds from the duff and soil of sugar pine forests. *Forest Science* 2: 36-42.

Quick, C. R. 1959. Ceanothus seeds and seedlings on burns. *Madrono* 15: 79-81.

Reeves, G. H., L. E. Benda, K. M. Burnett, P. A. Bisson, and J. R. Sedell. 1995. A disturbance-based ecosystem approach to maintaining and restoring freshwater habitats of evolutionarily significant units of anadromous salmonids in the Pacific Northwest. *Proceedings American Fisheries Society Symposium* 17: 334-349.

Ruggiero, L. F., K. B. Aubrey, A. B. Carey, and M. H. Huff, tech. coord. 1991. *Wildlife and Vegetation of Unmanaged Douglas-fir Forests.* General Technical Report PNW-GTR-285, USDA Forest Service, Pacific Northwest Research Station, Portland OR.

Ruth, R. H. 1970. *Effect of Shade on Germination and Growth of Salmonberry.* Research Paper 96, USDA Forest Service, Pacific Northwest Research Station, Portland OR.

Schrader, B. A. 1998. *Structural Development of Late Successional Forests in the Central Oregon Coast Range: Abundance, Dispersal and Growth of Western Hemlock (*Tsuga heterophylla*) Regeneration.* PhD dissertation, College of Forestry, Oregon State University, Corvallis.

Sensenig, T. 2002. *Growth, Fire History and Stand Development of Old and Young Forests in Southwestern Oregon.* PhD dissertation, Oregon State University, Corvallis.

Smith, D. M., B. C. Larson, M. J. Kelty, and P. M. S. Ashton. 1996. *The Practice of Silviculture: Applied Forest Ecology.* John Wiley and Sons, New York.

Staebler, G. R. 1960. Theoretical derivation of numerical thinning schedules for Douglas-fir. *Forest Science* 6: 98-109.

State of Oregon. 1971. *Forest Practices Act.* Oregon Revised Statutes Section 527 (as amended 1991), Salem.

Stein, W. I. 1995. *Ten-year Development of Douglas-fir and Associated Vegetation after Site Preparation on Coast Range Clearcuts.* Research Paper PNW-RP-473, USDA Forest Service, Pacific Northwest Research Station, Portland OR.

Steinblums, I. J., H. A. Froehlich, and J. K. Lyons. 1984. Designing stable buffer strips for stream protection. *Journal of Forestry* 82: 49-52.

Tappeiner, J. C. II, and J. C. Zasada. 1993. Establishment of salmonberry, salal, vine maple and bigleaf maple seedlings in the coastal forests of Oregon. *Canadian Journal of Forest Research* 23: 1775-1780.

Tappeiner, J. C. II, T. B. Harrington, and J. D. Walstad. 1984. Predicting recovery of tanoak and Pacific madrone after cutting or burning. *Weed Science* 32: 415-417.

Tappeiner, J. C. II, P. M. McDonald, and T. F. Hughes. 1986. Survival of tanoak (*Lithocarpus densiflorus*) and Pacific madrone (*Arbutus menziesii*) seedlings in forests of southwestern Oregon. *New Forests* 1: 43-55.

Tappeiner, J. C. II, J. C. Zasada, P. Ryan, and M. Newton. 1991. Salmonberry clonal and population structure: the basis for a persistent cover. *Ecology* 72: 609-618.

Tappeiner, J. C. II, D. W. Huffman, D. Marshall, T. A. Spies, and J. D. Bailey. 1997. Density, ages, and growth rates in old-growth and young-growth forests in coastal Oregon. *Canadian Journal of Forest Research* 27: 638-648.

Tesch, S. D., and E. J. Korpella. 1993. Douglas-fir and white fir advanced regeneration for renewal of mixed conifer forests. *Canadian Journal of Forest Research* 23: 1427-1437.

Tesch, S. D., K. Baker-Katz, E. J. Korpella, and J. W. Mann. 1993. Recovery of Douglas-fir seedlings and saplings wounded during overstory removal. *Canadian Journal of Forest Research* 23: 1684-1694.

Walstad, J. D., and P. J. Kuch. 1987. *Forest Vegetation Management for Conifer Production.* John Wiley and Sons, New York.

Waring, R. H., and J. F. Franklin. 1979. Evergreen coniferous forests of the Pacific Northwest. *Science* 204: 1380-1386.

Withrow-Robinson, B., D. Hibbs, and J. Beuter. 1995. *Poplar Chip Production for Willamette Valley Grass Seed Sites.* Research Contribution 11, Forest Research Laboratory, Oregon State University, Corvallis.

Zasada, J. C., J. C. Tappeiner II, B. D. Maxwell, and M. A. Radiwan. 1994. Seasonal changes in shoot and root production and carbohydrate content of salmonberry (*Rubus spectabilis*) rhizome segments from the central Oregon Coast Range. *Canadian Journal of Forest Research* 24: 272-277.

Major Forest Diseases of the Oregon Coast Range and their Management

Walter G. Thies and Ellen Michaels Goheen

Introduction

This chapter addresses the role of major forest diseases found in the Oregon Coast Range. We briefly review pathogen biologies and their ecological roles, address management strategies, and identify the gaps in knowledge that must be filled to support the implementation of ecosystem management in coastal forests of Oregon.

Forest tree diseases are among the most important disturbance agents operating in the forest ecosystems of the Oregon Coast Range. Even though the effects of tree diseases are rarely spectacular at any one point in time, at landscape scales and over long time frames they are substantial. In contrast to dramatic events that affect only a relatively small area for a relatively short time, such as the eruption of Mt. St. Helens in 1980 or the Tillamook fires of the early 1930s and 1940s, tree pathogens have ongoing effects in virtually all forests. To meet ecosystem management objectives, it is critical to understand the ecological roles of pathogens, their historical and current distributions, their impacts on resources, how management influences them, and how they affect management.

We define tree disease as a sustained disturbance to the normal function, structure, or form of a tree, as provoked by biological, physical, or chemical factors of the environment. Thus disease results from the sometimes-complex interactions between the host tree, the pathogen (the organism that causes disease), and the environment. A change in any one of these factors can influence the severity of disease.

Native disease organisms have coevolved with their hosts in the Coast Range ecosystems. They are natural, and necessary, parts of healthy ecosystems and are always present at some level that helps initiate change and renewal. These host-pathogen combinations are genetically buffered, and impacts tend to be limited. However, human activities that disrupt these native host-pathogen interactions can have serious consequences. For example, creating numerous infection courts in the form of freshly cut stumps has led to increased incidence of annosus root disease. Movement of a host beyond its area of adaptation can lead to high levels of disease in relatively short periods of time. This is one of several factors associated with the current outbreak of Swiss needle cast in the Coast Range.

In contrast to the native pathogens, recently introduced pathogens, such as *Cronartium ribicola*, the cause of white pine blister rust, or *Phytophthora lateralis*, the cause of Port-Orford-cedar (*Chamaecyparis lawsoniana*) root disease, have not coevolved with their hosts. As a result, these host-pathogen relationships have developed essentially without the genetic buffers found in coevolved systems. Their impacts on sites conducive to disease can often be extreme. Management strategies designed to reduce their impacts are usually multifaceted and aggressive.

Whatever their origins, diseases may cause individual trees to suffer growth loss, reduced vigor, dieback of tops and branches, deformities, predisposition to attack by insects, and mortality. At the scale of stands and landscapes, disease may exert a profound influence on forest structure, species composition, stocking, function, and the rate and direction of vegetative succession. For example, wood decays, along with wind breakage and bark-beetle activity, are extremely important to the successional progression of Douglas-fir (*Pseudotsuga menziesii*) stands toward western hemlock (*Tsuga heterophylla*) climax forests in the Oregon Coast Range. Diseases can create standing dead, broken, hollow, decayed, and down wood; create canopy openings of various sizes; and shift nutrient and water balances. These changes then influence species diversity of forbs, shrubs, and trees.

Forest-tree diseases influence animals by creating or altering habitat. Examples include heartwood decays, which create the conditions necessary for primary cavity nesters to excavate cavities in wood, and Port-Orford-cedar root disease, which kills large, streamside Port-Orford-cedars quickly, thus eliminating an important component of riparian habitat, with consequences for stream-temperature fluctuations and long-term recruitment of in-stream structures.

The influence of disease on species composition, structure, and stocking may be subtle or direct. Root diseases in particular create openings in conifer stands where herbaceous and woody angiosperms may become established. As root diseases spread into surrounding stands, vegetative composition and stocking slowly change. Laminated root rot (*Phellinus weirii*) may be responsible for a volume reduction of as much as 50 percent at rotation age in affected areas. In recent years, Swiss needle cast has influenced the growth and vigor of Douglas-fir within the coastal fog zone on the northern Oregon coast, even eliminating its competitive advantage over other species in some locations. Changes in species composition due to diseases can be even more direct: white pine blister rust has eliminated or greatly reduced the presence of western white pine (*Pinus monticola*) on sites that are at high risk for the rust.

Forest-tree diseases are best known for their deleterious effects, and much of the forest-pathology literature addresses disease impacts in terms of losses. However, the effects of pathogens are not always bad. The way effects are perceived relates directly to management objectives. High incidence of disease may be consistent with maintaining or enhancing late-successional character in forest stands or the management of special forest products, such as mushrooms, bark, huckleberries, moss, and boughs. On the other hand, even minor amounts of disease may be unacceptable when fiber production is being emphasized or when human safety is a concern.

The idea that forest diseases cause a loss or reduction in value is a social, not a biological, concept. The changes brought about by forest-tree pathogens are neither good nor bad, except as they relate to human values. For example, a Douglas-fir stand that is heavily affected by laminated root rot might be considered by some people to be undesirable if developed facilities, such as buildings or campgrounds, are at risk from windthrown diseased trees, or if harvestable fiber is significantly reduced through mortality and slower growth of infected trees. But that same stand might be considered by other people to be highly desirable because of its distinct large openings, scattered mortality (open stand), and abundant large woody debris. These characteristics may provide big-game forage, small-animal habitat, a source for special forest products, more favorable conditions for retaining water from snowfall, increased diversity in the plant community, and a generally more varied and visually interesting area for recreation. Furthermore, biological diversity may be increased by pathogens, by selectively reducing the population of some tree species as well as opening the stand, and thus both freeing resources and creating the disturbance necessary to allow additional species to successfully compete. By influencing the structure and composition of the overstory vegetation, pathogens profoundly affect the understory vegetation, the below-ground ecosystem, and the animals that are important parts of these communities.

Many pathogens are present in coastal forests, but only a few play what appear to be critical roles. While the root diseases are the most important, a dwarf mistletoe, a canker disease, a foliar disease, and several stem decays are important to land managers in the Coast Range. In the sections that follow, we describe the basic biologies of these important pathogens, including information on conducive environments and strategy for spread, our understanding of the ecological roles of these forest diseases, and common management strategies.

For additional general information about forest disease, see Boyce (1961), Hadfield and others (1986), Scharpf (1993), Filip and others (1995), and Hansen and Lewis (1997).

Important Root Diseases

Root diseases are extremely important natural agents of disturbance in the forest ecosystems of the Oregon Coast Range. For ecosystem management objectives to be met, the biology, incidence, severity, and impact of root disease must be understood and taken into account. The influence of root diseases on forest structure, composition, function, and yield is profound. Root diseases are important gap formers; they create openings of varied sizes in the forest, depending on the pathogens and hosts involved. These gaps are then colonized by a variety of forbs, shrubs, and trees.

The degree to which root disease is perpetuated on a site is a function of the relative susceptibilities of the regeneration. If highly susceptible species continue to colonize root-disease centers, the pathogens will be maintained on the site and host trees are unlikely to reach large size. On the other hand, if immune species occupy these areas, the pathogen will die, and eventually susceptible species may again grow on the site without root disease. If intermediately susceptible species colonize the centers, pathogens will be perpetuated on the site, and the disease will continue to influence tree development according to the susceptibility of the resident and colonizing species. Susceptible species may grow more slowly, develop increased levels of butt decay, and have an increased chance of windthrow and death. Depending on site conditions, species diversity and production of special forest products may either increase or decrease.

Root diseases influence forest structure by reducing the likelihood that some trees will grow to large sizes or, at the very least, by slowing their growth. Stocking levels may be reduced in discrete areas or across the stand, depending on the distribution of inoculum and the tree species present. The standing dead trees created may be important for wildlife habitat. However, the nature of the decay associated with some root diseases means that many of these snags will remain standing for only a short time compared to trees killed by other agents. The down woody material resulting from root disease is itself important for wildlife habitat, water-holding capacity of the soil, and nutrient cycling.

Accepting the effects of root disease may be consistent with meeting resource objectives in the immediate future; mortality and spread rates are generally low and impacts are not dramatic in the short term. However, over time and space, the impacts of root disease are sufficient to require that they be considered in long-term planning. Accurately describing the pathogen(s), inoculum concentration and distribution, and population of host species of an area can be critical to understanding long-term effects. In the Oregon Coast Range, there is a legacy of many acres where susceptible species were planted on sites with high inoculum levels. The decision about whether an infected stand is entered for a regeneration harvest or thinning treatments, interplanted with resistant species, or left alone will depend on the future condition desired for the stand. In some cases the desired condition may be unattainable without extreme intervention. For example, it would be impossible to grow large trees rapidly (e.g., to create late-successional characteristics or to satisfy timber objectives) on Douglas-fir plantations with high levels of laminated root rot. To achieve these goals it would be necessary to plant other species or eliminate the pathogen.

Knowledge of the occurrence and level of root disease in an area is required to predict the impacts of root disease. Surveys, stand examinations, and resource inventories must collect at least a minimal level of information pertaining to root disease: tree species, size, and stocking; identity of root diseases present; proportion and species of trees affected; distribution of root disease in the area; and an estimate of the area affected. This information can help managers identify and select the best management options. It can also be used to predict future impacts with the aid of tools such as the Western Root Disease Model Extension to the Forest Vegetation Simulator (Frankel 1998).

Although many root diseases develop in coastal forests, five caused by fungal pathogens are of greatest concern for making forest management decisions and so have the highest priority for research: (1) laminated root rot caused by *Phellinus weirii*; (2) Port-Orford-cedar root disease caused by *Phytophthora lateralis*; (3) black stain root disease caused by *Leptographium wageneri*; (4) annosus root disease caused by *Heterobasidion annosum*; and (5) Armillaria root disease caused by *Armillaria ostoyae*.

Laminated root rot

Basic biology

Laminated root rot, caused by *Phellinus weirii*, is the most widespread and hardest-to-manage natural agent of disturbance that forest managers must deal with. It affects nearly all species of commercially important conifers in coastal stands. Douglas-fir, western hemlock, true firs (*Abies* spp.), and spruces (*Picea* spp.) are highly to moderately susceptible; pines (*Pinus* spp.), western red cedar (*Thuja plicata*), and incense-cedar (*Calocedrus decurrens*) are tolerant to resistant. Hardwoods are immune. The disease is conservatively estimated to reduce annual productivity of Douglas-fir west of the crest of the Cascade Mountains by 32 million cubic feet. A survey of 12,000 acres in the Alsea Ranger District of the Siuslaw National Forest estimated volume losses (mortality) exceeding 20 percent and predicted volume losses of over 50 percent by end of rotation. Other high-site coastal stands suffer similar losses.

When infected trees die, the pathogen continues to live saprophytically in dead roots for 50 years or more. Infection in a young stand begins when roots of young trees contact residual infested stumps and roots from the preceding stand (Figure 8-1a in color section following page 148). The fungus grows from the infected wood onto the uninfected roots and colonizes them. It then spreads between living trees through root contact (Figure 8-1b in color section following page 148). As the fungus advances along a tree's roots, the roots distal to the fungus are killed, denying the tree water and nutrients it needs for growth. Crown symptoms may appear 5 to 15 years after initial infection, but are usually not seen until at least half of the root system is affected. These symptoms include retarded leader growth, short chlorotic foliage, and distress cone crops (Figures 8-1c, 8-1d in color section following page 148). As roots decay, a tree dies standing or is robbed of structural support and then windthrown (Figures 8-1e in color section following page 148, 8-1f in color section following page 180). Windthrow is more common with this disease than with those caused by other root pathogens in the Coast Range. Laminated-root-rot centers provide wildlife habitat in the form of standing dead trees and down woody material (Figure 8-1g in color section following page 180), but because of decayed roots, the standing dead trees remain upright for only a relatively short time. Several species of bark beetles, including Douglas-fir beetles (*Dendroctonus pseudotsuqae*), Douglas-fir pole beetles (*Pseudohylesinus nebulosus*), Douglas-fir engraver (*Scolytus unispinosus*), and fir engraver (*Scolytus ventralis*), are frequently associated with disease centers and recent windthrow.

Laminated root rot is distinguished from other root pathogens that cause similar crown symptoms by the characteristic decay of roots and butts (Figure 8-1e). The decayed wood separates readily at the annual rings (Figures 8-1h, 8-1i in color section following page 180), with pits and characteristic brown setal hyphae (Figures 8-1j, 8-1k in color section following page 180) on both sides of the sheets. A crusty light-colored mycelial sheath (ectotrophic mycelia) is found on root surfaces of young trees (Figures 8-1a, 8-1m in color section following page 180) and within root-bark crevices on old trees. A distinctive brown, crescent-shaped stain is found in fresh cross sections of infected trees and roots (Figures 8-1n, 8-1o in color section following page 180).

Phellinus weirii spreads little, if at all, by windblown spores, and it can grow only a short distance into the soil. Virtually all spread is by mycelia on or within roots. The fungus persists on the site in the large roots and stumps of dead or cut trees (Figure 8-1o) and appears to effectively wall itself off in the infested wood by a protective hyphal sheath, seen as dark zone lines in cut wood (Figure 8-1i). When a site infested with *P. weirii* is regenerated with susceptible tree species, the disease nearly always causes mortality in the new stand (Figure 8-1b). Seedlings planted in contact with inoculum may be killed during the first year. Initial mortality involves scattered individuals close to old infested stumps (Figure 8-1b), but by stand age 15 to 20 years, *P. weirii* may be spreading from the original infection foci and forming disease centers that expand radially about 1 foot per year. These centers eventually appear in the stand as variable-sized, understocked openings (Figure 8-1g) that contain dead standing trees, windthrown (Figure 8-1e) or broken (Figure 8-1p in color section following page 180) trees, and occasional uninfected trees. Symptomatic trees in various stages of decline occur around the margins of centers, which commonly fill in with hardwoods, shrubs, and resistant conifers (Figure 8-1g). Susceptible conifers may regenerate in the centers, but they commonly contact inoculum at an early age and die. Through the formation of disease centers or gaps in homogenous Douglas-fir stands, this single disease greatly increases species and structural diversity in westside stands in Washington and Oregon.

Management strategies

Phellinus weirii appears to have coevolved with its hosts and is well adapted to the conditions that favor susceptible species. Laminated root rot does not appear to be restricted by topography, climate, or soil conditions in the Coast Range, and it appears to be influenced minimally, if at all, by vigor-enhancing treatments such as fertilizing or thinning. High frequency of the disease on middle to upper slopes may reflect distribution of the most susceptible hosts.

Mitigating strategies that are being tested and used operationally include regenerating diseased sites with less-susceptible species and removing or inactivating inoculum. Infested roots left in the soil may harbor the pathogen for many years. In the absence of mitigating strategies, *P. weirii* can survive on a site as long as consecutive rotations of Douglas-fir or other highly susceptible species continue there. In unmanaged stands, infected trees often fall over, pulling much of the inoculum out of the soil. However, in a managed stand, stumps are created in which the pathogen survives for long periods; indeed, these may remain on the site longer than one rotation. Under a management regime that ignores the presence of *P. weirii*, the inoculum on a site, in the forms of residual infested stump and root wood, may increase over time.

Distribution of the pathogen in an infested stand has generally been perceived as clustered around disease centers. This perception has led to a mitigative strategy of cutting a buffer strip of host trees around centers. However, recent mapping of infested stands has provided evidence to suggest that the distribution of infected trees is sometimes much more diffuse than previously thought. A diffuse pattern of distribution would limit the effectiveness of buffers and may mean significantly more infected trees and higher long-term losses than expected due to reduced growth in diseased stands. Changing the tree species grown on the site to one that is less susceptible can be expected to reduce disease-caused mortality, but that may need to be done on a much larger scale than previously expected. For additional information concerning laminated root rot, see Wallis (1976), Hansen (1978), Tkacz and Hansen (1982), Hadfield (1985), Hadfield and others (1986), Thies and Sturrock (1995), and Frankel (1998).

Port-Orford-cedar root disease

Basic biology

Port-Orford-cedar root disease, caused by the introduced pathogen *Phytophthora lateralis*, is a severe threat to Port-Orford-cedar on sites favorable for the pathogen. It is also known to attack Pacific yew (*Taxus brevifolia*) in areas of high inoculum.

Port-Orford-cedar is native to a narrow coastal strip in southern Oregon and northern California. It grows on a wide variety of sites, including streambanks, bogs, coastal sand dunes, deep productive soils, and dry sites. Port-Orford-cedar needs a consistent supply of water and is an important species in riparian ecosystems. In such environments, it provides streamside shade, and its snags and logs are long-lasting components of terrestrial and aquatic wildlife habitat. It is often the only large-tree component (Figure 8-2a in color section following page 180) on soils that are high in iron and magnesium but low in calcium (i.e., serpentine-like soils). Also, Port-Orford-cedar has important cultural value to the indigenous people of the area. Despite a limited domestic market, the export market for Port-Orford-cedar has been substantial for decades, making it the most valuable conifer timber species in North America.

The pathogen was introduced to the native range of Port-Orford-cedar in 1952. Mortality soon became conspicuous near coastal towns and along major roads. *Phytophthora lateralis* is now present throughout the range of Port-Orford-cedar; infected and healthy trees are intermingled. Where the disease has been present for several decades, mortality has been extensive, especially along streams and drainage ditches. On drier microsites, mortality is less frequent and often absent.

The first external symptom of Port-Orford-cedar root disease is a slight discoloration of the host foliage, which may within months progress to yellow, then bright red, then dark red, and finally brown (Figure 8-2b in color section following page 180). The foliage may hang on the tree for several years. The disease is best identified by a cinnamon-colored stain in the inner bark and cambium that abruptly joins creamy-white, healthy inner bark in roots and lower boles of declining trees (Figure 8-

2c in color section following page 180). Once an infected tree starts to decline, it may be attacked by cedar bark beetles (*Phloeosinus* spp. *(Scolytidae))*.

The pathogen is spread mainly by human transportation of infected seedlings and infested soil into previously disease-free sites. Although livestock and wildlife may transport the pathogen in soil on their feet, that is believed to be only a minor factor in most situations. The most common method of spread is the movement of soil on machinery and other vehicles from an infested to a noninfested area during road building, logging, or travel through the forest.

The pathogen is an oomycete, or "water mold," and is spread by water-borne spores (zoospores), which can swim a short distance; they move in water films in the soil and are attracted to the uninjured root tips of cedar. They are totally dependent on free water for infection and spread and cannot survive in dry soil. Virtually all natural spread is downslope or downstream in water moved by gravity (Figure 8-2d in color section following page 180). After the zoospores germinate, mycelium penetrates the roots, kills the phloem, and spreads internally to the base of the tree and to other roots. If the roots of infected and adjacent trees are grafted, the mycelium can grow from root to root, with an estimated movement of about 6 inches per year; however, this is a minor contributor to disease spread. In addition to the zoospores, the pathogen also produces durable resting spores within infected phloem. As a dead root deteriorates, these spores are released to the soil, where they may remain viable for at least six years under favorable conditions.

Management strategies

The importance of Port-Orford-cedar within its range and the fact that *P. lateralis* is an aggressive exotic pathogen makes it critical that this disease be managed aggressively, regardless of resource objectives. Mitigation strategies involve preventing the movement of water or soil from an infested site to the root zone of healthy trees. Such strategies include closing roads and restricting operations to reduce movement of infested soil, cleaning vehicles and equipment before entering or leaving specified areas to remove soil that may contain spores, and building berms along roadsides to reduce splash and runoff. Other mitigation strategies involve the host, such as retaining Port-Orford-cedar on microsites not favorable to the pathogen; removing Port-Orford-cedar from roadside strips to prevent infestation of disease-free stands; and, where the pathogen is already present, removing the host to allow the fungus die in its absence. Port-Orford-cedar with some resistance to *P. lateralis* have been identified; however, they are only tolerant and slower to die, not immune. Researchers are working aggressively to identify additional resistant individuals throughout the range of the species and to breed for enhanced resistance. The fungicide metalaxyl is an effective protectant that can halt early infection. Although this preventive treatment has been demonstrated on ornamentals, the scale of application makes its use impractical in the forest. For additional information concerning Port-Orford-cedar root disease, see Zobel and others (1985), Hadfield and others (1986), Roth and others (1987), and Kliejunas (1994).

Black stain root disease

Basic biology

Black stain root disease is caused by the fungus *Leptographium wageneri* (the sexual stage is *Ophiostoma wageneri*, which is rarely found). This disease is unique among the root diseases affecting Oregon's forests, because the pathogen is spread by root-feeding insects and because it is a vascular wilt disease that does not actually cause decay of root or stem tissue. Black stain root disease is caused by three recognized varieties of *L. wageneri*; only one variety is of concern in the Oregon Coast Range, *L. wageneri* var. *pseudotsugae*. This disease was largely unrecognized in the Pacific Northwest prior to 1969, but it is now found in most parts of Washington and Oregon. The disease is common in west-side stands and causes a significant amount of mortality in stands in southwestern Oregon, where about 25 percent of the Douglas-fir plantations have the disease and about 10 percent of those are suffering severe losses.

Black stain root disease may cause many of the same symptoms as other root diseases, but it often kills host trees more quickly. The disease is characterized by a black stain, which may appear as streaks, in the sapwood (Figure 8-3a in color section following page 180). The stain is usually limited to one or two growth rings and does not extend radially towards the center of the tree as do other insect-introduced stains (Figure 8-3b in color section following page 180). Due to general discoloration, this characteristic stain becomes indistinct fairly quickly after a tree dies; therefore, the highest probability of accurately identifying the disease occurs when the trees are still symptomatic or recently dead.

Douglas-fir is by far the most common host of black stain root disease in the Pacific Northwest, especially in coastal forests, but other conifer species, especially pines, may also be affected. The disease appears in 10- to 25-year-old Douglas-fir plantations or in younger, dense natural stands. Douglas-fir over 30 years old are less likely to be infected and killed.

Disease centers appear as small (usually 0.1 acre or less) groups of dead and symptomatic trees. Dead trees remain standing unless they are also affected by other root diseases, and centers may occur as clusters of standing dead trees within a stand (Figure 8-3c in color section following page 180). The clustered distribution may reflect insect vectoring of the fungus. Insect vectoring occurs when adult root-feeding bark beetles and weevils carry spores into the trees (Figure 8-3d in color section following page 180). Fruiting bodies of the fungus form in the egg-laying galleries of the insects; as mature insects emerge, sticky spores adhere to the insects and are dispersed by them. The spore-carrying bark beetles and weevils feed and breed in roots of less-vigorous trees, so the occurrence of black stain root disease in disturbed areas probably reflects the insect's preference for stressed or injured trees.

Although long-distance spread involves insects, the fungus, once introduced into a tree, can move to new trees across root contacts or by growing up to 6 inches through the soil. Tree-to-tree spread does not depend on the low vigor of host trees, and centers may expand radially at a rate of nearly 5 feet per year. However, the fungus is relatively nonpersistent; survival appears to be limited to a year after the tree dies.

Management strategies

Black stain root disease is usually found in areas of site disturbance or tree injury, such as in pre-commercially thinned stands, along roads and skid trails, or where rotary-blade brush cutters have been used to clear roadsides. The insects that carry the fungus prefer injured or stressed trees; therefore, minimizing severe wounds to trees or disturbances such as soil compaction are important strategies for managing this disease. For example, where soils are particularly prone to compaction, high-lead and skyline yarding would be preferred over tractor logging. If tractor logging is necessary, the risk of serious compaction can be reduced if skid trails are designated, the area covered by skid roads is minimized, trees are felled to the yarding lead, and yarding is restricted to the dry season. New road construction through Douglas-fir plantations may injure trees and encourage the spread of black stain root disease. If construction occurs or if old roads are reopened, removing injured trees and those whose roots are covered by sidecast fill will reduce the risk of disease; this strategy would preclude the use of equipment that damages trees, such as rotary-blade brush cutters, on roads adjacent to Douglas-fir plantations.

When thinning in high-risk areas, particularly those within 1 mile of known black stain root disease, schedule thinning in juvenile stands between late June and early September. This is the period after the insects that carry the pathogen have already emerged and established themselves for breeding in other dead, dying, or down trees. Thinning at that time will attract fewer insects into thinned stands. Additionally, juvenile wood on the ground before September will dry sufficiently so that it will not still be attractive to insects the next spring.

Planning for a diverse mix of species is the best approach in areas where black stain root disease is a concern. For example, in plantations where the disease is already causing mortality, resistant or immune species can be retained during thinning. Where no resistant species are present, the spread of the disease can be minimized by avoiding active infection centers and creating a 30-foot no-thin buffer around them.

As with other root diseases, high levels of inoculum of black stain root disease in relatively pure stands of susceptible species may preclude adequate stocking to meet management objectives. The nonstocked or low-stocked areas created by the disease may eventually develop into stands that are diverse in species composition and structure, but development will be slower than if preventive strategies had been implemented. For additional information concerning black stain root disease see Goheen and Hansen (1978), Hansen (1978), Witcosky and Hansen (1985), Hadfield and others (1986), Hansen and others (1988), and Hessburg and others (1995).

Annosus root disease

Basic biology

Annosus root disease is caused by the fungus *Heterobasidion annosum* (previously *Fomes annosus*). Two forms of the fungus are recognized in Oregon: the "P-group," which affects chiefly pine, and the "S-group," which affects chiefly western hemlock, true firs, and spruce. All west-side conifer species can be infected by *H. annosum*, but they differ in degree of susceptibility and damage; hardwoods are rarely attacked. In Coast Range stands, western hemlock and true firs are commonly affected, and infection levels are high—often 20 percent or more of the trees are affected. Trees are less frequently killed outright by annosus root disease than by other root diseases. However, trees weakened by this disease may also be attacked by bark beetles and killed. Infection results in reduced strength and wood value due to decay, stain, and breakage; windthrow and stem breakage occur frequently. These effects are usually most noticeable after trees are at least 120 years old; effects in young stands tend to be minor unless trees have been damaged. *Heterobasidion annosum* appears to contribute significantly to pathogenic rotation in coastal stands, affecting stand structures and age-class distributions by causing the decline of older western hemlock, Sitka spruce, and true firs. Douglas-fir are only slightly susceptible and are rarely damaged in coastal stands; however, Douglas-fir is a very common host in the intermountain region.

Annosus root disease is difficult to diagnose because many trees do not show outward symptoms. Those that do show symptoms typical of other root diseases: chlorotic or thinning foliage, reduced growth, and production of distress cone crops. The most reliable way to diagnose annosus root disease is to find fruiting bodies or conks of the fungus, usually on trees in advanced stages of decline. Conks are often small with a dark brown to black upper surface, a white to cream pore surface with rounded pores, and a white-cream sterile margin (without pores). Conks usually occur in hollows of infected stumps (Figure 8-4a in color section following page 180), and they are frequently found in the duff layer on the exterior of roots and root collars of infected trees (Figure 8-4b in color section following page 180). Decay is stringy with large white streaks and scattered black flecks (Figures 8-4c, 8-4d in color section following page 180). In more advanced stages on some hosts, decayed wood often separates at the annual rings (Figure 8-4d). This kind of advanced decay is similar to the delaminating decay caused by *P. weirii*; however, with annosus root disease, setal hyphae are not formed, and the wood is usually pitted on only one side of the lamina. This delaminated decay is more common in true firs than in western hemlock. In hemlock, advanced decay more commonly appears as a white, stringy or spongy rot containing numerous black flecks.

Heterobasidion annosum infects its hosts in two ways: by mycelial growth across root-to-root contacts and by windblown spores. Spore infection of freshly cut stumps is the primary way that new infection centers start. In coastal stands, stumps can become infected at any time of the year. After growing vegetatively down through the stump and root system, the fungus can move to adjacent trees across root contacts, but it does not grow through the soil. Nevertheless, infection centers can expand at 1 foot per year. The fungus can remain viable in large stumps for at least 50 to 60 years, but in stumps that are less than 6 inches in diameter inside the bark, the fungus dies out within a few years. The fungus can also infect wood exposed when living trees are wounded. The rate of spread is two to three times greater upward than downward in a tree when the fungus colonizes an injury (Figure 8-4e in color section following page 180).

The incidence of annosus root disease increases with incidence of stand entry; that is, damage is greater in stands that have had multiple entries than in stands that have been entered only once. One reason is that with more entries, more stumps are created that become infected and initiate disease centers. Wounds created during road building, harvesting, and yarding are additional entrance points for the pathogen.

Management strategies

Heterobasidion annosum is common in coastal forests, and spores can be detected readily by trapping them from the air. However, damage is minimal in unentered stands, since disease incidence relates directly to the presence of wounded trees or fresh stumps. Effects of *H. annosum* can be minimized where timber production is the main objective, by managing damage-prone species on short rotations (40 to 120 years), by favoring more-resistant species on appropriate sites, and by limiting tree wounding.

In developed recreational areas, risk of tree failure can be reduced by monitoring highly susceptible species, especially if wounded, for annosus root disease and removing the tree if there is an unacceptable amount of decay.

The impacts of annosus root disease can be minimized by treating recently cut stump surfaces with Sporax®, a registered borax fungicide with the active ingredient sodium tetraborate decahydrate. Borax fungicides chemically alter the surfaces of stumps, making them inhospitable for *H. annosum* spore germination and infection. Effectiveness of the fungicide is reduced in areas of high rainfall, which makes these compounds less useful in coastal forests than in interior forests.

With increasing emphasis on longer rotations, thinning, and management of late-successional forests, impacts of annosus root disease can be expected to increase. Extensive decay and stem breakage of older susceptible trees may be desirable for achieving some objectives, such as the development of wildlife habitat. However, where long-term retention of trees is desired, annosus root disease poses a challenge. Carefully managing mixtures of species not prone to the disease may be the solution. For additional information concerning annosus root disease, see Goheen and others (1980), Hadfield and others (1986), Otrosina and Scharpf (1989), Frankel (1998), and Woodard and others (1998).

Armillaria root disease

Basic biology

Armillaria root disease of conifers, caused by the fungus *Armillaria ostoyae,* is widespread in the Coast Range and can affect all native species of conifers; however, it causes limited damage. In general, *A. ostoyae* behaves quite differently east of the crest of the Cascade Range, where it is a widespread root pathogen and at times very aggressive, killing many host trees. While this type of behavior can occur in the Coast Range, it is less common.

Armillaria root disease is often associated with trees under stress. It is frequently found on sites with compacted soils, along skid roads, where trees have been poorly planted (Figure 8-5a in color section following page 180), or where there has been a poor match of stock to site. It is most frequent and has the greatest impacts in stands younger than 30 years old, particularly Douglas-fir and noble fir (*Abies procera*) plantations (Figure 8-5b in color section following page 180). Pines, hemlocks, and cedars are tolerant or resistant. Armillaria root disease on hardwoods is most likely caused by species of *Armillaria* other than *A. ostoyae.*

In Coast Range forests, conifers with Armillaria root disease may exhibit thin or chlorotic foliage, distress cone crops, and a flow of pitch just above ground level (Figure 8-5c in color section following page 180). Honey-colored mushrooms may be produced in the fall at the base of infected trees (Figure 8-5d in color section following page 180). The decay is a yellow stringy root and butt rot, especially in nonresinous conifers such as western hemlock and true firs. The flow of pitch is the tree's response to the fungus growing beneath the bark. Extensive growth of the fungus eventually girdles and kills the tree. Chopping into the bark reveals white to cream sheets of fungus called mycelial fans (Figure 8-5e in color section following page 180). These fans can be found beneath the bark of roots and root collars, but never on the outside of the bark, of symptomatic trees. The fungus also often produces black shoestring-like structures called rhizomorphs (Figure 8-5a) that can be found in the soil around infected roots and under bark at the base of dead trees. Rhizomorphs are root-like in form, up to 0.05 inch in diameter, and have a white, nonwoody core.

Armillaria ostoyae is often confused with other species of *Armillaria* that are found on already-dead hosts or trees dying from other causes. Specifically, white fans found beneath the bark of trees that died because of suppression may be mistaken for those caused by *A. ostoyae*. The presence of host reactions such as symptomatic foliage, basal resinosis, and thick, robust mycelial fans must be used to positively identify *A. ostoyae*.

Armillaria ostoyae can spread by spores, but that is uncommon. Most spread is by the mycelium, which can survive for up to 35 years in old-growth stumps and roots before being replaced by other organisms. Survival of the fungus is influenced by stump size, tree species, and habitat type. Larger stumps provide a more substantial food base and longer survival. Small stumps of precommercial size are not effective inoculum sources. The fungus spreads to trees in a new stand and between trees by growing across root contacts and, to a much lesser extent, by rhizomorphs, which can grow a short distance through the soil to infect roots of susceptible trees. Once in a root, the fungus spreads both proximally and distally. Some trees resist the fungus at the time of infection by walling it off with callus tissue. However, if these trees subsequently become stressed or are cut, the fungus may grow through the callus tissue and spread rapidly in the roots.

Armillaria root disease centers may take a number of years to develop. Spread rates are about 1 foot per year. Some centers spontaneously become inactive, but others attain very large size. The primary effect of Armillaria root disease is that it kills the infected trees; in some cases it may initially cause butt rot and reduced growth. Infected trees may be windthrown, but they more often die standing. Armillaria root disease centers that occur in older stands appear to retain standing dead trees for relatively long periods, so they do not quickly result in large quantities of down woody material. Bark beetles may attack trees weakened by the disease and thus add dead trees to the expanding centers.

Management strategies

Strategies designed to maintain vigorous growth that are likely to reduce the effects of Armillaria root disease include using resistant species where possible, and minimizing the number of stand entries. Close examination of affected sites may be necessary to determine the conditions responsible for tree stress so that these may be avoided on similar sites in the future.

Because Armillaria root disease usually occurs early in the life of the stand, it is important to have a quality planting effort that uses resistant species (where appropriate) and stock that is carefully matched to the site. Depending on the distribution of the disease and the proportion of susceptible hosts, stocking goals may be seriously affected. Thinning stands with low levels of infection may increase tree vigor enough to compensate for disease-caused mortality. Favoring resistant species during these entries will help reduce future losses to disease. To ensure adequate stocking, avoid thinning in stands that have high levels of disease. Interplanting with resistant species may be needed, although stand development may be slowed and additional time may be needed to achieve the desired future condition. For additional information concerning Armillaria root disease, see Morrison (1981), Hadfield and others (1986), Shaw and Kile (1991), and Frankel (1998).

Other Forest Diseases

Western hemlock dwarf mistletoe

Basic biology

Dwarf mistletoes are parasitic, seed-bearing vascular plants that infect conifer species. These parasites survive only on live hosts and obtain all their water and most of their food from their hosts. Western hemlock dwarf mistletoe (*Arceuthobium tsugense* subsp. *tsugense*) is common in the Oregon Coast Range. It occurs principally on western hemlock and associated noble fir and Pacific silver fir (*Abies amabilis*). Other species may also become infected; grand fir (*Abis grandis*) is an occasional host and Sitka spruce (*Picea sitchensis*), western white pine, Douglas-fir, and mountain hemlock (*Tsuga mertensiana*) (Figure 8-6a in color section following page 180) are rare hosts.

Dwarf mistletoes have separate male and female plants. Female plants produce berries (Figure 8-6b in color section following page 180) in the fall which, when mature, forcibly discharge their seed an average distance of about 15 feet. Seeds released from the tops of trees can be carried more than 100 feet by wind. Long-distance dispersal, though rare, is accomplished by birds and mammals.

The seeds are sticky. When they land on host needles they stick, but with the addition of moisture, they eventually slide down the needle and lodge at the base of the needle on the twig. The germinating seed mechanically penetrates the relatively young, thin host tissues. Within two years of the dwarf-mistletoe infection, the infected tree branch develops a localized spindle-shaped swelling (Figure 8-6c in color section following page 180). After another one to two years, small yellowish-green buds break through the bark and develop into aerial shoots that are usually 2 to 3 inches long, segmented and nonwoody, with scale-like leaves. Flowering occurs in late summer, and insects are the primary pollinators. The fruit matures by fall of the year following flower production.

Host branches usually react to dwarf-mistletoe infection by producing a "witches' broom" (Figure 8-6c)—an abnormal proliferation of many small twigs on the branch that appears as a clustered mass of twigs and foliage. Young infections cause small brooms that are not readily observed, but older brooms may weigh several hundred pounds (Figure 8-6d in color section following page 180). Brooms monopolize water and nutrients at the expense of the rest of the tree. Swelling of the bole may result from direct infection or growth of the parasite from an infected branch. Wood-decay fungi often enter bark cracks and openings in these swellings.

Although witches' brooms are the most conspicuous evidence of infection, aerial shoots are the most positive diagnostic characteristic. Shoots are larger and more abundant on open-grown trees or in upper crowns exposed to sunlight (Figure 8-6e in color section following page 180). Shoots may be absent or sparse and poorly developed on older infections, on lower branches, and especially in the very dense stands common in the Coast Range. Even when aerial shoots are absent, basal cups (a residual structure on the branch at the point of shoot attachment) may be present on the spindle-shaped swellings of infected branches.

Western hemlock dwarf mistletoe infection causes branch and top dieback (Figure 8-6e), reduced growth rates, stem deformities, and increased decay and mortality. Effects are greater with increasing numbers of infections in a tree. Effects can be very significant on heavily infected older trees (trees more than 150 years old) or trees of any age that have numerous infections in the upper third of their crowns (Figures 8-6e, 8-6f in color section following page 180).

Western hemlock dwarf mistletoe plays several ecological roles. Where the disease is severe, it contributes to a pathogenic rotation for western hemlock by causing substantial decline and mortality of old infected trees. It limits the development of a western-hemlock understory in infected stands and gives nonhosts distinct growth advantages over the infected western hemlocks. Furthermore, dwarf mistletoe-induced brooms and large infected branches provide preferred habitat for several wildlife species, including marbled murrelets.

The rate of spread of western hemlock dwarf mistletoe in the forest depends on complex and interrelated factors, including the composition of tree species, structure of the stand, and spacing of the trees. Understory hosts that are constantly exposed to dwarf mistletoe seeds from infected overstory trees will themselves have many infections. Dwarf-mistletoe plants spread fastest and develop best in relatively open, multistoried, pure western hemlock stands where selective logging has left infected residual trees in a wide range of size classes. A regeneration harvest that leaves many small infected residual trees will also result in rapid spread of the mistletoe.

The lowest infection rates are associated with either very slow or very rapid tree growth. Trees that grow slowly develop dense, compact foliage that intercepts the seeds; thus fewer mistletoe seeds are dispersed between trees. Trees that grow rapidly shed internal infected branches rapidly, and tree height growth outpaces dwarf-mistletoe movement within the tree. The spread of western hemlock dwarf mistletoe is slowest in dense, single-storied stands of pure western hemlock and in stands with large components of nonhost species.

Management strategies

Management strategies for stands with western hemlock dwarf mistletoe depend on resource objectives. If the goal is to grow well-formed western hemlocks of average or better height, volume, and life span, then it is appropriate to use management strategies to maintain dwarf mistletoe at low frequency levels and limit its spread to young hosts. Strategies might include harvesting infected trees or stands, favoring nonhosts in infected areas, or living with the disease if intensity is light, by managing on short rotations (40 to 120 years). To reduce the likelihood of infection, remove infected overstory trees before western-hemlock regeneration is 10 feet tall. Well-scattered infected regeneration will result in spread to intervening regeneration during the course of stand development; the higher the proportion of infected regeneration, the faster the spread. Many infections will be missed if regeneration is examined too soon after overstory removal, because one- to two-year-old infections may not be visible. Trees with the highest overall levels of dwarf mistletoe infection and those with infections in the upper portions of their crowns have the highest priority for removal.

To meet other management objectives, maintaining some level of western hemlock dwarf mistletoe in the stand may be acceptable or even desired. These situations represent significant planning challenges, particularly where stands must be managed on long rotations. It is a fine balance to maintain some dwarf mistletoe but not so much that long-term impacts are severe enough to prevent desired future conditions from being met. Some combination of treatments will have to be used, although our experience regarding the choices is limited. Prescriptions that promote nonhosts, make use of strategically placed group-selection cuts, and maintain low levels of infection in even-aged stand components may hold promise. For additional information concerning western hemlock dwarf mistletoes, see Smith (1969, 1971, and 1977) and Hawksworth and Wiens (1996).

White pine blister rust

Basic biology

White pine blister rust is a branch- and stem-canker disease caused by the rust fungus *Cronartium ribicola.* It affects five-needle pines. Damage includes topkill, branch dieback (Figure 8-7a in color section following page 180), predisposition to attack by other agents (including bark beetles), and mortality. Additional symptoms include chlorotic needle spots (Figure 8-7b in color section following page 180), spindle-shaped branch swellings (Figure 8-7c in color section following page 180), aecia on branch or stem cankers (Figures 8-7d, 8-7e in color section following page 212), and heavy resin flow from stem cankers (Figure 8-7f in color section following page 212).

The rust is native to Asia but was introduced to western North America via Europe on a shipment of seedlings in 1910. It then quickly spread throughout most of the range of five-needle pines. These pines have been all but eliminated in some coastal stands, and their numbers have been seriously reduced in others. In many areas, the only remaining natural hosts are large trees that have had their tops and numerous branches killed by the fungus. Trees in this condition are frequently predisposed to attack by mountain pine beetle (*Dendoctonus ponderosae*). Because five-needle pines have economic value, are important elements of forest diversity, and have such special virtues as resistance to laminated root rot, interest in managing white pine blister rust is high.

The life cycle of the fungus is complex: it takes four to five years to complete and involves five spore forms. The fungus is a parasite that requires a live host to survive in nature and an alternate host in the genus *Ribes* (gooseberries and currants) to complete its life cycle. Yellow-orange pustules or blisters on pines produce aeciospores (Figure 8-7d) annually in the late spring. These thick-walled, long-lived spores are fairly resistant to radiation and desiccation; as a result they are capable of long-distance travel. Aeciospores land on and infect the leaves of *Ribes*, which support annual infections that produce urediospores in midsummer (Figure 8-7g in color section following page 212). Urediospores infect other *Ribes* leaves and build up the population of the rust in local areas. Teliospores are produced on the underside of infected *Ribes* leaves in late summer and early fall. These germinate in place and produce basidiospores. Basidiospores are spread via wind to pine needles, where they land and germinate. These thin-walled, short-lived spores require cool temperatures and moist conditions for survival and germination; clouds and fog in late summer and early fall increase the chance of successful movement from *Ribes* back to pine hosts. The remaining spore form, pycniospores or spermatia, are produced on pines at the margins of the cankers. Although these spores are incapable of initiating new infections, they probably have a sexual function.

The distribution, frequency, and association of the hosts, as well as micro- and macroclimatic conditions, are important factors in the spread of the rust. Dense populations of *Ribes* increase the probability of high production of basidiospores and infection of local pine populations. Certain *Ribes* species are more susceptible to infection than others. Large-scale weather events that bring moist conditions to a region can increase levels of infection and long-distance travel of fragile spores. Topographic features combine with microclimate, such as saddles or mountaintops where summer fog is frequent, to influence infection locally.

Management strategies

Current management strategies in coastal forests include planting five-needle pines with various levels of resistance to *C. ribicola* on appropriate sites. Tree-improvement programs to screen apparently resistant pines and to breed for higher levels of resistance have existed for some time. Because of the fungus' ability to mutate, special care is taken to maintain a variety of resistance mechanisms in the breeding programs. Other strategies to manage white pine blister rust include the use of risk-rating systems to match levels of resistance in planting stock to local site conditions. Also, the lower branches of planted five-needle pines may be pruned to prevent stem infections and to make the microclimates in plantations less favorable for the pathogen. For additional information concerning white pine blister rust, see Van Arsdel and others (1956), Boyce (1961), Kimmey and Wagener (1961), McDonald and others (1984), and Hunt (1998).

Swiss needle cast

Basic biology

Swiss needle cast, caused by the fungus *Phaeocryptopus gaeumannii,* is native to North America, attacks only Douglas-fir, and occurs throughout the Douglas-fir region. Probably introduced to Europe from North America, the disease and the fungus that causes it were first described in 1925 on Douglas-fir trees planted in Switzerland. During the next 20 years, the fungus spread throughout Europe. The disease was not reported in the United States until 1938, but its presence on herbarium specimens of Douglas-fir collected in California in 1916 and common occurrence on Douglas-fir in the western United States led to the conclusion that the fungus is native but not damaging in the range of Douglas-fir. In the 1970s, however, Swiss needle cast emerged as an important disease of Christmas trees in the Pacific Northwest, and in the 1980s it was identified as causing damage in young forest plantations in some coastal areas. Current concern about the disease started in the 1980s, when needle loss was severe in young Douglas-fir plantations in the Tillamook area. Damage is now severe in scattered plantations and natural stands along the Oregon and Washington coasts. Aerial surveys of coastal Oregon forests, conducted by the Oregon Department of Forestry, identified 130,000 acres in 1996, 145,000 acres in 1997, and 173,000 acres in 1998 that showed obvious symptoms of the disease. Most stands with severe discoloration occurred within about 18 miles of the coast, in the fog belt or spruce/hemlock zone.

The disease cycle begins in spring when spores are released from fruiting bodies on older diseased needles. Spores are carried on air currents to newly emerged needles. When adequate moisture is present, the spores germinate and the fungus enters the needle through the stomata. The major period for infection is weather dependent, but is usually from mid-May through the end of June, coinciding with needle emergence and early needle elongation; however, infection may occur at other times of the year. Symptoms do not usually appear the first year after infection, except in severely diseased trees. The fungus ramifies through the needle, and eventually the fruiting bodies form in the stomata in the fall and winter and mature in the spring. They appear as two soot-like streaks on the underside of the needle, one along each side of the midrib (Figure 8-8a in color section following page 212). The individual fruiting bodies are black and spherical (up to 0.04 in. in diameter) and easily seen with a hand lens (Figure 8-8b in color section following page 212). Black specks not associated with stomata are not Swiss needle cast.

The fungus attacks only the host's needles, causing them to yellow and drop prematurely. Some trees may retain only the current year's needles, and others may retain three or more years' foliage (Figures 8-8c, 8-8d in color section following page 212), in contrast to the four-year complement retained by healthy Douglas-fir. Trees that have been damaged for many years have epicormic branches and narrow, decrepit crowns. However, the only reliable sign or symptom of the disease is the presence of the fruiting bodies. Although trees seldom die from Swiss needle cast alone, the disease does reduce growth, which reduces wood production and affects stand development.

Diseased stands appear yellow to brownish-yellow when viewed from a distance (Figure 8-8e in color section following page 212). Discoloration is most noticeable from late winter until bud break. When needle loss is severe, the additional light reaching the forest floor causes lush growth of understory plants.

Swiss needle cast has long been considered an innocuous disease caused by a weak native pathogen. The current outbreak is not well understood and is the subject of a significant research effort. One or more of the following factors may be responsible: (1) Douglas-fir has been planted to pure stands and at high densities in areas of the fog belt where it was previously only a minor component of the natural stand; (2) some of the plantations have been established with off-site seed; (3) as the disease intensifies, the presence of more spores increases the probability of infection; and (4) recent changes in climatic conditions may favor intensification of the disease.

Management strategies

Although there are no research-proven management strategies, land managers can take some reasonable precautions:

(1) In areas where the disease is causing stand discoloration, discriminate against Douglas-fir when either planting or thinning.

(2) Use seed from the local seed zone and elevation—do not deviate by even 500 feet in elevation or 1 mile in latitude.

(3) If the stand is of commercial size, harvest and convert to species other than Douglas-fir.

Fungicides have been used effectively to control Swiss needle cast in Christmas-tree plantations; however, due to its difficulty of application and high cost, fungicide is not considered an effective operational treatment in forest stands at this time. For additional information concerning Swiss needle cast, see Nelson and others (1989), Kanaskie (1994a, 1994b, 1998), Hood (1997), Filip (1998), and Maguire and others (1998).

Common stem decays

Wood decay is an important ecological process. As stands mature, decay organisms may cause tree death or breakage and thus create canopy gaps. Decay fungi decompose wood into components that, once on the ground, increase soil moisture-holding capacity and nutrient availability. Decayed wood provides opportunities for nesting, resting, and foraging by many wildlife species. For example, wood softened by decay fungi is more easily excavated by primary cavity-nesting species such as woodpeckers and nuthatches. Vacated cavities, broken tops, and hollowed logs are used by many other wildlife species.

At the same time, decays also cause loss in economic value and are responsible for the loss of millions of board feet of timber each year in North America. Decay of sound wood reduces wood quality and renders timber unmerchantable. Breakage and windthrow reduce stocking. While these losses occur most frequently in old-growth forests, second-growth stands may be seriously affected.

Stem decays are widely distributed in coastal forests and are caused by many species of fungi. Many of these fungi break down only the nonliving heartwood portion of living trees. Others decompose heartwood or sapwood of dead trees only; decompose slash, roots, and organic matter in the soil; or cause wood in use by humans to deteriorate.

Brown rots develop as the fungus selectively uses carbohydrates (primarily cellulose), leaving behind the brownish lignin component of the wood. Brown rotted wood is usually dry and fragile; it tends to crumble readily or break down into cubes. Most brown rots form solid columns of decay, though some form pockets.

White rots are produced by fungi that attack both the carbohydrate and the lignin components of the wood. They may form in pockets or cause the wood to be stringy. Every tree species is susceptible to at least one heart-rot fungus while alive; many fungal species are also associated with dead trees. As a rule, heart-rot fungi do not penetrate sound trees but invade through openings into the heartwood. Any opening into the heartwood or exposure of dead sapwood next to heartwood is a potential infection site for decay fungi. Wounds caused by fire, weather, animals, or human activities are common points of entry for decay fungi. Natural openings in trees, such as branch stubs, open knots, and dead branches, also provide entry. Some decay fungi enter the tree through injured roots or through basal fire scars. Others kill the root wood before entering the heartwood.

Decay fungi often go unnoticed. One reason is that early stages of decay may be very difficult to detect. While some incipient decays can be seen as distinct color changes in wood, others are much less obvious or virtually invisible. If conks are present, decay may be apparent; however, very small, viable, and sporulating conks may escape casual inspection, and annual conks may not be apparent at all during part of the year.

The decays causing the greatest impact in coastal forests are all caused by fungi and include red ring rot (caused by *Phellinus pini*), rust-red stringy rot (caused by *Echinodontium tinctorium*), schweinitzii butt rot (caused by *Phaeolus schweinitzii*), and brown crumbly rot (caused by *Fomitopsis pinicola*). Other decays of concern include brown trunk rot (caused by *Fomitopsis officinalis*), brown cubical rot (caused by *Laetiporus sulphureus*), and yellow-brown top rot (caused by *Fomitopsis cajanderi*).

Basic biology

Red ring rot, caused by *Phellinus pini*, is widespread throughout the Pacific Northwest. In western North America, it is the most common cause of heart rot. Hosts include Douglas-fir, pines, hemlocks, spruces, true firs, and western redcedar, and rarely incense-cedar. The decay is a cellulose- and lignin-destroying white pocket rot that usually occurs in the heartwood and occasionally enters living sapwood. Incipient decay is usually red-purple in Douglas-fir, light purplish gray in spruce, pink to reddish in pines, and colorless in cedar. Advanced decay appears as spindle-shaped white pockets with firm wood in between (Figures 8-9a, 8-9b in color section following page 212). Pockets may coalesce and in the later stages may form distinct rings or crescents (Figure 8-9c in color section following page 212), hence the name "red ring rot." Eventually a void will be created. Zone lines are sometimes produced.

Fruiting bodies occur on tree boles (Figure 8-9d in color section following page 212). They are perennial, hoof-shaped to bracket-like, and produced on the stem at branch stubs and knots (Figure 8-9e in color section following page 212). They range from 2 to 12 inches in width. The upper surface is rough, dark gray to brownish-black, and concentrically furrowed. The interior and lower surface are cinnamon-brown. Pores are irregular in shape (nearly round to maze-like or angular). Swollen knots filled with fungal tissue (punk knots) may be present on stems. Stems may be slightly flattened or "dished out" in areas associated with conks and punk knots.

Windborne basidiospores are disseminated from fruiting bodies. Infection of living trees is not well understood but is believed to result from basidiospores germinating on branch stubs and producing mycelium that grows through the branch stub into the heartwood.

Red ring rot is more severe in southern Oregon than in other parts of the Coast Range. In general it is also more severe in older stands, in pure stands of a host species, on steep slopes, and on shallow soils. Conks occur higher on trees in older stands. Larger conks usually indicate more decay. The fungus does not typically invade dead wood and quickly dies in cut lumber.

Rust-red stringy rot, caused by *Echinodontium tinctorium* (also known as Indian paint fungus), is distributed throughout the western United States on true firs and western hemlock. While found throughout the Oregon Coast Range, it is more abundant in the south than in the north. This fungus is the most serious heart-rot organism on these tree species. In some old-growth stands, losses of 25 to 50 percent or more of the gross volume have been recorded. The rot is most common in the mid-trunk region, but it may extend into the butt or down from the top.

The fruiting bodies are distinctive. The woody, perennial, hoof-shaped conks range from 2 to more than 8 inches in width and are quite common on infected trees (Figures 8-10a, 8-10b in color section following page 212). The upper surface is dull, black, and rough. The undersurface is usually level (Figure 8-10a) but made up of hard, coarse teeth or spines (Figure 8-10b); this toothed surface is usually gray when fresh (Figure 8-10c in color section following page 212), but turns black when old. The interiors of conks are bright orange-red (Figure 8-10b), and the pigment is water soluble. Conks develop on the bole, usually on the underside of branches or branch stubs (Figure 8-10b). Punk knots may be present; these are not usually swollen.

The decay appears as a brown stringy rot (Figures 8-10b, 8-10c), but technically *E. tinctorium* is a white-rot organism, since the fungus decomposes lignin and, to a lesser extent, cellulose. Early stages of decay weaken the springwood, causing separation during seasoning of lumber and resulting in ringshake. In the first stages of noticeable decay, the wood becomes soft, with a light yellow to brown or water-soaked stain. This discoloration gradually deepens to pale reddish-brown. Fine rusty-red or black zone lines may appear at this stage. Advanced decay is a distinct brownish, reddish-brown, or rusty red and is soft and stringy. Rotted wood separates along the annual rings. In late stages of decay, the wood is reduced to a fibrous, stringy mass. Logs may be hollow in very advanced stages (Figures 8-10c, 8-10d in color section following page 212). This disease is of particular concern in developed recreational sites because of the risk of windthrow or breakage.

Conks produce viable spores throughout the year; however, conditions for infection are optimal during the spring and fall. Spores infect small (less than 0.1 inch in diameter) exposed branchlet stubs just before the branchlets are overgrown. Suppressed and slow-growing trees tend to have more shade-killed branchlets and to heal these branchlet stubs very slowly, thus allowing more opportunities for fungal infection and colonization. After spores have germinated and mycelium develops within the branch, fungal growth continues until the branchlets are overgrown. The fungus then enters a dormant state as a resting spore (chlamydospore), which can survive for 50 years or more without causing decay. Dormant infections are activated immediately by mechanical injuries, frost cracks, or formation of large branch stubs that allow air into the trunk interior. To cause the infection to become active, injuries must be within approximately 1 foot of a dormant infection. Wound size is unimportant for activating these infections, but the larger the injury, the more likely it is to be near one or more dormant infections that will be activated, leading to decay development. After extensive decay has formed, conks are produced, usually at branch stubs and occasionally at wounds where the fungus has a continuous pathway from the interior to the exterior of the tree (Figure 8-10b). Spores can be produced for several years after conks are formed and may even be produced for as long as 10 years after host trees have died or been felled. More conks are produced on felled trees than on standing dead trees, and more are found on felled but intact trees than on trees bucked into logs.

Schweinitzii butt rot (also known as red brown butt rot), caused by *Phaeolus schweinitzii* (also known as velvet-top fungus), occurs throughout the world. The fungus causes a brown cubical heartwood rot of the roots and butt (Figure 8-11a in color section following page 212), destroying the cellulose in stems up to about 9 feet above the ground. Along the Pacific Coast, the fungus is most common on Douglas-fir. However, its host range is wide and includes pines, true firs, larch (*Larix* spp.), spruces, incense-cedar, western redcedar, and, rarely, hemlocks. Infection results in high levels of decay in lower stems and predisposition to windthrow and breakage. Schweinitzii butt rot is of particular concern in developed recreational sites because of the risk of windthrow and breakage.

The annual fruiting bodies of the fungus may appear either on the butt of the tree, usually in the first 9 feet above the ground, or on the soil at the base of an infected tree. Depending on the site of occurrence, they may be either bracket-form (on exposed wood) (Figure 8-11b in color section following page 212) or mushroom-like (on soil where they have emerged from infected roots) (Figure 8-11c in color section following page 212). Fresh fruiting bodies have a soft velvety top with concentric rings, usually reddish-brown and encircled by a yellowish margin (Figure 8-11c). The spore-producing lower surface varies from dark green to light brown. Old fruiting bodies darken and dry to a corky or leathery texture (Figure 8-11d in color section following page 212). A large infected Douglas-fir may have a distinctly swollen butt (Figure 8-11e in color section following page 212). The incipient decay is usually a difficult-to-detect yellowish to light green stain that extends longitudinally an average of 2 feet beyond visibly decayed wood. The wood becomes soft before the advanced stage is reached. Wood with advanced decay is soft, brittle, and brown, and it breaks into large irregular cubes that are easily crumbled between the fingers (Figure 8-11f in color section following page 212). Very thin layers of resinous white mycelium may be found in shrinkage cracks.

Fruiting bodies are produced in summer and fall, and spores are windborne. The fungus generally enters trees through root wounds and fire scars. Tree-to-tree spread is also possible across root contacts and grafts.

Brown crumbly rot, caused by *Fomitopsis pinicola* (also known as red belt fungus), is one of the most common wood decays in coniferous forests in western North America. The rot is a brown cubical rot; cellulose is destroyed (Figure 8-12a in color section following page 212). The host list includes pines, true firs, Douglas-fir, western hemlock, Sitka spruce, and western redcedar. While *F. pinicola* mainly decomposes dead and down timber, it can cause heart rot in living trees. Decay from this fungus in living true firs is associated with large open bole swellings caused by dwarf mistletoe. The fungus is important as a slash rotter and in the deterioration of salvageable, killed timber, down trees, and stored logs.

Fruiting bodies are commonly seen on dead and fallen conifers. They range from 4 to 18 inches in width and are leathery to woody, perennial, bracket-shaped structures (Figure 8-12b in color section following page 212). Young fruiting bodies are round white fungal masses. As they mature, the upper surfaces turn dark gray to black; fresh lower pore surfaces remain white to creamy. A conspicuous reddish margin develops between the two surfaces; hence the common name "red belt fungus" (Figure 8-12b).

Decay develops rapidly in the sapwood and then progresses to the heartwood. Early stages of decay may appear as a faint yellow-brown to brown stain. As decay advances, it becomes light reddish brown and forms a crumbly mass broken into rough, rather small cubes (Figure 8-12a). Small patches of lighter color may give a mottled appearance on the ends of logs. Nonresinous mycelial felts form in the shrinkage cracks. These felts are thicker than those associated with most similar rots, but not as thick as those formed by the quinine fungus (see below).

Spores are windborne. Living trees are infected through wounds or dead or broken tops. They are introduced into dying and dead trees by wood-boring insects.

Brown trunk rot, caused by *Fomitopsis officinalis* (also known as the quinine fungus), is a brown, cubical heartwood rot of tree tops and trunks. The fungus is frequently associated with upper portions of old-growth conifers; it is rare in the butt. It occurs on Douglas-fir, pines, Sitka spruce, western hemlock, and, very rarely, true firs. Its incidence in second-growth forests is not well known but appears to be low.

Fruiting bodies are not common. When present, they are hard, perennial, and hoof-shaped, becoming long and cylindrical after several years of development (Figures 8-13a, 8-13b in color section following page 212). They develop at branch stubs (Figure 8-13b) or on wounds, and they range in size from 1 to 24 inches long. Conks are chalky white to grayish, with ridged and cracked surfaces (Figure 8-13a). Conk interiors are soft and white with a bitter taste.

The decay is frequently difficult to detect in its early stages. Incipient decay color varies by host species. In Douglas-fir, discoloration is unusual but when present it is purplish. In ponderosa pine (*Pinus ponderosa*), incipient decay is commonly red-brown or brown. In advanced stages, the wood breaks down into a crumbly mass of yellow-brown (Figure 8-13c in color section following page 212) to reddish-brown cubical chunks. In late stages, thick, whitish mycelial felts are common in the shrinkage cracks between the cubes (Figure 8-13c). Felts become 0.2 inch thick and may cover a large area in one continuous sheet (Figure 8-13d in color section following page 212). Felts have a characteristic bitter taste and resinous pockets or resinouscrusty areas.

Spores are windborne. They are believed to enter trees through broken tops, branch stubs, and occasionally fire scars.

Brown cubical rot, caused by *Laetiporus sulphureus* (also known as the sulfur fungus), occurs in a variety of conifers, including Douglas-fir, true firs, pines, western hemlock, Sitka spruce, and western redcedar. It causes considerable decay in true firs. This fungus also attacks hardwoods. Although brown cubical rot is not considered a major slash-decay organism, it is often seen on stumps, logs, and dead trees.

The conks are among the most conspicuous of all wood-decay fungi. Many appear in the fall on wounds at or near the base of living trees (Figure 8-14 in color section following page 212), on stumps, and on fallen logs. They are annual, clustered, shelf-like fruiting bodies that are soft, fleshy, and brilliant yellow-orange to red-orange when fresh. The pore surface is bright sulfur-yellow. Old fruiting bodies are hard, brittle, and chalky white.

Hidden decay is usually present but detectable only microscopically. The earliest detectable stage is a light-brown stain. In advanced decay, the wood breaks down into medium-sized reddish-brown cubes. Cracks of cubes are often completely filled with nonresinous white mycelial felts. Decay is most often associated with the butts of trees.

Spores are windborne. The fungus generally enters through basal fire scars and other wounds on conifers.

Yellow-brown top rot (also known as rose fomes rot), caused by *Fomitopsis cajanderi* (also known as the rose-colored conk), is a brown cubical heartwood rot in living trees that is often limited to the upper portion of the tree. It is frequently found in both standing and down dead trees. Decay develops rapidly, is important in piled logs and pulpwood, and may continue to develop in wood in service. Hosts include Douglas-fir, grand fir, western white pine, and spruce.

Fruiting bodies are perennial, woody, and bracket-like to hoof-shaped (Figures 8-15a, 8-15b in color section following page 212), with a pinkish to rose undersurface and inner tissue (Figure 8-15c in color section following page 212). The upper surface is brown to black and is usually cracked and roughened. Conks often appear stacked on one another (Figure 8-15a).

Wood may be moderately affected before any discoloration or change in texture becomes evident. A faint brownish or yellow-brown stain, sometimes marked by greenish-brown zone lines, may be seen in the early stages. As decay advances, yellowish to reddish-brown, soft, irregular cubes are formed. Thin mycelial felts that vary from white to faint rose may develop in the cracks between the cubes.

Spores are windborne. The fungus enters trees through broken tops or branch stubs.

Management strategies

Management strategies for wood decays must be thoughtfully planned to balance the tradeoffs between undesirable losses of forest products and desirable decayed wood for wildlife habitat and other ecological functions. In many coastal stands that are intensively managed, the removal of large trees has decreased the average tree age and amount of decay. Areas with extensive young stands may now lack decay and defect.

To achieve desired conditions, managers can limit unwanted decay while deliberately introducing other decay. The amount of decay in a stand can be reduced by keeping rotations short and avoiding tree injury during stand entries, such as for thinning. In the case of rust-red stringy rot, decay levels can also be kept low by not retaining suppressed hosts under infected overstory trees. Decay levels can be enhanced by intentionally damaging trees, retaining defective trees, and inoculating trees with wood-decay fungi.

When trees in a stand are to be selected for retention, objectives must be carefully defined and the marking crew must be trained to identify the decay. This will prevent situations where sound, decay-free trees are selected for retention and then treated to initiate decay, while trees that have existing decay are harvested and sent to the mill.

Planting susceptible tree species on sites that had severely decayed trees should not lead to serious decay in the new stand unless the trees are grown for more than 150 years. However, since current management policy on federal lands favors longer rotations and retention of existing stands of old-growth trees, increased occurrence of decay can be expected on these lands. This shift in turn increases the importance of considering decays in future management decisions, especially the management of developed recreational sites where visitor safety is an issue. For additional information concerning common stem decays, see Boyce (1961), Aho (1982), Aho and others (1983), and Gilbertson and Ryvarden (1987).

Special Concerns for Disease Management

Planning

Although no universal system exists for making land-management decisions, some variation of the following decision ladder may be useful for including disease management in the selection of appropriate management scenarios for forests in the Oregon Coast Range:

(1) *Establish management objectives.* The choice of available options will likely vary depending on the objectives. Examples include fiber production, wildlife-habitat enhancement, development of special forest products, or some combination of these.

(2) *Get a historical perspective.* Determine the historical appearance of the stand, including tree species present (not just what was planted), relative stand structure, composition, and age; diseases that were and are present and their distribution; and availability (or probable future availability) of inoculum from nearby stands.

(3) *Propose management strategies.* Use available models to evaluate the effects of the proposed strategies over time. For example, the western root disease model (Frankel 1998) can be used to address laminated root rot, annosus root disease, and Armillaria root disease and their interactions with bark beetles. Simulations designed around one of these three root diseases can be used to approximate the others. Simulated stands can be "grown" to evaluate various cultural treatments and test options. In the absence of models, use local data to project the impacts of various scenarios. Because growing a forest takes a long time, it is important to consider cumulative effects of various actions and inactions.

(4) *Integrate.* Integrate the decision about managing disease with the management of other resources on the site. Try to integrate the activities on the subject site with those planned for adjacent sites and to project impacts to the next larger scale. For example, try to project how the actions planned for a stand will, through time, affect other activities in the watershed.

Managers must expect occasional failures, because even the best plans have some built-in uncertainty. In general, diseases, especially root diseases with their hidden inoculum, cannot be controlled completely. Where controls are available, they are generally too expensive to use on large areas. Therefore, when we speak of management to reduce disease impact, we are really only talking about reducing risk. We are more likely to succeed if we adopt management objectives based on biological realities, clearly quantify those objectives, and develop management strategies with the goal of preventing unacceptable future outcomes.

Managing forests is a long-term enterprise, and changes will occur that are well beyond our reasonable planning horizon. Examples include changes in local weather conditions, the unexpected appearance of other disturbance agents that alter the outcome of our applied management strategy, policy shifts that alter desired outcomes, and changes in demand for particular products.

Topics for research

Current knowledge of forest diseases of the Oregon Coast Range suffers from many serious gaps. A better understanding of the basic biology of a number of pathogens is needed, as are improved techniques for determining and modeling their effects on tree mortality and growth. There is a further need to increase understanding of interactions among pathogens as well as between pathogens and other disturbance agents. Especially critical is understanding the roles of pathogens in natural stands and how various forest management techniques and strategies affect pathogen survival, spread, and intensification. Managers need to be able to judge the effects of diseases on all management objectives, not just on timber production. The following topics, which are not prioritized, are important for future research:

Biology

The roles of tree diseases, especially the more virulent root diseases and stem decays, in creating stand openings, promoting species and structural diversity, and recruiting short- and long-term snags and down wood. Development of models of disease dynamics on stand and landscape scales would be especially useful.

The historic effects of diseases in unmanaged stands, both unentered stands and those that were previously managed.

The introduction, establishment, and spread of exotic pathogens. Such information could form a basis for preventing future introductions.

The biology of stem-decay organisms, their role in creating wildlife habitat, and their interactions with various wildlife species.

The function of forest diseases at the landscape level; past efforts at understanding disease impacts have been concentrated at the individual tree or stand level.

Management

The effects of various regeneration techniques, spacing treatments, vigor-enhancement treatments, pruning, rotation lengths, and other stand treatments on disease intensification and spread, especially for laminated root rot, black stain root disease, Armillaria root disease, annosus root disease, stem decays, and Swiss needle cast.

The use of the genetic resistance of hosts to manage forest-tree diseases, especially exotic diseases.

Methods for predicting the distribution of major pathogens, especially root diseases, in terms of stand and site conditions, including the range of conditions that would lead to either an increase or a decrease in disease severity.

Strategies for simultaneously managing multiple disturbance agents on the same site.

Nondestructive techniques for identifying and monitoring forest diseases early and in living trees.

Summary and Conclusions

Forest tree diseases profoundly influence forest structure, composition, and function. They are among the most important disturbance agents operating in the forest ecosystems of the Oregon Coast Range. Although their effects are not spectacular at any given time, their impacts at landscape scales and over long time frames are considerable. Much is understood regarding the biologies and impacts of many of the important diseases found in the Oregon Coast Range. Decades of research have been dedicated to developing an understanding of how pathogens survive, invade their hosts, spread, affect resources, influence forest management practices, and are in turn influenced by forest management. The knowledge gained indicates not only that diseases will continue to influence forest processes, but that their impacts will, in all likelihood, increase as forest managers move toward managing late-seral species on long rotations, particularly if stands are entered repeatedly. Stumps created during stand entries are important infection courts for annosus root disease, for example; thus, stand entries greatly increase the impacts of this disease. Wounds created during stand entries also provide infection courts for a variety of decay fungi. Long rotations allow diseases such as annosus root disease to develop to the point where stem decay and breakage are frequent. Late-seral species tend to be more decay-prone in general. Long rotations also increase opportunities for spread of dwarf mistletoe from overstory trees to developing understories, thus leading to high infection levels in the understory and greater impacts on tree growth and survival. And there is much that still needs to be learned.

In recent years attention has begun to be focused on understanding the roles diseases play in forest ecosystems. Emphasis has shifted to studies of gap dynamics, habitat development, and the influence of disease on forest succession and forest structure. Tools now available make it easier to study cellular and genetic aspects of the pathogens. Models have been developed to assist with predicting long-term impacts. There is renewed interest in the effects of introduced pathogens. The slow spread of most diseases, the protracted period before results of experimentation can be analyzed, and the complexity of ecosystems and their processes make gaining a solid understanding difficult. Therefore, interdisciplinary team approaches to questions of disturbance processes are critical.

Understanding the ecological roles of diseases, their historical and current distributions, their impacts on resources, and the ways management influences them and they affect management, is crucial to meeting ecosystem-management objectives. Unless diseases are included as a planning consideration, ecosystem-management strategies are unlikely to succeed in achieving desired future conditions.

Epilogue

Although the pathogen names used in this chapter are currently valid, readers should be aware of potential changes. As molecular and serological techniques improve, our ability to discern genetic differences between organisms and our understanding of natural groupings change, and new names and descriptions of organisms are published. Each name is considered valid when it is published, but to gain general acceptance it must meet with the approval of scientists in the field and be used in the literature. Thus a single fungal pathogen may be known by many different names over the years, each accepted as valid for some period of time. As a result, the literature for a particular disease can be confusing. For the 15 diseases described in this chapter, the names of eight of the causal agents have changed at least once in the past two decades. Given ongoing studies, we anticipate that, within the next 10 years, there will be changes in the names of the fungal pathogens causing laminated root rot, black stain root disease, annosus root rot, Swiss needle cast, and red ring rot.

Photo credits

The authors express their appreciation to everyone who offered us photographs for use in Chapter 8. About half of the photos used originated with others, and all were taken by government employees. The following individuals provided photographs that appear here: USDA Forest Service—Jerry Beatty, Don Goheen, Ellen Goheen, Jim Hadfield, Mike Larsen, Earl Nelson, John Pronos, Walt Thies, and Doug Westlind; USDI Bureau of Land Management—Walt Kastner; California Department of Forestry—Dave Adams; Oregon Department of Forestry—Alan Kanaskie; Oregon State University—Greg Filip, Everett Hansen and Jeff Witcosky; Canadian Forestry Service—Rona Sturrock. While nearly all the photographs were taken by the above individuals, some were from agency archives where the identity of the original photographer is unclear. We regret any oversight in our failure to properly acknowledge a contributed photograph.

Literature Cited

Aho, P. E. 1982. *Indicators of Cull in Western Oregon Conifers*. General Technical Report PNW-144, USDA Forest Service, Pacific Northwest Research Station, Portland OR.

Aho, P. E., G. Fiddler, and G. M. Filip. 1983. *How to Reduce Injuries to Residual Trees During Stand Management Activities*. General Technical Report PNW-156, USDA Forest Service, Pacific Northwest Research Station, Portland OR.

Boyce, J. S. 1961. *Forest Pathology*. McGraw Hill Book Co., New York.

Filip, G. M. 1998. *Swiss Needle Cast Cooperative Annual Report*. Forest Research Laboratory, Oregon State University, Corvallis.

Filip, G. M., A. Kanaskie, and A. Campbell III. 1995. *Forest Disease Ecology and Management in Oregon*. Manual 9, OSU Extension Service, Oregon State University, Corvallis.

Frankel, S. J. 1998. *Users Guide to the Western Root Disease Model, Version 3.0*. General Technical Report PSW-GTR-165, USDA Forest Service, Pacific Southwest Research Station, Albany CA.

Gilbertson, R. L., and L. Ryvarden. 1987. *North American Polypores, Volumes 1 and 2, Fungiflora.* **Publisher???** Oslo, Norway.

Goheen, D. J., and E. M. Hansen. 1978. Black stain root disease in Oregon and Washington. *Plant Disease Reporter* 62: 1098-1102.

Goheen, D. J., G. M. Filip, C. L. Schmitt, and T. F. Gregg. 1980. *Losses from Decay in 40- to 120-year-old Oregon and Washington Western Hemlock Stands.* USDA Forest Service, Forest Pest Management, Pacific Northwest Region, Portland OR.

Hadfield, J. S. 1985. *Laminated Root Rot: A Guide for Reducing and Preventing Losses in Oregon and Washington Forests.* USDA Forest Service, Pacific Northwest Region, Portland OR.

Hadfield, J. S., D. J. Goheen, G. M. Filip, [and others]. 1986. *Root Diseases in Oregon and Washington Conifers.* USDA Forest Service, Pacific Northwest Region, Portland OR.

Hansen, E. M. 1978. Incidence of *Verticicladiella wageneri* and *Phellinus weirii* in Douglas-fir adjacent to and away from roads in western Oregon. *Plant Disease Reporter* 62: 179-181.

Hansen, E. M., and K. J. Lewis. 1997. *Compendium of Conifer Diseases.* APS Press, The American Phytopathological Society, St. Paul MN.

Hansen, E. M., D. J. Goheen, P. F. Hessburg, J. J. Witcosky, and T. D. Schowalter. 1988. Biology and management of black stain root disease in Douglas-fir, pp. 63-80 in *Leptographium Root Diseases of Conifers,* T. C. Harrington and F. W. Cobb, Jr., ed. APS Press, The American Phytopathological Society, St. Paul MN.

Hawksworth, F. G., and D. Wiens. 1996. *Dwarf Mistletoes: Biology, Pathology, and Systematics.* Agricultural Handbook 709, USDA Forest Service, Washington DC.

Hessburg, P. F., D. J. Goheen, and R. V. Bega. 1995. *Black Stain Root Disease of Conifers.* Forest Insect and Disease Leaflet 145, USDA Forest Service, Portland OR.

Hood, I. A. 1997. Swiss needle cast, pp. 14-15 in *Compendium of Conifer Diseases,* E. M. Hansen and K. J. Lewis, ed. APS Press, The American Phytopathological Society, St. Paul MN.

Hunt, R. S. 1998. Pruning western white pine in British Columbia to reduce white pine blister rust losses: 10-year results. *Western Journal of Applied Forestry* 13: 60-63.

Kanaskie, A. 1994a. *Effect of an Aerial Application of Bravo 720 (Chlorothalonil) on Needle Retention in Four Douglas-fir Plantations near Tillamook, Oregon.* Forest Health Report 94-3, Oregon Department of Forestry, Salem.

Kanaskie, A. 1994b. *Effects of Nitrogen and Phosphorous Fertilization on Height Growth Increment, Needle Loss, and Chlorosis of Douglas-fir in Coastal Northwest Oregon.* Forest Health Report 94-2, Oregon Department of Forestry, Salem.

Kanaskie, A. 1998. *Swiss Needle Cast of Douglas-fir in Coastal Western Oregon.* Forest Health Note October 1998, Oregon Department of Forestry, Salem.

Kimmey, J. W., and W. W. Wagener. 1961. *Spread of White Pine Blister Rust from Ribes to Sugar Pine in California and Oregon.* Technical Bulletin 1251, USDA Forest Service, Portland OR.

Kliejunas, J. T. 1994. Port-Orford-cedar root disease. *Fremontia* 22: 3-11.

Maguire, D. A., A. Kanaskie, R. Johnson, G. Johnson, and B. Voelker. 1998. Growth impact study: progress report from phases I and II, pp. 9-13 in *Swiss Needle Cast Cooperative Annual Report,* G. M. Filip, compil. Forest Research Laboratory, Oregon State University, Corvallis.

McDonald, G I., E. M. Hansen, C. A. Osterhas, and S. Samman. 1984. Initial characterization of a new strain of *Cronartium ribicola* from the Cascade Mountains of Oregon. *Plant Disease* 68: 800-804.

Morrison, D. J. 1981. *Armillaria Root Disease: A Guide to Disease Diagnosis, Development, and Management in British Columbia.* BC-X-203, Canadian Forestry Service, Pacific Forest Research Center, Victoria BC.

Nelson, E. E., R. R. Silen, and N. L. Mandel. 1989. Effects of Douglas-fir parentage on Swiss needle cast expression. *European Journal of Forest Pathology* 19: 1-6.

Otrosina, W. J., and R. F. Scharpf, tech. coord. 1989. *Research and Management of Annosus Root Disease* (Heterobasidion annosum) *in Western North America.* General Technical Report PSW-116, USDA Forest Service, Pacific Southwest Forest and Range Experiment Station, Albany CA.

Roth, L. F., R. D. Harvey, Jr., and J. T. Kliejunas. 1987. *Port-Orford-Cedar Root Disease.* Forest Pest Management Report R6-FPM-PR-010-91, USDA Forest Service, Pacific Northwest Region, Portland OR.

Scharpf, R. F. 1993. *Diseases of Pacific Coast Conifers.* Agricultural Handbook 521, USDA Forest Service, Pacific Southwest Research Station, Albany CA.

Shaw, C G. III, and G. A. Kile. 1991. *Armillaria Root Disease.* Agriculture Handbook 691, USDA Forest Service, Washington DC.

Smith, R. B. 1969. Assessing dwarf mistletoe on western hemlock. *Forest Science* 15: 277-285.

Smith, R. B. 1971. Development of dwarf mistletoe infections on western hemlock, shore pine, and western larch. *Canadian Journal of Forest Research* 1: 35-42.

Smith, R. B. 1977. Overstory spread and intensification of hemlock dwarf mistletoe. *Canadian Journal of Forest Research* 7: 632-640.

Thies, W. G., and R. N. Sturrock. 1995. *Laminated Root Rot in Western North America.* General Technical Report PNW-GTR-349, USDA Forest Service, Pacific Northwest Research Station, Portland OR, in cooperation with Natural Resources Canada, Canadian Forestty Service, Pacific Forest Research Center, Victoria BC.

Tkacz, B. M., and E. M. Hansen. 1982. Damage by laminated root rot in two succeeding stands of Douglas-fir. *Journal of Forestry* 80: 345-356.

Van Arsdel, E. P., A. J. Riker, and R. F. Patton. 1956. Effects of temperature and moisture on the spread of white pine blister rust. *Phytopathology* 46: 307-308.

Wallis, G. W. 1976. *Phellinus (Poria) weirii Root Rot: Detection and Management Proposals in Douglas-fir Stands*. Forestry Technical Report 12, Canadian Forestry Service, Victoria BC.

Witcosky, J. J., and E. M. Hansen. 1985. Root colonizing insects associated with Douglas-fir in various stages of decline due to black stain root disease. *Phytopathology* 75: 399-402.

Woodard, S., J. Stenlid, R. Karjalainen, and A. Huttermann. 1998. *Heterobasidion annosum: Biology, Ecology, Impact, and Control*. CAB International, Wallingford UK.

Zobel, D. B., L. F. Roth, and G. M. Hawk. 1985. *Ecology, Pathology, and Management of Port-Orford-Cedar (Chamaecyparus lawsoniana)*. General Technical Report PNW-184, USDA Forest Service, Pacific Northwest Forest and Range Experiment Station, Portland OR.

Figure 8-7d. **White pine blister rust**: Bright yellow aecia form on infected branches in the spring 3 to 4 years after infection. Aeciospores carried by air currents from *Ribes* spp. infect the leaves.

Figure 8-7e. **White pine blister rust**: Bright yellow aecia form on infected branches in the spring 3 to 4 years after infection. Infections on the lower portion of the stem are always fatal to the tree.

Figure 8-7f. **White pine blister rust**: A heavy resin flow is often associated with white pine blister rust cankers. White pine are killed when stem cankers form near the ground and girdle the tree. Small trees may be killed rapidly.

Figure 8-7g. **White pine blister rust**: *Ribes* spp. are the alternate hosts for white pine blister rust. Orange urediospores are produced on the underside of the leaves of *Ribes* spp. during the summer. These spores carry the fungus to other *Ribes* spp. plants. Teliospores, produced late in the summer on the same leaves, germinate in place and produce basidiospores that are carried by air currents to white pine.

Figure 8-8a. **Swiss needle cast:** Fruiting bodies appear on the underside of 1- or 2-year-old needles as two soot-like streaks, one along each side of the midrib.

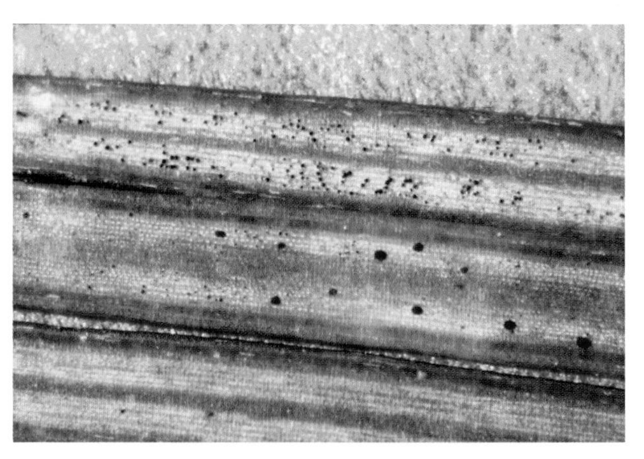

Figure 8-8b. **Swiss needle cast**: Underside of Douglas-fir needles as seen with a hand lens. The two light bands are stomata. Individual fruiting bodies of *Phaeocryptopus gaeumannii*, the fungus causing Swiss needle cast, are black and spherical (up to 0.04 inches in diameter) and are easily seen with a hand lens. Top needle has many *P. gaeumannii* fruiting bodies. Middle needle has a few fruiting bodies of *P. gaeumannii* and fruiting bodies of another less troublesome needle spot called fly speck. Bottom needle is essentially healthy.

Figure 8-8c. **Swiss needle cast**: Needle loss on individual branches can be severe, leaving only emerging needles and a few residual 1-year-old needles.

Figure 8-8d. **Swiss needle cast**: The disease can severely reduce the needle complement on an entire tree, thus reducing the tree's photosynthetic capability and growth.

Figure 8-8e. **Swiss needle cast**: Severely diseased stands appear yellow to brownish-yellow when viewed from a distance. Discoloration is most noticeable from late winter until bud break.

Figure 8-9a. **Red ring rot**: Typical white pocket rot caused by *Phellinus pini*. The occurrence of few to many spindle-shaped white pockets with firm wood in between give the decay its common name, "white-speck."

Figure 8-9b. **Red ring rot**: Closeup of *Phellinus pini*-caused white pocket rot. Here the decay is more advanced than in Figure 8-9a, and the pockets are longer with less firm wood between pockets.

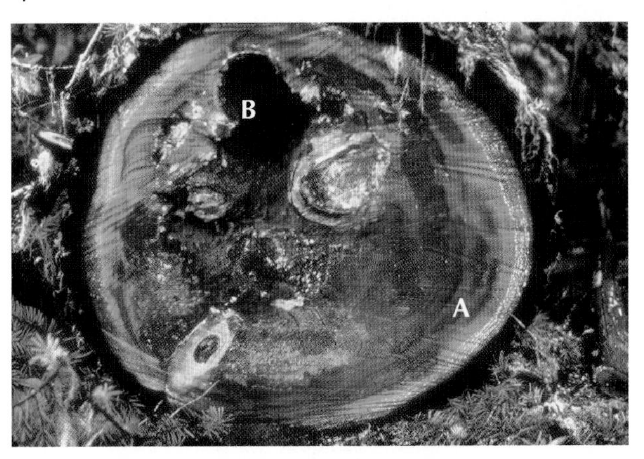

Figure 8-9c. **Red ring rot**: Cross section of Douglas-fir showing advanced decay caused by *Phellinus pini*. Pockets of advanced decay have coalesced into rings or crescents (A) with a complete void in the upper part of this cross section (B).

Figure 8-9d. **Red ring rot**: *Phellinus pini* conks on western hemlock. The occurrence of conks on older trees, numerous conks on a tree or large conks indicate extensive decay.

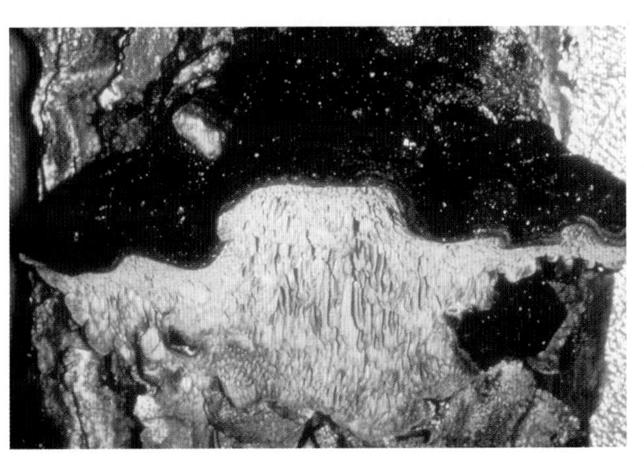

Figure 8-9e. **Red ring rot**: *Phellinus pini* conks are perennial, hoof-shaped to bracket-like; they are produced on branches and on stems at branch stubs and knots. The upper surface of the conk is rough, dark gray to brownish-black, and concentrically furrowed. The conk interior and pore layer are cinnamon-brown. Pores are irregular in shape. Conks range in size from 2 to 10 inches across.

Figure 8-10a. **Rust red stringy rot**: *Echinodontium tinctorium* fruiting body on true fir bole. Conks are perennial, hoof-shaped, woody, rough in texture, and dark gray to black. The undersurface is made up of coarse, hard spines or teeth. The toothed surface is gray when fresh and darkens as it ages.

Figure 8-10b. **Rust red stringy rot**: Cross section of tree bole through an *Echinodontium tinctorium* conk. Note the conk's orange-red interior, its coarse irregular teeth, and the fibrous advanced decay of the wood.

Figure 8-10c. **Rust red stringy rot**: Late stage of decay caused by *Echinodontium tinctorium*. The wood is reduced to a fibrous, stringy mass; later the stem may become completely hollow.

Figure 8-10d. **Rust red stringy rot**: Late stage of decay caused by *Echinodontium tinctorium*. In advanced decay, stems may be completely hollow.

Figure 8-11a. **Schweinitzii butt rot**: *Phaeolus schweinitzii* causes a brown cubical decay in the roots and butts of a number of conifers. Infected trees often break or are windthrown.

Figure 8-11b. **Schweinitzii butt rot**: While most fruiting bodies of *Phaeolus schweinitzii* emerge from roots or at the base of trees, occasionally conks fruit on boles within approximately 9 feet off the ground.

Figure 8-11c. **Schweinitzii butt rot**: Large, annual, mushroom-like fruiting bodies of *Phaeolus schweinitzii* are produced in late summer and fall. New conks are velvety and green to red to brown, with a yellow to yellow-green margin; hence the common name "velvet-top fungus."

Figure 8-11d. **Schweinitzii butt rot**: As *Phaeolus schweinitzii* conks age, they darken to brown or brownish-black and become leathery or brittle. *Phaeolus schweinitzii* is also referred to as the "cow pie fungus."

Figure 8-11e. **Schweinitzii butt rot**: Large, mature *Phaeolus schweinitzii*-infected Douglas-firs may have distinctly swollen butts.

Figure 8-11f. **Schweinitzii butt rot**: *Phaeolus schweinitzii* causes a brown cubical decay in the roots and butts of conifers. Note: The decayed wood easily breaks into cube-like pieces.

Figure 8-12a. **Brown crumbly rot**: *Fomitopsis pinicola* causes a brown cubical decay of dead trees, stumps, and down wood. It may also occur on dead portions of living trees.

Figure 8-12b. **Brown crumbly rot**: When young, *Fomitopsis pinicola* conks appear as white, round, fungal masses; as they mature, the upper surface turns gray to black, the pore surface remains white to cream, and a reddish margin forms between the two layers; hence the common name "red-belt fungus." Conks are perennial and leathery to woody.

Figure 8-13a. **Brown trunk rot**: Conks of *Fomitopsis officinalis* are perennial, hoof-shaped, and chalky white to gray; the surface is cracked or ridged. They range in size from a few inches to more than 2 feet long. Conk interiors are soft and white and have a bitter flavor, giving the fungus the common name "quinine fungus."

8-13b. **Brown trunk rot.** Conks of *Fomitopsis officinalis* are perennial and occasionally appear dark to brown.

8-13c. **Brown trunk rot**: Decay caused by *Fomitopsis officinalis* is a yellow-brown to reddish-brown cubical rot. *Fomitopsis officinalis* is most often associated with old-growth conifers and is especially common in trees with old broken tops.

8-13d. **Brown trunk rot**: *Fomitopsis officinalis*-caused decay is a yellow-brown to reddish-brown cubical rot. Thick, white mycelial mats grow in the shrinkage cracks between cubes. Note: upper right corner of mat is starting to peel away from the wood.

8-14. **Brown cubical rot**: *Laetiporus sulphureus* fruiting bodies are annual, clustered or shelf-like and bright yellow-orange. They appear in late summer and autumn on wounds at or near the base of living trees, on stumps, and on down wood. As they age, they dry out to a chalky white. The fungus is also known as the "sulfur fungus" or "chicken of the woods."

8-15a. **Yellow-brown top rot**: The upper surface of conks produced by *Fomitopsis cajanderi* is brown to black and often rough.

Figure 8-15b. **Yellow-brown top rot**: Conks produced by *Fomitopsis cajanderi* may appear anywhere on standing and down dead trees.

Figure 8-15c. **Yellow-brown top rot**: The pore layer and interior of conks produced by *Fomitopsis cajanderi* are pink to violet, giving the fungus the common name the "rose-colored conk."

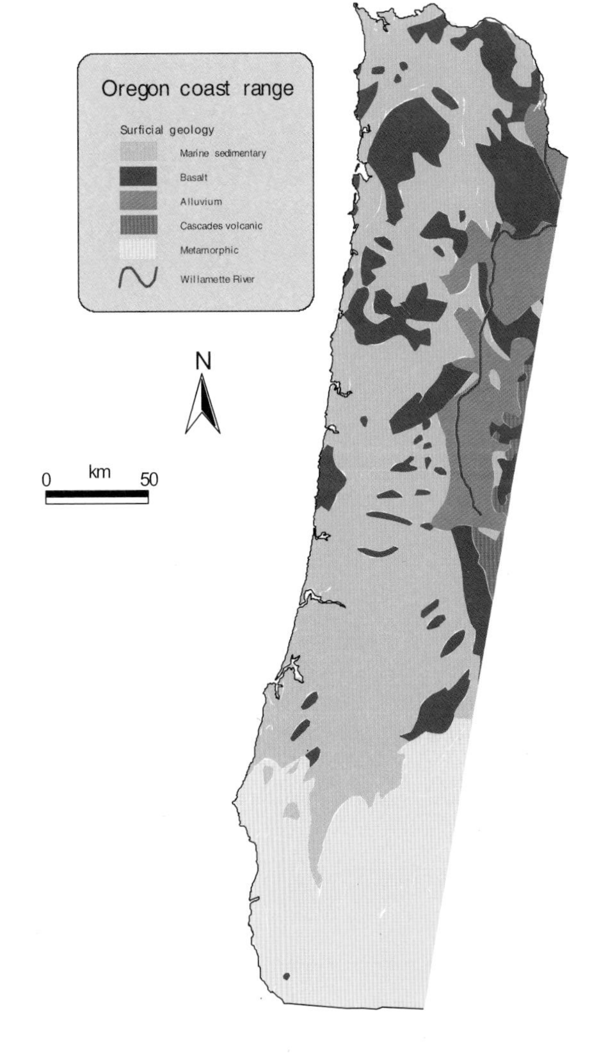

Figure 9-1. Geology of the Oregon Coast Range.

Figure 9-2. Average monthly precipitation and evapotranspiration for Cottage Grove and Vernonia in the Oregon Coast Range.

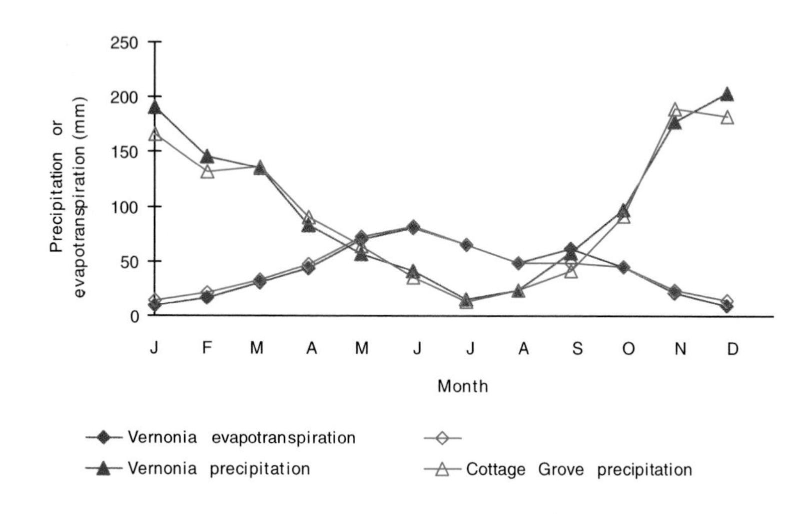

Figure 9-3. Median monthly streamflow for the Upper Nestucca, Alsea, and North Fork Coquille Rivers in the Oregon Coast Range.

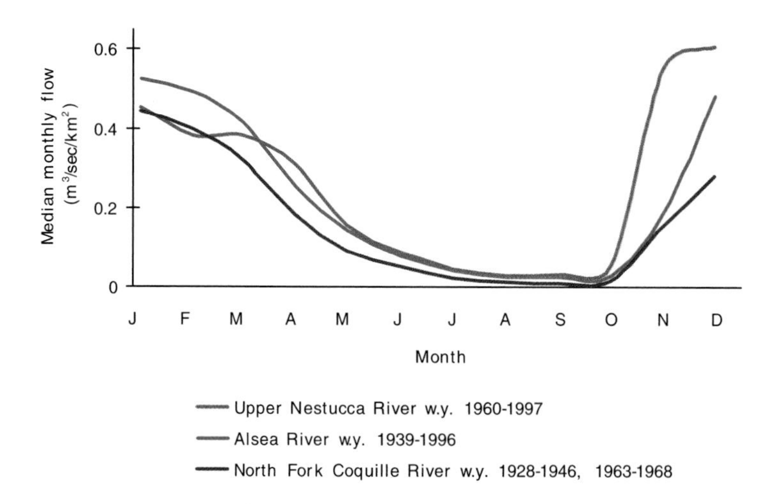

Figure 9-4. Recurrence interval for one-hour precipitation intensity for four locations within the Mapleton District of the Siuslaw National Forest in the central Oregon Coast Range.

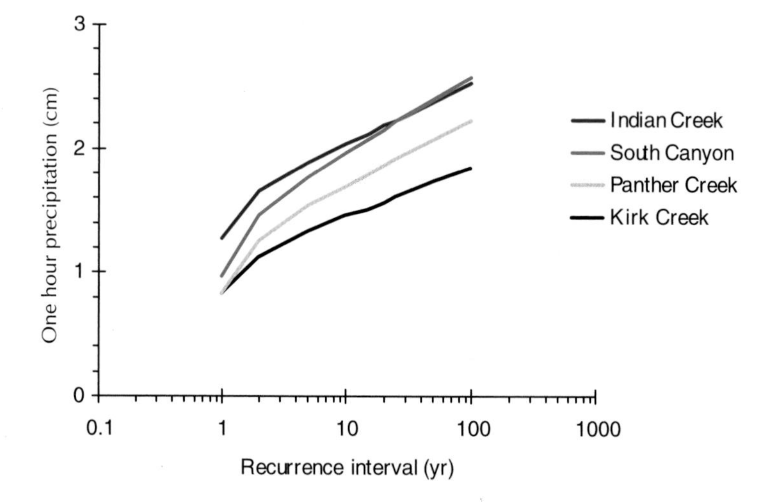

Landslides, Surface Erosion, and Forest Operations in the Oregon Coast Range

Arne E. Skaugset, Gordon H. Reeves, and Richard F. Keim

Introduction

Accelerated erosion from forest management activities, particularly timber harvest, has long been a focus of environmental concern. Associated with the accelerated loss of soil is an accelerated loss of soil nutrients, which ultimately may reduce the long-term productivity of intensively managed forest soils. As erosion processes in forested landscapes of western Oregon became better understood, especially the patchiness of erosion by shallow, translational landslides, the focus changed to effects on downstream values. With the listing of salmonids as threatened or endangered species, concern has become more sharply focused on aquatic habitat in headwater streams draining managed forest landscapes.

After a brief section providing the physical setting of the Oregon Coast Range, we discuss the erosion processes found in forested landscapes of western Oregon, with an emphasis on the factors and environmental conditions that initiate shallow, translational landslides. This section concludes with a synopsis of what is known about the relationship between these landslides and aquatic habitat.

The next section discusses how contemporary and legacy forest practices in western Oregon forests have affected accelerated erosion. The emphasis is on the observed relationship between forest management practices and the increased occurrence of shallow, translational landslides. This section concludes with a discussion of how forest management activities might affect the occurrence, size, and composition of debris-flow deposits.

The last section of the chapter presents the practices currently used to mitigate the effects of forest management activities. The regulation of forest practices to mitigate erosion has evolved over the last several decades from the application of water-quality standards to the use of best management practices (BMPs). Currently, these practices are in a transition from site-specific BMPs to watershed- or landscape-scale practices that fit into a construct of ecosystem management. This section presents site-specific BMPs that have been developed to reduce accelerated erosion, with emphasis on those practices used to locate high-risk, landslide-prone terrain and to mitigate against the occurrence of shallow, translational landslides. The section concludes with a discussion of BMPs that could be added to the current prescriptions, and how these should be used on the landscape to ensure that watersheds function optimally for aquatic habitat while allowing for forest management and the production of solid wood products.

Physical Setting of the Oregon Coast Range

Geology

The Oregon Coast Range is composed of rocks uplifted by the collision of two plates. The ocean floor, which is the Juan de Fuca Plate, is subducting under the terrestrial North American Plate, causing the Coast Range to rise by as much as 5 millimeters per year (Kelsey et al. 1994, 1996). The presence of small faults and folds throughout the overriding plate results in spatially variable uplift; the rate of uplift varies by as much as an order of magnitude (Kelsey and Bockheim 1994). At the same time, erosion is lowering the range by an average of 0.03 to 0.11 millimeters per year (Reneau et al. 1989; Reneau and Dietrich 1991).

The rocks of the Coast Range are predominantly layered and interbedded sandstones and mudstones formed from ocean sediments before uplift, and there are many intrusions of basalt into this matrix (Kelsey et al. 1996). Some of these basalts were already in place when the sedimentary platform formed on the former ocean floor and the two rock types uplifted together (Orr et al. 1992). Most of the basalt in the northern Coast Range and some of the coastal headlands, however, originated from terrestrial volcanoes east of the Cascade Range (Orr et al. 1992). Because basalt is less susceptible to erosion than sedimentary rock, many of these areas of intrusive basalt are now the highest elevations in the Coast Range. This differential susceptibility to erosion also means that sedimentary areas are more highly dissected than the basalt areas. The proportions of weaker, interbedded mudstones, as well as variations in strike and dip, affect the stability and erosivity of the bedrock.

Much of the eastern boundary of the Coast Range is the Willamette Valley, which is a fault-block valley of alluvial and lacustrine sediments. The volcanic Cascade Range borders the Coast Range to the southeast, and the volcanic and metamorphic rocks of the Siskiyou Mountains form the southern boundary (Figure 9-1 in color section following page 212). The Coast Range is bounded on the west by the Pacific Ocean.

Hydrology

The Coast Range has a Mediterranean climate, in which most of the precipitation falls as rain between October and April (Figure 9-2 in color section following page 212). During these months, precipitation exceeds evapotranspiration, causing runoff as streamflow. From May through September, warmer temperatures and long days promote vigorous plant growth, and evapotranspiration exceeds rainfall. This seasonal shift in weather results in a signature annual hydrograph for rivers of the Coast Range, in which nearly all streamflow occurs in the fall through the spring (Figure 9-3 in color section following page 212). Extended droughts in the summer can reduce streamflow to extremely low levels.

Precipitation falls predominantly as rain in the contributing areas for most streams, although some mountain peaks in the Coast Range can receive frequent snowfall during the winter. Most rain falls during frontal storms with intensities of 0.5 to 1.0 centimeters per hour (Figure 9-4 in color section following page 212). It is the extended periods of rainfall that result in peak streamflows during the wettest months; however, peak flows are largest when an intense rainfall occurs during such periods of extended rainfall or when rain falls on snow.

Landslides and Surface Erosion in Unmanaged Coast Range Forests

Erosion processes can be divided into three general groups: surface erosion, channel scour, and mass movement (Brown 1980). *Surface erosion* is defined as the movement of individual soil particles. Swanson and others (1982) list the surface-erosion processes that occur in unmanaged, forested watersheds as dry ravel, snow- and ice-induced movement, and soil movement accompanying root throw. Sediment budgets show that surface erosion is negligible in unmanaged, forested watersheds of western Oregon (Dietrich and Dunne 1978; Swanson et al. 1982).

Channel scour occurs within the stream channel, as soil and rock material from the streambed and stream banks becomes suspended sediment and bed load, which are available to the stream to transport (Brown 1980; Swanson et al. 1982). In gently sloping watersheds in the absence of debris flows, channel scour is the dominant process responsible for eroding stream banks and transporting streambed and bank material out of a watershed and further downstream.

Mass soil movement is the type of erosion that dominates sediment budgets for unmanaged, humid, temperate, forested watersheds (Swanston 1978). Mass soil-movement processes that occur in unmanaged, forested watersheds of the Coast Range include: soil creep, earthflows, deep-seated landslides, and shallow-translational landslides (Dietrich and Dunne 1978; Swanston 1978; Swanson et al. 1982).

Soil creep (Figure 9-5a) is a slow (millimeters per year) downslope movement of the entire soil profile

(a)

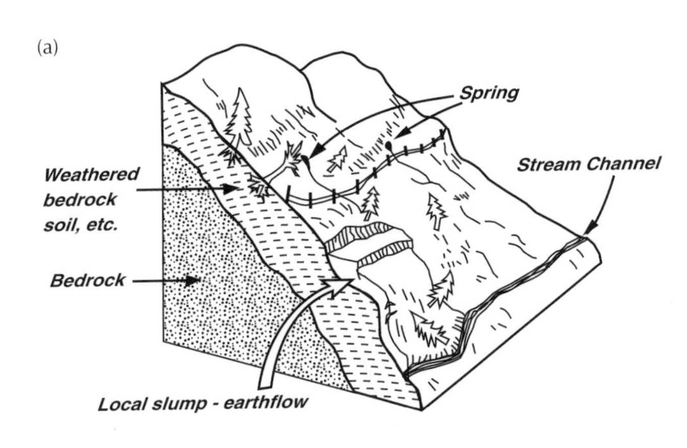

(b)

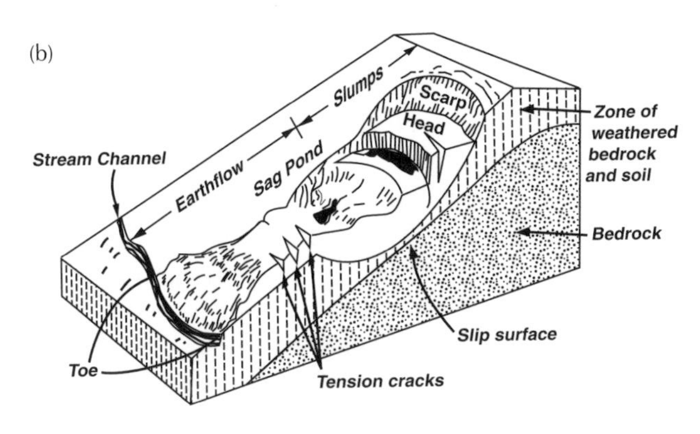

(c)

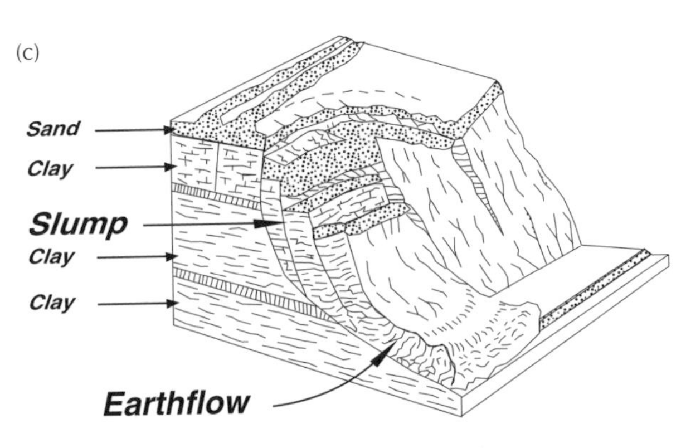

Figure 9-5. Three different types of mass soil movement processes that occur in the Oregon Coast Range including (a) soil creep, (b) earthflows, and (c) rotational landslides (Brown 1980).

in response to gravity. It is very small soil movements over a very large scale, and can involve entire forested slopes or a small watershed. However, the soil deformations are barely noticeable and difficult to measure; for the most part, they do not represent discrete failures (Swanston 1978). For gentle slopes (< 50%), soil creep is the dominant erosion process responsible for transporting soil from hillslopes to streambanks in unmanaged, forested watersheds.

In contrast, *earthflows* (Figure 9-5b) are large with distinguishable margins and a discrete failure surface. Although their speed may be faster than that of soil creep, it is still slow (centimeters to meters per year). The soil flows or glides over the underlying bedrock as a series of distinct blocks. The movement is sufficient to produce discrete failures (Swanston 1978).

Deep-seated landslides can be either rotational slumps (Figure 9-5c) or block-glide-type failures. These landslides are often large but locally discrete failures in which a block of soil and regolith either moves by rotation over a broadly concave failure surface or slides along bedding planes between layers of bedded deposits. Deep-seated landslides occur over periods lasting from minutes or hours to several days. They are often characterized by a steep scarp at the head of the failure and earthflow features at the toe.

Shallow, translational landslides in the Coast Range consist primarily of debris slides and debris flows (Figure 9-6). The primary difference between a debris slide and a debris flow is the moisture content of the slide mass. A *debris slide* is on the dry side of the continuum and consists of soil and rock that is displaced outward and downward along a failure surface, sliding out over the original ground surface (Cruden and Varnes 1996). Debris slides deposit at the foot of the failure scarp and do not mobilize and move farther downslope. Additional water content creates a slurry of soil, rock, water, and organic material called a *debris flow*. Debris flows follow a stream channel until they set up in debris-flow deposits (Cruden and Varnes 1996). (Although the terms *debris avalanche* and *debris torrent* are often used interchangeably with the more general term "debris flow" [Cruden and Varnes 1996], "debris avalanche" is a dated term and is currently not used in this context.) In this chapter, "debris slide" refers to the landslide-initiation sites and "debris flow" to features that move down stream channels and ultimately deposit lower in the system.

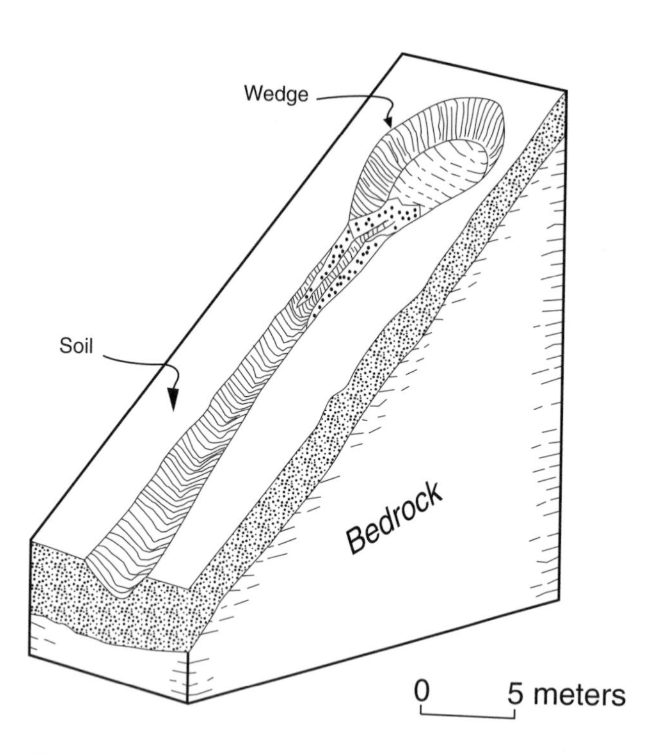

Figure 9-6. A debris flow with associated downslope scour (Humphrey 1981).

Shallow, translational landslides are initially small, locally discrete failures that can move very rapidly. They occur on steep slopes (> 50%) and involve the movement of shallow layers of soil downslope. Within seconds or minutes, they can move tens to hundreds of meters. They flow down stream channels and deposit in lower-gradient, higher-order stream channels.

Debris flows are the mass-movement processes of most interest and concern relative to both public safety and impacts on aquatic resources (Pyles et al. 1998). Landslide inventories show them to be ubiquitous across the landscape in the Coast Range (Robison et al. 1999; see Table 9-1). Although earthflows and rotational slumps may dominate sediment budgets within a given watershed or along a particular stream reach (Swanson and Swanston 1977), debris flows dominate erosion processes across the Coast Range. As the dominant erosion process in western Oregon, they overwhelm the sediment budget of unmanaged watersheds (Dietrich and Dunne 1978; Swanson et al. 1982). Debris flows are critical to public safety because of their ubiquity, size, speed, and extent (Pyles et al. 1998), and they dominate concerns about aquatic habitat quality (Sullivan et al. 1987; Swanson et al. 1987; Reeves et al. 1995; see Chapter 4). Given their size and ubiquitous distribution across the landscape, debris flows are most prone to being influenced by forest management activities, particularly forest roads and timber harvest.

Geomorphic context for debris flows

Debris flows occur on only a small proportion of the landscape (Ice 1985). Nearly all initiate on slopes steeper than 50 percent, and most occur on slopes between 80 and 99 percent (Ketcheson and Froehlich 1978). Debris flows are hypothesized to initiate in unchanneled bedrock hollows (Dietrich et al. 1987; Reneau and Dietrich 1987; Montgomery and Dietrich 1994), which are almost always thought of as being synonymous with topographic convergence or topographic swales (Montgomery and Dietrich 1994). In fact, unchanneled bedrock hollows may exist on relatively planar slopes that exhibit little or very subtle topographic expression as well as slopes with distinctive topographic swales. Landslide inventories have reported that as many as half of the inventoried landslides initiate on either planar slopes or slopes with very little topographic expression (Ketcheson and Froehlich 1978; May 1998; Robison et al. 1999).

Bedrock hollows are characterized by bedrock surfaces that are concave both across-slope and downslope (Sidle et al. 1985) and parabolic in cross section (Burroughs 1984), with the long axes extending downslope (Swanson et al. 1987). This shape causes subsurface groundwater to converge toward the hollow (O'Loughlin 1986; Dietrich et al. 1987; Reneau and Dietrich 1987; Montgomery and Dietrich 1994), as does the accumulation of colluvium (Sidle et al. 1985). The bedrock depressions are themselves created by repeated landsliding (Alger and Ellen 1987), and they undergo cycles of failure followed by filling with colluvium (Sidle et al. 1985; Benda and Dunne 1997a). The material that refills the hollows comes from hillside erosion that is primarily due to soil creep (Swanston and Swanson 1976), to the biogenic activities of animals and plants (Swanson et al. 1982), and to surface erosion (Swanson et al. 1987). The susceptibility of the hollows to landsliding during rainstorms increases over time because of this colluvial in-filling (Dietrich et al. 1995; Benda and Dunne 1997a; Dunne 1998).

Bedrock hollows filled with colluvium are found all over the world (Hack and Goodlett 1960; Marron

Table 9-1. Research results from landslide inventories for forests, clearcuts, and road rights-of-way carried out in Oregon, Washington, British Columbia, and New Zealand.

Location and researcher	*Forests*			*Clearcuts*			*Logging Road*		*Right-of-way*
	Slide rate (#/ha)	*Slide volume (m^3)*	*Erosion rate (m^3/ha/yr)*	*Slide rate (#/ha)*	*Slide volume (m^3)*	*Erosion rate[1] (m^3/ha/yr)*	*Slide rate (#/ha)*	*Slide volume (m^3)*	*Erosion rate[1] (m^3/ha/yr)*
Coast Range Mtns. OR									
Ketcheson & Froehlich 1978	1/11	31	0.05	1/8.5	36	0.20 (3.7)	1/1.6	386	4.7 (49)
Swanson et al. 1977	1/13	32	0.09	1/18	111	0.17 (1.9)			
Schroeder & Brown 1984	1/25			1/3.5					
Cascade Mtns. OR									
Swanson & Dyrness 1975	1/157	837	0.10	1/33	1,344	0.45 (4.3)	1/2.8	1,351	5.3 (51)
Morrison 1975	1/176	1,990	0.13	1/25	439	0.33 (2.6)	1/0.81	1,428	43 (343)
British Columbia, Canada									
O'Loughlin 1972	1/848	3,041	0.03	1/146	1,150	0.07 (2.2)	1/38	3,451	0.79 (25)
Olympic Mtns. WA									
Fiksdal 1974	1/77	4,655	0.20	0	0	0	1/0.81	599	33 (165)
Siskiyou Mtns. OR									
Amaranthus et al. 1985	1/483	2,425	0.07	1/27	891	0.47 (6.8)	1/3.6	2,015	7.7 (12)
New Zealand[2]									
O'Loughlin & Pearce 1976	1/150*	550*	0.28*	1/6.5	562	6.10 (21.7)			

[1] The number in parentheses indicates the increase in erosion rate compared to the forest.

[2] New Zealand data marked with an * are estimated.

1985; Tsukamoto and Minematsu 1987; Crozier et al. 1990). Their density varies from 0.1 to 2.0 hollows per hectare depending upon geology, topography, and climate (Benda et al. undated). Colluvial hollows are predictably the locations of landslides, especially where shallow soils overlie bedrock (Montgomery et al. 1998), as in the Coast Range.

Estimated return intervals of landslides for an individual hollow vary from a few hundred years to 15,000 years or more (Swanson et al. 1987; Benda and Dunne 1997a), although the effect of fluctuations in climate on failure rates is unknown. Locally, differences in surficial rocks, such as a concentration of mudstone layers or differences in the physical integrity of sandstone, may affect the rate of accumulation of colluvium (Reneau et al. 1989) and the concentration of subsurface water (Anderson et al. 1997).

Mechanistic context for debris flows

Using the geomorphic context for debris flows described above, we can identify the locations that are most likely to be initiation sites. In any given landslide-producing storm, less than 10 percent of sites prone to debris slides will fail (Dunne 1998; Robison et al. 1999). However, with the current level of knowledge, it is not possible to predict which sites prone to debris slides will fail during the next landslide-producing storm. However, slope-stability analysis can be used to gain some insight into the likelihood that a site will or will not fail.

Slope-stability analysis

Slope-stability analysis is a way to assign a numerical value to the stability of a slide-prone site. It is a quantitative treatment of the stresses acting on a hillslope or valley and the strength of the soil involved. In carrying out a slope-stability analysis, a factor of safety for the slide-prone site is calculated. The factor of safety is the ratio of the forces driving failure and those resisting it for a given slide mass. The forces driving failure are due to the downslope component of the weight of the soil within the slide mass. The forces resisting failure comprise soil strength and any additional strength due to plant roots.

The methods available for analyzing slope stability range from fairly simple and straightforward to exceedingly complex. The analysis methods can involve only a simple balance of forces or moment equilibrium, or both could be required. The slide mass involved can be analyzed as a one-, two-, or three-dimensional body. Finally, the failure surface can be modeled as a simple circular surface or as a more complicated irregular surface. The simplest analysis is a stress balance on a one-dimensional slide mass with a regular surface; much more complex is a rigorous force-and-moment analysis on an irregular three-dimensional surface. To learn more about slope-stability analysis methods, see Nash (1987).

The simplest and most straightforward method of slope-stability analysis is the infinite-slope method. It is used most often to quantitatively investigate the stability of shallow, landslide-prone forest soils (Gonsior and Gardner 1971; O'Loughlin 1972, 1974; Wu et al. 1979; Sidle and Swanston 1982). The infinite-slope method is a ratio of the soil shear strength and the shear stress of a representative slice of soil that has a unit width and length; thus only soil depth varies, making the analysis one-dimensional. The generalized equation for the infinite-slope method is provided below for a soil in which strength is represented by both cohesion and friction and groundwater flow is represented by steady-state seepage parallel to the slope (Abramson et al. 1996).

$$FOS = \frac{c' + h\cos^2\beta[(1-m)\gamma_m + m\gamma']\tan\phi'}{h\sin\beta\cos\beta[(1-m)\gamma_m + m\gamma_{sat}]}$$

In this equation, *FOS* is the factor of safety; and c' and ϕ' are soil-strength parameters; γ_m, γ' and γ_{sat} are the moist, buoyant, and saturated unit weights of the soil; h is soil depth; and m is a dimensionless depth of water ranging from zero to one. The other terms in the equation are illustrated in Figure 9-7. In general, the contribution of plant roots to soil strength takes the form of an apparent cohesion, *ca*, or root cohesion, *cr*, in the numerator of the equation that is summed with the soil cohesion term, (Sidle et al. 1985).

All slope-stability-analysis methods, including the infinite-slope method, require the following information: (1) the location of the failure surface, (2) the mode of failure, which dictates the appropriate analysis method, (3) the loading on the critical failure surface, (4) the soil strength parameters, including the effect of roots, and(5) the pore water pressure on the failure surface at failure.

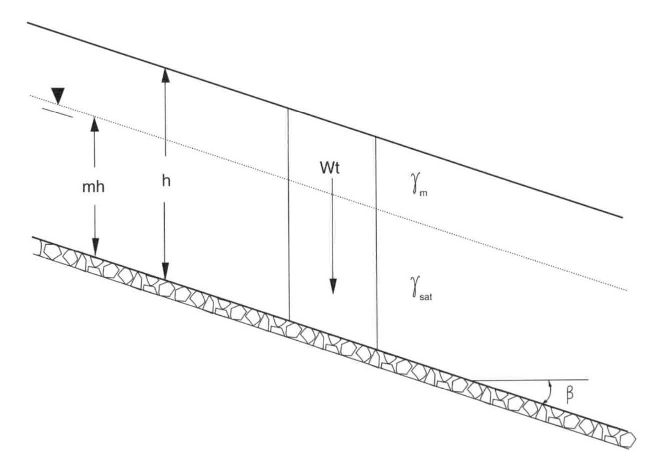

Figure 9-7. Definition sketch for the infinite slope method for slope stability analysis.

Slope-stability analysis has been carried out for naturally occurring debris flows in shallow, residual soils (Swanston 1970; Gonsior and Gardner 1971; O'Loughlin 1972, 1974; Wu et al. 1979; Sidle and Swanston 1982; Buchanan and Savigny 1990). The infinite-slope method was used for most of these analyses. The exceptions were those of Swanston (1970) who used the "ordinary method of slices," and Buchanan and Savigny (1990), who used Bishop's modified method. Both analyses combined conventional soil mechanics with limit-equilibrium methods. Soil strength was determined using standard engineering tests. Positive pore pressures were estimated based on the direct response of a groundwater table to rainfall.

Although a standard analysis of the stability of a site prone to debris slides using the infinite-slope method provides a general framework for considering mechanisms of failure, it is of little practical value in determining absolute stability. Its value is limited because some of the critical parameters in a slope-stability analysis can be measured or known only with a very low degree of precision. These parameters include soil strength, root strength, and pore water pressures at failure. The lack of precision for these critical parameters results in a less precise estimate of the absolute stability of an individual site. Therefore, slope-stability analysis has a very limited application for identifying stable versus unstable sites.

Considerable research has been carried out in recent years on two of the critical components of slope-stability analysis for forested sites prone to debris slides that have shallow, coarse-textured or granular soils. These are the estimates of soil strength and pore water pressure at failure resulting from the hydrologic response of hillslopes to rainfall.

Soil strength

The strength of the forest soils for steep, landslide-prone terrain in the Coast Range (or for any soil) is not a fixed quantity. It is a function of the effective normal stress on a failure surface when failure occurs. These soils are generally considered to be cohesionless or granular soils and are classified as fine sands or fine sandy loams. They are shallow, 1.0-meter-deep (± 0.5 meter), loose soils with low, dry densities and high void ratios; thus the confining and normal stresses at typical soil depths are low.

The relationship between soil strength and the effective normal stress is called the Mohr-Coulomb strength envelope. Figure 9-8 shows a typical Mohr-Coulomb strength envelope. This relationship takes the following form:

$$\tau = c' + \sigma_n' \tan \phi'$$

where τ is the soil strength (kPa); c' is the effective soil cohesion (kPa); σ_n' is the effective normal stress on the failure surface (kPa); and ϕ' is the effective internal angle of friction (°).

The effective normal stress on a failure surface is the total normal stress minus the pore water pressure, which is a function of the depth of the groundwater table over the failure surface. The total normal stress is that component of the total soil weight above the failure surface that is perpendicular to the failure surface. Normal stress is

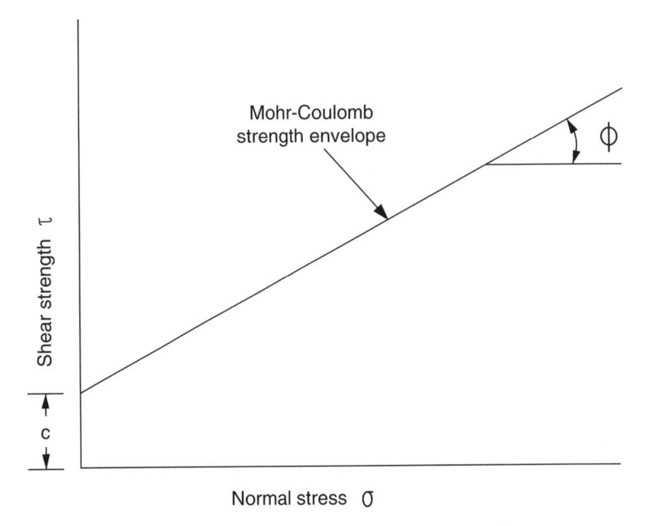

Figure 9-8. A Mohr-Coulomb strength envelope.

expressed per unit area. Therefore, the Mohr-Coulomb strength envelope is presented in terms of total stress as:

$$\tau = c + (\sigma_N - \mu)\tan\phi$$

where c is the total soil cohesion (kPa); σ_N is the total normal stress on the failure surface (kPa); and μ is the pore water pressure, which is the product of the depth of the water times the density of water (kPa).

Knowing the location of the groundwater table at failure is critical to the stability analysis of steep, landslide-prone slopes, because it is the only way to calculate the effective stress on the failure surface. The location of the groundwater table and the pore pressures on the failure surface at failure are doubly critical because the groundwater response to rainfall is the ultimate triggering mechanism for almost all landslides on forestlands in Oregon.

Soil strength is not a theoretical quantity; values of soil strength must be determined empirically. There are a variety of soil-strength tests, ranging from relatively simple *in situ* field tests to complex and time-consuming laboratory tests. The accuracy and precision of the estimates of soil strength derived from these tests are generally proportional to the time and effort put into the test. The most common and accurate soil-strength test is a laboratory test called the triaxial test. It uses a cylindrical soil sample that is nominally 5 centimeters in diameter and 15 centimeters long. The sample is placed in a cell where different pressures or stresses can be applied to determine the strength of the soil sample at different depths. Once the appropriate stresses or confining pressures have been applied, a force or an increase in load is applied to the long axes of the sample until it fails. The triaxial test has several advantages: (1) it allows a failure surface to develop along the weakest failure plane in the soil sample instead of forcing an arbitrary failure surface through the sample, (2) a wide array of stress states can be imposed on the soil sample, and (3) the positive pore pressure in the sample can be modeled and measured.

Estimates of soil strength that have been used for slope-stability analysis of steep, landslide-prone forested slopes with shallow soils have generally come from conventional tests of soil strength (Swanston 1970; O'Loughlin 1972; Sidle and Swanston 1982; Wu et al. 1988; Buchanan and Savigny 1990). These conventional tests are carried out using standard geotechnical engineering practices, which include the triaxial test as well as others. Unfortunately, the resulting estimates of soil strength reflect stress states and test conditions that are *not representative* of shallow soils on steep, landslide-prone forested slopes at the time of failure.

The conventional tests for soil strength are not representative because they are carried out at high confining pressures using a standard stress path. Although these conditions are representative of the stress states and the type of loading commonly found in geotechnical engineering, the *in situ* conditions of a high confining stress (i.e., deep soils) and an increase in weight or total load (i.e., standard stress path) do not occur on forested slopes. The failure surface for most debris slides is shallow, roughly 1.0 meter (± 0.5 meter). In addition, the soils are loose and cohesionless, with low bulk densities (1.1 to 1.4 megagrams per cubic meter). As a result, the *in situ* confining stresses for these soils at these depths are low, in the range of 6 kPa ± 3 kPa. Another factor making conventional estimates inappropriate for predicting rainfall-induced landslides is that the stress path is not an increase in total load, but rather a decrease in the effective stress. Shallow, translational landslides occur when the effective stress is reduced by the water table that occurs in response to rainfall.

Soil strength at low confining stresses (i.e., at the *in situ* confining stresses for loose, shallow, cohesionless soils) is not a linear function of effective stress. For standard soil-strength testing at conventional stress states, the internal angle of friction for granular soils increases with decreasing confining stress (Lambe and Whitman 1969; Mitchell 1976; Holtz and Kovacs 1981). Therefore, triaxial soil-strength tests must be carried out using low confining stresses for shallow, cohesionless soils from steep, landslide-prone forested slopes. Morgan (1995) has compiled published soil-strength values for Coast Range soils; Figure 9-9 shows a non-linear strength envelope fitted through the data points. As shown in the figure, the slope of the line or the internal angle of friction, ϕ', decreases and the intercept of the line, c', increases with increasing average normal effective stress, which is represented by p' in the figure. Linear regressions of the data show values as high as 45° for an effective stress of less than 35 kPa, which corresponds roughly to a soil depth of approximately 4 to 5 meters, and as

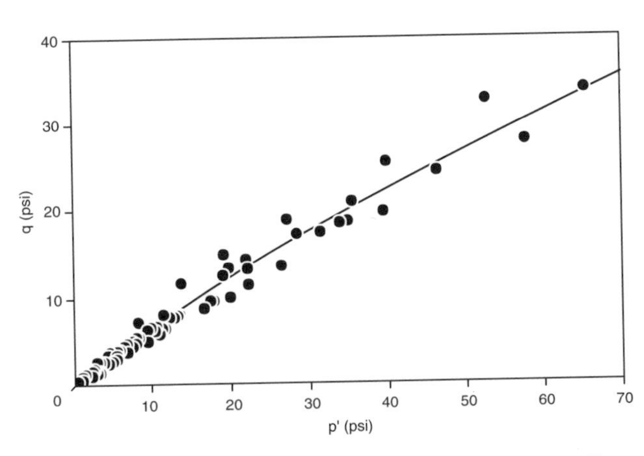

Figure 9-9. A graph of the failure envelope using all the information available for soil strength test data from shallow, cohesionless forest soils on steep, landslide-prone sites in the Oregon Coast Range. The index p′ is the normal stress on the soil sample and q is an index of the shear stress or strength (Collins 1995).

low as 29° for effective stresses greater than 200 kPa. The corresponding values of c' range from nearly zero to approximately 25 kPa. After conducting triaxial tests of loose, cohesionless soils from the Coast Range, Collins (1995) also reported a nonlinear strength envelope and recommended that strength tests for these soils be carried out using triaxial tests at *in situ* effective stress levels. Similarly, testing a shallow, landslide-prone soil from the San Francisco Bay area, Anderson and Sitar (1995) found the strength envelopes to be nonlinear and steeper at lower confining pressures.

The effect of the stress path on soil strength has also been investigated. Collins (1995) conducted strength tests using a field stress path, which initiates failure by reducing effective stress without increasing total load, and compared the results with those for the same soils using a standard stress path. Differences in the two stress paths for cohesion and the internal angle of friction were negligible compared with the differences that occurred simply as a result of variability between test specimens and tests. So although testing at *in situ* confining pressures is recommended, it is not necessary to use a non-standard stress path in the test.

Root strength

The effect of roots on soil strength is modeled by adding an apparent-cohesion term, *ca*, or a root-cohesion term, *cr*, to the Mohr-Coulomb strength envelope. Values for the apparent cohesion due to roots have been derived using various research methods for a variety of soils and plant species. These methods include determining the tensile strength of roots (Burroughs and Thomas 1977), carrying out both laboratory and *in situ* direct shear tests (Endo and Tsuruta 1969; Waldron and Dakessian 1981, 1982; O'Loughlin et al. 1982; Waldron et al. 1983), and landslide back-analysis (Swanston 1970; O'Loughlin 1974; Sidle and Swanston 1982). Values of apparent cohesion due to roots vary from 1.0 to 17.5 kPa (Sidle et al. 1985).

Groundwater response to rainfall on hillslopes

Using slope-stability analysis to distinguish unstable sites prone to debris slides from stable ones requires knowing the location of the groundwater table at failure, because its location at the point of incipient instability defines the degree of stability or instability. The response of the groundwater regime in a given debris-slide-prone site is the key to predicting the point of incipient instability as a function of low-frequency, long-return-period storms.

Our understanding of the relationship between rainfall and the development of a groundwater table on steep, landslide-prone forested slopes has developed over the last several decades (Campbell 1975). The current model consists of a two-phase system representing a residual soil, or colluvium, overlying bedrock (Figure 9-10). The soil has high permeability and is quite porous compared to the impervious bedrock. Rainfall infiltrates the soil and moves vertically until it encounters bedrock (or other less permeable layers); the subsurface flow then becomes parallel to the underlying slope. According to this model, groundwater will accumulate and the transient groundwater table will get thicker toward the base of slopes (Figure 9-10). Based on this model and the relationships observed between rainfall and piezometric head, predictive models have been developed to locate the position of groundwater tables for a given amount of rainfall or return periods of storms (Swanston 1967; Burroughs 1984). These relationships also have been included in slope stability models to help predict the location of landslide-prone terrain (Schroeder and Swanston 1987; Hammond et al. 1992).

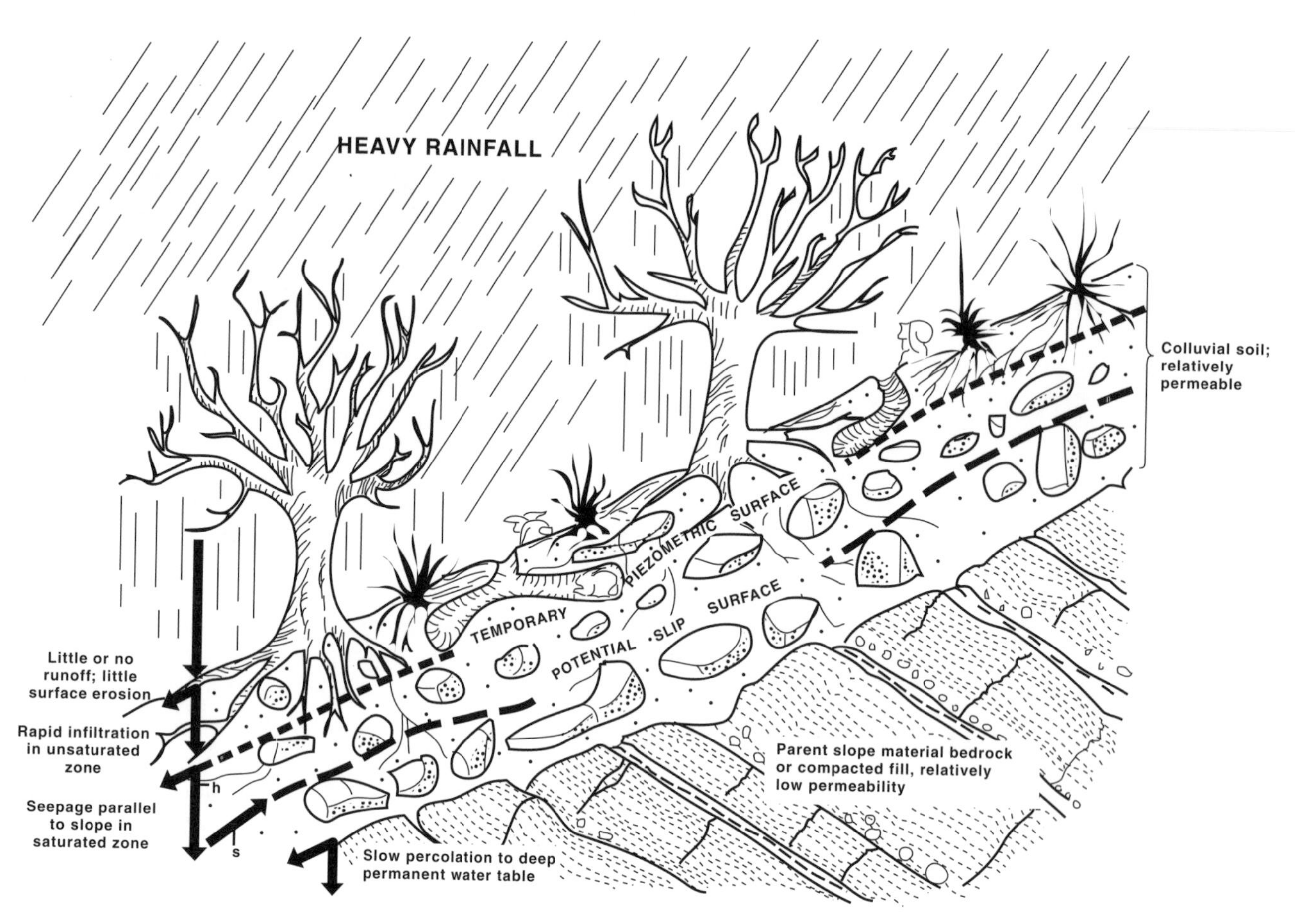

Figure 9-10. A conceptual model for a two-phase system of groundwater response to rainfall for steep, forested hillslopes (Campbell 1975).

However, research has shown that this conceptual model is oversimplified. Harr (1977), Johnson and Sitar (1990), and Montgomery et al. (1997) observed that the occurrence of positive pore pressure during large storms was highly localized on a hillslope. Zones of positive pore pressure did not occur at the base of slopes as expected, but at mid-slope locations associated with convergent topography. Therefore, Johnson and Sitar (1990) hypothesized that topographic convergence causes flow to move laterally to swales or hollows, and that subsurface flow in these locations does not come from the surface soil layers directly upslope. They further hypothesized that subsurface flow may also occur through the bedrock layer.

In the Oregon Coast Range, Anderson et al. (1997) and Montgomery et al. (1997) have observed a three-phase system, with a highly pervious layer of fractured bedrock separating the soil and impervious bedrock. Subsurface flow through the fractured bedrock contributed significantly both to runoff generation for a small bedrock hollow and to the generation of positive pore pressure. Zones of positive pore pressure were discontinuous and highly localized, and their occurrence coincided with areas of groundwater exfiltration from the bedrock into the overlying soil. Subsurface flow tended to infiltrate into and exfiltrate out of the fractured bedrock layer several times, which accounted for the discontinuous pattern of positive pore pressure. When the bedrock hollow subsequently failed during a natural storm, that failure coincided with the location of subsurface flow exfiltration from the bedrock to the overlying soil (Dietrich and Sitar 1997).

More sophisticated physically based models of subsurface flow that predict the location of groundwater tables may help to differentiate unstable and stable landforms on forested hillslopes; for example, TOPMODEL (Bevin et al. 1995) or CLAWS (Duan 1996). However, these models currently cannot deal with the site-specific complexity illustrated by Montgomery et al. (1997) or cannot parameterize it. But even though preferential flow paths and pipes or macropores in soil are notoriously difficult to measure and to represent in models (Bonell 1998), they govern the behavior of subsurface flow in hollows during landslide-producing storms.

An alternative that avoids this problem is the use of simple empirical models to simulate subsurface flow (Beschta 1998). They can be used to simulate the location of groundwater tables in sites prone to debris slides in response to rainfall (Bransom 1997). The Antecedent Precipitation Index or API model has been used to simulate runoff from small, forested watersheds in response to rainfall (Fedora and Beschta 1989). The greatest benefit of the API model is that the parameterization of the complex subsurface-drainage environment can be accomplished using a single empirical coefficient (Bransom 1997). However, this model approach does require that local piezometric data be developed to generate the empirical coefficient, which can be a formidable task.

So far this discussion of the initiation of landslides has dealt only with static pore pressures (those resulting from precipitation and the aggregation of subsurface flow). Another consideration is dynamic pore pressures, specifically the excess pore pressures that are generated as a result of the incipient movement of the landslide. The pore pressures during mobilization of the debris flow are termed excess pore pressures because they cannot be predicted or accounted for by traditional means, which consider the drained initiation state and not the undrained mobilization state (Anderson and Sitar 1995).

As a part of the research on the influence of confining pressures and stress path on soil strength, the response of the pore pressure in the soil sample to undrained loading was investigated (Bransom 1990; Keough 1994; Morgan 1995). The proportionality constant between axial undrained loading and pore-pressure response, called Skempton's "A" pore-pressure parameter, was positive for these strength tests, meaning that these soils are compressive (Morgan 1995). This means that their volume decreases with incipient axial loading and shear. Although we lack information about how a compressive soil responds to incipient shear strain as a result of reduced effective stress, it is likely that such a situation would increase pore pressures. This response has been observed in physical modeling of Reid and others (1997); the maximum excess pore pressure during debris-flow initiation and mobilization was much greater than the static pore pressure.

Debris flows and aquatic habitat

Landslides initiate on steep, forested hillslopes in response to the presence of a groundwater table, which means that the soils have high water contents and high positive pore pressures. At such times the saturated-water content of forest soils in steep, sites prone to debris slides is higher than their liquid limit (Bransom 1990); once initial failure occurs, the soils tend to behave as liquids and quickly mobilize into debris flows.

Debris flows initiate on steep, soil-mantled hillslopes at the upstream end of the drainage system and run down through steep zero-order swales and sometimes first- and second-order stream channels. Eventually they deposit in lower-gradient first- and second-order streams while some reach third- and higher-order stream channels (Swanson and Lienkaemper 1978). Debris flows can remove all the sediment and large wood from the steep swales and first-order stream channels, leaving a bedrock channel (Benda 1990). However, the stream channel upstream of a debris-flow deposit is generally neither totally erosional nor totally depositional; it retains some sediment and channel substrate material although the channel is highly disturbed and the large wood is removed.

Debris-flow deposits vary in size and type, depending on the size of the flow and the gradient and width of the stream channel (Benda 1990; May 1998). At one end of the continuum is the *debris jam*, a large, valley-spanning accumulation of organic debris that dams the stream, forcing the accumulation of a large wedge of sediment. However, debris-flow deposits can also form fans of large wood and sediment that force a larger stream against the opposite bank, thus increasing channel sinuosity without blocking the stream. At the other end of the continuum, debris fans of sediment and large wood may deposit across floodplains and not interact with the receiving stream channel except at very high flows (Benda 1990; May 1998). Of the 53 debris-flow deposits in the central Coast Range that occurred following the February 1996 storm, only 12 (23 percent) were valley-spanning debris dams. Of the remainder, 36 (66 percent) interacted with the receiving channel to various degrees without completely blocking it, and 5 (11 percent) were washed away during the storm and not found (May 1998).

Debris flows can be disastrous to fish and aquatic habitat. When the debris flow occurs, there is the likelihood of direct mortality to fish and other aquatic species. Later, the form of the stream may be greatly altered and simplified. The large wood will be gone, the stream will have low hydraulic retention, the overhanging canopy will be more open, and the surrounding riparian area will be highly disturbed (Everest and Meehan 1981; Lamberti et al. 1991). Downstream erosion may be accelerated; large, abrupt inputs of sediment may mean more fine sediment, which affects spawning-bed quality, and more transport-resistant sediment, which can cause aggradation of stream channels and filling in of pools (Everest et al. 1987; Sullivan et al. 1987).

Some research results suggest, however, that the relationship between debris flows and aquatic habitat may not be entirely negative. Despite short-term localized damage, Everest and Meehan (1981) found that debris-flow deposits created resting habitat for spawning adult coho salmon and rearing habitat for juvenile coho without interfering with upstream spawning migration. In this case, the net effect of debris flows on fish and aquatic habitat was positive. In the Elk River watershed in southwest Oregon, the most productive areas for both rearing and spawning habitat were associated with large "flats" or areas of relatively low gradient. These areas were persistent features at tributary junctions and apparently resulted from debris-flow deposits that occurred in conjunction with catastrophic wildfires over 100 years ago (G. Reeves, personal communication). This scenario would link high-quality aquatic habitat with the occurrence of debris flows.

Recently, Reeves et al. (1995) offered a long-term view that portrays aquatic habitat as dynamic over the geographical scale of watersheds and landscapes and over the temporal scale of decades and centuries. In response to disturbance, aquatic habitat, like terrestrial habitat, undergoes successional stages and its quality at a given point in a watershed will succeed from young to old forms over time. The mechanism for resetting the age of aquatic habitat has historically been catastrophic disturbance, specifically stand-replacement wildfires followed by intense, landslide-producing storms that cause the widespread occurrence of debris flows (Benda and Dunne 1997a, b; Benda et al. 1998). The aquatic habitat is of highest quality and the fish communities are most diverse when the disturbances are of intermediate age, neither recently disturbed nor senescent (Reeves et al. 1995; see also Chapter 3). This disturbance-based theory of the succession of aquatic habitat recognizes debris flows as the mechanism that brings critical habitat ingredients, namely large wood, boulders, and sediment, from the hillslopes and into the streams.

Landslides and Surface Erosion in Managed Coast Range Forests

Surface erosion

Because of the low intensities of precipitation in the Coast Range and the high infiltration capacities of forest soils, surface erosion resulting from overland flow is virtually nonexistent. However, minor amounts of surface erosion occur, especially on steep slopes because of ravel from exposed soils resulting from local disturbances, such as windthrow and burrowing animals. Managed forests offer more opportunity for surface erosion because soil disturbance is more drastic, widespread, and chronic. Surface erosion resulting from forest management activities can be divided into two general categories: surface erosion from within harvest units, where the predominant management activities include the removal of trees and site preparation, and surface erosion from drastically disturbed and compacted soil surfaces, such as skid trails for ground-based harvesting systems, roads, and landings.

Surface erosion from harvest units

Surface erosion coming from within harvest units takes two main forms. The first, dry ravel, is predominantly associated with steep slopes and cohesionless or skeletal soils in western Oregon. The second form, infiltration-limited surface erosion (wet erosion) is associated with rainfall and overland flow.

Dry ravel

Timber harvesting can accelerate the rate of dry ravel, especially if broadcast burning is used as a site-preparation treatment. Swanson et al. (1989) summarized the research findings on dry ravel. Increases in either slope or slash burning cause the rate of dry ravel to increase greatly. On steep, broadcast-burned harvested areas in the Oregon Coast Range, the surface-erosion rate was approximately eight times the rate on similar but gentler terrain and approximately 13 times the rate for steep, unburned slopes. On steep, burned slopes, two-thirds of the first year's erosion occurred within 24 hours after the burn (Bennett 1982). In the Oregon Cascades, dry ravel on a steep, unburned slope increased markedly after harvest, then within 3 to 5 years settled to a level approaching that of an undisturbed forest (Swanson et al. 1989). Therefore, even with marked increases in dry ravel on harvested sites, the window of accelerated erosion is very small compared to expected rotation lengths, and the total amount of erosion produced is minuscule compared to other erosion processes.

Infiltration-limited surface erosion

Surface erosion associated with infiltration-limited overland flow is virtually nonexistent for harvested areas in western Oregon. The simple removal of trees does not change the infiltration capacity of forest soils sufficiently to cause infiltration-limited surface erosion. While the absence of trees does not increase surface erosion, the manner in which they were removed and the site-preparation techniques used after harvest may increase its risk. Aerial-logging systems, such as skyline systems or helicopter logging, expose and disturb less surface soil than ground-based harvest systems, such as tractors and skidders (Chamberlin et al. 1991; see Chapter 6).

Chemical site preparation leaves the duff, litter, and slash intact and causes the least increase in the risk of surface erosion. In contrast, site preparation by broadcast burning removes logging slash, may remove litter and duff layers, and may create hydrophobic conditions, all factors that increase the risk of surface erosion. The risk is increased most by mechanical site preparation, in which logging slash is pushed into windrows using a bulldozer or put into piles using a hydraulic excavator or a bulldozer. In such cases, not only is the slash layer removed, but the surface soil is also disturbed and compacted.

Regardless of these factors, it is still difficult to generate surface erosion in Coast Range forests. Even moderate soil disturbance, compaction, and broadcast burning do not reduce infiltration capacities of forest soils to a point where they are less than rainfall rates. Johnson and Beschta (1978) found that infiltration capacities on areas harvested using various methods, as well as infiltration capacities on tractor skid roads, were several times higher than expected rainfall rates. For areas that had been harvested and broadcast burned, McNabb and others (1989) found that the single lowest measurement of infiltration capacity was still twice the expected 100-year-return-period rainfall intensity.

Research results from paired watershed studies reinforce these findings. During the Alsea Watershed study, one small sub-watershed was 90 percent clearcut and harvested using a high-lead cable system; there was no increase in suspended sediment as a result of the clearcut logging alone (Brown and Krygier 1971). In a paired watershed study in the H. J. Andrews Experimental Forest, the rate of soil loss attributed to surface erosion from clearcut harvested areas was very small (Fredriksen 1970).

Therefore, erosion associated with infiltration-limited overland flow in western Oregon forested watersheds occurs only on sites that are drastically disturbed and compacted. These sites include roads, landings, and the portions of skid trails that have experienced multiple passes (six turns or more) with ground-based harvesting equipment. On such sites, the soils are bare and have been compacted sufficiently to make the infiltration capacity less than expected rainfall intensities.

Surface erosion from roads and landings

Forest roads have long been a focus of concern regarding accelerated erosion. The construction of a forest road, especially one with any degree of side slope, results in a high degree of disturbance. When the road is completed, the remaining structure has two locally oversteepened slopes (the cut slope and the fill slope), a surface that must support loaded log trucks, and a roadside ditch. All of these surfaces are bare soil until vegetation becomes reestablished. The forest-road surface is intentionally compacted to develop its strength and is a virtually impervious surface. The roadside ditch is designed to carry both runoff from the road surface and subsurface flow intercepted by the cut slope to a cross-drain culvert or other drainage structure. Moreover, the environmental effects of forest roads are much more persistent than the effects of harvesting. Because forest roads are constructed, maintained, and used over an extended period of time, they generate surface erosion over this period.

The effects of construction in accelerating erosion have been studied extensively. These research sources are paired watershed studies for the Alsea Watershed (Brown and Krygier 1971), the H. J. Andrews Experimental Forest (Fredriksen 1970), and Caspar Creek in northern California (Krammes and Burns 1973). However, the information obtained is of marginal use, for a number of reasons. The studies were not specific to erosion processes, and the researchers lumped erosion processes together and measured overall effects at the mouth of the watershed. Also, the three studies were carried out in landslide-prone terrain, and landslides undoubtedly contributed a significant amount to the reported erosion. Therefore, the results say little about surface erosion from forest roads. Finally, the existing roads were constructed according to methods used 30 to 40 years ago, and road-location and -construction practices have improved greatly during the intervening years.

In contemporary managed forests, it is the environmental effect of this legacy of forest roads that is of most concern. In western Oregon, there are literally tens of thousands of kilometers of forest road, and all represent some potential to produce chronic surface erosion. Several studies have investigated the environmental factors that affect surface erosion from forest roads. One of the most important predictors of surface erosion from these roads is the level of traffic (Reid and Dunne 1984; Bilby et al. 1989). Reid and Dunne (1984) report that a heavily traveled road will yield, by an order of magnitude, more fine sediment than a similar road that has only light traffic. The thickness and quality of the surfacing material of the road also affect sediment production. Surface erosion is less where a thick layer of higher-quality erosion-resistant rock was used for the surface course than where a thinner layer of lesser-quality rock was used (Bilby et al. 1989). The thick surface seems to minimize the amount of fines that is pumped from the subgrade to the surface, and the high-quality rock seems to be less prone to erode. However, both surfacing thickness and quality were secondary to traffic intensity.

In the central Coast Range, Luce and Black (1999) investigated how road attributes affected surface erosion in the absence of traffic. They found that sediment production was best predicted using a term that is the product of the length and the square of the slope of the road segment. Total sediment production from road segments was independent of the height of the road cutslope. However, when the road cutslope and roadside ditch were vegetated, the amount of surface erosion was reduced by a factor of seven. Similarly, Foltz (1996) found water velocities in unvegetated roadside ditches to be two to three times faster than those in vegetated or grass-lined ditches.

Modeling the process of surface erosion from roads has shown that it is influenced by the formation of ruts (Foltz and Burroughs 1990, 1991; Foltz 1993). Well-defined, deep, continuous ruts cause long, concentrated flow paths, in contrast to the short flow paths across the road. Maintenance of the road surface to prevent and fix rut formation lowered the erosion rate drastically, because the length of the flow paths was reduced. Although grading and shaping the road surface made material available to erode in the next storm, overland flow was not able to remove the material efficiently because of the shorter flow path. However, traffic on a road increased surface erosion long before ruts form. Even a slight increase in the flow path caused an increase in surface erosion from the road.

Effect of forest management activities on debris slides and flows

Mass movement processes (most debris slides and flows) dominate the sediment budgets of pristine watersheds in western Oregon (Dietrich and Dunne 1978; Swanson et al. 1982). It is these same processes that have the greatest potential to be adversely affected by forest management activities and to result in the largest net increases in accelerated erosion. Landslides associated with forest management activities are divided into two groups: in-unit landslides (landslides that occur in harvest units and are not associated with forest roads) and road-related landslides. The management-related causes of landslides and their mitigation are different for each group. For in-unit landslides, the causes relate to vegetation management; for road-related landslides, the causes are related to site disturbance and changes in hillslope hydrology.

In-unit landslides

Since the earliest research in this area, investigators have believed that there is a basic cause-and-effect relationship between timber harvesting and landslide occurrence (Croft and Adams 1950; Bishop and Stevens 1964). The early research tool used to investigate this relationship was landslide inventories. In particular, inventories for harvested terrain were compared with inventories for similar but unmanaged terrain. Landslide inventories have been carried out for much of the Pacific Rim, including the Pacific Northwest, coastal British Columbia, southeast Alaska, Japan, and New Zealand. The results have been compiled to show background landslide and erosion rates and the effect of timber harvesting on these rates (Sidle et al. 1985). Table 9-1 contains such a compilation. In every landslide inventory but one, the erosion rate attributed to debris flows increased after harvest, by 1.9 to 21.7 times. Similarly, all but two studies documented an increase in the rate of occurrence of landslides after harvest.

However, care must be used when interpreting these data. Most of the studies used landslide data collected from aerial photographs. Unfortunately, given differential visibility in aerial photographs of debris flows in forests versus clearcuts, photo-based inventories could bias the results of the research (Pyles and Froehlich 1987). The Oregon Department of Forestry (ODF) conducted a systematic, ground-based landslide inventory after a series of storms in in 1996 in western Oregon. The survey showed conclusively that landslide-inventory data from aerial photographs underrepresented both the number of landslides and total landslide erosion (Robison et al. 1999). Therefore, it is best to base conclusions on inventories that are systematic and ground based, and that give every landslide an equal opportunity to be sampled.

The ODF landslide inventory was intended to monitor the effect of forest practices on landslides (Robison et al. 1999). The inventory covered eight 26-square-kilometer study areas in the Cascade Mountains and Coast Range of western Oregon. Five of the study areas were labeled "red zones" because they had a high incidence of landslides resulting from the 1996 storms. Table 9-2 presents the landslide densities for four of the red zones stratified by the age class of the forest stand (data for one red zone, the Tillamook study area, is omitted because only one forest age class was present).

Although this study represents only the response of steep, landslide-prone, forested landscapes to a single landslide-producing storm, its results encapsulate the trends shown by a compilation of systematic, ground-based landslide inventories (Pyles et al. 1998). Therefore, the trends seen in the ODF data seem to be consistent across many types of landslide inventories from a variety of terrain types.

The ODF study shows that there is, on average, an increased incidence of landslides in the first decade after clearcut harvest compared with adjacent late-seral forests. The average increase in landslide density was 42 percent, based on a background rate of 5.2 landslides per square kilometer and a post-harvest landslide rate of 7.4 landslides per square kilometer. This average increase is less than the average increase indicated in Table 9-1, but it is consistent with the increase observed in ground-based landslide inventories (Pyles et al. 1998).

Table 9-2. Landslide density for forest age classes in four "red zone" study areas in the Coast and Cascade Ranges after the February 1996 storm (Robison et al. 1999).

Study area	*0-9 yrs (#/km²)*	*10-29 yrs (#/km²)*	*30-100 yrs (#/km²)*	*>100 yrs (#/km²)*	*Change[1] (%)*
Elk	22.2	20.5	15.4	26.1	-15
Vida	13.7	3.3	2.7	8.8	56
Mapleton	21.1	1.9	6.1	13.5	58
Scottsburg	20.1	10.5	7.3	5.7	250
Average	19.2	9.1	7.9	13.5	42

[1] This column is the percent change in the landslide density for the 0-9 year age class compared with the >100 year age class.

The increased incidence of landslides in the first decade after harvest is followed by a decrease in landslide occurrence in the 10- to 30-year and 30- to 100-year forest age classes. Because the ODF study was the first to sample multiple forest age classes, there is no other database for comparison. However, this result was previously hypothesized (Froehlich 1978; Swanson and Fredriksen 1982), and the merit of the hypothesis has been discussed vigorously without resolution.

The average increase in landslide density attributed to timber harvest represents widely varying values for individual study areas, ranging from a 15 percent decrease in the incidence of landslides in one study area to a 250 percent increase in another. These two extremely different results come from study areas that are within 10 kilometers of each other and have the same geology, landforms, and management regime; the landslides occurred during the same storm. Such variability in landslide response has been noted when only ground-based inventories were considered (Pyles et al. 1998); therefore, it seems to be normal.

These results indicate the importance of naturally occurring factors that influence landslide occurrence as compared with timber harvest. One of these factors is the variability in the total amount and intensity of precipitation that trigger landslides. Insufficient precipitation data were collected to satisfactorily determine whether the cause of the extreme variability in landslide occurrence was driven by precipitation. However, sufficient precipitation-intensity data were collected during the February 1996 storm to show that precipitation amounts and intensities that trigger landslides do vary significantly across distances represented by the ODF landslide inventory (M. Clark, personal communication).

For the four multi-age-class study areas, the results of the ODF landslide inventory show an average increase in landslide density as a result of timber harvest. However, there are no statistically significant differences in landslide density among the four age classes (Robison et al. 1999). The small sample size (only four treatments and four study areas) and the high variability in response undoubtedly contribute to this lack of significance. The result is consistent with those from the only other landslide study in which the effect of timber harvest was tested statistically (Martin 1997). This study investigated the effectiveness of leaving patches of intact forest in areas with steep, convergent topography within a clearcut harvest area (headwall-leave areas) in mitigating the occurrence of landslides (Martin 1997). The incidence of landslides in forested areas, clearcut areas, and headwall-leave areas was found to be statistically insignificant because of high variability. The results from the ODF landslide inventory for erosion volume correspond to those for landslide density, including the lack of statistical significance.

Although landslide density and erosion volume are appropriate variables given the monitoring questions that were asked in the ODF study, they provide little insight into what the response would mean for a specific managed forest. In other words, what does a 42 percent increase in landslide density look like? An approach that will yield more insight is to present the landslide-density data as the number of failed slide-prone sites per unit area or the percentage of slide-prone sites that failed. In calculating these numbers, some implied assumptions are made. First, the term "bedrock hollow" will be used synonymously for a debris-slide-prone site. Since bedrock hollows can exist on planar slopes with little or very subtle topographic expression, the assumption is made that all the landslides inventoried by the ODF initiated in bedrock hollows. Second, even though the reported value of the average spatial density of 100 hollows per square kilometer is for the central Oregon Coast Range (Benda 1990), it is assumed that this value will adequately represent the ODF study areas. It is recognized that both of these assumptions are more than likely in error; however, the consequence of the error will result in the numbers being more conservative.

Using the value of 100 hollows per square kilometer, it is possible to convert landslide-density values to the number and percentage of failed hollows per square kilometer (Table 9-3, for 0- to 9- and >100-year forest age classes). The highest landslide density, for the greater-than-100-year age class in the Elk study area in the central Coast Range—26 per square kilometer—represents a failure rate of only about 10 percent of the hollows. Therefore, on a landscape basis, fewer than 10 percent of the hollows failed during the landslide-producing storm, whether the forest was harvested or not. The average increase in landslide density, 42 percent, means an increase of two landslides per square kilometer if the entire square kilometer were

Table 9-3. The percent of debris slide-prone sites that failed in two forest age classes in four "red zone" study areas in the Coast and Cascade Ranges of Oregon after the February 1996 storm.

Study area	*0-9 yrs (%)*	*>100 yrs (%)*	*Change[1] (%)*
Elk	8.6	10.2	-1.6
Vida	5.3	3.4	1.9
Mapleton	8.2	5.3	2.9
Scottsburg	7.8	2.2	5.5
Average	7.5	5.3	2.1

[1] This column is the change in the percent of debris slide-prone sites that failed for the 0-9 year age class compared with the >100 year age class. Data based on an assumed density of debris slide-prone sites of 100/km^2.

harvested. That is, the number of failed hollows would increase by approximately 2.2 percent in the harvested area. The change in the percentage of hollows that failed ranged from a low of -1.5 percent, which means fewer hollows failed after harvest, to a high of 5.6 percent. These values represent an extreme condition: a rare, landslide-producing storm on a 100-percent-clearcut harvest on highly unstable terrain. These values are actually high for an entire managed forest because a whole forest is rarely in a 0- to 9-year age class. For areas with older age classes interspersed, the increase in the percentage of hollows that failed would be lower.

Road-related landslides

Forest roads have long been considered the dominant source of accelerated erosion caused by forest management activities. The early paired watershed studies first showed the importance of road-related landslides in harvest-related accelerated erosion (Fredriksen 1970; Brown and Krygier 1971; Beschta 1978). Inventories of road-related landslides further documented the importance of forest roads to the increase in landslide erosion associated with forest management (O'Loughlin 1972; Fiksdal 1974; Morrison 1975; Swanson and Dyrness 1975; Swanson et al. 1977; Amaranthus et al. 1985). As shown in Table 9-1, the erosion rate from the rights-of-way of forest roads is 12 to 343 times the erosion rate from similar unmanaged forested terrain. As discussed above, many of the landslide inventories represented in Table 9-1 were photo-based; the same concerns about bias in the data hold for road-related landslides as for in-unit landslides. However, because the difference indicated by the inventories of road-related landslides is much greater than background variability, it is easier to infer that a problem exists.

During the last two to three decades, considerable effort and expense have been put into improving the environmental performance of forest roads, especially with regard to landslides. Improvements include wholesale changes in the way steep, landslide-prone forested terrain is harvested and how forest roads are located, constructed, and maintained, including (1) widespread use of long-span, high-lift cable systems, like skyline systems, to reduce the amount of road needed and increase the flexibility of road location; (2) use of ridge-top locations for forest road systems; (3) use of high-gradient roads to reach and stay on ridge-tops; (4) use of full-bench, end-haul road-construction practices; and (5) improved road-drainage and road-maintenance practices.

The degree to which these improved practices have reduced the negative environmental effects of roads is not fully known. However, an inventory of road-related landslides in the central Coast Range compared the effects of forest roads constructed using old practices with those constructed using improved ones (Sessions et al. 1987). The results showed that the improved practices reduced both the number and the size of road-related landslides. Unfortunately, the validity of these results remained in doubt because, at the time of the study, the roads constructed using the improved practices had not been tested by a landslide-producing storm. The February 1996 storm provided the opportunity to evaluate the effects of contemporary forest-road systems.

Two inventories of road-related landslides were carried out following the February 1996 storm. As a part of the ODF study discussed previously, all road-related landslides within the eight 26-square-kilometer study areas were inventoried (Robison et al. 1999). The other road-related landslide inventory was carried out on Blue River and Lookout Creek in the central Oregon Cascades (Wemple 1998).

The results from both inventories show that contemporary forest-road systems remain a significant source of erosion from debris slides/flows (Skaugset and Wemple 1999). Road drainage was associated with approximately half of the debris flows that initiated at road fills, confirming the importance of road drainage to accelerated erosion. The cutslope height of the roads was correlated with

Table 9-4. Comparisons of inventories of road-related landslides for the Oregon Coast and Cascade Ranges from the literature with the post-1996 ODF flood monitoring study and the PNW inventory of Blue River and Lookout Creek watersheds.

Oregon Coast Range Comparisons

Researcher	*Road length (km)*	*Number of landslides*	*Average landslide volume (m^3)*	*Landslide density (#/km)*	*Erosion rate (m^3/km)*
Swanson et al. 1977	80.0	89	386	1.10	731
Sessions et al. 1987	240.0	93	179	0.39	119
Mapleton (Robison et al. 1999)	29.6	29	120	1.10	203
Tillamook (Robison et al. 1999)	28.6	10	124	0.35	74

Oregon Cascades Comparisons

Researcher	*Road R/W area (km)*	*Number of landslides*	*Average landslide volume (m^3)*	*Landslide density (#/km)*	*Erosion rate (m^3/km)*
Swanson & Dyrness 1975	2.10	73	1,351	6.0	640
Morrison 1975	0.61	75	1,942	20.0	3,200
Vida (Robison et al. 1999)	0.53	22	175	7.0	330
Blue R./Lookout Cr. (Wemple 1998)	11.20[1]	43	482	0.8	43

Researcher	*Road R/W area (km)*	*Number of landslides*	*Average landslide volume (m^3)*	*Landslide density (#/km)*	*Erosion rate (m^3/km)*
Swanson & Dyrness 1975	171[1]	—	—	0.43	7.90
Morrison 1975	49[1]	—	—	1.50	40.00
Vida (Robison et al. 1999)	42	—	—	0.52	4.20
Blue R./Lookout Cr. (Wemple 1998)	898	—	—	0.05	0.54

[1] Assumed a 12.5-meter road prism width based on average road width for "red zone" study areas in the ODF study (Robison et al. 1999).

both the occurrence and the volume of road-related landslides.

On a landscape level, road age and topographic position were factors in erosion-related damage to roads. Older roads (those constructed during the 1960s or earlier) were the source of significantly more erosion by landslides than more recent roads. Midslope roads produced more erosion than either ridge or valley-bottom roads.

Despite improved road standards, the density of landslides from logging roads in 1996 was roughly the same as in older inventories of roads with lower construction and drainage standards (Table 9-4). This demonstrates that, regardless of the standards, if roads are constructed in steep, landslide-prone terrain, there will be landslides. However, the data also show that the average volume of the road-related landslides in 1996 decreased compared with rates from the older inventories (Table 9-4), in some cases markedly. These results, and the lower road density that is now standard, mean that while road-related landslides still occur, the total accelerated erosion from roads should be less (Skaugset and Wemple 1999).

The effect of forestry on debris flows

Since the presentation of a disturbance-based approach to the management and recovery of aquatic ecosystems (Reeves et al. 1995; see Chapter 3), increasing attention has been paid to the effects of forestry on the size and composition of debris flows and their deposits. There is some evidence that the primary effect of forest management is on how far debris flows run and where they deposit. Ketcheson and Froehlich (1978) report that debris flows from clearcuts run farther and set up in larger deposits than debris flows from forests. May (1998) found that, of 53 debris flow deposits in anadromous fish-bearing streams, a disproportionate share came from and ran through harvest units.

There is very little information regarding the direct effect of timber harvest on the size and composition of debris flows. Empirical models that describe where debris flows will run and deposit show that deposition is governed by stream-channel geometry, stream gradient, and stream-valley width (Benda and Cundy 1990; Fannin and Rollerson 1993). However, the models describe only the runout and deposition of debris flows and not the composition of the deposit. May (1998) and Robison and others (1999) suggest that although the size of trees along the periphery of the debris flow may affect the length of runout, the primary de-terminants are channel geometry and gradient.

May (1998) investigated 53 deposits from debris flows in the central Coast Range resulting from the February 1996 storm. Forest management appeared to have no effect on the distribution of large wood in debris-flow deposits, and the size of the wood did not correspond to the size of the trees on the hillslope at the time of the debris flow. This result reflects a legacy of large wood stored in low-order stream channels. Forest management did affect the lengths of large wood pieces in the debris-flow deposits. Although short pieces of wood dominated the length distribution for all management classes, pieces were longer in debris flows that ran through mature forests.

Roads were found to be the one forest management activity that undeniably affected the composition, size, and run-out length of debris flows. Debris flows that initiated from roads had the largest initial slide volumes, traveled the farthest, and contained the most sediment (May 1998). In the ODF study areas, road-related landslides were, by count, the minority of all landslides, yet they accounted for a disproportionate amount of the length of highly impacted channels. This may be a result of the larger initial slide volumes of road-related landslides, which would cause them to travel farther (Robison et al. 1999).

Prevention and Mitigation of Accelerated Erosion in a Managed Coast Range Forest

The recognition of the link between forest management and accelerated erosion has prompted the development of alternative forest management practices that prevent or mitigate erosion. Since the initiation of Forest Practice Rules in 1972, and the research results from paired watershed studies, the development of alternative practices to mitigate accelerated erosion has intensified.

Three management or regulatory tools can drive changes in forest practices to prevent or mitigate accelerated erosion: water-quality standards, BMPs, and watershed management. Water-quality standards, which are the regulatory tool of choice for point sources of pollution, are not practical for non-point sources (Brown 1980). Nevertheless, debate continues regarding their efficacy for non-point sources of pollution from forestry practices.

Most progress made with regard to the prevention and mitigation of accelerated erosion has come through the establishment of BMPs. They are forest practices that are prescribed for a specific operation and are effective; that is, they reduce accelerated erosion from forestry activities and are technically feasible and economically viable.

The last tool is watershed management, or the management of forest practices on a scale that is larger than an individual forest stand or operation (the scale for BMPs). Watershed management prescribes the distribution of forest practices on a watershed both spatially and through time.

Contemporary forest management is currently in transition between using BMPs and adopting watershed management. Best management practices have been implemented for virtually all forest management activities, as evidenced by the robust sets of forest-practice rules adopted by all of the western timber-producing states. Some changes in BMPs and a few aspects of forest practices continue, but by and large BMPs are firmly in place. Conversely, although some rudimentary watershed-scale management practices are required, for example, harvest unit size and "green up" or adjacency constraints for clearcuts, watershed management as a regulatory tool is in its infancy.

Best management practices and accelerated erosion

Many BMPs have been developed to mitigate the effect of forest management activities on accelerated erosion. Their breadth of coverage in the forest-practice rules of western timber-producing states is a testament to the sheer number of site-specific and activity-specific BMPs that have been developed. It is beyond the scope of this chapter to review all BMPs developed to prevent or mitigate accelerated erosion. However, their efficacy at the watershed scale can be illustrated by reviewing research results from the Caspar Creek Watershed study. This watershed study is an excellent example of the advances made to prevent and mitigate accelerated erosion from timber harvesting by using BMPs.

The Caspar Creek study is unique because it pairs a study carried out in the 1960s, before the advent of forest-practice rules, with another carried out in the 1990s using contemporary BMPs. The experimental watersheds are the North Fork (473 hectares) and the South Fork (424 hectares) of Caspar Creek, located in northern California in the redwood region approximately 7 kilometers from the Pacific Ocean. The watersheds, originally logged between 1860 and 1904, grew back to young-growth redwood (*Sequoia sempervirens*) and Douglas-fir (*Pseudotsuga menziesii*) before contemporary treatments began. Both receive approximately 1,200 millimeters of precipitation annually coming as rain during the winter months, and both are underlain by marine sedimentary rock (Wright et al. 1990; Lewis 1998).

During the original study, the North Fork was used as a control and the South Fork was harvested. Roads were constructed in the South Fork in 1967 and harvesting took place between 1971 and 1973. Approximately 65 percent of the timber volume was selectively logged using tractors; 75 percent of the roads were within 60 meters of the stream. As a result, 15 percent of the watershed was compacted (Wright et al. 1990).

Harvesting for the second study was carried out between 1989 and 1992. In this study, the South Fork was the control and the North Fork was harvested. Because of limits on the size of clearcuts and adjacency constraints, only about 48 percent of the North Fork area was harvested. The harvesting was clearcut silviculture, using, for the most part, skyline-logging systems. Half of the harvest units were broadcast burned and half were not. Ground disturbance for roads, landings, skid trails, and fire lines was limited to 3.2 percent of the watershed. Streams were buffered by leaving a protection zone from 24 to 60 meters in width (Lewis 1998).

Rice and others (1979) found increases in the suspended sediment load in the South Fork watershed of 1,403 kilograms per hectare per year after road construction in 1967 and 3,254 kilograms per hectare per year for the five years after timber harvest began (1971 to 1976). When the roles of the two watersheds were reversed and the North Fork was harvested, Lewis (1998) reported an increase in suspended sediment of 188 kilograms per hectare per year for the six-year period after harvest began (1990 to 1996). When differences in sampling methods and streamflow are accounted for, the relative increases in excess suspended sediment load are 212 percent to 331 percent for the South Fork study and 89 percent for the North Fork study (Lewis 1998). These data suggest that harvesting using BMPs reduced the potential increase in suspended sediment load by 2.4 to 3.7 times.

The difference in the rate of accelerated erosion between the South Fork and North Fork watersheds is attributed to a difference in the harvesting practices or BMPs used. In both cases, roughly the same volume of timber was removed. The South Fork was selectively harvested to remove 65 percent of the volume while the North Fork was 48 percent clearcut. Three kinds of BMPs were used in the North Fork watershed, associated with the harvesting system, the road system, and stream protection.

The South Fork was harvested using a ground-based tractor system where location of the skid roads for the tractors was at the discretion of the tractor operators, meaning that a tractor was driven to virtually every log that was harvested. This harvesting method would have left a quarter to a third of the watershed disturbed (Froehlich 1976; Froehlich et al. 1981) and 15 percent of the South Fork was compacted by roads, landings, and skid trails (Wright et al. 1990, Lewis 1998). The North Fork was harvested using skyline-yarding systems, which are high-lift, long-span, cable-logging systems that result in less soil disturbance (3 to 4 percent; Rice et al. 1972), and virtually no compaction.

The second BMP that is credited with reducing accelerated erosion is improved road systems. The tractor-logging system for the South Fork required that roads be constructed low in the watershed and close to the stream. Over 70 percent of the road system in the South Fork was built within 60 meters of the stream. The skyline systems used in the North Fork watershed require fewer roads, and these roads are constructed on ridges or high in the watershed away from streams. Only 3.2 percent of the North Fork was in roads, landings, and skid trails and was considered compacted (Lewis 1998).

The third BMP used to reduce accelerated erosion was stream protection. The South Fork was harvested without any formal buffer strips for stream protection and with no equipment-exclusion zones around streams. In contrast, when the North Fork was logged, selectively logged buffer strips from 24 to 60 meters in width were left adjacent to the streams, and heavy equipment was excluded from these streamside areas.

Manuals are available for BMPs for forest practices and erosion. The most comprehensive are those associated with a given state's forest-practice rules (see Oregon Department of Forestry 1995). Even states that lack regulatory programs for forestry have manuals that describe BMPs for forestry. In addition, Weaver and Hagans (1994) have authored a good manual on BMPs to reduce erosion from forest roads.

Best management practices and debris slides and flows

With recognition of the link between forest management activities, especially timber harvesting, and debris flows (Bishop and Stevens 1964; Swanston and Swanson 1976) came the need to prevent or mitigate harvest-related debris flows. Therefore, BMPs have been developed over the years to prevent or mitigate the occurrence of debris flows associated with timber harvesting.

The first step in applying BMPs is recognition of where landslides are most likely to occur. Identification of the attributes of landslide-prone terrain has been an output of most landslide inventories (Bishop and Stevens 1964; O'Loughlin 1972; Megahan et al. 1978; Gresswell et al. 1979). Repeatedly, landslide inventories have shown that debris flows initiate on steep slopes and in convergent topography. In general, debris slides do not occur on slopes less than 50 percent, and most occur on slopes greater than 80 percent. In addition, debris flows are identified with topographic features or landforms that are convergent or concave, such as incipient stream channels, headwalls, hollows, swales, and deep, v-notched gullies.

The identification of key landscape attributes that are correlated with initiation sites has allowed the development of formal risk ratings for debris flow failures. These risk ratings are determined on the basis of aggregations of geomorphic and topographic features that are numerically weighted and summed. One such risk-rating system (Environmental Protection Agency 1980) was the basis for the Mapleton headwall risk-rating system that is used to rate the risk associated with debris-flow-initiation sites within the Mapleton District of the Siuslaw National Forest (Swanson and Roach 1987). The system stratifies headwalls into hazard classes that describe their likelihood of failure (Martin 1997). The ODF has used a definition of a high-risk site as a geomorphic hazard rating for potential initiation sites of debris flows. Their criteria for defining a high-risk site include: (1) active landslides, (2) uniform slopes steeper than 80 percent, (3) headwalls (convergent slopes) steeper than 70 percent, (4) abrupt slope breaks where the steeper slope exceeds 70 percent (mid-slope benches), and (5) inner gorges with side slopes steeper than 60 percent.

Recognizing slope steepness and convergence and applying geomorphic hazard ratings must be done in the field, standing on the site of interest and evaluating its potential to be a debris-slide or -flow-initiation site. However, computer algorithms can use digital elevation models (DEMs) with one-dimensional stability-analysis and rainfall-runoff relationships to predict the locations of colluvial hollows (Montgomery and Dietrich 1994; Wu and Sidle 1995). Evaluation of the efficacy of these models has shown that model output is generally consistent with empirical findings (Dietrich and Sitar 1997; Montgomery et al. 1998; Sidle and Wu 1999). The efficacy of these models illuminates the importance of topography to the initiation sites for debris-flow failures, and modeling allows the locations of potential high-risk sites to be evaluated over a large land area with a minimum investment in time and resources.

Of course, the efficacy of the computer algorithms to predict the location of potential debris-slide-initiation sites is only as good as the data they use or the DEMs being interrogated. Large scale DEMs, such as 30-meter DEMs, can obscure micro topography and underestimate slope steepness, both characteristics that are important to landslide-hazard assessment. Robison and others (1999) found a poor correlation between site-specific slope measurements and those derived from 30-meter DEMs and they also found that, on an area basis, 30-meter DEMs overrepresented moderate slope terrain (< 50%) and underrepresented very steep terrain (> 70%).

Once initiation sites of potential debris-flow failures have been identified, a number of BMPs can be prescribed to mitigate the increase in failure rate associated with timber harvesting. These BMPs are designed to maintain more root biomass after timber harvest than would be present after a typical clearcut silviculture prescription. The site-specific prescriptions range from a ban on harvesting to clearcut silviculture that accelerates the growth of young trees. At one end of the continuum is a complete ban on timber harvest on large areas that contain very steep slopes. This is followed by headwall-leave areas, which are patches of uncut trees that are left on the high-risk sites or hollows and on very steep slopes within a harvest unit. These leave areas can range in size from several hectares to several tens of hectares. Other possible prescriptions include minimizing site disturbance during harvest to maintain understory shrubs or brush to provide reinforcement after harvest. For the same reason, site preparation by either broadcast burning or silvicultural chemicals would be discouraged after harvest. Drastic disturbance and gouging of the soil on these sites during harvest are also discouraged to minimize effects on subsurface preferential-flow paths and soil strength. On some sites, an intermediate brush-release spray also would be discouraged.

At the opposite end of the prescription continuum is very intensive treatment of the site, including clearcut harvest, site preparation, planting superior stock, and brush treatment. A window of vulnerability after harvest is recognized as a part of this treatment (Ziemer 1981). However, instead of providing more reinforcement during that period of vulnerability, this prescription narrows the window of vulnerability by encouraging rapid reforestation of the site and fast growth of the stock through intensive forest management.

The efficacy of only one of these BMPs for treatment of initiation sites has been tested. Martin (1997) found numerical differences for the failure rates of headwalls in unmanaged forest (36 percent), clearcut (38 percent), and headwall-leave area within a clearcut (41 percent). These numerical differences were not statistically significant. The efficacies of other BMPs have not been tested. That is, the effects of differences in site disturbance, site preparation, planting-stock quality, and intermediate treatments have not been tested; all have been lumped into the category of "clearcut" in landslide inventories.

During the ODF study, investigators checked for compliance with the Oregon Forest Practice Rules at sites where debris-flow failures initiated. The specific forest practices considered included linear gouging during cable yarding, construction of tractor skid trails, and accumulations of logging slash. In no case had the forest practice rules been violated, and none of these factors was associated with the initiation of debris flows (Robison et al. 1999).

A final concern is the effect of timber harvesting on debris-flow composition, especially that timber harvesting would result in decreases in the amount and size (diameter and length) of large wood in debris flows. What little research has been done on this subject does not support that hypothesis. In fact, harvested terrain exhibits a legacy of large wood in debris flows (May 1998). However, if large wood is not allowed to grow adjacent to debris-flow paths, then those paths that fail will not fill with large wood. Given that timber harvest has generally reduced the amount of large wood in and available to streams and that large wood is an integral component of aquatic habitat and the storage and routing of sediment, it is important that debris flows associated with timber harvest include large wood (see Chapter 3). Therefore, although site-specific BMPs have not been developed for debris-flow composition, it would seem reasonable to require both an increase in the size and extent of protected riparian zones and the establishment or nonharvesting of conifers along selected debris-flow paths.

Watershed management and debris flows

A growing body of research and thought proposes a new paradigm for the management of landslide-prone forested terrain. The new paradigm does not prescribe BMPs alone to mitigate the effect of timber harvesting on debris flows and aquatic habitat. Rather, it is based on the prescription of forest management activities on a larger scale in both space and time. This watershed-scale prescription of forest management activities is designed to make timber harvest more nearly mimic a natural disturbance regime. The theoretical framework and the basis for this new paradigm is presented by Reeves et al. (1995) and in Chapter 4.

Anadromous salmonids evolved in ecosystems that were dynamic in both space and time. The dynamic nature of aquatic habitat is the result of a natural disturbance regime that is, in turn, the result of stand-replacing wildfires, which leave the terrain vulnerable to landslides, whose frequency, size, and spatial distribution on a landscape scale create a range of channel conditions. When a landslide-producing storm occurs subsequent to a wildfire, the resulting widespread occurrence of debris flows transports large quantities of wood and sediment into the stream channel. The episodic delivery of sediment to stream channels causes them to shift between periods of aggradation and degradation. Because the quality of aquatic habitat is a function of channel conditions, it will also shift, with the optimum in complex aquatic habitat occurring between the extremes of aggraded and degraded states.

Several attributes characterize a natural disturbance regime in the Oregon Coast Range (Benda 1994). First is disturbance, such as a stand-replacement wildfire, which has an average size of approximately 30 square kilometers and a return rate of 200 to 300 years. Using these figures, Benda (1994) reports that in a natural regime, about 15 to 25 percent of the forested landscape will be in an early-successional state at any given time. The wildfires leave standing and down large wood, a biological legacy that becomes incorporated into debris flows after the fire. Together these attributes lead to the pattern of channel conditions and associated aquatic habitat conditions in which salmonids evolved.

The disturbance regime created by contemporary forest management practices differs in these attributes. First, the size of the disturbance made by a typical harvest unit is much smaller, and its recurrence is much more frequent, such as every 50 to 100 years. As a result, 35 percent or more of contemporary managed forests may be in an early-successional state at any given time. Second, given the purpose of timber harvest, managed forests historically have not left the amount or size of standing and down wood that wildfires have.

The landscape conditions in a contemporary managed forest may not be capable of providing the range of channel and habitat conditions that ultimately result in high-quality aquatic habitat. For these conditions to develop, the forested landscape must be managed in such a manner that the disturbance regime more closely resembles the natural disturbance regime of the Coast Range, including the episodic delivery of large amounts of sediment and large wood.

Several prescriptions at the landscape scale might bring a managed disturbance regime to more closely resemble a natural disturbance regime: (1) lengthen the rotation ages to ± 150 years; (2) reduce the proportion of the landscape in young stands (Robison et al. 1999); (3) aggregate harvest units into one area in a watershed instead of scattering them across the landscape; and (4) leave debris-flow paths with a biological legacy of large wood. The debris-flow paths that have the strongest likelihood of delivering large wood to streams can be identified (Benda 1990) and conifers should be established, allowed to become large, and left to enhance the biological legacy of subsequent debris flows.

Summary

Erosion is a natural process that occurs even in pristine forested terrain in western Oregon. The dominant erosional form is mass movement and, in particular, debris flows. Surface erosion is virtually nonexistent. Forest management activities, especially timber harvest, cause accelerated erosion in the forms of surface erosion from forest roads and an increased occurrence of landslides from clearcut harvest units and forest roads.

The BMPs used over the last three decades have reduced the amount of accelerated erosion that occurs as a result of forest management. These BMPs include protection zones around streams or buffer strips that are used to limit the amount of disturbance to streambed and banks. Contemporary harvest systems include more long-span, high-lift cable systems that require fewer roads and leave less ground disturbance than older systems. Finally, the density of forest roads has been reduced, and the roads are better located, built, drained, and maintained. Even so, the opportunity for accelerated erosion remains.

One objective of BMPs for surface erosion and debris flows from roads is to limit accelerated erosion to the extent practicable and feasible. In contrast, a different perspective is being taken for debris flows from within harvest units. According to new data and a new line of thinking regarding the role of natural disturbances in the formation of aquatic habitat, debris flows are the vectors that bring large wood and sediment, the basic building blocks for aquatic habitat, into the stream. Rather than trying to prevent all debris flows from harvest units, it may be more important and realistic to institute forest management schemes in which resulting debris flows mimic natural processes with regards to timing and large-wood composition as much as possible.

Forest management can affect aquatic habitat by changing the timing and composition of debris flows. Timber harvest changes the disturbance pattern of a forested landscape by introducing rotation ages that are shorter than the interval between naturally occurring wildfires and by spreading timber harvest out in space and time. Furthermore, timber harvest removes more large wood and leaves less standing and down large wood than wildfires. It is possible to develop BMPs that will ultimately result in debris-flow paths retaining sufficient large wood. However, landscape management, not BMPs, is needed to provide a more natural timing of the occurrence of debris flows.

Literature Cited

Abramson, L. W., T. S. Lee, S. Sharma, and G. M. Boyce. 1996. *Slope Stability and Stabilization Methods*. John Wiley & Sons, New York.

Alger, C. S., and S. D. Ellen. 1987. Zero-order basins shaped by debris flows, Sunol, California, USA, pp. 111-119 in *Erosion and Sedimentation in the Pacific Rim*, R. L. Beschta, T. Blinn, G. E. Grant, G. G. Ice, and F. J. Swanson, ed. IAHS publication No. 165, Wallingford U.K.

Amaranthus, M. P., R. M. Rice, N. R. Barr, and R. R. Ziemer. 1985. Logging and forest roads related to increased debris slides in southwestern Oregon. *Journal of Forestry* 83: 229-233.

Anderson, S. A., and N. Sitar. 1995. Analysis of rainfall-induced debris flows. *Journal of Geotechnical Engineering* 121(7): 544-552.

Anderson, S. P., W. E. Dietrich, D. R. Montgomery, R. Torres, M. E. Conrad, and K. Loague. 1997. Subsurface flow paths in a steep, unchanneled catchment. *Water Resources Research* 33: 2637-2653.

Benda, L. E. 1990. The influence of debris flows on channels and valley floors in the Oregon Coast Range, USA. *Earth Surface Processes and Landforms* 15: 457-466.

Benda, L. E. 1994. *Stochastic Geomorphology in a Humid Mountain Landscape*. PhD dissertation, University of Washington, Seattle.

Benda, L. E., and T. W. Cundy. 1990. Predicting deposition of debris flows in mountain channels. *Canadian Geotechnical Journal* 27: 409-417.

Benda, L. E., and T. Dunne. 1997a. Stochastic forcing of sediment supply to channel networks from landsliding and debris flow. *Water Resources Research* 33: 2849-2863.

Benda, L. E., and T. Dunne. 1997b. Stochastic forcing of sediment routing and storage in channel networks. *Water Resources Research* 33: 2865-2880.

Benda, L. E., D. J. Miller, T. Dunne, G. H. Reeves, and J. K. Agee. 1998. Dynamic landscape systems, pp. 261-288 in *River Ecology and Management: Lessons from the Pacific Coastal Ecoregion*, R. J. Naiman and R. E. Bilby, ed. Springer-Verlag, New York.

Benda, L., C. Veldhuisen, D. Miller, and L.R. Miller. Undated. *Slope Instability and Forest Land Managers, a Primer and Field Guide*. Earth Systems Institute, Seattle WA.

Bennett, K. A. 1982. *Effects of Slash Burning on Surface Soil Erosion Rates in the Oregon Coast Range*. MS thesis, Oregon State University, Corvallis.

Beschta, R. L. 1978. Long-term patterns of sediment production following road construction and logging in the Oregon Coast Range. *Water Resources Research* 14(6): 1011-1016.

Beschta, R. L. 1998. Forest hydrology in the Pacific Northwest: additional research needs. *Journal of the American Water Resources Association* 34(4): 729-763.

Bevin, K. J., R. Lamb, P. Quinn, R. Romanowicz, and J. Freer. 1995. TOPMODEL, pp. 627-668 in *Computer Models of Watershed Hydrology*, V. P. Singh, ed. Water Resources Publications, LLC, Highlands Ranch CO.

Bilby, R. E., K. Sullivan, and S. H. Duncan. 1989. The generation and fate of road-surface sediment in forested watersheds in southwestern Washington. *Forest Science* 35(2): 453-468.

Bishop, D. M., and M. E. Stevens. 1964. *Landslides on Logged Areas in Southeast Alaska*. Research Paper NOR-1, USDA Forest Service, Northern Forest Experiment Station, Juneau AK.

Bonell, M. 1998. Selected challenges in runoff generation research in forests from the hillslope to headwater drainage basin scale. *Journal of the American Water Resources Association* 34(4): 765-785.

Bransom, M. 1990. *Soil Engineering Properties and Vegetative Characteristics for Headwall Slope Stability Analysis in the Oregon Coast Range*. MS thesis, Oregon State University, Corvallis.

Bransom, M. 1997. *Geohydrologic Conditions on a Steep Forested Slope: Modeling Transient Piezometric Response to Precipitation*. PhD dissertation, Oregon State University, Corvallis.

Brown, G. W. 1980. *Forestry and Water Quality*. Oregon State University Bookstores, Inc., Corvallis.

Brown, G. W., and J. T. Krygier. 1971. Clearcut logging and sediment production in the Oregon Coast Range. *Water Resources Research* 7(5): 1189-1198.

Buchanan, P., and K. W. Savigny. 1990. Factors controlling debris avalanche initiation. *Canadian Geotechnical Journal* 19(2): 167-174.

Burroughs, E. R. 1984. Landslide hazard rating for portions of the Oregon Coast Range, pp. 265-274 in *Symposium on Effects of Forest Land Use on Erosion and Slope Stability*, C. L. O'Loughlin and A. J. Pearce, ed. Environment and Policy Institute, University of Hawaii, Honolulu.

Burroughs, E. R., and B. R. Thomas. 1977. *Declining Root Strength in Douglas-fir After Falling as a Factor in Slope Stability*. Research Paper INT-190, USDA Forest Service, Intermountain Forest and Range Experiment Station, Ogden UT.

Campbell, R. H. 1975. *Soil Slips, Debris Flows, and Rainstorms in the Santa Monica Mountains and Vicinity, Southern California*. Professional Paper 851, U.S. Geological Survey, Washington DC.

Chamberlin, T. W., R. D. Harr, and F. H. Everest. 1991. Timber harvesting, silviculture, and watershed processes, pp. 181-205 in *Influences of Forest and Rangeland Management on Salmonid Fishes and Their Habitats*, W. R. Meehan, ed. Special Publication 19, American Fisheries Society, Bethesda MD.

Collins, T. S. 1995. *Evaluation of Triaxial Strength Tests for Soils of the Oregon Coast Range*. Engineering Report, Department of Civil Engineering, Oregon State University, Corvallis.

Croft, A. R., and J. A. Adams. 1950. *Landslides and Sedimentation in the North Fork of Ogden River, May 1959*. Research Paper No. 21, USDA Forest Service, Intermountain Forest and Range Experiment Station, Ogden UT.

Crozier, M. J., E. E. Vaughan, and J. M. Tippett. 1990. Relative instability of colluvium-filled bedrock depressions. *Earth Surface Processes and Landforms* 15: 329-339.

Cruden, D. M., and D. J. Varnes. 1996. Landslide types and processes, pp. 36-75 in *Landslides Investigation and Mitigation*, A. K. Turner and R. L. Schuster, ed. Special Report 247, Transportation Research Board, National Research Council, National Academy Press, Washington DC.

Dietrich, W. E., and T. Dunne. 1978. Sediment budget for a small catchment in mountainous terrain. *Z. Geomorph. N.F. Suppl. Bd.* 29: 191-206.

Dietrich, W. E., and N. Sitar. 1997. Geoscience and geotechnical engineering aspects of debris-flow hazard assessment, pp. 656-676 in *Debris-flow Hazards Mitigation: Mechanics, Prediction, and Assessment*, C. Chen, ed. Proceedings of First International Conference, San Francisco CA.

Dietrich, W. E., S. L. Reneau, and C. J. Wilson. 1987. Overview: "zero-order basins" and problems of drainage density, sediment transport and hillslope morphology, pp. 27-37 in *Erosion and Sedimentation in the Pacific Rim*, R. L. Beschta, T. Blinn, G. E. Grant, G. G. Ice, and F. J. Swanson, ed. IAHS Publication No. 165, Wallingford U.K.

Dietrich, W. E., R. Reiss, M. L. Hsu, and D. R. Montgomery. 1995. A process-based model for colluvial soil depth and shallow landsliding using digital elevation data. *Hydrological Processes* 9: 383-400.

Duan, J. 1996. *A Coupled Hydrologic-geomorphic Model for Evaluating Effects of Vegetation Change on Watersheds*. PhD dissertation, Oregon State University, Corvallis.

Dunne, T. 1998. Critical data requirements for prediction of erosion and sedimentation in mountain drainage basins. *Journal of the American Water Resources Association* 34: 795-808.

Endo, T., and T. Tsuruta. 1969. The effect of the tree's roots upon the shear strength of soil, pp. 167-182 in *1968 Annual Report, Hokkaido Branch, Forest Experiment Station*. Sapporo, Japan.

Environmental Protection Agency. 1980. *An Approach to Water Resources Evaluation of Non-point Silvicultural Sources (A Procedural Handbook)*. Environmental Research Laboratory, Office of Research and Development, Athens GA.

Everest, F. H., and W. R. Meehan. 1981. Forest management and anadromous fish habitat productivity, pp. 521-530 in *Transactions of the 46th North American Wildlife and Natural Resources Conference*. Wildlife Management Institute, Washington DC.

Everest, F. H., R. L. Beschta, J. C. Scrivener, K. V. Koski, J. R. Sedell, and C. J. Cederholm. 1987. Fine sediment and salmonid production: a paradox, pp. 98-142 in *Streamside Management: Forestry and Fishery Interactions*, E. O. Salo and T. W. Cundy, ed. Contribution No. 57, College of Forest Resources, University of Washington, Seattle.

Fannin, R. J., and T. P. Rollerson. 1993. Debris flows: some physical characteristics and behaviour. *Canadian Geotechnical Journal* 30: 71-81.

Fedora, M. A., and R. L. Beschta. 1989. Storm runoff simulation using an antecedent precipitation index (API) model. *Journal of Hydrology* 112: 121-133.

Fiksdal, A. J. 1974. *A Landslide Survey of the Stequaleho Creek Watershed*. Supplement to Final Report UW-7404, Fisheries Research institute, University of Washington, Seattle.

Foltz, R. B. 1993. *Sediment Processes in Wheel Ruts on Unsurfaced Forest Roads*. PhD dissertation, University of Idaho, Moscow.

Foltz, R. B. 1996. *Roughness Coefficients in Forest Road-side Ditches*. Poster paper H12B-9 published as a supplement to EOS Transactions, AGU 77(46), November 12, 1996. Fall Meeting of the American Geophysical Union, San Francisco CA.

Foltz, R. B., and E. R. Burroughs, Jr. 1990. Sediment production from forest roads with wheel ruts, pp. 226-275 in *Watershed Planning and Analysis; Proceedings of a Symposium, July 9-11, 1989*. American Society of Civil Engineers, Durango CO.

Foltz, R. B., and E. R. Burroughs, Jr. 1991. A test of normal tire pressure and reduced tire pressure on forest roads: sedimentation effects, in *Proceedings of the June 5-6, 1991 Conference on Forest and Environment Engineering Solutions*. American Society of Agricultural Engineers, St. Joseph MI.

Fredriksen, R. L. 1970. *Erosion and Sedimentation Following Road Construction and Timber Harvest on Unstable Soils in Three Small Western Oregon Watersheds*. Research Paper PNW-104, USDA Forest Service, Pacific Northwest Forest and Range Experiment Station, Portland OR.

Froehlich, H. A. 1976. The influence of different thinning systems on damage to soil and trees, pp. 333-344 in *Proceedings XVI IUFRO World Congress, Division IV.* Norway.

Froehlich, H. A. 1978. The influence of clearcutting and road building activities on landscape stability in western United States, pp. 165-173 in *Proceedings of the 5th North American Forest Soils Conference*. Colorado State University, Fort Collins.

Froehlich, H. A., D. E. Aurlich and R. Curtis. 1981. *Designing Skid Trail Systems to Reduce Soil Impacts from Tractive Logging Machines*. Research Paper 44, Oregon State University, School of Forestry, Corvallis.

Gonsior, M. J., and R. B. Gardner. 1971. *Investigation of Slope Failures in the Idaho Batholith*. Research Paper INT-97, USDA Forest Service, Intermountain Forest and Range Experiment Station, Ogden UT.

Gresswell, S., D. Heller, and D. N. Swanston. 1979. *Mass Movement Response to Forest Management in the Central Oregon Coast Range*. Resources Bulletin PNW-84, USDA Forest Service, Pacific Northwest Forest and Range Experiment Station, Portland OR.

Hack, J. T., and J. C. Goodlett. 1960. *Geomorphology and Forest Ecology of Mountain Region in the Central Appalachians*. Professional Paper 347, U.S. Geological Survey, Washington DC.

Hammond, C., D. Hall, S. Miller, and P. Swetik. 1992. *Level 1 Stability Analysis (LISA) Documentation for Version 2.0.* General Technical Report INT-285, USDA Forest Service, Intermountain Forest and Range Experiment Station, Ogden UT.

Harr, R. D. 1977. Water flux in soil and subsoil on a steep forested slope. *Journal of Hydrology* 33: 37-58.

Holtz, R. D., and W. D. Kovacs. 1981. *An Introduction to Geotechnical Engineering*. Prentice-Hall, Inc., Englewood Cliffs NJ.

Humphrey, N. F. 1981. *Pore Pressures in Debris Failure Initiation*. Project Completion Report, OWRT Project Number A-108-WASH, Department of Geological Sciences, University of Washington, Seattle.

Ice, G. G. 1985. *Catalog of Landslide Inventories for the Northwest*. Technical Bulletin No. 456, National Council of the Paper Industry for Air and Stream Improvement, New York.

Johnson, K. A., and N. Sitar. 1990. Hydrologic conditions leading to debris-flow initiation. *Canadian Journal of Geotechnical Engineering* 27: 789-801.

Johnson, M. G., and R. L. Beschta. 1978. Logging, infiltration capacity, and surface erodibility in western Oregon. *Journal of Forestry* 78(6): 334-337.

Kelsey, H. M., and J. G. Bockheim. 1994. Coastal landscape evolution as a function of eustasy and surface uplift rate, Cascadia margin, southern Oregon. *Geological Society of America Bulletin* 106: 840-854.

Kelsey, H. M., D. C. Engebretson, C. E. Mitchell, and R. L. Ticknor. 1994. Topographic form of the Coast Ranges of the Cascadia Margin in relation to coastal uplift rates and plate subduction. *Journal of Geophysical Research* 99: 12245-55.

Kelsey, H. M., R. L. Ticknor, J. G. Bockheim, and C. E. Mitchell. 1996. Quaternary upper plate deformation in coastal Oregon. *Geological Society of America Bulletin* 108: 843-860.

Keough, D. 1994. *An Investigation of Index and Strength Properties of Landslide Susceptible Soils in the Oregon Coast Range*. Engineering Report, Department of Civil Engineering, Oregon State University, Corvallis.

Ketcheson, G., and H. A. Froehlich. 1978. *Hydrologic Factors and Environmental Impacts of Mass Soil Movements in the Oregon Coast Range*. WRRI-56, Oregon Water Resources Research Institute, Oregon State University, Corvallis.

Krammes, J. S., and D. M. Burns. 1973. *Road Construction on Caspar Creek Watersheds, 10-year Report on Impact*. Research Paper PSW-93, USDA Forest Service, Pacific Southwest Forest and Range Experiment Station, Berkeley CA.

Lambe, T. W., and R. V. Whitman. 1969. *Soil Mechanics*. John Wiley & Sons, Inc., New York.

Lamberti, G. A., S. V. Gregory, L. R. Ashkenas, R. C. Wildman, and K. M. Moore. 1991. Stream ecosystem recovery following a catastrophic debris flow. *Canadian Journal of Fisheries and Aquatic Sciences* 48: 196-208.

Lewis, J. 1998. Evaluating the impacts of logging activities on erosion and suspended sediment transport in the Caspar Creek watersheds, pp. 55-69 in *Proceedings of the Conference on Coastal Watersheds: the Caspar Creek Story*, R. R. Ziemert, tech. coord. General Technical Report. PSW-GTR-168, USDA Forest Service, Pacific Southwest Forest and Range Experiment Station, Berkeley CA.

Luce, C. H., and T. A. Black. 1999. Sediment production from forest roads in western Oregon. *Water Resources Research* 35(8): 2561-2570.

Marron, D. C. 1985. Colluvium in bedrock hollows on steep slopes, Redwood Creek drainage basin, northwestern California. *Catena Supplement* 6: 59-68.

Martin, K. 1997. *Forest Management on Landslide Prone Sites: the Effectiveness of Headwall Leave Areas and Evaluation of Two Headwall Risk Rating Methods*. Engineering Report, Department of Civil Engineering, Oregon State University, Corvallis.

May, C. 1998. *Debris Flow Characteristics Associated with Forest Practices in the Central Oregon Coast Range*. MS thesis, Oregon State University, Corvallis.

McNabb, D. H., F. Gaweda, and H. A. Froehlich. 1989. Infiltration, water repellency, and soil moisture content after broadcast burning a forest site in southwest Oregon. *Journal of Soil and Water Conservation* 44(1): 87-90.

Megahan, W. F., N. F. Day, and T. M. Bliss. 1978. Landslide occurrence in the western and central Northern Rocky Mountain Physiographic Province in Idaho, pp. 116-139 in *Proceedings of the 5th North American Forest Soils Conference*. Colorado State University, Fort Collins.

Mitchell, J. K. 1976. *Fundamentals of Soil Behavior*. John Wiley & Sons, Inc., New York.

Montgomery, D. R., and W. E. Dietrich. 1994. A physically based model for the topographic control on shallow landsliding. *Water Resources Research* 30: 1153-1171.

Montgomery, D. R., W. E. Dietrich, R. Torres, S. P. Anderson, J. T. Heffner, and K. Loague. 1997. Hydrologic response of a steep, unchanneled valley to natural and applied rainfall. *Water Resources Research* 33(1): 91-109.

Montgomery, D. R., K. Sullivan, and H. M. Greenberg. 1998. Regional test of a model for shallow landsliding. *Hydrologic Processes* 12: 943-955.

Morgan, D. J. 1995. *Strength Parameters and Triaxial Strength Testing of Soils of the Oregon Coast Range*. Engineering Report, Department of Civil Engineering, Oregon State University, Corvallis.

Morrison, P. H. 1975. Ecological and geomorphological consequences of mass movements in the Alder Creek watershed and implications for forest land management. BA thesis, University of Oregon, Eugene.

Nash, D. 1987. Chapter 2: A comparative review of limit equilibrium methods of stability analysis, pp. 11-75 in *Slope Stability*. M. G. Anderson and K. S. Richards, ed. John Wiley & Sons Inc., New York NY.

O'Loughlin, C. L. 1972. *An Investigation of the Stability of the Steepland Forest Soils in the Coast Mountains, Southwest British Columbia*. PhD dissertation, University of British Columbia, Vancouver.

O'Loughlin, C. L. 1974. The effect of timber removal on the stability of forest soils. *Journal of Hydrology (NZ)* 13(2): 121-134.

O'Loughlin, C. L., and A. J. Pearce. 1976. Influence of Cenozoic geology on mass movement and sediment yield response to forest removal, North Westland, New Zealand. *Bulletin of the International Association of Engineering and Geology* 14: 41-46.

O'Loughlin, C. L., L. K. Rowe, and A. J. Pearce. 1982. Exceptional storm influences on slope erosion and sediment yield in small forest catchments, North Westland, New Zealand, pp. 84-91 in *Proceeding of the First National Symposium on Forest Hydrology, Melbourne, Australia, May 1982*. Publication No. 82/6, The Institute of Engineers, Australia, National Conference, Barton, ACT, Australia.

O'Loughlin, E. M. 1986. Prediction of surface saturation zones in natural catchments by topographic analysis. *Water Resources Research* 22: 794-804.

Oregon Department of Forestry. 1995. *Forest Practices Field Guide*. Forest Practices Section, Salem.

Orr, E. L., W. N. Orr, and E. M. Baldwin. 1992. *Geology of Oregon*. Fourth edition. Kendall/Hunt, Dubuque IA.

Pyles, M. R., and H. A. Froehlich. 1987. Discussion of rates of land sliding as impacted by timber management activities in northwestern California. *Bulletin of the Association of Engineering Geologists* 24(3): 425-431.

Pyles, M. R., P. W. Adams, R. L. Beschta, and A. E. Skaugset. 1998. *Forest Practices and Landslides. A Report Prepared for Governor John A. Kitzhaber*. Department of Forest Engineering, Oregon State University, Corvallis.

Reeves, G. H., L. E. Benda, K. M. Burnett, P. A. Bisson, and J. R. Sedell. 1995. A disturbance-based ecosystem approach to maintaining and restoring freshwater habitats of evolutionarily significant units of anadromous salmonids in the Pacific Northwest. *American Fisheries Society Symposium* 17: 334-349.

Reid, L. M., and T. Dunne. 1984. Sediment production from forest road surfaces. *Water Resources Research* 20(11): 1753-1761.

Reid, M. E., R. G. LaHusen, and R. M. Iverson. 1997. Debris-flow initiation experiments using diverse hydrologic triggers, pp. 1-11 in *Debris-flow Hazards Mitigation: Mechanics, Prediction, and Assessment*, C. Chen, ed. Proceedings of First International Conference, San Francisco CA.

Reneau, S. L., and W. E. Dietrich. 1987. Size and location of colluvial landslides in a steep forested landscape, pp. 38-48 in *Erosion and Sedimentation in the Pacific Rim*, R. L. Beschta, T. Blinn, G E. Grant, G. G. Ice, and F. J. Swanson, ed. IAHS publication No. 165, Wallingford U.K.

Reneau, S. L., and W. E. Dietrich. 1991. Erosion rates in the southern Oregon Coast Range: evidence for an equilibrium between hillslope erosion and sediment yield. *Earth Surface Processes and Landforms* 16: 307-322.

Reneau, S. L., W. E. Dietrich, M. Rubin, D J. Donahue, and A. T. Jull. 1989. Analysis of hillslope erosion rates using dated colluvial deposits. *Journal of Geology* 97: 45-63.

Rice, R. M., L. S. Rothacher, and W. F. Megahan. 1972. Erosional consequences of timber harvesting: an appraisal, pp. 321-329 in *Proceedings of a Symposium on "Watersheds in Transition,"* S. D. Csallany, T. G. McLaughlin, and W. D. Striffler, ed. Fort Collins CO.

Rice, R. M., F. B. Tilley, and P. A. Datzman. 1979. *A Watershed's Response to Logging and Roads: South Fork of Caspar Creek, California, 1967-1976*. Research Paper, PSW-146, USDA Forest Service, Pacific Southwest Forest and Range Experiment Station, Berkeley CA.

Robison, E. G., K. Mills, J. Paul, L. Dent, and A. Skaugset. 1999. *Oregon Department of Forestry Storm Impacts and Landslides of 1996: Final Report*. Forest Practices Technical Report No. 4, Oregon Department of Forestry, Salem.

Schroeder, W. L., and G. W. Brown. 1984. Debris torrents, precipitation, and roads in two coastal Oregon watersheds, pp. 117-122 in *Symposium on the Effects of Forest Land Use on Erosion and Slope Stability, May 7-11*. Honolulu HI.

Schroeder, W. L., and D. N. Swanston. 1987. *Application of Geotechnical Data to Resource Planning in Southeast Alaska*. General Technical Report PNW-198, USDA Forest Service, Pacific Northwest Resource Station, Portland OR.

Sessions, J., J. C. Balcom, and K. Boston. 1987. Road location and construction practices: effects on landslide frequency and size in the Oregon Coast Range. *Western Journal of Applied Forestry* 2(4): 119-124.

Sidle, R. C., and D. N. Swanston. 1982. Analysis of a small debris slide in coastal Alaska. *Canadian Geotechnical Journal* 19(2): 167-174.

Sidle, R. C., and W. Wu. 1999. Simulating effects of timber harvesting on the temporal and spatial distribution of shallow landslides. *Z. Geomorph. N.F.* 43(2): 185-201.

Sidle, R. C., A. J. Pearce, and C. L. O'Loughlin. 1985. *Hillslope Stability and Land Use*. Water Resources Monograph 11, American Geophysical Union.

Skaugset, A. E., and B. C. Wemple. 1999. The response of forest roads on steep, landslide-prone terrain in western Oregon to the February 1996 storm, pp. 193-203 in *Proceedings of the International Mountain Logging and 10th Pacific Northwest Skyline Symposium, March 28-April 1, 1999*, J. Sessions and W. Chung, ed. Corvallis OR.

Sullivan, K., T. E. Lisle, C. A. Dolloff, G. E. Grant, and L. M. Reid. 1987. Stream channels: the link between forests and fishes, pp. 39-97 in *Streamside Management: Forestry and Fishery Interactions*, E. O. Salo and T. W. Cundy, ed. Contribution No. 57, College of Forest Resources, University of Washington, Seattle.

Swanson, F. J., and C. T. Dyrness. 1975. Impact of clearcutting and road construction on soil erosion by landslides in the western Cascades Range, Oregon. *Geology* 3(7): 393-396.

Swanson, F. J., and R. L. Fredriksen. 1982. Sediment routing and budgets: Implications for judging impacts of forestry practices, pp. 129-137 in *Sediment Budgets and Routing in Forested Drainage Basins*, F. J. Swanson, R. J. Janda, T. Dunne, and D. N. Swanston, ed. General Technical Report, PNW-141, USDA Forest Service, Pacific Northwest Forest and Range Experiment Station, Portland OR.

Swanson, F. J., and G. W. Lienkaemper. 1978. *Physical Consequences of Large Organic Debris in Pacific Northwest Streams*. General Technical Report PNW-69, USDA Forest Service, Pacific Northwest Forest and Range Experiment Station, Portland OR.

Swanson, F. J., and C. J. Roach. 1987. *Mapleton Leave Area Study*. Administrative Report, USDA Forest Service, Pacific Northwest Research Station, Corvallis OR.

Swanson, F. J., and D. N. Swanston. 1977. Complex mass-movement terrains in the western Cascade Range, Oregon. *Geological Society of America, Reviews in Engineering Geology* 3: 113-124.

Swanson, F. J., M. M. Swanson, and C. Woods. 1977. *Inventory of Mass Erosion in the Mapleton Ranger District, Siuslaw National Forest*. Final Report, Siuslaw National Forest and Pacific Northwest Forest and Range Experiment Station, Forestry Sciences Laboratory, Corvallis OR.

Swanson, F. J., R. L. Fredriksen, and F. M. McCorison. 1982. Material transfer in a western Oregon forested watershed, Chapter 8, pp. 233-266 in *Analysis of Coniferous Forest Ecosystems in the Western United States*. R. L. Edmonds, ed. US/IBP Synthesis Series 14, Hutchinson Ross Publishing Company, Stroudsburg PA.

Swanson, F. J., L. E. Benda, S. H. Duncan, G. E. Grant, W. F. Megahan, L. M. Reid, and R. R. Ziemer. 1987. Mass failures and other processes of sediment production in Pacific Northwest forest landscapes, pp. 9-38 in *Streamside Management: Forestry and Fishery Interactions*, E. O. Salo and T. W. Cundy, ed. Contribution No. 57, College of Forest Resources, University of Washington, Seattle.

Swanson, F. J., J. L. Clayton, W. F. Megahan, and G. Bush. 1989. Erosional processes and long-term site-productivity, pp. 67-81 in *Maintaining the Long-term Productivity of Pacific Northwest Forest Ecosystems*, D. A. Perry, R. Meurisse, B. Thomas, R. Miller, J. Boyle, J. Means, C. R. Perry, and R. F. Powers, ed. Timber Press, Portland OR.

Swanston, D. N. 1967. *Soil-water Pieziometry in a Southeast Alaska Landslide Area*. Research Note PNW-68, USDA Forest Service, Pacific Northwest Forest and Range Experiment Station, Portland OR.

Swanston, D. N. 1970. *Mechanics of Debris Avalanching in Shallow Till Soils of Southeast Alaska*. Research Paper PNW-103, USDA Forest Service, Pacific Northwest Forest and Range Experiment Station, Portland OR.

Swanston, D. N. 1978. Effect of geology on soil mass movement activity in the Pacific Northwest, pp. 89-115 in *Forest Soils and Land Use, Proceedings of the Fifth North American Forest Soils Conference*, C. T. Youngberg, ed. Colorado State University, Fort Collins.

Swanston, D. N., and F. J. Swanson. 1976. Timber harvesting, mass erosion, and steepland forest geomorphology in the Pacific Northwest, pp. 199-221 in *Geomorphology and Engineering*, D. R. Coates, ed. Dowden, Hutchinson, and Ross, Stroudsburg PA.

Tsukamoto, Y., and H. Minematsu. 1987. Hydrogeomorphological characteristics of a sero-order basin, pp. 61-70 in *Erosion and Sedimentation in the Pacific Rim*, R. L. Beschta, T. Blinn, G. E. Grant, G. G. Ice, and F. J. Swanson, ed. IAHS Publication No. 165, Wallingford U.K.

Waldron, L. J., and S. Dakessian. 1981. Soil reinforcement by roots: calculation of increased soil shear resistance from root properties. *Soil Science* 132(6): 427-435.

Waldron, L. J., and S. Dakessian. 1982. Effect of grass, legume, and tree roots on soil shearing resistance. *Soil Science Society of America Journal* 46: 894-899.

Waldron, L. J., S. Dakessian, and J. A. Nemson. 1983. Shear resistance enhancement of 1.22-meter diameter soil cross sections by pine and alfalfa roots. *Soil Science Society of America Journal* 47: 9-14.

Weaver, W. E., and D. K. Hagans. 1994. *Handbook for Forest and Ranch Roads*. Mendocino County Resource Conservation District, Ukiah CA.

Wemple, B. C. 1998. *Investigations of Runoff Production and Sedimentation on Forest Roads*. PhD dissertation, Oregon State University, Corvallis.

Wright, K. A., K. H. Sendek, R. M. Rice, and R. B. Thomas. 1990. Logging effects on streamflow: storm runoff at Caspar Creek in northwestern California. *Water Resources Research* 26(7): 1657-1667.

Wu, T. H., W. P. McKinnell III, and D N. Swanston. 1979. Strength of tree roots and landslides on Prince of Wales Island, Alaska. *Canadian Geotechnical Journal* 16: 19-33.

Wu, T. H., P. E. Beale, and C. Lan. 1988. In-situ shear test of soil-root systems. *ASCE Journal of Geotechnical Engineering* 114(12): 1376-1394.

Wu, W., and R. C. Sidle. 1995. A distributed slope stability model for steep forested basins. *Water Resources Research* 31: 2097-2110.

Ziemer, R. R. 1981. Roots and the stability of forested slopes, pp. 343-361 in *Erosion and Sediment Transport in Pacific Rim Steeplands*. IAHS Publication No. 132, Wallingford U.K.

Moving toward Sustainability

Stephen D. Hobbs, John P. Hayes, Rebecca L. Johnson, Gordon H. Reeves, Thomas A. Spies, and John C. Tappeiner II

Introduction

The decade of the 1990s was a period of tremendous change in how forest and stream resources were managed in the Oregon Coast Range. These changes were brought about by a number of factors. One of the most significant was increased concern about the environment and the welfare of fish and wildlife species. Application of the Endangered Species Act to the northern spotted owl (*Strix occidentalis*) and its listing as threatened, followed by subsequent listings of the marbled murrelet (*Brachyramphus marmoratus*) and coho salmon (*Oncorhynchus kisutch*), had a profound impact on management, particularly on federal forestland. Other federal laws such as the Clean Water Act, Clean Air Act, National Forest Management Act, and others also have driven change. Certainly one of the biggest agents of change was the Northwest Forest Plan for Forest Service and Bureau of Land Management lands. First announced in 1993 and implemented in 1994, this federal action dramatically redirected how these lands would be managed, shifting the focus from timber harvest to protection of species and habitat. Although the Northwest Forest Plan contains provisions for an active timber harvest program, albeit at levels significantly lower than those reached in the 1970s and 1980s, interpretation of the plan and subsequent litigation have prevented attainment of the harvest levels originally envisioned.

There have been other significant regional changes as well. The high-technology industry has become a growing force in Oregon's economy, while there has been reduction and consolidation among companies in the wood-products sector. Moreover, wood-processing companies have retooled to use smaller logs and have implemented technological advances in manufacturing of wood products. Reductions in availability of harvestable timber from federal lands have forced companies to rely more on wood from their own lands or from other private or nonfederal public sources (county or state forestlands).

Oregon's demographic landscape has also changed over the last decade and will continue to change in the future. Portland and surrounding areas have experienced tremendous growth, as has the Interstate 5 corridor between Portland and Salem. Urbanization and the growing political power of urban populations will increasingly influence the management of forests and streams in the Coast Range. Population growth and the booming economy of the 1990s have increased tourism and recreation in the Coast Range, particularly along the immediate coast. As visitors travel to and from the coast, they form opinions about forest management activities based on the landscapes they see. There is also increasing evidence that Oregonians are unwilling to accept declines in the quality of forest aesthetics, recreational opportunities, salmon, wildlife, or other forest- and stream-related values as a result of the production of wood products. These opinions and impressions have political implications, but, as will be discussed later in this chapter, public beliefs about how forest resources should be managed are not necessarily consistent with society's consumption of wood.

The movement toward sustainability of forest and stream resources will take time, but many of the necessary components are already in place. For example, some steps have been taken in this direction by implementation of the Oregon Plan, a plan for the restoration of salmon and watersheds in western Oregon (State of Oregon 1997), the formation of watershed councils, and planning for the management of state forestland in northwestern Oregon (Bordelon et al. 2000). There is clearly an increasing need for information sharing and collaboration among diverse landowners and others on issues of resource management and utilization that transcend property boundaries. More recently, the Oregon Department of Forestry issued *Oregon's First Approximation Report for Forest Sustainability* (Oregon Department of Forestry 2000). This report represents the first effort by any state to evaluate forestlands using internationally agreed criteria and indicators for the conservation and sustainable management of temperate and boreal forests. These criteria and indicators were developed through the United Nations in response to the Statement of Forest Principles and Chapter 11 of Agenda 21, which came out of the 1992 United Nations Conference on Environment and Development (Earth Summit) held in Rio de Janeiro. Using what has come to be known as the Montreal Process, a working group met in Geneva in 1994 to develop the criteria and indicators (Montreal Process Working Group, 1998a). These were finalized in 1995 in Santiago, Chile, when Australia, Canada, Chile, China, Japan, the Republic of Korea, Mexico, New Zealand, the Russian Federation, and the United States endorsed the criteria and indicators (Montreal Process Working Group, 1998b).

In this chapter we provide a broad overview of the challenges and opportunities associated with integrated resource management and achieving sustainability in the Coast Range. We concur with Clark and others (1999) that integrated resource management is a process rather than an end, and we suggest this process is necessary to achieving sustainability. As you read the chapter, this concept should be seen as the common thread woven through the discussion of different influences affecting integrated resource management and sustainability; these influences are the focus of our attention. We take a big-picture perspective, because policies affecting socioeconomic and ecological issues in the Coast Range will be increasingly affected by the interaction of local, state, regional, national, and international socioeconomic and political influences. We have not attempted to identify specific sustainability objectives for the Coast Range. What constitutes sustainability will evolve through time as environmental, economic, and social conditions change and new information becomes available. The issues are complex, and although a specific solution to achieve sustainability is not developed in the chapter, important challenges, prerequisites, guidelines, and ideas are discussed.

Integrated Resource Management, Sustainability, and Biological Diversity

Integrated resource management

Helms (1998) defines integrated resource management as "the simultaneous consideration of ecological, physical, economic, and social aspects of lands, water, and resources in developing and implementing multiple-use, sustained-yield management." In its simplest form, integration means considering all the parts or bringing the parts together to create a whole entity. Integration is more about a process for achieving goals than the actual outcomes themselves (Clark et al. 1999). Moreover, within the land management context, integration can be viewed as a process for simultaneously considering and meeting diverse human uses and values *and* the biophysical attributes of the environment. It is important to understand that integration is basically a human process involving mutual learning, shared decision making, accommodation, and cooperation (Clark et al. 1999).

Integrated resource management is a loosely defined decision-making process by which stakeholders consider the many values and factors that affect sustainability. This can take many forms and can consider different spatial and temporal scales. What is important is that people agree to broad goals, respect other opinions, are flexible in their thinking and actions, and engage in meaningful dialogue. Integrated resource management should include collaboration, cooperation, and an interdisciplinary approach in the process (Clark et al. 1999), necessitated in part by the fact that many influences affecting sustainability transcend boundaries.

Sustainability

Much has been written over the last decade about sustainability. Terms such as sustainable forestry, sustainable ecosystems, sustainable development, and sustainable management abound in the literature. For example, sustainable forestry is generally accepted to mean forestry that "...must be ecologically sound, economically viable, and socially desirable" (Aplet et al. 1993). Helms (1998) defines sustainable forestry as, "the practice of meeting the forest resource needs and values of the present without compromising the similar capability of future generations." A discussion of the various and evolving definitions of sustainability viewed from the anthropocentric, ecocentric, and contextual perspectives is given by Borchers (1996). In a more recent report, Floyd and others (2001) address forest sustainability; they include case studies and an annotated bibliography. Several common ideas underlie the concepts of sustainability, sustainable development, and sustainable forestry:

(1) Economic, social, political, and ecological/ biophysical environmental factors are integral parts of sustainability.

(2) Human activities today should not limit options for future generations.

(3) There are limits on the products and values that can be derived through time from forest and stream ecosystems without jeopardizing ecosystem integrity and resilience.

(4) There is a balance between what ecosystems can safely produce (without jeopardizing ecosystem integrity and resilience) and the demands humans make on them (Bormann et al. 1994).

Interest in the issue of sustainability has not been restricted to the United States; considerable international attention has been focused on the subject (Schlaepfer 1997; Montreal Process Working Group 1998a, 1998b). In 1995, the Santiago Declaration, issued by Montreal Process member countries (Montreal Process Working Group 1998a), contained seven national-level criteria on what should be considered important to the conservation and sustainable management of temperate and boreal forests: (1) conservation of biological diversity; (2) maintenance of productive capacity of forest ecosystems; (3) maintenance of forest ecosystem health and vitality; (4) conservation and maintenance of soil and water resources; (5) maintenance of forest contribution to global carbon cycles; (6) maintenance and enhancement of long-term multiple socioeconomic benefits; and (7) legal, institutional, and economic framework for forest conservation and sustainable management.

These criteria and the 67 indicators that accompany them were developed because of growing concern over the future of temperate and boreal forests and their continuing ability to meet the needs of future generations. Although all the criteria are equally important, we will examine just one of them, the conservation of biological diversity, as an example of how a criterion might be measured and used in an evaluation of Coast Range forests.

The conservation of biological diversity

The conservation of biological diversity is a key building block for sustainability, and achieving it is one of the most challenging problems facing policymakers, community leaders, and scientists today. Like sustainability, there are different definitions of the conservation of biological diversity. Helms (1998) defines biological diversity (biodiversity) as, "the variety and abundance of life forms, processes, functions, and structures of plants, animals, and other living organisms, including the relative complexity of species, communities, gene pools, and ecosystems at spatial scales that range from local through regional to global." Biological diversity can be viewed at three levels, all of which are necessary: genetic diversity, species diversity, and community-level diversity (Primack 1993).

Integration of ecological and socioeconomic values into forest management requires methods to describe the different dimensions of forest ecosystems and to visualize the consequences of different management actions. Characterizing the complexity of biological diversity (the variety of life and ecosystems) in terms of measures and indicators is a relatively new and rapidly developing field. Conservation biologists have devised various strategies for sustaining biological diversity. These strategies have been classified into two general types: "fine-filter," which deals with individual species, and "coarse-filter," which deals with communities and landscapes (Noss 1987). The "filter" in these cases is the conservation strategy and how well it captures different parts of biological diversity in its safety net.

At its most fundamental level, conservation of biological diversity is based on sustaining populations of all native, and in some cases desired nonnative, species in a region, including important genotypes. A fine-filter approach may be the most direct method to achieving this goal because it focuses on species such as those at risk (northern spotted owl) or others that play important roles in ecosystem functioning (beaver). However, it is the least comprehensive method because it is not possible to use a fine-filter approach for all species of a region. We know too little about the ecology of many individual species, and there are simply too many species to create databases and management plans for each one.

Biological diversity also exists at higher levels of biological organization, the levels of communities and ecosystems. These entities not only have value in their own right but also provide a foundation for conservation of individual species dependent on the habitats and ecosystems in which they occur. The coarse-filter approach is based on maintaining a diversity of vegetation types and the natural disturbance regimes that created them. Without attention to disturbance, succession, and stand development, conservation strategies will ultimately fail. For example, many forest types such as oak woodlands require relatively frequent patchy, surface fires to thin out invading conifers and maintain their characteristic composition and structure. The disadvantage of the coarse-filter approach is that it may not capture some individual species with narrow habitat requirements or some endangered species that require special conservation efforts. In addition, for many landscapes the disturbance history of the ecosystems may not be known well enough to provide a good template for active management. Consequently, a combination of fine- and coarse-filter approaches is typically needed in regional conservation strategies.

Current conservation strategies in the Coast Range are a mix of fine- and coarse-filter approaches. Habitat-conservation plans for the northern spotted owl and species-viability assessments that were done for the Northwest Forest Plan by the Forest Ecosystem Management Assessment Team (FEMAT 1993) are examples of fine-filter approaches, methods that focus on conservation of individual species. Coarse-filter approaches are exemplified by the aquatic-conservation strategy of the Northwest Forest Plan and structure-based management approaches being implemented on state forests in northwest Oregon (Bordelon et al. 2000). It is unclear how the relative mix of these strategies will change, or how they will be combined in future management and policy efforts that focus on sustainable forestry. Recent plans for federal lands have placed considerable emphasis on fine-filter approaches such as the "survey and manage" component of the Northwest Forest Plan. This has been driven by concern that coarse-filter approaches may place too much risk on individual, poorly known species such as some species of mollusks and fungi. However, there is limited knowledge of the biology of many species, which makes it impractical to rely on a species-by-species approach. In addition, recent studies of fire history in the Coast Range (Wimberly et al. 2000), while validating the relatively high abundance of old-growth forests over the last several thousand years, also demonstrate the dynamic nature of these ecosystems. Long-term strategies to conserve biological diversity in this region will need to include landscape planning across all ownerships. This planning should include strategies for using disturbance to provide for all stages of forest succession.

Forest management increasingly considers biological diversity. For example, in the last 10 years, new forest policies that directly or indirectly address biodiversity have been implemented on all ownerships in the Coast Range (the Northwest Forest Plan, revisions to the Oregon Forest Practices Act, plans for state forests). These new efforts are a blend of fine- and coarse-filter approaches. Coarse-filter approaches have been applied at multiple spatial scales. At the stand level, harvesting, reforestation, and thinning have been modified on many sites to follow natural disturbance regimes more closely or to provide habitat components, such as individual old trees, snags, logs, and complex vertical and horizontal structure in upland and stream ecosystems (see Chapters 5 and 7). At the landscape level, silvicultural treatments on federal and state lands are now scheduled with the goal of maintaining connectivity and providing some large blocks of interior forest and a range of age classes. Watersheds provide a basis for multi-ownership planning that involves a mixture of voluntary and regulatory approaches to achieve watershed goals.

Although we have learned much about conserving biological diversity in the Coast Range, we must remember that our knowledge and management practices contain a dose of uncertainty. While this should not prevent us from taking actions to conserve biological diversity and produce other values from our forests, it should serve as a caution against assuming we know all the answers, or that only one approach is the best. The management approaches we use today may not be the best ones in the future. The challenge for managers and the public is to find a balance between the long-term strategies needed to manage these forest ecosystems and the course corrections that will be needed along the way as scientific information, social values, economics, and institutions change.

Changing Realities in the Oregon Coast Range

During the last several decades, there has been a shift to increased regulation of forest and stream resources, through state and federal law and increased awareness of nonconsumptive-resource values. This resulted from three factors acting in concert. First, research increased understanding of ecosystems, including biophysical processes, species-environment interactions, and the effects of management activities on species and ecosystems. This knowledge raised awareness of species decline, environmental degradation, and the importance of ecosystem resilience, integrity, and biodiversity. Second, rising public concern for the environment and increased appreciation for use of nonconsumptive forest resources raised the visibility of environmental issues. Finally, the environmental movement has been a powerful political force that has significantly influenced environmental and land management policies through the legislative process and litigation. These trends are likely to continue in the immediate future and will affect how forests and streams are managed.

Forest conditions

The Coast Range is an ever-changing mosaic of forest conditions driven in part by socioeconomic and political forces that influence landowner actions. Major forestland-owner groups are the federal land management agencies (Forest Service and Bureau of Land Management), the forest industry, the State of Oregon (principally the Oregon Department of Forestry), and nonindustrial private forest owners. Collectively, the Forest Service and Bureau of Land Management manage approximately 26 percent of the forestland in the Coast Range, and the State of Oregon manages 11 percent. Most of the state forestland is managed by the Oregon Department of Forestry. The forest industry owns an estimated 40 percent of the forestland, while nonindustrial private forest and woodland owners manage an estimated 21 percent. All other owners hold less than 2 percent of the forestland (Bettinger et al. 2000).

The nature of these forests is likely to change considerably in the next several decades. These changes will occur because of (1) policies and practices of landowners and the applicable laws and regulations under which they operate, (2) the legacy of past forest practices, and (3) society's changing views of forests and uses of forestlands. For example, there will be increasing contrast between federal and industrial forestlands. Federal forests will be characterized by increasing amounts of older stands, while industrial forests will be considerably younger, with harvests on relatively short rotations. It has been hypothesized that existing policies will result in increased edge effects and decreased size of core areas over time (Spies et al. in press).

In addition, disturbance agents such as Swiss needle cast disease may play a role in shaping future forests of the Coast Range. It is difficult to predict future specific conditions of forests with a high degree of confidence because of unforeseen changes in policy or economic conditions, but the following general trends seem likely to occur over the next several decades.

Federal forests

Implementation of the Northwest Forest Plan in 1994 dramatically changed management of Forest Service and Bureau of Land Management lands in the range of the northern spotted owl in Washington, Oregon, and California. Tuchmann and others (1996) provide an excellent summary of events leading up to the Northwest Forest Plan and a description of how the plan was developed and implemented. The pendulum has swung from an emphasis on timber harvest to an emphasis on protection of properly functioning ecosystems and restoration of degraded ecosystems and populations of at-risk species. The plan is one of the first and most comprehensive attempts to use principles of ecosystem management and conservation biology to guide management of a large area of federal lands. Prior to the plan, timber harvesting over many decades had decreased the amount of old-growth forests by more than 50 percent (Bolsinger and Wadell 1993) and had threatened species and ecological processes associated with those ecosystems (FEMAT 1993). Many innovative elements of ecosystem management and conservation biology were included in the plan. For example, landscape- and watershed-scale designs were developed to maintain or restore large patches of interior old-growth habitat, connectivity, and aquatic ecosystems, and adaptive-management areas were created to demonstrate and test new approaches to forest management.

Although the Northwest Forest Plan was intended to stabilize the flow of federal timber to mills at a lower level than historic harvests, since its implementation relatively little timber has in fact been harvested from federal lands. Several factors have decreased the level of timber production from what was originally intended, especially the late addition to the plan of special protection for rare plants, fungi, and animals. Insufficient time has passed to fairly judge whether goals related to protection of the environment will be achieved, although it seems reasonable to speculate that improvements in habitat conditions for many species will be realized, particularly for those associated with older forests. If left unaltered, the plan will dramatically increase the amount of older forests on federal lands over the next 100 years (Bettinger et al. 2000; Spies et al. in press). However, some of these older forests may not provide the habitat associated with original old-growth forest without stand-density management (see Chapter 7).

Approximately 80 percent of federal forestlands in the Coast Range are either in reserve status (no timber harvest) or other land-allocation designations in which only thinning is allowed to meet ecological goals such as maintaining or producing late-successional and riparian stands (Bettinger et al. 2000). Many of these lands currently support young forest stands. The future of these stands and the extent to which they achieve current objectives depend on the treatments they receive in the next several decades. Thinning would enable them to produce large trees and eventually achieve some of the characteristics of old-growth forests (see Chapter 7). If trees were left at high densities, these stands would produce high yields of wood. However, tree sizes would be small, understory development would be minimal, and these stands would not develop in the same way or at the same rate as current old-growth stands. As a consequence, it is questionable whether these stands would achieve the same level of structural diversity as is found in current old-growth stands.

State of Oregon forests

The vast majority of state forestlands in the Coast Range are managed by the Oregon Department of Forestry. These forestlands are managed for an array of economic, environmental, and social benefits, including timber harvest and hence revenue for local taxing districts, counties, and the State of Oregon. Most of the stands are less than 85 years in age, although stands of older timber can be found throughout many of the lands. The Oregon Department of Forestry's "structure-based management" approach to the management of state forestlands in the northern Oregon Coast Range is an attempt to achieve landscapes with diverse forest structure, including old-forest habitat characteristics, along with wood production, through a combination of stand-density management, regeneration methods, and different patch sizes and placement integrated at multiple spatial scales (Bordelon et al. 2000). For example, one of the goals is to maintain 20 to 30 percent of the forested landscape in older forest structure. It will take time to evaluate whether this new approach is successful in achieving the multiple objectives identified in the plan.

Industrial forests

Industrial forestlands are intensively managed for the production of wood fiber and are dominated by young even-aged stands typically less than 100 years old. However, an increasing number of companies are actively seeking the development of new practices that will enhance other forest and stream resource values while allowing for profitable timber harvest. Many companies have invested in research, such as the Coastal Oregon Productivity Enhancement (COPE) Program (see Chapter 1), to develop silvicultural and other practices that allow timber harvest while enhancing fish, wildlife, and other resource values. The management of nontimber resources is a growing trend and is best exemplified by the American Forest and Paper Association's

Sustainable Forestry Initiative (Berg and Cantrell 1999). Members are required to follow principles, objectives, and performance measures consistent with the International Standards Organization 14001 Environmental Management Systems Standard.

Silvicultural practices will vary among industrial owners; however, on many ownerships there is likely to be intensive control of noncrop vegetation, early thinning, and fertilization to achieve desired tree sizes (often in the range of 16 to 24 inches in diameter at breast height) within 40 years. Some companies may grow trees on longer rotations for large logs and special markets. However, current technology enables the efficient manufacturing of relatively small logs into beams, siding, and other products. The role that market-based incentives, such as third-party certification or "green certification," will play in influencing future silvicultural practices and other forest operations is unknown. There is growing interest in the retail business to sell wood products certified as produced by environmentally responsible methods or from sustainable forests.

Nonindustrial private forests

Although largely dominated by Douglas-fir (*Pseudotsuga menziesii*), nonindustrial private forests are diverse, with a wide range in stand age and structure. Nonindustrial private forest (NIPF) owners have many objectives and own their forestland for many reasons. Consequently, no single stereotype can be used to characterize NIPF owners. Johnson and others (1997) surveyed NIPF owners in western Oregon and Washington and reported that 73 percent of respondents identified the enjoyment of owning green space as an important or very important reason for owning forestland. Fifty-five percent identified timber production as an important or very important reason for owning forestland. Interestingly, more than a third of the respondents indicated that timber production was less important or not important at all. However, the study found size of ownership was associated with certain owner attitudes; owners with larger amounts of forestland regarded timber production as an important reason for owning the land. Moreover, owners with large acreages indicated they might harvest sooner than planned because of potential regulations.

Historically, since 1975, harvest from NIPF lands has been relatively stable, but this changed in the early 1990s. Starting in 1992, annual timber harvest from NIPF lands started to increase, with an unusually high peak harvest reached in 1993 (Lettman and Campbell 1997). In the following years, harvest levels declined to more normal levels seen prior to that year. Favorable stumpage prices and increased demand created by declining timber harvest from federal lands probably contributed to these increased harvests on NIPF lands.

Between 1961 and 1994 over 600,000 acres of nonindustrial forestland were acquired by the forest industry (Zheng and Alig, 1999). The significance of this shift in ownership and whether this trend continues today are unknown. However, it is likely these newly acquired lands will be managed on the even-age, single-species, short-rotation basis typical of most industrial ownerships.

Other influences

Currently Swiss needle cast disease, as well as root diseases, are recognized as biological agents that will have important impacts on future forests (see Chapter 8). If so, attention may be increasingly given to growing stands of mixed species composition as a buffer against diseases that could have a major impact on the forest. Species mixtures of Douglas-fir, western hemlock (*Tsuga heterophylla*), western red cedar (*Thuja plicata*), and red alder (*Alnus rubra*) are being considered as possible mixed stands on some private and public forestlands. Use of cottonwood (*Populus* spp.) plantations and the introduction of exotic species grown on 10- to 20-year rotations for intensive wood production could possibly increase on valley-bottom sites. However, the extent to which this actually happens will depend on economic factors and an evaluation of the potential environmental consequences. Whether genetically engineered trees are used will depend largely on public acceptance.

As the socioeconomic characteristics of Oregon's population change, public opinion about how forests are managed will likely have an increasing influence on forest practices and the laws that regulate them. This may be largely driven by an increasing urban population whose attitudes about forests and their utilization are changing. Using voting data from the 1998 Oregon Ballot Measure 64 (Oregon Forest Conservation Initiative) and research by others, Kline and Armstrong (2001) discuss how changes in Oregon's population are associated with increased concern for forest-related values other than the production of wood products. Citing other research, they note that Oregonians are becoming more disconnected from the forest industry and that an increasingly affluent and urban population may have a more environmental orientation. In addition, increasing numbers of urban dwellers may move to more rural settings seeking an improved quality of life, a process referred to as exurbanization (Egan and Luloff 2000); this movement also may affect forest policies because of the values the migrants bring with them to local communities (Egan and Luloff 2000).

Stream conditions

Coastal Oregon streams and rivers have undergone large changes in the past 150 years. Historically, streams and rivers had large concentrations of wood in all parts of the stream network. Early explorers and settlers reported massive jams, which were often difficult or impossible to pass over. Low-gradient valley bottoms had well-developed off-channel areas, often with multiple channels that grew larger or smaller in size depending on the season. These areas were often sites of large beaver concentrations and probably were the sites of greatest anadromous salmon and trout production in the watershed.

The valley bottoms were among the first areas to be settled by Euro-Americans, who cleared vegetation, drained wetlands, and diked channels to create homesteads and farms. They cleared large wood and boulders out of channels to facilitate movement up and down streams and rivers of all sizes. As timber harvest developed, splash dams were placed on many coastal streams. These were structures that spanned the channel and ponded water behind them. Cut logs were dropped into the pond or floated to the pond from upstream. At high flows, the dam was opened or blown up and the logs carried downstream to processing mills. The consequences of these activities included a decrease in the quantity and quality of fish habitat and a likely decline in fish production in streams with these dams.

Modern activities have also impacted Oregon coastal streams. The lower portion of almost every river has been diked or drained for agricultural use or development. Trees were also cleared from these areas. Trees were harvested to the edge of streams in upstream areas until 1972 when the first State Forest Practices Act was implemented. In the 1950s and 1960s there were also programs developed and directed by fish biologists to remove wood from channels, based on the belief that wood impeded the movement of fish. These activities and the lingering impact of earlier activities have left the majority of Oregon coastal streams in a degraded state, and thus they do not furnish favorable habitat for anadromous salmon and trout.

Many streams on public and private lands in the Coast Range are currently in poor condition. Thorn et al. (2000) found that streams lacked woody debris and pools and had elevated levels of fine sediments and degraded riparian conditions. Only about 6 percent of stream reaches surveyed were considered of high quality. Urban, nonforested, and agricultural lands represented the poorest habitat conditions in the survey area, which included streams south of the Columbia River flowing into the Pacific Ocean. These areas of low-gradient streams have historically provided the most productive freshwater habitat for salmonids (Thorn et al. 2000).

The degraded condition of coastal Oregon streams is a major factor associated with the current decline in anadromous salmon and trout. Currently, coho salmon along the entire Oregon Coast, steelhead on the south Oregon Coast, and chum salmon in the lower Columbia River, are listed as threatened under the Endangered Species Act. Fall Chinook salmon and coastal cutthroat trout were evaluated by the National Marine Fisheries Service, but their numbers were sufficiently large that a listing was averted. A suite of factors, including variable ocean conditions, a decline in the quantity and quality of freshwater and estuary habitats, and the impact of genetic practices, are associated with the status of these fish (Nehlsen et al. 1991), but habitat alteration is the most common factor associated with the decline of individual species and populations.

Whether stream conditions in the Coast Range will improve is uncertain. While the Oregon Plan has been a positive step toward improving fish habitat, much remains to be done, particularly for low-gradient streams. Streams on federal lands are likely to improve the most, primarily as a result of the riparian requirements in the Northwest Forest Plan. However, improvements in some areas may be limited because of the reluctance of managers and decision makers to undertake or allow silvicultural activities in riparian areas as described in Chapter 7, and by the lack of economic incentives to do these things.

The Independent Multidisciplinary Science Team (1999) concluded that current rules of the Oregon Forest Practices Act and measures of the Oregon Plan are insufficient to achieve the mission of the Oregon Plan (recovery of wild salmonid stocks in coastal Oregon streams). Current riparian rules are insufficient to improve riparian areas because of the restricted size of riparian management zones and the number of trees that may be removed from them. Also, currently little consideration is given in current Oregon Forest Practice rules to riparian areas along non-fish-bearing streams. However, as of this writing, changes to the riparian rules that would result in improved conditions are under consideration.

Streams in the lower portions of river systems, which historically have been areas of high fish production, are even less likely to produce fish in the future than streams in the forested upper and middle portions of the network. Today these areas are primarily in urban and agricultural settings, and the potential for improving fish-habitat conditions is limited. However, with innovative approaches to planning and economic incentives that encourage landowners to improve fish habitat, this could change.

Some private landowners have made substantial changes in the management of selected watersheds, but these cover only a relatively small portion of the Coast Range. Since the condition of streams is determined to a large extent by ownership patterns, there will likely be large gaps in the distribution of streams with conditions favorable to anadromous salmonids and other native fish. Many streams and watersheds simply will not contribute to the recovery of depressed fish populations unless significant steps are taken, such as those recommended by the Independent Multidisciplinary Science Team (1999). One of that team's recommendations was that actions at individual sites were in and of themselves not adequate, and that a landscape perspective utilizing landscape analysis should be used as a framework for policy formulation. This reinforces points made later in this chapter that landscape-level planning and cross-boundary cooperation will be necessary to achieve sustainability.

Challenges Facing Policymakers

Forest and stream management policies have been subject to major changes, largely due to a better understanding of species and ecosystems and changing socioeconomic pressures that have played out in our political system. In the last 50 years, public concern has shifted from providing cheap building materials and other wood products for a rapidly growing population to protecting nonconsumptive goods and experiences. But while an increasing percentage of our society demands protection of forest and stream resources, our appetite for wood products has dramatically increased (Force and Fizzell 2000). The contradictory position of consuming more while simultaneously demanding increased conservation through preservation and regulation makes the formulation of meaningful policy difficult for any level of government. More ominous are the ramifications of human global-population growth and associated increases in the consumption of the world's natural resources.

Globalization, population growth, and demands on resources

World economies have become increasingly interconnected in the twentieth century and will become more so in the twenty-first. Economic interdependencies span the range of geopolitical scales from local to international. Improvements in communications, computing, data management, and transportation systems, as well as reduced trade barriers, have facilitated development of a global economy. Changes in the economies of trading partners can influence interest rates, investment opportunities, and other components of the United States economy. Global trends and international events will increasingly affect even the most remote communities in the United States (Cinnamon et al. 1999). What happens in communities of the Coast Range is affected by what happens at the state, regional, national, and global scales. For example, as demand for wood products increases in the Pacific Northwest, the cost of imported wood could influence local markets and the well-being of forestland owners in the Coast Range. Forest resources are traded on a worldwide market, and what occurs in one region has a ripple effect through other regions (Sedjo et al. 1999). An example of this point is the implementation of the Northwest Forest

Plan and its subsequent effects on timber supply and prices and on the transfer of environmental impacts to other regions (Sedjo et al. 1999).

Key among the factors that will affect demand for wood is human-population growth (Brooks 1997). Although a worldwide shortage of raw wood material has been avoided by the substitution of other materials and increased technological efficiencies (Lippke and Bishop 1999), there are concerns about the relationship between world-population growth and the demand for wood products. How the supply and demand for these products is distributed among developed and developing countries is also of concern, particularly as it relates to potential environmental impacts in developing countries.

Global-population growth and increased gross domestic product and per capita income in both developed and developing countries will continue to drive demand for wood products, perhaps at an unsustainable rate. Compounding the problem is the fact that forests are being converted to other uses, particularly in developing countries. The Food and Agriculture Organization (1999) estimated that between 1990 and 1995, the world's forests decreased by 139.1 million acres. This included an increase of 21.7 million acres in developed countries and a decrease of 160.8 million acres in developing countries. Unfortunately, many developing countries lack actively enforced laws requiring adequate reforestation or other forest practices to ensure resource renewal and environmental protection.

Another important concern is the potential impact of environmental policy in the United States and how it may influence resource utilization in other countries (Sedjo 1993). In 1993, Aplet and others (1993) cautioned against exporting domestic environmental problems and suggested greater care should be taken in balancing domestic supply and consumption of wood fiber. Six years after Sedjo (1993) and Aplet and others (1993) published their papers, Cohen (1999) echoed the same warning: "American planners, managers, and citizens must consider the global perspective, even if they are concerned only to protect American resources and interests, because the United States is and will be intimately linked to the rest of the world. In future American land use and forestry, purely domestic factors will increasingly have to be balanced against demographic, economic, environmental, and cultural influences that originate outside of American boundaries." This point is particularly well illustrated by changes in federal forest management policy in the Pacific Northwest during the early to mid-1990s, which had impacts well beyond the region (Sedjo et al. 1999). This issue is particularly important given the recent projection that "America's appetite for timber will continue to grow, and consumption will exceed domestic harvest over the next 50 years (Adams 2002)."

The public values, private-property rights debate

The widely recognized tension that exists between advocates for the public good and advocates for the rights of private-property owners is a fundamental issue that must be recognized and addressed more satisfactorily than is currently the case. In the Pacific Northwest as in other parts of the country, the Clean Air, Clean Water, and Endangered Species Acts, as well as changing state regulations, have intensified the debate. Central to this issue are the questions of what constitutes a "taking" and how to interpret the Fifth Amendment of the Constitution, which states, "...nor shall private property be taken for public use, without just compensation." Much has been written about this issue (Achterman 1993; DeCoster 1994; Flick 1994; Cubbage 1995; Flick et al. 1995; Lewis 1995; Zhang 1996; Meidinger 1997). The key question is whether public values and the public good take precedence over private-property rights in disputes that involve resources that know no property boundaries. Should management activities that negatively impact other ownerships or public values be protected by federal, state, or municipal law? This issue has often been resolved through legislation and court actions in favor of the public good. Despite these outcomes, the strong beliefs held by advocates for private-property rights have not been dampened.

Fundamental to this issue is the potential dissonance between ownership and natural boundaries. The needs of many fish and wildlife species (northern spotted owl, coho salmon) transcend property boundaries, as do physical environmental attributes (clean air and water) and values (aesthetics). The ecological literature is relatively clear in its condemnation of situations where ecological and social boundaries do not match (Meidinger 1998). Meidinger describes how boundaries make coordination difficult, slow the exchange of information, and spur organizations to realize benefits while externalizing their costs. However, Meidinger, citing other authors, also describes the positive aspects of such boundaries. Boundaries can impede the effects of inappropriate policies or disturbance, thus allowing time for appropriate adjustments to be made (Naiman and Decamps 1990; Morehouse 1995). Boundaries also clearly fix responsibility for management actions and subsequent consequences of those decisions (Ellickson 1993). Finally, boundaries help define who needs information (Williamson 1985). It follows that to successfully manage multiple resources for sustainability, a broad landscape perspective must be taken that respects the positive and minimizes the negative aspects of ownership boundaries. This point has been emphasized by Spies and others (in press) in their study of the Coast Range. They examined potential changes in the Coast Range that would result if current policies were projected 100 years into the future across multiple ownerships. Based on simulation results of current policies and land ownership, they hypothesize that future biophysical processes will be influenced more by boundaries and ownership patterns than in the past. The study is the only one of its kind in the Coast Range, and one of only a few similar studies in the United States.

In landscapes with intermingled private and public ownerships, such as in the Coast Range, some level of cooperation and consultation among landowners will be necessary to achieve sustainability. The alternative is landscape-level management through increased command-and-control systems. Command and control is often used to refer to governmental control of resource management activities through laws and administrative regulations enabled by law (Achterman 1993). There is no doubt that some level of command and control is necessary and that we have benefited greatly from it. The issue, however, is at what point increasing command and control becomes deleterious. Consider the much broader definition of command and control offered by Holling and Meffe (1996). They view it as control to reduce variation in different aspects of human health and happiness, including within the context of the management of natural resources. Their thesis is that increased command and control in natural resource management ultimately reduces ecosystem resilience by reducing the range of natural variability, and ultimately decreasing sustainability. They suggest that economically driven enterprises and environmental movements, focused primarily on regulation and prohibition, can also represent forms of command and control.

Significant benefits to animal and plant species, ecosystems, and the economic well-being of communities have resulted from current laws focused on the management, utilization, and restoration of forest and stream resources. Likewise, economically driven enterprises have enabled communities to prosper and have contributed to an increasing standard of living. The environmental community has also contributed by raising public awareness of important environmental issues. Command and control becomes a problem when its influence, whether driven by preservation or profit, decreases the range of natural variability of ecosystems and landscapes. From a landscape perspective, pressures exerted by different forms of command and control should be considered in aggregate. One key to achieving sustainability is to identify an appropriate mix of command-and-control measures that does not further reduce the range of natural variability. However, this issue is further complicated by the question of how long we can expect to maintain the range of natural variability in the face of a rapidly growing population.

Feelings about private-property rights and the public good are equally strong among individuals on either side of the debate. It is clear to us that, if progress is to be made toward achieving sustainability through integrated resource management, there needs to be a greater understanding among all participants for the views of those on opposing sides of the private-property rights – public values debate. As long as private-property owners perceive their rights as being in jeopardy, and as long as those advocating for the public good perceive that public values ought always to take precedence over private-property rights, progress will be an uphill battle. We suggest much can be learned about finding common ground and equitable solutions by studying Knight and Landres (1998), Clark and others (1999), Hummel and Freet (1999), Johnson and others (1999), and Yaffee (1999). These authors offer insight to the problems and solutions associated with achieving goals requiring the cooperation and collaboration of diverse interest groups.

Conflicting policies and legal requirements

From a big-picture perspective, the policies and laws that govern management activities on private and public lands have largely served us well. For example, the Clean Water, Clean Air, and Endangered Species Acts have improved environmental quality and have brought some species populations (Bald Eagle, Peregrine Falcon, grizzly bear, and gray wolf) back to viable levels in some parts of the country. On private and state lands, the Oregon Forest Practices Act requires prompt reforestation following timber harvest and better protection of riparian areas, although there is debate about whether current riparian rules are adequate (see Independent Multidisciplinary Science Team 1999). In addition, Oregon's land-use laws require careful planning.

Despite the many positive outcomes attributed to these and other laws, managers are often frustrated by the myriad of regulations and administrative procedures they must deal with. Depending on circumstances, laws can be confusing and overly complex, presenting managers with a conflicting potpourri of requirements or differing interpretations by agencies. Their implementation may be confounded by deeply held beliefs in private-property rights, the public good, historical precedent, differences in agency mandates, and intervention by the courts. Public agencies often face coordination difficulties and may sometimes find themselves at cross-purposes in pursuing their respective missions (Sedjo et al. 1999). Three recent publications discuss these issues in detail (Meidinger 1997, 1998; Mealey 2000). Mealey (2000), discussing the integration of science and policy, saw three barriers to successfully achieving ecosystem health: prevailing political and administrative cultures, differences between science and management, and legal requirements. Using the Interior Columbia Basin Ecosystem Management Project (Quigley et al. 1996; Quigley and Cole 1997) as a model, he describes how interpretation of the Endangered Species, Clean Air, and Clean Water Acts by different agencies prevented the implementation of management policies recognized as necessary to improve ecosystem health. Although Mealey (2000) addressed the barriers in terms of ecosystem health, we feel the argument he makes also has applicability to integrated resource management and achieving sustainability.

To examine some of the challenges that would be faced in implementing new ideas, consider the paper written by Reeves and others (1995), in which they present a new disturbance-based approach to improving freshwater habitat for anadromous salmonids (also see Chapter 4). The authors contend that, over the long term, a static system of reserves is not in the best interest of anadromous salmonids. Rather, they suggest that at a regional scale, areas of good habitat should shift among watersheds over a period of many years as a result of natural or anthropogenic disturbances. However, because wildfire has been largely eliminated as a natural disturbance factor, the authors propose the use of timber harvesting to induce the natural processes associated with fish-habitat rejuvenation. Timber harvest activity would be concentrated in fewer watersheds rather than dispersed among many watersheds. Ecologically appropriate riparian buffers would be left along fish bearing and selected non-fish-bearing streams. Rotation time between harvests on any given site would be greatly increased, and centers of harvest activity would shift among watersheds over a period of many years. However, harvest activity would be exclusive of any reserve system where human activity is minimized, such that a spatial and temporal pattern of timber harvest would more closely mimic historic patterns of natural disturbance, and, in the long term, improve fish habitat. Reeves and others (1995) also discuss some of the obstacles that would have to be overcome to implement their approach. For example, they discuss the need for people to think in longer time periods and to realize that disturbance should be seen as an important, positive influence on aquatic ecosystems. We note that public acceptance might be difficult as people have strong attachments to place and might be unlikely to accept short-term disturbance to their favorite places in exchange for uncertain ecological benefits far into the future.

Many obstacles would have to be overcome to implement an idea like that proposed by Reeves and others (1995). Key among these would be the issue of whether state and federal agencies have the legal authority to allow short-term disturbance in the interest of achieving long-term ecological goals, particularly if such actions might result in the incidental taking of a species listed as threatened or endangered, or result in temporary failure to meet legal standards, such as water quality standards. If agencies do have the authority, their cultures might impede appropriate action. These are issues also raised by Mealey (2000). Likewise, corporations might be reluctant or unable to coordinate or otherwise share timber harvest planning information because of antitrust laws or concern about providing information to competitors or drawing the attention of regulatory agencies. Another significant obstacle is a lack of trust among stakeholders. This is something not often discussed upfront among stakeholders, but it is present, and it does influence how or even whether new information and ideas will be implemented. Equally important, this mistrust may influence how laws and regulations are interpreted when some latitude is permitted in the actions that can be taken to meet the intent of the law or regulation. As science changes our understanding of ecosystems and the sociology of resource management, laws and administrative procedures that prevent or slow the utilization of better information should be reexamined.

Four policy- and institution-related changes are currently underway that may influence stewardship across boundaries (Meidinger 1998). These we see as essential for successful integrated resource management and achieving sustainability. Meidinger (1998) is careful to note that in some cases it is too early to tell whether these policy changes will have, on balance, positive or negative effects on stewardship across boundaries. The changes he describes are (1) privatization of policy making, (2) rule to discretion, (3) decentralization, and (4) politicization of information.

Privatization of policy making refers to a growing trend of public policy being made outside government processes by nongovernmental organizations. He gives examples such as the Applegate Partnership and the Forest Stewardship Council. In his discussion of rule to discretion, Meidinger (1998) outlines the fact that most institutions rely on rules which are often either over- or under-inclusive and hence do not allow for much flexibility in how they are interpreted and enforced. He also makes the point, also noted by Mealey (2000), that rules may not be compatible, which places managers in a difficult position when it comes to compliance. Meidinger (1998) sees a tentative shift toward more discretionary decision making in which the different parties negotiate innovative ways to meet the spirit and general intent of the law to achieve overall superior results without meeting every detailed requirement.

Decentralization is another positive change seen by Meidinger (1998). This trend reflects a shift from centralized to decentralized policy development and enforcement. For example, states are assuming a greater role in policy development and enforcement, often through voluntary-compliance programs developed by stakeholders. A good example is the Oregon Plan (State of Oregon 1997).

The fourth trend Meidinger (1998) discusses is the politicization of information that bears on the relationship between research and policy. This trend revolves around increased reliance on science-based information and the sharing of it in the formulation of policy. In the Coast Range, models and data developed by the Coastal Landscape Analysis and Modeling Study (CLAMS; Bettinger et al. 2000; Spies et al. in press) could provide the basis for evaluating how different policies might affect forest and stream resources.

Some would argue that central to the problem is the lack of a clear national policy about the nation's forests and associated stream resources around which laws governing management could be fashioned. It is easy to understand why the current system of complex laws and regulations has evolved when there is no guiding national policy for the use and protection of forest and stream resources. In a recent report by the National Research Council (1998) on nonfederal forests, a major problem identified was the absence of a national policy for these forests. If a national policy were developed collaboratively with the states, providing only general direction, it could allow individual states to legislate their own regulations to meet the spirit or intent of the national policy. Others would suggest a national policy is unnecessary, and that any natural resource policies should be solely the responsibility of individual states.

Although there are no easy or readily apparent solutions to this dilemma of conflicting policies and laws, policymakers should consider the barriers identified by Mealey (2000) and the changes affecting cross-boundary stewardship described by Meidinger (1998). One initial step to reduce the complexity of laws and the confusion often surrounding their interpretation would be the formation of a commission or other suitable body to evaluate this issue.

Establishing a framework for discussion

There is growing recognition among public and private resource managers and policymakers that varying levels of cooperation, collaboration, and partnership across ownerships will be necessary to increase the probability of achieving sustainability. In addition, there are two things we feel will increase the chance for success. First, the ability to monitor, assess, and predict biophysical changes and subsequent socioeconomic effects across broad landscapes will improve, as will confidence in the information produced. Better predictive capability at multiple spatial and temporal scales will dramatically change our ability to evaluate the potential consequences of proposed policy changes. Secondly, increased voluntary and incentive-based cooperation among landowners and public agencies on important resource issues will be necessary if increased command-and-control measures are to be avoided in the future. Such cooperation will require innovation, flexibility, and leadership at all levels of government and in the private sector.

We do feel there are opportunities to make progress on issues surrounding the management of forest and stream resources, such as salmon populations, water quality, and timber supply in the Coast Range. In particular, we suggest that formal or informal partnerships among stakeholders to solve contentious issues offer promise.

Establishing a framework for discussion to solve difficult resource management and policy issues will take hard work, patience, and dedication. The principal challenge lies in integrating and reconciling often vastly different opinions among stakeholders. However, we submit that differing views on management of forest and stream resources may be a strength rather than a weakness. The division of influence (political clout) among interest groups is also important to the balance of power in the arena of natural resource management and policy development. Just as biological diversity is widely seen as an indicator of ecosystem resilience, socioeconomic and political diversity in a democratic society are also seen as strengths. Diversity of opinion and strongly held views should not necessarily be seen as insurmountable obstacles to meaningful and productive discussions that can result in acceptable solutions to difficult problems. There are many examples of effective collaboration, cooperation, and partnership (Giusti 1994; Glick and Clark 1998; Hummel and Freet 1999; Johnson et al. 1999). Solutions derived from processes that involve a wide array of stakeholders representing diverse views on a particular issue will be more widely accepted by special-interest groups and the public at large. The key is providing the necessary framework for such discussions to occur.

We have developed a list of characteristics or requirements we feel are necessary for meaningful discussion to occur among stakeholders with differing opinions. This is not an exhaustive list, and although it is drawn from our own experiences in Oregon, it has much in common with more thorough treatments of the topic by others (Shands 1994; Shindler and Neburka 1997; Williams and Ellefson 1997; Landres 1998; Yaffee 1998, 1999; Clark et al. 1999; Hummel and Freet 1999). The key requirements include:

(1) *Leadership*. Someone needs to take the first step.

(2) *Openness*. There should be a common willingness among stakeholders to enter into a dialogue.

(3) *Common vision*. There must be a common, big-picture goal or a clearly defined problem.

(4) *Effective forum*. There should be a forum for discussion and decision making (watershed councils, associations, committees, task forces).

(5) *Well-defined decision-making process*. A clear process for conducting meetings and making decisions needs to be agreed upon by the participants.

(6) *Trust*. There needs to be a high level of trust. Often this only comes with time and positive experiences.

(7) *Ownership*. Stakeholders need to feel they have influence (that their opinion counts) and to recognize that all opinions and perspectives have value.

(8) *Clearly defined spatial scope*. The issues need to be bounded by a clearly defined landscape. We suggest the watershed is the smallest scale that should be used.

(9) *Adequate information base*. Information necessary for meaningful discussion must be available.

(10) *Respect and recognition of the issues of public good and private-property rights*. Both are strongly held beliefs considered fundamental to American culture and legitimized by law.

(11) *Interdisciplinary focus*. Most issues related to forest and stream management will require interdisciplinary collaboration.

(12) *Disaggregation and synthesis.* The issue needs to be broken down into its component parts to facilitate planning, organization, division of responsibility, and accomplishment of the goal. This approach is commonly used in science. In the collaborative problem-solving context, the re-aggregation of information developed from the solution of component problems may be necessary to address more complex issues (Norris 1995).

Summary

In the last decade, tremendous changes have taken place that have influenced the management of forest and stream resources in the Oregon Coast Range. Implementation of the Northwest Forest Plan dramatically changed how Forest Service and Bureau of Land Management lands within the range of the northern spotted owl are managed. During this same period, the technology, recreation, and service industries in Oregon grew, while the wood-products industry experienced consolidation. Oregon's population has also grown, particularly in the larger metropolitan areas along the I-5 corridor. Urban growth has brought changing attitudes about how forest and stream resources should be managed. Increasing numbers of Oregonians hold strong feelings about the importance of recreation opportunities, aesthetics, and the well-being of fish and wildlife species, and are less willing to accept declines in these values in return for the production of wood products. Simultaneously, demand for wood products has grown. These contrary forces make the formulation of resource management policy and progress toward achieving sustainability challenging. However, the use of integrated resource management, as a loosely defined decision-making process by which stakeholders consider the many values and factors that affect sustainability, does offer promise as a mechanism for achieving progress in a region that will continue to experience change.

Under current state and federal policies and current management practices in the private sector, forests in the Coast Range will change over the next 100 years. For example, there will be increasing contrast between federal and industrial forestlands. Federal forests will be characterized by increasing numbers of older stands, while industrial forests will be considerably younger, with harvests on relatively short rotations. It has been hypothesized that existing policies will result in increased edge effects and decreased size of core areas over time (Spies et al. in press).

Overall, streams of the Coast Range are in relatively poor condition in terms of habitat quality for anadromous salmonids and water quality for all land uses, particularly urban, agricultural, and other nonforest areas. These areas typically have low-gradient streams that historically had high fish productivity. Although recent steps to improve fish-habitat quality, such as the Oregon Plan, have been taken, much remains to be done if significant improvements are to be realized.

The process of integrated resource management can provide policymakers with an opportunity to make significant progress toward achieving sustainability. However, the hurdles they will face are many and complex. We do not suggest they are insurmountable, but they do need to be considered. This will mean thinking beyond traditional boundaries and constituencies. Globalization, population growth, and increasing demand for natural resources interact in ways that will increasingly affect communities in the Coast Range. The world is rapidly becoming more interconnected and the welfare of its countries interdependent. Major shifts in forest and stream resource management policy should no longer be made in isolation, without consideration of the implications beyond state, regional, or national borders.

As part of the move toward sustainability, stakeholders will have to better resolve the public values – private-property rights debate. Although there have been numerous precedents set in favor of the public good or values, the strongly held beliefs on both sides of the issue can be obstacles that should be dealt with in the integrated resource management process.

Likewise, the myriad of laws, regulations, and administrative procedures that apply to the management of forest and stream resources is complex and confusing. This is a significant problem that requires attention. Ways need to be found to simplify laws and streamline procedures to facilitate movement toward sustainability without sacrificing important rights and safeguards. Equally important, public agencies and private organizations need greater flexibility in how they choose to interpret or meet the spirit and intent of specific laws that govern management activities.

Through the integrated resource management process, private and public organizations can discuss and seek resolution to crucial management and policy issues. This will take hard work, patience, and dedication. The challenge lies in integrating and reconciling what are often vastly different opinions among stakeholders. However, diversity of opinions should be viewed as a strength rather than an obstacle to meaningful discussion and acceptable solutions to difficult problems. There are numerous examples of effective collaboration, cooperation, and partnership among diverse interest groups. The key is to establish an appropriate framework within which meaningful discussion and progress can occur.

Our ability to achieve sustainability of forest and stream resources in the Coast Range will be influenced by many factors, some of which are beyond the control of local communities. Nonetheless, we are optimistic that significant progress toward achieving sustainability has occurred and will continue to occur. A key ingredient to success will be continued improvement in our ability to integrate, evaluate, and communicate the interplay of biophysical, socioeconomic, and policy factors as they affect the quality of life for current and future generations.

Literature Cited

Achterman, G. L. 1993. Private property rights – public values, pp. 8-18 in *Communications, Natural Resources, and Policy*, B. Shelby and S. Arbogast, ed. Proceedings, 1993 Starker Lectures, College of Forestry, Oregon State University, Corvallis.

Adams, D. M. 2002. Harvest, inventory, and stumpage prices. *Journal of Forestry* 100(2): 26-31.

Aplet, G. H., N. Johnson, J. T. Olson, and V. A. Sample. 1993. Conclusion: prospects for a sustainable future, pp. 309-314 in *Defining Sustainable Forestry*, G. H. Aplet, N. Johnson, J. T. Olson, and V. A. Sample, ed. Island Press, Washington DC.

Berg, S., and R. Cantrell. 1999. Sustainable forestry initiative: toward a higher standard. *Journal of Forestry* 97(11): 33-35.

Bettinger, P., K. N. Johnson, J. Brooks, A A. Herstrom, and T A. Spies. 2000. *Phase I Report on Developing Landscape Simulation Methodologies for Assessing the Sustainability of Forest Resources in Western Oregon*. CLAMS Simulation Modeling Report, Oregon Department of Forestry, Salem.

Bolsinger, C. L., and K. L. Waddell. 1993. *Area of Old-growth Forests in California, Oregon, and Washington*. Resource Bulletin PNW-RB-197, USDA Forest Service, Pacific Northwest Research Station, Portland OR.

Borchers, J. G. 1996. A hierarchical context for sustaining ecosystem health, pp. 63-80 in *Search for a Solution: Sustaining the Land, People, and Economy of the Blue Mountains*, R. G. Jaindl and T. M. Quigley, ed. American Forests, Washington DC.

Bordelon, M A., D. C. McAllister, and R. Holloway. 2000. Sustainable forestry: Oregon style. *Journal of Forestry* 98(1): 26-34.

Bormann, B. T., M. H. Brookes, E. D. Ford, A. R. Kiester, C D. Oliver, and J. F. Weigand. 1994. *Volume V: A Framework for Sustainable Ecosystem Management*. General Technical Report PNW-GTR-331, USDA Forest Service, Pacific Northwest Research Station, Portland OR.

Brooks, D. J. 1997. The outlook for demand and supply of wood: implications for policy and sustainable management. *Commonwealth Forestry Review* 76(1): 31-36.

Cinnamon, S. K., N. C. Johnson, G. Super, J. Nelson, and D. Loomis. 1999. Shifting human use and expected demands on natural resources, pp. 327-343 in *Ecological Stewardship: A Common Reference for Ecosystem Management, Volume III*, W. T. Sexton, A. J. Malk, R. C. Szaro, and N. C. Johnson, ed. Elsevier Science Ltd., Oxford UK.

Clark, R. N., G. H. Stankey, P. J. Brown, J. A. Burchfield, R. W. Haynes, and S. F. McCool. 1999. Toward an ecological approach: integrating social, economic, cultural, biological, and physical considerations, pp. 297-318 in *Ecological Stewardship: A Common Reference for Ecosystem Management, Volume III*, W. T. Sexton, A. J. Malk, R. C. Szaro, and N. C. Johnson, ed. Elsevier Science Ltd., Oxford UK.

Cohen, J. E. 1999. Human population growth and tradeoffs in land use, pp. 677-702 in *Ecological Stewardship: A Common Reference for Ecosystem Management, Volume II*, R. C. Szaro, N. C. Johnson, W. T. Sexton, and A. J. Malk, ed. Elsevier Science Ltd., Oxford UK.

Cubbage, F W. 1995. Regulation of private forest practices: What rights? Which policies? *Journal of Forestry* 93(6): 14-20.

DeCoster, L. A. 1994. Private property rights and other myths. *Journal of Forestry* 92(5): 28-29.

Egan, A. F., and A. E. Luloff. 2000. The exurbanization of America's forests: research in rural social science. *Journal of Forestry* 98(3): 26-30.

Ellickson, R. C. 1993. Property and land. *Yale Law Journal* 102: 1315-1400.

Flick, W. A. 1994. Changing times: forest owners and the law. *Journal of Forestry* 92(5): 30-33.

Flick, W. A., A. Barnes, and R. A. Tufts. 1995. The evolution of regulatory taking. *Journal of Forestry* 93(6): 21-24.

Floyd, D. W., S. L. Vonhof, and H. E. Seyfang. 2001. Forest sustainability: a discussion guide for professional resource managers. *Journal of Forestry* 99(2): 8-27.

Food and Agricultural Organization. 1999. *State of the World's Forests*. Food and Agriculture Organization of the United Nations, http://www.fao.org/forestry/FO/SOFO/SOFO99/sofo99-e.stm (March 16, 2001).

Force, J. E., and G. Fizzell. 2000. How social values have affected forest policy, pp. 16-22 in *Proceeding of the Society of American Foresters 1999 National Convention*, Society of American Foresters, Bethesda MD.

FEMAT (Forest Ecosystem Management Assessment Team). 1993. *Forest Ecosystem Management: An Ecological, Economic, and Social Assessment*. Report of the Forest Ecosystem Management Assessment Team, U.S. Government Printing Office 1973-793-071. U.S. Government Printing Office for USDA Forest Service; USDI Fish and Wildlife Service, Bureau of Land Management, and Park Service; U.S. Department of Commerce, National Oceanic and Atmospheric Administration and National Marine Fisheries Service; and the U.S. Environmental Protection Agency. Portland OR.

Giusti, G. A. 1994. Partnerships across boundaries, pp. 42-52 in *Proceedings, Society of American Foresters 1993 National Convention*, Society of American Foresters, Bethesda MD.

Glick, D. A., and T. W. Clark. 1998. Overcoming boundaries: the greater Yellowstone ecosystem, pp. 237-256 in *Stewardship Across Boundaries*, R. L. Knight and P. B. Landres, ed. Island Press, Washington DC and Covelo CA.

Helms, J. A. 1998. *The Dictionary of Forestry*. Society of American Foresters, Bethesda MD.

Holling, C. S., and G. K. Meffe. 1996. Command and control and the pathology of natural resource management. *Conservation Biology* 10(2): 328-337.

Hummel, M., and B. Freet. 1999. Collaborative processes for improving land stewardship and sustainability, pp. 97-129 in. *Ecological Stewardship: A Common Reference for Ecosystem Management, Volume III*, W. T. Sexton, A. J. Malk, R. C. Szaro, and N. C. Johnson, ed. Elsevier Science Ltd., Oxford UK.

Independent Multidisciplinary Science Team. 1999. *Recovery of Wild Salmonids in Western Oregon Forests: Oregon Forest Practices Act Rules and the Measures in the Oregon Plan for Salmon and Watersheds*. Technical Report 1999-1 to the Oregon Plan for Salmon and Watersheds, Governor's Natural Resources Office, Salem.

Johnson, K. M., A. Abee, G. Alcock, D. Behler, B. Culhane, K. Holtje, D. Howlett, G. Martinez, and K. Picarelli. 1999. Management perspectives on regional cooperation, pp. 155-179 in *Ecological Stewardship: A Common Reference for Ecosystem Management, Volume III*, W. T. Sexton, A. J. Malk, R. C. Szaro, and N. C. Johnson, ed. Elsevier Science Ltd., Oxford UK.

Johnson, R. L., R. J. Alig, E. Moore, and R. J. Moulton. 1997. NIPF landowners' view of regulation. *Journal of Forestry* 95(1): 23-28.

Kline, J. D., and C. Armstrong. 2001. Autopsy of a forestry ballot initiative. *Journal of Forestry* 99(5): 20-27.

Knight, R. L., and P. B. Landres (ed.). 1998. *Stewardship Across Boundaries*. Island Press, Washington DC and Covelo CA.

Landres, P. B. 1998. Integration: a beginning for landscape-scale stewardship, pp. 337-345 in *Stewardship Across Boundaries*, R. L. Knight and P. B. Landres, ed. Island Press, Washington DC and Covelo CA.

Lettman, G. J., and D. Campbell. 1997. *Timber Harvesting Practices on Private Forest Land in Western Oregon*. Oregon Department of Forestry, Salem.

Lewis, G. 1995. Private property rights: the conflict and the movement. *Journal of Forestry* 93(6): 25-26.

Lippke, B. R., and J. T. Bishop. 1999. The economic perspective, pp. 597-638 in *Maintaining Biodiversity in Forest Ecosystems*, M. L. Hunter, ed. Cambridge University Press, Cambridge UK.

Mealey, S. P. 2000. The influence of science and technological change on forest management since 1900, pp. 23-35 in *Proceedings of the Society of American Foresters 1999 National Convention*, Society of American Foresters, Bethesda MD.

Meidinger, E. E. 1997. Organizational and legal challenges for ecosystem management, pp. 361-379 in *Creating a Forestry for the 21st Century: the Science of Ecosystem Management*, K. A. Kohm and J. F. Franklin, ed. Island Press, Washington DC and Covelo CA.

Meidinger, E. E. 1998. Laws and institutions in cross-boundary stewardship, pp. 87-110 in *Stewardship Across Boundaries*, R. L. Knight and P. B. Landres, ed. Island Press, Washington DC and Covelo CA.

Montreal Process Working Group. 1998a. *The Montreal Process*. http://www.mpci.org/whatis/evolution_e.html (September 25, 2000).

Montreal Process Working Group. 1998b. *The Montreal Process*. http://www.mpci.org/meetings/santiago/santiago1_e.html (September 25, 2000).

Morehouse, B. 1995. A functional approach to boundaries in the context of environmental issues. *Journal of Borderlands Studies* 10: 53-74.

Naiman, R. J., and H. Decamps. 1990. *The Ecology and Management of Aquatic-terrestrial Ecotones*. UNESCO Press, Paris.

National Research Council. 1998. *Forested Landscapes in Perspective: Prospects and Opportunities for Sustainable Management of America's Nonfederal Forests*. National Academy Press, Washington DC.

Nehlsen, W., J. E. Williams, and J. A. Lichatowich. 1991. Pacific salmon at the crossroads: stocks at risk from California, Oregon, Idaho, and Washington. *Fisheries* 16(2): 4-21.

Norris, L. A. 1995. Ecosystem management – a science perspective, pp. 15-18 in *Proceedings Ecosystem Management in Western Interior Forests, May 3-5, 1994, Spokane WA*, Department of Natural Resource Sciences, Washington State University, Pullman WA.

Noss, R. F. 1987. From plant communities to landscapes in conservation inventories: a look at the Nature Conservancy (USA). *Biological Conservation* 41: 11-37.

Oregon Department of Forestry. 2000. *Oregon's First Approximation Report for Forest Sustainability*. Oregon Department of Forestry, Salem.

Primack, R. B. 1993. *Essentials of Conservation Biology*. Sinauer Associates Inc., Sunderland MA.

Quigley, T. M., and H. Bigler Cole. 1997. *Highlighted Scientific Findings of the Interior Columbia Basin Ecosystem Management Project*. General Technical Report PNW-GTR-404, USDA Forest Service, Pacific Northwest Research Station, USDI Bureau of Land Management, Portland OR.

Quigley, T. M., W. Haynes, and T. Graham. 1996. *Integrated Scientific Assessment for Ecosystem in the Interior Columbia Basin and Portions of the Klamath and Great Basins*. General Technical Report PNW-GTR-382, USDA Forest Service, Pacific Northwest Research Station, Portland OR.

Reeves, G. H., L. E. Benda, K. M. Burnett, P. A. Bisson, and J. R. Sedell. 1995. A disturbance-based ecosystem approach to maintaining and restoring freshwater habitats of evolutionary significant units of anadromous salmonids in the Pacific Northwest. *American Fisheries Society Symposium* 17: 334-349.

Schlaepfer, R. 1997. *Ecosystem-based Management of Natural Resources: Steps Towards Sustainable Development*. IUFRO Occasional Paper No. 6, International Union of Forest Research Organizations, Vienna, Austria.

Sedjo, R. A. 1993. Global consequences of US environmental policies. *Journal of Forestry* 91(4): 19-21.

Sedjo, R. A., D. E. Toweill, and J. E. Wagner. 1999. Economic interactions at local, regional, national, and international scales, pp. 345-357 in *Ecological Stewardship: A Common Reference for Ecosystem Management, Volume III*, W. T. Sexton, A. J. Malk, R. C. Szaro, and N. C. Johnson, ed. Elsevier Science Ltd., Oxford UK.

Shands, W. E. 1994. Decision processes and leadership, pp. 75-80 in *Proceedings Society of American Foresters 1993 National Convention*, Society of American Foresters, Bethesda MD.

Shindler, B., and J. Neburka. 1997. Public participation in forest planning. *Journal of Forestry* 95(1): 17-19.

Spies, T. A., G. H. Reeves, K. M. Burnett, W. C. McComb, K. N. Johnson, G. Grant, J. L. Ohmann, S. L. Garman, and P. Bettinger. In press. Assessing the ecological consequences of forest policies in a multi-ownership province in Oregon, in *Integrating Landscape Ecology into Natural Resource Management*, J. Lui and W. W. Taylor, ed. Cambridge University Press, Cambridge UK.

State of Oregon. 1997. *The Oregon Plan: Coastal Salmon Restoration Initiative*. http://www.oregon-plan.org/Final.html (April 19, 2000).

Thorn, B., K. Jones, P. Kavanagh, and K. Reis. 2000. *1999 Stream Habitat Conditions in Western Oregon*. Monitoring Program Report Number OPSW-ODFW-2000-5, Oregon Department of Fish and Wildlife, Portland.

Tuchmann, E. T., K. P. Connaughton, L. E. Freedman, and C. B. Moriwaki. 1996. *The Northwest Forest Plan: A Report to Congress*. USDA Forest Service, Pacific Northwest Research Station, Portland OR.

Williams, E. M., and P. V. Ellefson. 1997. Going into partnership to manage a landscape. *Journal of Forestry* 95(5): 29-33.

Williamson, O. 1985. *The Economic Institutions of Capitalism*. Free Press, New York.

Wimberly, M. A., T. A. Spies, C. J. Long, and C. Whitlock. 2000. Simulating historical variability in the amount of old forests in the Oregon Coast Range. *Conservation Biology* 14(1): 167-180.

Yaffe, S. L. 1998. Cooperation: A strategy for achieving stewardship across boundaries, pp. 299-324 in *Stewardship Across Boundaries*, R. L. Knight and P. B. Landres, ed. Island Press, Washington DC and Covelo CA.

Yaffee, S. L. 1999. Regional cooperation: a strategy for achieving ecological stewardship, pp.131-153 in *Ecological Stewardship: A Common Reference for Ecosystem Management, Volume III*, W. T. Sexton, A. J. Malk, R. C. Szaro, and N. C. Johnson, ed. Elsevier Science Ltd., Oxford UK.

Zhang, D. 1996. State property rights laws: What, where, and how? *Journal of Forestry* 94(4): 10-15.

Zheng, D., and R. J. Alig. 1999. *Changes in the Non-federal Land Base Involving Forestry in Western Oregon, 1961-1994*. Research Paper PNW-RP-518, USDA Forest Service, Pacific Northwest Research Station, Portland OR.

Conversion Factors

English-to-Metric	*Metric-to-English*
Length	
in x 2.540 = cm	cm x 0.393 = in
ft x 0.304 = m	m x 3.280 = ft
chain x 20.116 = m	m x 0.049 = chain
mi x 1.609 = km	km x 0.621 = mi
Area	
in^2 x 6.451 = cm^2	cm^2 x 0.155 = in^2
ft^2 x 0.092 = m^2	m^2 x 10.763 = ft^2
ac x 0.404 = ha	ha x 2.471 = ac
ft^2/ac x 0.299 = m^2/ha	m^2/ha x 4.356 = ft^2/ac
Volume	
in^3 x 16.387 = cm^3	cm^3 x 0.061 = in^3
ft^3 x 0.0238 = m^3	m^3 x 35.314 = ft^3
yd^3 x 0.764 = m^3	m^3 x 1.308 = yd^3
bd ft[a] x 0.00236 = m^3	m^3 x 423.729 = bd ft
fl oz x 29.573 = ml	ml x 0.033 = fl oz
qt x 0.946 = liter	liter x 1.056 = qt
U.S. gal x 3.785 = liter	liter x 0.264 = U.S. gal
Mass	
oz x 28.349 = g	g x 0.035 = oz
lb x 0.453 = kg	kg x 2.204 = lb
Temperature	
0.555(∞F - 32) = ∞C	
1.8∞C + 32 = ∞F	
Pressure	
bar x 10 = Mpa	

[a] Based on nominal measurement of 1 bd ft = 1 in x 1 ft x 1 ft.

Glossary

aecia. Cuplike or blisterlike structures bearing aeciospores in the rust fungi.[3]

anthropocentric. Focused on or driven by the needs and desires of humans.

autogenic. Originating from within a system.[3]

Best Management Practices (BMPs). A set of alternative forest practices that can reduce non-point source pollution from forest management practices, especially timber harvesting. To be a BMP, the practice must also be technically and economically feasible.

biological diversity. *See* biodiversity.

biodiversity. The variety and abundance of life forms, processes, functions, and structures of plants, animals, and other living organisms, including the relative complexity of species, communities, gene pools, and ecosystems at spatial scales that range from local through regional to global.[3]

cavity excavator. A species that partially or totally excavates cavities in dead wood for use as nesting or roosting sites.

clearcutting. The cutting of essentially all the trees, producing a fully exposed microclimate for the development of a new age class. Depending on management objectives, a clearcut may or may not have reserve trees left to attain goals other than regeneration.[3]

coarse-filter approach. Managing an area to provide a broad range of habitat conditions to provide habitat for multiple species; this approach contrasts with fine-filter approaches in which habitat is managed for individual species.

Coast Range province. The physiographic region, characterized by steep mountains and narrow valleys, extending from about the middle fork of the Coquille River in the south, northward to the Columbia River, and bounded by the Willamette and other interior valleys to the east and the Pacific Ocean on the west.

conservation biology. The applied science of maintaining the earth's biological diversity.[6]

conservation of biological diversity. Action to conserve and maintain genes, species, and ecosystems with a view to the sustainable management and use of biological resources.[9]

contextual. The interrelated conditions in which something exists or occurs.[8]

core area. That area of habitat essential in the breeding, nesting, and rearing of young, up to the point of dispersal of the young.[2]

debris flow. A shallow, translational landslide that transitions into a flowing mass of soil, rocks, shallow water, and organic debris and flows downslope, usually conveyed within a stream channel.

debris flow deposit. The mass of soil, rock, and organic material that deposits out into a stream channel when a debris flow runs out of momentum.

debris jam. *See* debris flow deposit.

debris slide. A shallow, translational landslide in which the landslide mass deposits immediately below the failure scarp and does not travel any further downslope.

decision support tools. Generally a computer-assisted system of decision-making rules designed to find solutions to problems based on parameters or constraints set by the user.

discrete failure. Evidence of the occurrence of a mass-movement feature by the presence of a tension crack or scarp where soil has visibly become displaced either horizontally (*tension crack*), vertically (*scarp*), or some combination of both.

disturbance regime. A pattern of repeated disturbance events in which each disturbance event is followed by a recovery period, e.g., a debris flow followed by recovery of riparian vegetation and stream characteristics prior to the next debris flow.

dry ravel. Dry surface erosion or the particle-by-particle movement of soil that occurs during dry periods.

earthflow. The downslope transport of soil and mantle materials either by a flowage mechanism or by gliding displacement of a series of blocks over a discrete failure surface.

ecocentric. Focused on environmental needs. Welfare of the environment is the primary concern.

ecoregion. A large ecosystem defined by predictable patterns of climate, topography, soils, and vegetation.

ecosystem. A spatially explicit, relatively homogeneous unit of the earth that includes all interacting organisms and components of the abiotic environment within its boundaries. An ecosystem can be of any size, e.g., a log, pond, field, forest, or the earth's biosphere.[5]

ecosystem health. There is no single definition of ecosystem health. The concept can be defined only within the context of the desired values that a particular seral stage of a forest ecosystem, or a particular forest landscape, is supposed to provide. One could say that if an ecosystem has its integrity, it is "healthy," and vice versa.[7]

ecosystem integrity. The maintenance of an ecosystem within the range of conditions or seral stages in which the processes of autogenic succession operate normally to return the ecosystem to, or toward, its predisturbance condition. Ecosystem integrity is very different from the integrity of a particular seral stage or condition, such as the integrity of the old-growth condition. An ecosystem that has been regressed from an old-growth condition to an earlier seral stage may not have experienced any loss of ecosystem integrity, but there will have been a loss in the integrity of the old-growth condition of that ecosystem.[7]

ecosystem resiliency. The capacity of an ecosystem to maintain or regain normal function and development following disturbance.[4]

edge. The more or less well-defined boundary between two or more elements of the environment, e.g., a field adjacent to a woodland or the boundary of different silvicultural treatments.[3]

edge effect. The modified environmental conditions or habitat along the margins (edges) of forest stands or patches.[3]

effective stress. The stress in the soil that is the difference between the total stress and the positive pore pressure.

end haul. A construction technique for building forest roads in steep terrain where the road spoil material is loaded into dump trucks and transported to and dumped at a stable location.

epicormic shoot. A shoot arising spontaneously from an adventitious or dormant bud on the stem or branch of a woody plant, often following exposure to increased light levels or fire.[3]

extensive forestry. The practice of forestry on a basis of low operating and investment costs per acre.[3]

feller-buncher. A machine for mechanical felling of trees with a shear or saw device. Some feller-bunchers accumulate several small trees before the bunch is placed in a selected position. Configurations include rubber-tired and tracked drive-to-tree machines, tracked swing-boom machines, and tracked leveling swing-boom machines for steep terrain.

fine-filter approach. Managing habitat for each species independently; this approach contrasts with coarse-filter approaches in which habitat is managed to provide a broad range of habitat conditions for multiple species.

forest management. The practical application of biological, physical, quantitative, managerial, economic, social, and policy principles to the regeneration, management, utilization, and conservation of forests to meet specific goals and objectives while maintaining the productivity of the forest.[3]

forwarder. A rubber-tired, articulated vehicle used for transporting shortwood or cut-to-length logs clear of ground. The forwarder is equipped with a grapple loader for loading and unloading.

fragmentation. The detaching or separation of expansive tracts of land into spatial fragments.

full bench. A construction technique for building forest roads where the entire running surface of the road is placed on an excavated bench and none is placed on fill material.

genetically buffered. State in which a coevolved host-pathogen combination has selected for genotypes that have reached near-equilibrium. Introduction of a new host or pathogen genotype to the system may upset the balance and result in an increased incidence of disease.

globalization. The process of increasing interdependence among the nations, economies, and people throughout the world.

habitat. The resources and conditions necessary to support presence, survival, and reproduction of a species through time.

headwall. A steeply sloping, concave landscape feature that is found at the distal end of first-order streams in steep, landslide-prone terrain. These are also sometimes referred to as swales, hollows, or zero-order basins.

height:diameter ratio. The ratio of the total tree height to the stem diameter at breast height when both measurements are in the same units.

hypogeous fungus. A fungus that fruits beneath the ground, such as a truffle.

infection courts. The site of infection by a pathogen.[3]

in situ stress. The stress-state place on a discrete element of soil that results from a combination of the overburden stress of the soil and the confining process.

integrated resource management. A process usually involving multiple stakeholders for the simultaneous consideration of ecological, physical, economic, and social aspects of lands, waters, and resources in developing and implementing management plans, often involving multiple-use or sustained-yield management, or the solution of complex forest resource problems.[4]

intensive forestry. The practice of forestry to obtain a high level of volume and quality of outturn per unit of area through the application of the best techniques of silviculture and management. Compared to extensive forestry, intensive forestry requires greater inputs of labor and capital in terms of quantity, quality, and frequency.[3]

intermediate support. A standing tree (or trees) located between the landing and tail tree to which a support jack (or J-bar) is attached to support a multispan skyline.

inter-riparian gradient. Changes in microclimate and vegetation from streamside to ridgetop; the transriparian gradient.

intra-riparian gradient. Changes in riparian conditions as one moves from headwater streams to major rivers and the ocean.

keystone species. A species that exerts influences on the ecology of an area at a level disproportionate to its abundance.

landscape. A spatial mosaic of several ecosystems, landforms, and plant communities across a defined area irrespective of ownership or other artificial boundaries and repeated in similar form throughout.[3]

landscape analysis. The process of collecting and interpreting landscape information about the spatial arrangement of matrices, patches, and processes and how they affect each another.[1]

landscape composition. The amount and types of habitat patches occurring in a landscape.

landscape configuration. The spatial pattern of habitat patches in a landscape.

landscape structure. Tthe composition and spatial configuration of patches of habitat in a landscape.

legacy. Components of a pre-disturbance ecosystem that persist in a post-disturbance ecosystem.

limiting stress. In slope stability terms, this is the stress state at which incipient failure occurs.

multiple-use forestry. Any practice of forestry fulfilling two or more objectives of management. Multiple uses may be integrated at one site or segregated from each other.The Multiple-Use Sustained Yield Act of 1960 identified the following multiple uses: timber, range, watershed, wildlife and fish, and outdoor recreation.[3]

non-point sources of pollution. Any source of material that is found in natural watercourses that is not a point source of pollution.

oomycetes (watermolds). A class of organisms that reproduce asexually by means of spores capable of swimming a short distance in a film of water. Oomycetes are ecologically favored by the presence of free water in soil and on foliage.

overland flow. Precipitation that does not infiltrate the mineral soil surface and runs over the surface of the soil to the nearest stream channel without infiltrating at any point.

partial cut. A general term used for tree removal other than clearcutting, in which selected trees are harvested.

pathogenic rotation. The stand age at which pathogens are reducing product value faster than it is being produced, usually causing the stand to be harvested at a younger age than it would be if pathogens were not active.

point source pollution. A pollutant carried to natural waters through a "discrete conveyance," implying a pipe, channel outfall, or dumping point for industrial, municipal, and agricultural wastes.

positive pore pressure. The pressure that exists in a soil on the potential failure surface that is a product of the depth of the failure surface below the groundwater surface and the density of the water.

primary cavity user. A species that partially or totally excavates cavities in dead wood for use as nesting or roosting sites; a cavity excavator.

province. A subdivision of a region that has relatively similar geology, climate, and vegetation.

regolith. The unconsolidated mantle of weathered rock and soil material on the earth's surface; loose earth materials above solid rock.

ringshake. A fissure or crack in a log or stem that follows a growth ring for some distance.[3]

riparian area. The region where direct interaction of the terrestrial and aquatic environments occurs.

root throw. The downslope movement of soil that accompanies the uprooting and downslope movement of wind-thrown trees.

rotational slump. The downslope movement of a block of soil and regolith materials over a broadly concave failure surface.

secondary cavity user. A species that uses cavities in dead wood but does not excavate its own cavity.

sediment budget. A method to quantify the transport and storage of soil and sediment in watersheds or smaller landscape units.

seral stage. A temporal and intermediate stage in the process of succession.[3]

shear strain. In slope stability, it is the movement of one layer of soil along another layer or plane. When landslides occur, the incipient movement of one layer of soil relative to the other is called shear strain.

silvicultural system. A planned series of treatments for tending, harvesting, and re-establishing a stand.[3]

silviculture. The art and science of controlling the establishment, growth, composition, health, and quality of forests and woodlands to meet the diverse needs and values of landowners and society on a sustainable basis.[3]

single-grip harvester. A machine that fells, delimbs, tops, and crosscuts (bucks) the tree at the stump area.

soil creep. The process of slow downslope movement of soil mantle materials in response to gravitational stress sufficient to cause permanent deformation but too small to result in discrete failure.

spoil material. The excess regolith, soil and rock material, excavated from the hillslope and either sidecast or stored in a disposal area during the process of constructing roads and landings.

springwood. That part of the annual ring of wood that is less dense and composed of large-diameter, thin-walled, secondary xylem cells laid down early in the growing season.[3]

stand. In silviculture, a contiguous group of trees sufficiently uniform in age-class distribution, composition, and structure, and growing on a site of sufficiently uniform quality, to be a distinguishable unit.[3]

stand density. A quantitative measure of tree crowding within a stand usually expressed as the number of trees per unit area.

stand structure. In silviculture, the horizontal and vertical distribution of components of a forest stand including the height, diameter, crown layers, and stems of trees, shrubs, herbaceous understory, snags, and down woody debris.[3]

stress. The intensity of a force expressed as force per unit area.

structure-based management. Landscape- level management focused on providing a wide array of benefits from forests by achieving a specific mix of all stand development stages through silvicultural operations.

structural diversity. One of the measures of biological diversity in forest ecosystems. It refers to the variation in tree size and canopy layering, the variety of different life forms of vegetation (trees, herbs, shrubs, mosses, climbers, epiphytes, etc.), and the relative size and abundance of standing dead trees (snags) and decaying logs on the ground (coarse woody debris). Structural diversity refers to these features within a particular local ecosystem (alpha structural diversity), or to variations in them between local ecosystems across the local landscape (beta structural diversity).[7]

succession. The gradual supplanting of one community of plants by another.[3]

sustainability. The capacity of forests, ranging from stands to ecoregions, to maintain their health, productivity, diversity, and overall integrity, in the long run, in the context of human activity and use.[3]

sustainable forest management. The practice of meeting the forest resource needs and values of the present without compromising the similar capability of future generations.[3]

sustained yield. The yield that a forest can produce continuously at a given intensity of management. Sustained-yield management implies continuous production so planned as to achieve, at the earliest practical time, a balance between increment and cutting. Sustained yield can also mean the achievement and maintenance in perpetuity of a high-level annual or regular periodic output of the various renewable resources without impairment of the productivity of the land.[3]

swing yarder. A specialized yarder with boom movement which provides greater flexibility in landing logs.

tail tree. A standing tree at the outer end of a skyline yarding system that elevates and supports one end of the skyline.

thinning. The removal of selected trees from an immature stand to stimulate or maintain desired growth of residual trees.

total normal stress. The stress on a potential failure surface that results from the total weight of the soil and the weight of the water above the failure surface.

translational slide. A landslide in which the slide mass moves out or down and out along a planar or gently undulating surface and has little of the rotary movement or backward-tilting characteristics of a slump.

transriparian gradient. Changes in microclimate and vegetation from streamside to ridgetop; the inter-riparian gradient.

tree jacking. A technique of mechanically assisted directional felling in which hydraulic tree jacks are placed in the back cut of the tree to push it against its natural lean.

tree pulling. A technique of mechanically assisted directional felling in which a winch or yarder drum pulls a tree into the desired lay.

urbanization. The transformation of rural areas or communities to communities with city-like characteristics achieved through increased human population.

variegation. The condition of a landscape in which patches vary with respect to suitability and permeability to movement along a continuum, and the quality of habitat in patches varies through time.

vertical structure. The amount, type, and distribution of vegetation from the forest floor to the top of the canopy.

watershed. A region or land area drained by a single stream, river, or drainage network.[3]

wedging. A nonmechanical directional felling technique in which a faller pounds wedges in the back cut of a tree to push it against its natural lean.

wolf tree. A common term used for open-grown trees with branches along the entire bole.

yarding. Yarding, or primary transportation, is defined as the initial movement of logs from the felling site to the roadside in preparation for further transport. The term "yarding" is generally associated with cable systems and "skidding" is generally associated with ground-based systems.

Sources

1. Adapted from Diaz, N. M., and S. Bell. 1997. "Landscape analysis and design." pp. 255-269 in K. A. Kohm and J. F. Franklin, eds. Creating a Forestry for the 21st Century. © 1997 by Island Press, Washington D.C. and Covelo, CA.
2. FEMAT (Forest Ecosystem Management Assessment Team). 1993. *Forest Ecosystem Management: An Ecological, Economic, and Social Assessment.* Report of the Forest Ecosystem Management Assessment Team, U.S. Government Printing Office 1973-793-071. Portland OR.
3. Helms, J.A. (ed.). 1998. *The Dictionary of Forestry.* Society of American Foresters. Bethesda MD.
4. Adapted from 3.
5. From Henderson's *Dictionary of Biological Terms*, 11th Edition, 1995, Eleanor Lawrence, Editor. Used by permission.
6. Hunter, M. L., Jr. 1996. *Fundamentals of Conservation Biology*. Blackwell Science, Cambridge, MA.
7. Kimmins, J. P. 1997. Forest Ecology: A Foundation for Sustainable Management. Prentice Hall, Upper Saddle River NJ.
8. Adapted from *Merriam-Webster's Collegiate® Dictionary*, 10th Edition, © 2001 at www.Merriam-Webster.com by Merriam-Webster, Incorporated.
9. Paraphrased from United Nations Environmental Programme. 2000. Agenda 21. Chapter 15. http://www.unep.org/Documents/Default.asp?DocumentID=52&ArticleID=63. Accessed 12-07-01.

List of Species

Common Name	Scientific Name
Amphibians	
clouded salamander	*Aneides ferreus*
Del Norte salamander	*Plethodon elongatus*
Dunn's salamander	*Plethodon dunni*
Larch Mountain salamander	*Plethodon larselli*
northwestern salamander	*Ambystoma gracile*
Pacific giant salamander	*Dicamptodon tenebrosus*
rough-skinned newt	*Taricha granulosa*
Shasta salamander	*Hydromantes shastae*
Siskiyou Mountains salamander	*Plethodon stormi*
southern torrent salamander	*Rhyacotriton variegatus*
tailed frog	*Ascaphus truei*
Van Dyke's salamander	*Plethodon vandykei*
Reptiles	
common garter snake	*Thamnophis sirtalis*
northern alligator lizard	*Elgaria coerulea coerulea*
western fence lizard	*Sceloporus occidentalis*
Mammals	
American beaver	*Castor canadensis*
American marten	*Martes americana*
Baird's shrew	*Sorex bairdii*
black bear	*Ursus americanus*
creeping vole	*Microtus oregoni*
deer mouse	*Peromyscus maniculatus*
Douglas-squirrel	*Tamiasciurus douglasii*
elk	*Cervus elaphus*
fog shrew	*Sorex sonomae*
long-legged myotis	*Myotis volans*
long-tailed vole	*Microtus longicaudus*
northern flying squirrel	*Glaucomys sabrinus*
Pacific jumping mouse	*Zapus trinotatus*
Pacific shrew	*Sorex pacificus*
red tree vole	*Arborimus longicaudus*
river otter	*Lutra canadensis*
silver-haired bat	*Lasionycteris noctivagans*
Townsend's big-eared bat	*Corynorhinus townsendii*
Townsend's chipmunk	*Tamias townsendii*
Townsend's vole	*Microtus townsendii*
Trowbridge's shrew	*Sorex trowbridgii*
western gray squirrel	*Sciurus griseus*
western long-eared myotis	*Myotis evotis*
western red-backed vole	*Clethrionomys californicus*
white-footed vole	*Arborimus albipes*
Birds	
American dipper	*Cinclus mexicanus*
band-tailed pigeon	*Columba fasciata*
belted kingfisher	*Ceryle alcyon*
black-capped chickadee	*Parus atricapillus*
black-throated gray warbler	*Dendroica nigrescens*
brown creeper	*Certhia americana*
chestnut-backed chickadee	*Parus rufescens*
dark-eyed junco	*Junco hyemalis*
evening grosbeak	*Coccothraustes vespertinus*
golden-crowned kinglet	*Regulus satrapa*
gray jay	*Perisoreus canadensis*
great blue heron	*Ardea herodias*
hairy woodpecker	*Picoides villosus*
Hammond's flycatcher	*Empidonax hammondii*
harlequin duck	*Histrionicus histrionicus*
hermit warbler	*Dendroica occidentalis*
Hutton's vireo	*Vireo huttoni*
mallard duck	*Anas platyrhynchos*
marbled murrelet	*Brachyramphus marmoratus*
northern flicker	*Colaptes auratus*
orange-crowned warbler	*Vermivora celata*
Pacific-slope flycatcher	*Epidonax difficilis*
peregrine falcon	*Falco peregrinus*
pileated woodpecker	*Dryocopus pileatus*
purple martin	*Progne subis*
red-breasted nuthatch	*Sitta canadensis*
red-breasted sapsucker	*Sphyrapicus ruber*
rufous hummingbird	*Selasphorus rufus*
Steller's jay	*Cyanocitta stelleri*
Swainson's thrush	*Catharus ustulatus*
Townsend's solitaire	*Myadestes townsendi*
varied thrush	*Ixoreus naevius*
Vaux's swift	*Chaetura vauxi*
warbling vireo	*Vireo gilvus*
willow flycatcher	*Empidonax traillii*
Wilson's warbler	*Wisonia pusilla*
winter wren	*Troglodytes troglodytes*
Fish	
American shad	*Alosa sapidissima*
candle fish	*Thaleichthys pacificus*
chinook salmon	*Oncorhynchus tshawytscha*
chum salmon	*Oncorhynchus keta*
Coast Range sculpin	*Cottus aleuticus*
coho salmon	*Oncorhynchus kisutch*
cutthroat trout	*Oncorhynchus clarkii*
Dolly Varden	*Salvelinus malma*

Fish (cntinued)

hake	*Merluccius productus*
lamprey	*Petromyzontidae*
longnose dace	*Rhinichthys cataractae*
minnow	*Cyprinidae*
northern pike minnow	*Ptychocheilus oregonensis*
Pacific lamprey	*Lampetra tridentata*
Pacific salmon and trout	*Oncorhynchus*
pink salmon	*Oncorhynchus gorbuscha*
rainbow trout	*Oncorhynchus mykiss*
red-side shiner	*Richardsonius balteatus*
reticulate sculpin	*Cottus perplexus*
riffle sculpin	*Cottus gulosus*
salmon and trout	*Salmonidae*
sculpin	*Cottus*
speckled dace	*Rhinichthys osculus*
steelhead	*Oncorhynchus mykiss*
stickleback	*Gasterosteidae*
striped bass	*Morone saxatilis*
sucker	*Catostomidae*

Trees & Shrubs

alder	*Alnus* spp.
bigleaf maple	*Acer macrophyllum*
ceanothus	*Ceanothus* spp.
chinquapin	*Castanopsis chrysophylla*
cottonwood	*Populus* spp.
currant	*Ribes* spp.
Douglas-fir	*Pseudotsuga menziesii*
gooseberry	*Ribes* spp.
grand fir	*Abies grandis*
huckleberry	*Vaccinium* spp.
hybrid poplar	*Populus deltoides x Populus nigra*
incense cedar	*Calocedrus decurrens*
larch	*Larix* spp.
mountain hemlock	*Tsuga mertensiana*
noble fir	*Abies procera*
oceanspray	*Holodiscus discolor*
Oregon ash	*Fraxinus latifolia*
Oregon white oak	*Quercus garryana*
Oregon grape	*Berberis aquifolium*
Pacific madrone	*Arbutus menziesii*
Pacific silver fir	*Abies amabilis*
Pacific yew	*Taxus brevifolia*
Ponderosa pine	*Pinus ponderosa*
Port Orford cedar	*Chamaecyparis lawsoniana*
red alder	*Alnus rubra*
red elderberry	*Sambucus racemosa*
salal	*Gaultheria shallon*
salmonberry	*Rubus spectabilis*
Sitka Spruce	*Picea sitchensis*
snowberry	*Symphoricarpos* spp.
tanoak	*Lithocarpus densiflorus*
thimbleberry	*Rubus parviflorus*
vine maple	*Acer circinatum*
wax-myrtle	*Myrica californica*
western hemlock	*Tsuga heterophylla*
western redcedar	*Thuja plicata*
western white pine	*Pinus monticola*
dwarf mistletoe	*Arceuthobium* spp.
western hemlock dwarf mistletoe	*Arceuthobium tsugense subspecies tsugense*

Fungi

annosus root and butt rot	*Heterobasidion annosum (Fomes annosus)*
black stain root disease	*Leptographium wageneri*
Douglas-fir black stain root disease	*Leptographium wageneri* var. *pseudotsugae*
honey mushroom	*Armillaria ostoyae*
Indian paint fungus	*Echinodontium tinctorium*
laminated root rot	*Phellinus weirii*
Port Orford cedar root rot	*Phytophthora lateralis*
quinine fungus	*Fomitopsis officinalis*
red belt fungus	*Fomitopsis pinicola*
red ring rot	*Phellinus pini*
rose-colored conk	*Fomitopsis cajanderi*
sulfur fungus, chicken of the woods	*Laetiporus sulphureus*
swiss needle cast	*Phaeocryptopus gaeumannii*
velvet-top fungus, cow pie fungus	*Phaeolus schweinitzii*
white pine blister rust	*Cronartium ribicola*

Insects

cedar bark beetle	*Phloeosinus* spp. *(Scolytidae)*
Douglas-fir beetle	*Dendroctonus pseudotsuqae*
Douglas-fir engraver	*Scolytus unispinosus*
Douglas-fir pole beetle	*Pseudohylesinus nebulosus*
fir engraver	*Scolytus ventralis*
midge	*Chironomidae*
mountain pine beetle	*Dendoctonus ponderosae*

Index

Figures, tables, and color plates are indicated with *f, t, fig.*